Advances in Bioorganometallic Chemistry

Advances in Bioorganometallic Chemistry

Edited by

Toshikazu Hirao
The Institute of Scientific and Industrial Research, Osaka University, Ibaraki, Japan

Toshiyuki Moriuchi
Division of Molecular Materials Science, Graduate School of Science, Osaka City University, Osaka, Japan

ELSEVIER

Elsevier
Radarweg 29, PO Box 211, 1000 AE Amsterdam, Netherlands
The Boulevard, Langford Lane, Kidlington, Oxford OX5 1GB, United Kingdom
50 Hampshire Street, 5th Floor, Cambridge, MA 02139, United States

Notices

Knowledge and best practice in this field are constantly changing. As new research and experience broaden our understanding, changes in research methods, professional practices, or medical treatment may become necessary.

Practitioners and researchers must always rely on their own experience and knowledge in evaluating and using any information, methods, compounds, or experiments described herein. In using such information or methods they should be mindful of their own safety and the safety of others, including parties for whom they have a professional responsibility.

To the fullest extent of the law, neither the Publisher nor the authors, contributors, or editors, assume any liability for any injury and/or damage to persons or property as a matter of products liability, negligence or otherwise, or from any use or operation of any methods, products, instructions, or ideas contained in the material herein.

British Library Cataloguing-in-Publication Data
A catalogue record for this book is available from the British Library

Library of Congress Cataloging-in-Publication Data
A catalog record for this book is available from the Library of Congress

ISBN: 978-0-12-814197-7

Publisher: Susan Dennis
Acquisition Editor: Emily M. McCloskey
Editorial Project Manager: Tasha Frank
Production Project Manager: Denny Mansingh
Cover Designer: Miles Hitchen

Cover image created by Toshikazu Hirao

Typeset by MPS Limited, Chennai, India

Contents

Part III
Metalloprotein, Hydrogenase, and Vitamin B12

List of Contributors

Satoshi Abe Department of Life Science and Technology, School of Life Science and Technology, Tokyo Institute of Technology, Yokohama, Japan

Roger Alberto Department of Chemistry, University of Zurich, Zurich Switzerland

J.B. Alexander Ross Department of Chemistry and Biochemistry, University of Montana, Missoula, MT, United States

Nadia Baartzes Department of Chemistry, University of Cape Town, Cape Town, South Africa

Timothy P. Curran Department of Chemistry, Trinity College, Hartford, CT, United States

Catherine Elleouet Université de Bretagne Occidentale, Brest; CNRS, UMR 6521, Chimie, Electrochimie Moléculaires et Chimie Analytique, CS 93837, Brest Cedex 3, France

Ian J.S. Fairlamb Department of Chemistry, University of York, York, United Kingdom

Natashya Falcone Department of Physical and Environmental Sciences, University of Toronto Scarborough, Toronto, ON, Canada; Department of Chemical Engineering and Applied Chemistry, University of Toronto, Toronto, ON, Canada

Richard H. Fish Lawrence Berkeley National Laboratory, University of California, Berkeley, CA, United States

Christian G. Hartinger School of Chemical Sciences, University of Auckland, Private Bag 92019, Auckland, New Zealand

Takashi Hayashi Department of Applied Chemistry, Osaka University, Suita, Japan

Toshikazu Hirao The Institute of Scientific and Industrial Research, Osaka University, Osaka, Japan

Yoshio Hisaeda Department of Chemistry and Biochemistry, Graduate School of Engineering, Kyushu University, Fukuoka, Japan

Toshihiro Ihara Division of Materials Science and Chemistry, Faculty of Advanced Science and Technology, Kumamoto University, Kumamoto, Japan

Katsuhiro Isozaki Institute for Chemical Research, Kyoto University, Kyoto, Japan

Heinz-Bernhard Kraatz Department of Physical and Environmental Sciences, University of Toronto Scarborough, Toronto, ON, Canada; Department of Chemical Engineering and Applied Chemistry, University of Toronto, Toronto, ON, Canada; Department of Chemistry, University of Toronto, Toronto, ON, Canada

Bernhard Kräutler Institute of Organic Chemistry and Centre of Molecular Biosciences University of Innsbruck, Innsbruck, Austria

Jason M. Lynam Department of Chemistry, University of York, York, United Kingdom

Basudev Maity Department of Life Science and Technology, School of Life Science and Technology, Tokyo Institute of Technology, Yokohama, Japan

Takashi Matsuo Division of Materials Science, Graduate School of Science and Technology, Nara Institute of Science and Technology (NAIST), Ikoma, Nara, Japan

Giuseppe Meola Department of Chemistry, University of Zurich, Zurich Switzerland

Toshiyuki Moriuchi Division of Molecular Materials Science, Graduate School of Science, Osaka City University, Osaka, Japan

Masaharu Nakamura Institute for Chemical Research, Kyoto University, Kyoto, Japan

Jun Okuda Institute of Inorganic Chemistry, RWTH Aachen University, Aachen, Germany

Akira Onoda Department of Applied Chemistry, Osaka University, Suita, Japan

François Y. Pétillon Université de Bretagne Occidentale, Brest; CNRS, UMR 6521, Chimie, Electrochimie Moléculaires et Chimie Analytique, CS 93837, Brest Cedex 3, France

Edward Rosenberg Department of Chemistry and Biochemistry, University of Montana, Missoula, MT, United States

Shinobu Sato Department of Applied Chemistry, Kyushu Institute of Technology, Kitakyushu, Japan

Daniel F. Sauer Institute of Inorganic Chemistry, RWTH Aachen University, Aachen, Germany

Ulrich Schatzschneider Institut für Anorganische Chemie, Julius-Maximilians-Universität Würzburg, Am Hubland, Würzburg, Germany

Philippe Schollhammer Université de Bretagne Occidentale, Brest; CNRS, UMR 6521, Chimie, Electrochimie Moléculaires et Chimie Analytique, CS 93837, Brest Cedex 3, France

Takafumi Shanoh Institute for Chemical Research, Kyoto University, Kyoto, Japan; Department of Energy and Hydrocarbon Chemistry, Graduate School of Engineering, Kyoto University, Kyoto, Japan

Ayesha Sharmin Department of Chemistry, Bangladesh University of Engineering and Technology (BUET), Dhaka, Bangladesh

Hisashi Shimakoshi Department of Chemistry and Biochemistry, Graduate School of Engineering, Kyushu University, Fukuoka, Japan

Yasuhito Shomura Institute of Quantum Beam Science, Graduate School of Science and Engineering, Ibaraki University, Hitachi, Japan

Gregory S. Smith Department of Chemistry, University of Cape Town, Cape Town, South Africa

Tameryn Stringer Department of Chemistry, University of Cape Town, Cape Town, South Africa

Hikaru Takaya Institute for Chemical Research, Kyoto University, Kyoto, Japan

Shigeori Takenaka Department of Applied Chemistry, Kyushu Institute of Technology, Kitakyushu, Japan

Takafumi Ueno Department of Life Science and Technology, School of Life Science and Technology, Tokyo Institute of Technology, Yokohama, Japan

Daniel Hernández Valdés Department of Chemistry, University of Zurich, Zurich Switzerland

Gerard van Koten Organic Chemistry and Catalysis, Faculty of Science, Utrecht University, Utrecht, The Netherlands

Nobuhiro Yasuda Japan Synchrotron Radiation Research Institute (JASRI), Kouto, Japan

Ryota Yoshida Institute for Chemical Research, Kyoto University, Kyoto, Japan; Department of Energy and Hydrocarbon Chemistry, Graduate School of Engineering, Kyoto University, Kyoto, Japan

Preface

Bioorganometallic Chemistry, which is defined as "Organometallic Chemistry at the Interface with Biology," is an innovative and rapidly growing interdisciplinary research area that focuses on the overlap between Biochemistry, Biological Processes, and Medicinal Organometallic Chemistry, as several pertinent examples. The history of Bioorganometallic Chemistry has been previously reviewed in Chapter 1, *Bioorganometallic Chemistry, Origins, and Founding Principles* by G. Jaouen, W. Beck, and M. J. McGlinchey.[1]

Furthermore, organotin compounds, which were used as pesticides on food crops, were studied with Cytochrome P450 enzymes; *Organometallic Chemistry at the Interface with Biocatalysis*, a paper published in 1975, was the first to demonstrate that organometallic complexes were metabolized in the presence of a biocatalytic oxidation enzyme, to provide predominately α- and β-hydroxybutyltin complexes.[2] If we reflect on the past, we become aware of the Ancient Origin of Bioorganometallic Chemistry, with the natural organometallic compounds that were identified previously, as exemplified by X-ray crystallography studies on vitamin B_{12} (Fig. 1).[3]

Vitamin B_{12} is a natural bioorganometallic compound in which the Methyl B_{12} derivative has played an important role in the methylation of environmentally important toxic metals, such as Hg^{2+} and Sn^{2+}.[4] Furthermore, Fe–Fe and Fe–Ni hydrogenases have also been identified as Ancient Bioorganometallic Complexes (Fig. 2).[5]

Importantly, the renaissance for the discipline *Bioorganometallic Chemistry* initially began when the first International Symposium on Bioorganometallic Chemistry (ISBOMC), organized by R. H. Fish and G. Jaouen, was held in Paris, France, in 2002. This seminal meeting provided the Bioorganometallic Chemistry community with a forum for the many topics now being researched around the world.

The symposiums held between 2002 and 2019 are as follows. ISBOMC 1: 2002, July; Paris, France; R. H. Fish and G. Jaouen; ISBOMC 2: 2004, July, Zurich, Switzerland, R. Alberto; ISMOMC 3: 2006, July, Milan, Italy, S. Maiorana and G. Jaouen; ISBOMC 4: 2008, July, Missoula, MT, USA, E. Rosenberg; ISBOMC 5: 2010, July, Bochum, Germany, N. Metzler-Nolte; ISBOMC 6: 2012, July, Toronto, Canada; B. Kraatz; ISBOMC 7: 2014, July, Vienna, Austria, B. Keppler and C. Hartinger; ISBOMC 8: 2016, September,

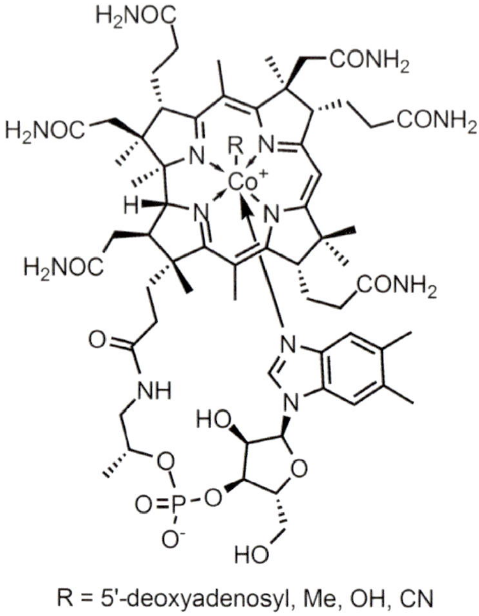

FIGURE 1 Structures of Vitamin B_{12}, R = CN; Methyl B_{12}, R = CH_3, Hydroxy B_{12}, R = OH, and Co-enzyme B_{12}, R = 5 ′-deoxyadenosyl.

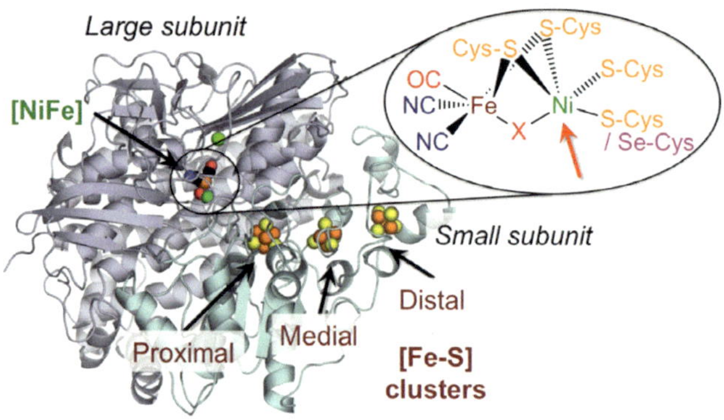

FIGURE 2 Structure of the Fe–Ni Hydrogenase.

Moscow, Russia, E. Milaeva and A. Nazarov; ISBOMC 9, 2019, July-August, York, United Kingdom, J. Lynam.

The broad variety within Bioorganometalllic Chemistry in the literature underlines the vibrancy of the field, allowing for the development of medicinal chemistry, probe applications, bio-inspired chemistry, functional materials, and biomaterials. Bioorganometallic Chemistry also includes synthesis, structures and reactivity of bioorganometallic complexes. Furthermore, biological applications, including the innovative topic of Metals in Medicine, has extended to subjects such as the use of organometallic complexes as new drugs for cancer therapy since the discovery of the inorganic complex, cisplatin.

More importantly, the new advances include, vitamin B_{12} chemistry, biosensors or probes, environmental biotransformations of organometallics, environmental analysis, and modified enzymes such as metallo-DNA, RNA and their analogues. Metalloproteins, which were obtained by the bioconjugation of biomolecules, including peptides, proteins, and apoenzymes, with organometallic complexes, will hopefully provide new types of artificial catalysts. The design of imaging probes based on the chemical and/or physical properties of organometallic complexes have been shown to be another important aspect of Bioorganometallic Chemistry. The survey and discussion in *Advances in Bioorganometallic Chemistry* consists of three parts:

Part I. Synthesis, Structure, and Reactivity of Bioorganometallic Compounds
Part II. Medicinal Chemistry and Imaging Probe
Part III. Metalloproteins, Hydrogenases, and Vitamin B_{12}

The structures and bio-inspired catalysis are described in the chapters of **Part I**. **Part II** mainly deals with metallo probes and drugs, including anticancer drugs. The chapters of **Part III** further display the function, structure, and bio-inspired catalysis aspects. Every chapter is based on the original work of the author(s). *Advances in Bioorganometallic Chemistry* has topics that synergize these new developments and includes the most important achievements that have been attained so far. Furthermore, future perspectives have great promise for a variety of new and exciting Bioorganometallic Chemistry developments.

Toshikazu Hirao (Editor)[1] and Toshiyuki Moriuchi (Editor)[2]
[1]The Institute of Scientific and Industrial Research, Osaka University, Ibaraki, Japan,
[2]Division of Molecular Materials Science, Graduate School of Science, Osaka City University, Osaka, Japan
Richard H. Fish
Lawrence Berkeley National Laboratory, University of California, Berkeley, CA, United States

REFERENCES

1. *Bioorganometallics: Biomolecules, Labeling, Medicine.* Edited by Jaouen G, Copyright 2006, Wiley-VCH Verlag GmbH & Co. Weinheim, Chapter 1.
2. Fish RH, Kimmel EC, Casida JE. *J Organometal Chem* 1975;**93**:C1.
3. Hodgkin DC. *Proc R Soc London, Ser A* 1965;**288**:294.
4. Dizikes LJ, Ridley WP, Wood JM. *J Am Chem Soc* 1978;**100**:1010.
5. Volbeda A, Charon M-H, Piras C, Hatchikian EC, Frey M, Fontecilla-Camps JC. *Nature* 1995;**373**:580.

Part I

Synthesis, Structure, and Reactivity of Bioorganometallic Compounds

Chapter 1

Organometallic Chemistry at the Interface with Biology

Richard H. Fish
Lawrence Berkeley National Laboratory, University of California, Berkeley, CA, United States

1.1 INTRODUCTION

In 1975, we published one of the first papers on bioorganometallic chemistry at the interface with biocatalysis, that focused on the oxidative metabolism of tributyltin acetate utilizing cytochrome P450 enzymes from rat liver microsomes, since these organotin compounds were being sprayed on food products for pest control, and therefore, the metabolic tributyltin compounds formed, and their toxicity, was important to determine for health related reasons.[1] In this overview of our bioorganometallic chemistry studies, we will start from 1975, and peruse the bioorganometallic chemistry projects we have conducted up to the present at LBNL/UC Berkeley, and with colleagues around the globe, that focus on aqueous synthesis, mechanisms, and the biological implications.

Thus, the following topics will be discussed that represents the important findings of our bioorganometallic chemistry studies: Tributyltin acetate and cyclohexyltriphenyltin metabolism studies with cytochrome P450 enzymes to provide predominately (α and β-hydroxybutyl)dibutyltin acetate, and the *trans*-4-hydroxycyclohexyltriphenyltin metabolites; respectively, as the major oxidation products; $[Cp^*Rh(H_2O)_3](OTf)_2$ reactions with nucleobases; for example, 1-methylthymine (1-MT), providing, at pH 10, a unique complex of anion and cation components, which was further stabilized by π−π interactions of the Cp*Rh moiety of $[(Cp^*Rh)_2(\mu\text{-}OH)_3]^+$ with the 1-MT linear complex, $[Rh(\eta^1(N^3)\text{-}1\text{-}MT)_2]^-$; synthesis and bioassay of [*cis*-Cp*Rh-hydroxytamoxifen]$(OTf)_2$, as a breast cancer antagonist drug, formed via a novel N-π rearrangement; 2D NMR and DFT structures of $[(\eta^6\text{-}Cp^*Rh\text{-}Tyr^\#)\text{-}$GPCR peptides$]^{2+}$ formed in the chemoselective reactions of the G-protein-coupled receptor peptides, $[Tyr^\#]$-GPCR, with $[Cp^*Rh(H_2O)_3](OTf)_2$, at pH 5.5, the synthon for many of our aqueous reactions; molecular docking studies of the $[(\eta^6\text{-}Cp^*Rh\text{-}Tyr^1)\text{-}Leu\text{-}Enkephalin]^{2+}$ complex to the X-ray crystallographic defined, μ-, ∂-, and κ − opioid receptors; biomimetic

Advances in Bioorganometallic Chemistry. DOI: https://doi.org/10.1016/B978-0-12-814197-7.00001-7

co-factors, N-substituted-1,4-dihydronicotinamide derivatives, formed by a chemoselective reaction of [Cp*Rh(bpy)H](OTf) with NAD^+ biomimics, in tandem with enzymes, horse liver alcohol dehydrogenase (HLADH) for achiral ketone reductions to chiral alcohols, and a cytochrome P450 BM3 Double Mutant enzyme, for biological oxidation reactions; Host, cyclic trimer, $[Cp^*Rh(2'\text{-deoxyadenosine})]_3(OTf)_3$, molecular recognition of Guests, such as, L-Tryptophan. Finally, our chapter, and others in this book, will demonstrate that the burgeoning bioorganometallic chemistry discipline represents exciting new vistas at the interface with biology.

1.2 BIOORGANOTIN METABOLISM CHEMISTRY: TRIBUTYLTIN ACETATE, AND CYCLOHEXYLTRIPHENYTIN IN REACTIONS WITH THE ENZYME, CYTOCHROME P450, FOR REGIO- AND STEREOSELECTIVITY HYDROXYLATION, AND THE POSSIBILITIES OF A FREE RADICAL MECHANISM

Our studies in the field of bioorganometallic chemistry began in 1973. We were interested in the metabolism of organotin compounds, which were being utilized as pesticides in agricultural applications. The focus was on the metabolism of tributyltin compounds in reactions with rat liver microsomes, cytochrome P450 enzymes for biological oxidation studies, for obvious environmental and health reasons. Thus, we discovered the first metabolites of tributyltin acetate with cytochrome P450 enzymes, and also studied their toxicology. Therefore, a new area of organometallic chemistry research, which we coined at the time, bioorganotin chemistry, was initiated with tributyltin acetate (Eq. 1.1), while the primary sites of hydroxylation were the α and β positions to the Sn atom.[1a–c] Thus, we surmised that the Sn–C bond controlled the regiochemistry. This pattern of hydroxylation suggested a free radical mechanism, where the Sn-C sigma bond electrons stabilized the developing carbon 2p orbital radicals, on predominately, both α and β carbon positions, and to a lesser extent, the γ and ∂ positions.

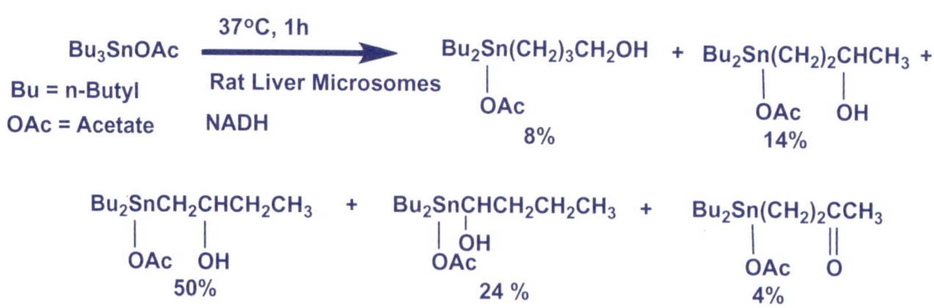

(1.1)

It was also important to study the chemo and stereoselectivity of these biological P450 enzyme reactions, and to understand more about a possible free radical mechanism; therefore, we conducted studies with cyclohexyltriphenyltin, and were able to define both aspects, with cytochrome P450 rat liver microsomes,[2] and their biomimetic models, Mn, and Fe porphyrin derivatives (Fig. 1.1).[3]

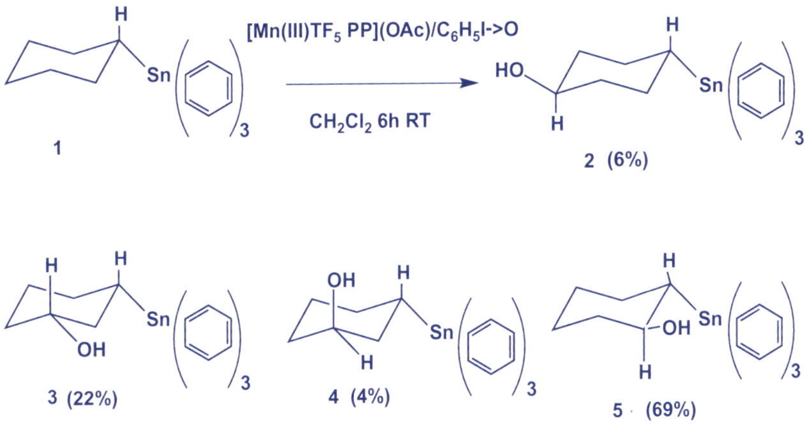

FIGURE 1.1 Biomimetic oxidation of cyclohexyltriphenyltin, **1**, with a cytochrome P450 model, [Mn(III)TF$_5$PP](OAc), and iodosobenzene as the oxygen transfer agent.[3]

The chemo and stereoselectivity in the hydroxylation reaction of cyclohexyltriphenyltin, **1**, with biomimetic precatalysts that mimic the active site of the cytochrome P-450 monooxygenase enzyme, iron(III) and manganese (III) tetrakis(pentafluorophenyl)porphyrin derivatives, [Fe(III) or [Mn(III) TF$_5$PP](Br,OAc), was studied utilizing the oxygen transfer agent, iodosylbenzene, and these results were compared to those obtained with the P-450 enzyme from rat liver microsomes (Table 1.1).[3] The [Mn(III)TF$_5$PP](OAc) biomimetic catalyst provided a 22% conversion of **1** to a mixture of *cis*- and *trans*-hydroxycyclohexyltriphenyltin compounds that included the *trans*-4 (6%), **2**; *cis*-3 (22%), **3**; *trans*-3 (3%), **4**; and *trans*-2 (69%), **5**, isomers. The chemoselectivity on a per hydrogen basis showed a C4:C3:C2: C1 ratio of 1:2:6:0 and a high stereoselectivity for equatorial over axial hydroxyl products with a EQ/AX ratio of 29. The corresponding [Fe(III) TF$_5$PP](Br) catalyst gave the same pattern of hydroxylation, as with the above-mentioned Mn catalyst. In comparison to the Cytochrome P-450 enzyme, which had a different chemoselectivity ratio on a per hydrogen basis for C4:C3:C2:C1 of 109:7:1:0, the biomimics appear to have less steric requirements at the putative reactive site, Mn = O, and was not encumbered by an enzyme pocket, which has limitations of both size and steric consequences.

TABLE 1.1 Comparison of the Chemoselectivity in the Hydroxylation of Compound 1 With Biomimic [Mn(III)TF$_5$PP](OAc)/Oxygen Transfer Agent, Iodosylbenzene, and a Cytochrome P-450 Enzyme From Rat Liver Microsomes

Compounds[a]	[Mn(III)TF$_5$PP](OAc), %	P-450 enzyme[b], %
2[c]	6	86
3[d]	22	7
4[d]	3	3
5	69	2

[a]*See Fig. 1.1 for structures* ***2-5.***
[b]*Cytochrome P450 enzyme from rat liver microsomes (see Ref. 1).*
[c]*The ketone of* ***2*** *was present in 1.9%.*
[d]*The ketone from* ***3*** *and* ***4*** *was present in 1.4%.*

Mechanistically the Sn-C sigma bond electrons also appeared to control the chemoselectivity of the hydroxylation reaction as shown in Scheme 1.1, by the fact that **3** and **5** were the major hydroxylation products, due to a stabilization of radical intermediates on carbons 2 and 3 by the carbon-tin σ-bond. Moreover, the hydroxyl rebound reaction to give products **2-5** appeared to be stereoselective for the sterically more favorable equatorial product.

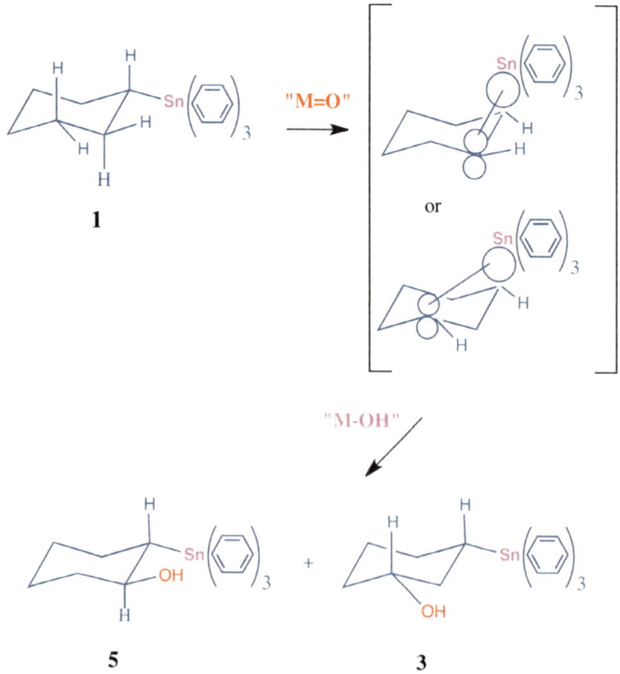

SCHEME 1.1 Carbon–Tin σ-bond stabilization of the developing carbon 2p orbital free radical during the oxidation reaction.

What the preceding experiments have clearly shown was that in the hydrophobic pocket of cytochrome P450, the cyclohexyl group must be orientated with the equatorial carbon-hydrogen on C4, being closer to the porphyrin Fe = O site of oxygen insertion (Scheme 1.2),[3] while the solution results do not have those restrictions, and thus, the chemoselectivity was opposite to the enzyme results; i.e., the C-Sn σ-bond overlapped with the developing carbon radicals on C2 and C3, which appears to dictate chemoselectivity (Scheme 1.1).[3]

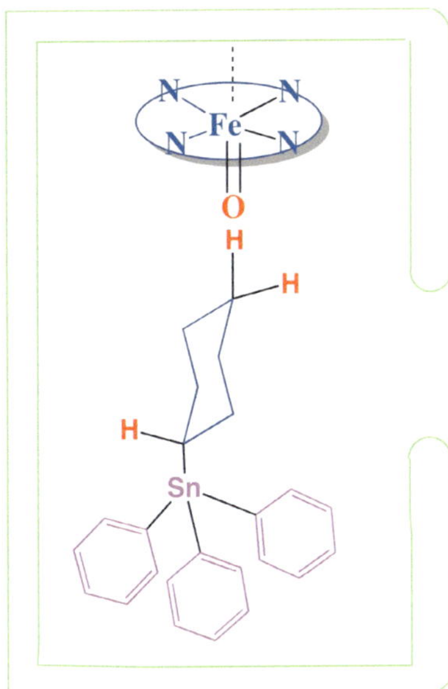

SCHEME 1.2 Putative conformation of compound **1** at the active site of the cytochrome P450 enzyme; C4-H equatorial hydrogen to C-OH, via a C• radical and HO• Rebound.

The metabolism studies of the above-mentioned organotin compounds were a prelude to the future use of these types of compounds as anticancer drugs, since this metabolism aspect would be an important component of any new organotin drug brought to clinical trials. A comprehensive review of the organotin anticancer drug discipline provided an overview of their veracity as new metal-based drugs.[4]

1.3 BIOORGANOMETALLIC CHEMISTRY OF [Cp*Rh(H_2O)$_3$](OTf)$_2$ WITH NUCLEOBASES: X-RAY STRUCTURES

In the early 1990s, the US National Institute of Health was initiating a project to map and sequence the human genome, and we were interested in how the Cp*Rh-N-heterocycle chemistry we had conducted during the 1980s and

1990s could be utilized for this purpose. Moreover, we thought it imperative to conduct this new bioorganometallic chemistry research in water, to make it more relevant to the biological systems. The focus of our collaborators was to design a method to anchor large DNA molecules to the gold electrode array on the stage of an epifluorescence microscope for mapping and sequencing λ DNA, as a DNA model 50,000 base pair oligonucleotide. We were able to utilize $[Cp^*Rh(H_2O)_3](OTf)_2$ and a synthesized 20 mer, consisting of a 10 mer of adenosine bases, and a 10 mer with the complementary sequence of the end λ DNA bases, for these purposes.[5] However, even though this approach was successful in binding λ DNA to the gold electrode array, we did not understand how the $[Cp^*Rh(H_2O)_3](OTf)_2$ was binding to the 20 mer.

Therefore, we embarked on a project to study the reactions of $[Cp^*Rh(H_2O)_3](OTf)_2$ with individual nucleobases, in water, and as a function of pH.

At that time, the early 1990s, relatively little was known about the pH dependent nature of the reactions of organometallic aqua complexes with DNA bases. We found that the pH dictated the structure of the Cp*Rh-DNA base complexes. For example, the reaction of 9-methyladenine (9-MA) with $[Cp^*Rh(H_2O)_3](OTf)_2$ provided a cyclic trimer at pH 6.8 (pD 7.2) (Fig. 1.2), whose structure, complex **6**, at that time, was unprecedented for an organometallic-DNA base complex.[6]

This was the paradigm for all the other Cp*Rh-DNA base structures we solved by single crystal X-ray analysis, with pH dictating the structure of the complexes that formed. This was also illustrated with 9-methyl/ethylhypoxanthine (9-MH/9-EH) in Fig. 1.3, where a different structure, based on the pKa of the hypoxanthine ligand, was obtained, complexes **7**, **8**, and **9**.

The X-ray structures of two other Cp*Rh-nucleobase complexes, 1-methyl cytosine and 1-methylthymine, as a function of pH, 5.4 and 10, respectively, were also elucidated by X-ray analysis, complexes **10** and **11** are shown in Fig. 1.4.[6]

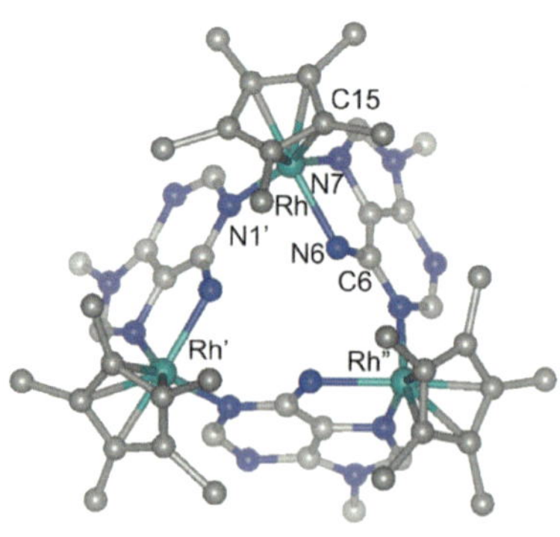

6

FIGURE 1.2 X-ray structure of trication, $[Cp^*Rh(\mu\text{-}\eta^1(N1)\text{:}\eta^2(N6, N7)\text{-}9\text{-}MA]_3(OTf)_3$ (triflate anions not shown for simplicity), **6**.

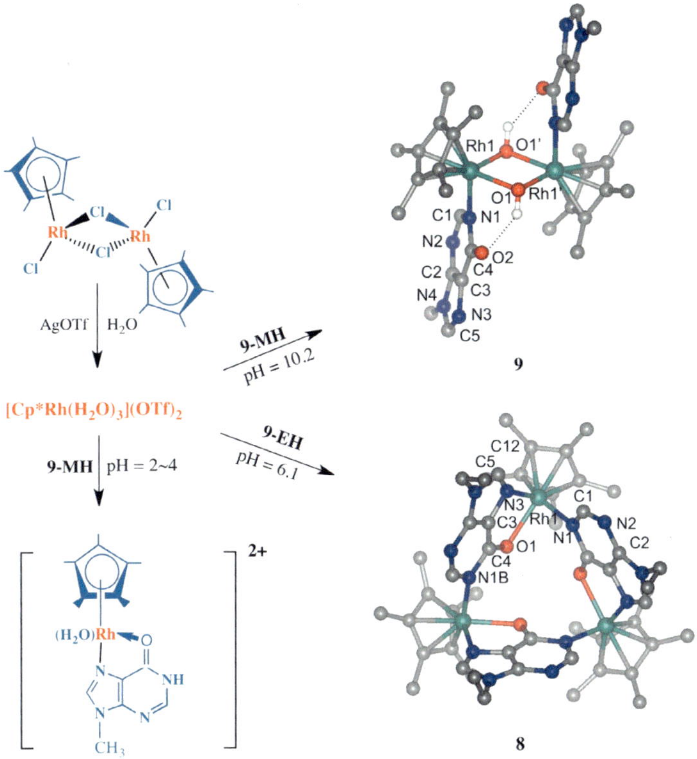

7

FIGURE 1.3 Cp*Rh-hypoxanthine structures, **7**, **8**, and **9**, dictated by pH.

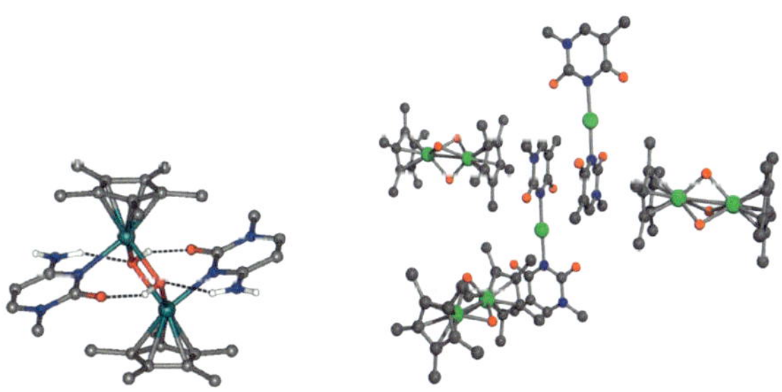

10 **11**

FIGURE 1.4 X-ray structures of *trans*-$[Cp^*Rh(\eta^1(N3)$ 1-methylcytosine$)(\mu\text{-}OH)]_2(OTf)_2$, **10**; 2 $[Rh(\eta^1(N^3)\text{-}1\text{-}MT)_2]^-$. $3[(Cp^*Rh)_2(\mu\text{-}OH)_3]^+$, **11**.

Since DNA may possibly be implicated as a site for a metal-based drug's mode of action, our studies on individual DNA bases, nucleotides, and nucleosides, in water, were important for structure–activity relationships, and the effect of pH on reactivity.

1.4 THE REACTION OF $[Cp^*Rh(H_2O)_3](OTf)_2$ WITH CO-FACTOR, NICOTINAMIDE ADENINE DINUCLIOTIDE (NAD^+)

We were also particularly intrigued by the common cyclic trimer structure that occurs with all 9-substituted adenine derivatives we have perused. This cyclic trimer, trication structure, $[Cp^*Rh(\mu\text{-}\eta^1(N1)\text{:}\eta^2(N6,N7)\text{-9-substituted adenine}]_3^{3+}$, with complex **6** as an example, forms via a self-assembly mechanism and provides this thermodynamically favored product, where the N-9-adenine substituent was alkyl, ribose, or a ribose phosphate ester. Thus, an adenine derivative that sparked our interest was nicotinamide adenine dinucleotide (NAD^+), an important co-factor in enzymatic reactions that utilizes its reduced form, 1,4-NADH, as a hydride source in chiral reduction reactions, in tandem with enzymes, such as HLADH.[7]

The ^1H NMR spectroscopy of NAD^+ and $[Cp^*Rh(H_2O)_3](OTf)_2$ showed very unambiguously that as you increase the pH from 3.0 to 6.0, two diastereomeric compounds are evident (Cp* signals at 1.670; 1.673 ppm, pH 6.0), and that these diastereomers have a very narrow stability range; i.e., at pH values less than 6.0, there is a presumed mixture of $[Cp^*Rh(NAD)]_3^{3+}$ and Cp*Rh aqua intermediates (at least eight Cp*Rh signals are evident) being formed, while at pH values greater than 6.0, the $[Cp^*Rh\text{-}NAD]_3^{3+}$ complex appeared to be decomposing to NAD^+ hydrolysis products, and the known $[(Cp^*Rh)_2(\mu\text{-}OH)_3](OTf)$.[7] Furthermore, the diagnostic chemical shifts that we have previously observed for all the enantiomeric/diastereomeric, cyclic trimer $[Cp^*Rh(\mu\text{-}\eta^1(N1)\text{:}\eta^2(N6,\ N7)\text{-9-substituted adeninato/5'-adenosinato/5'-adenosinato monophosphate methyl ester}]_3^{3+}$ derivatives we synthesized were evident in the ^1H NMR spectrum at pH 6.0 (Fig. 1.5). Thus, the H8 proton

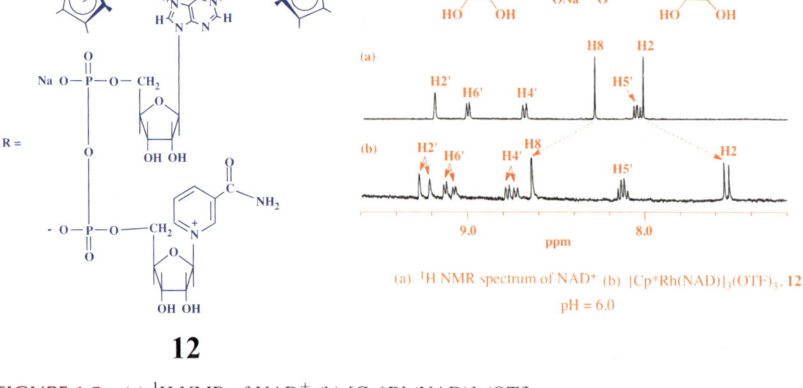

12

FIGURE 1.5 (a) ^1H NMR of NAD^+ (b) $[Cp^*Rh(NAD)]_3(OTf)_3$.

(8.28 ppm, **spectrum a**) of the adenine nucleus of free NAD^+ moves downfield ($\Delta\delta = 0.35$ ppm; 8.63, 8.64 ppm, **spectrum b**), while the free NAD^+ adenine H2 proton (8.01 ppm, **spectrum a**) moves dramatically upfield ($\Delta\delta = 0.48$ ppm; 7.54, 7.52 ppm, **spectrum b**). These diagnostic ^{1}H NMR spectroscopy results allow us to unequivocally assign a cyclic trimer structure to complex **12**; if complex **12** had a mononuclear structure, [Cp*Rh(η^2(N6, N7)-9-(5′-ribose pyrophosphate-5″-ribose-1″-nicotinamide)adeninato], the H8 and H2 protons would *both be downfield shifted* and no diastereomers would be evident, while the supramolecular, cyclic trimer structure of **12** provides a deshielding effect of H8 (downfield shift) and a shielding effect on H2 (upfield shift) with the presence of diastereomers (C3 symmetry). Interestingly, the nicotinamide protons also show two diastereomers with all pyridinium and ribose protons being doubled (Fig. 1.5). To our knowledge, this was the first bioorganometallic complex reported for NAD^+, itself.

1.5 MOLECULAR RECOGNITION STUDIES WITH CYCLIC TRIMER, [Cp*Rh(2′-DEOXYADENOSINE)]$_3$(OTf)$_3$, AS HOST, WITH BIOLOGICAL GUESTS, IN WATER

When we discovered the 9-substituted adenine, cyclic trimer structures, Fig. 1.2, as an example, having C3 symmetry, we found that the X-ray/computer generated molecular models conveyed a supramolecular, bowl structure to this potential host, and thought about the possibilities of non-covalent π−π, hydrophobic, and subtle hydrogen bonding interactions with biologically important guest molecules.[8] Indeed, this was the case and; moreover, we found that the [Cp*Rh(2′-deoxyadenosine)]$_3$(OTf)$_3$ complex, **13**, was the best host available (Fig. 1.6).

Therefore, we studied a variety of guest aromatic and aliphatic amino acids, substituted aromatic carboxylic acids, and aliphatic carboxylic acids including examples such as L-phenylalanine, L-tryptophan (**L-Trp**), phenylacetic and cyclohexylacetic (**CAA**) acids were studied by ^{1}H NMR spectroscopy (association constants [K_a] and free energies of complexation [ΔG^o]) for their

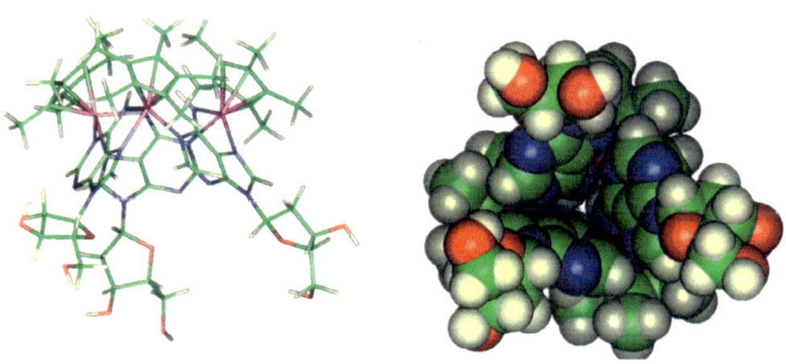

13

FIGURE 1.6 Shown on the left, the Dreiding model, while on the right, the CPK model of host, [Cp*Rh(2′-deoxyadenosine)]$_3$(OTf)$_3$, **13**.

non-covalent interactions with the host, [Cp*Rh(2′-deoxyadenosine)]$_3$(OTf)$_3$. Thus, the aromatic groups interacted by a classical π−π mechanism, while the aliphatic guests by classical hydrophobic interactions. The computer-generated molecular recognition process of **L-Trp** with [Cp*Rh(2′-deoxyadenosine)]$_3$(OTf)$_3$, **13**, is shown in the energy-minimized, space-filling host and the docking of **L-Trp** (Fig. 1.7).[8a,b] These overall results suggested that the molecular recognition of **L-Trp** with [Cp*Rh(2′-deoxyadenosine)]$_3$(OTf)$_3$ can be described in a way that places the **L-Trp** aromatic rings inside of the host cavity with the aromatic plane, or more specifically, the line which bisects the C-H(a) and C-H(a′) bonds parallel to the C3 axis of the host.[8a,b]

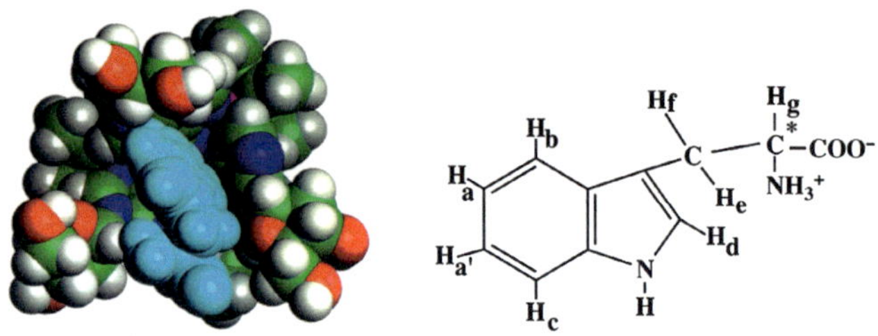

FIGURE 1.7 Molecular docking experiment with host, [Cp*Rh(2′-deoxyadenosine)]$_3$(OTf)$_3$, (**Left**), and guest, L-tryptophan, (**Right**).

1.6 BIOMIMICS OF NAD$^+$/1,4-NADH, AND CO-FACTOR REGENERATION WITH [Cp*Rh(bpy)H](OTf): ENZYME COMPATIBILITY WITH ORGANOMETALLIC CHEMISTRY

Furthermore, we were interested in understanding the mechanism of an important co-factor regeneration reaction that regioselectively reduced natural NAD$^+$ to 1.4-NADH using in situ formed [Cp*Rh(bpy)H](OTf).[9] However, we decided to simplify the natural NAD$^+$ structure by using a biomimic that stripped the ribose, pyrophosphate, and adenosine groups from the natural NAD$^+$ structure, while adding a benzyl group, N-benzylnicotinamide triflate, **14**. We also synthesized a NAD$^+$ biomimic that contained a ribose with a methyl ester phosphate group, N-ribose phosphate (methyl ester) nicotinamide, **15**. We then found that the biomimetic models were also regioselectively reduced by in situ generated [Cp*Rh(bpy)H](OTf) to form 1,4-NADH, biomimics, via the following proposed mechanism (Fig. 1.8).[9]

In a totally serendipitous manner, we decided to utilize the biomimetic models of NAD$^+$, **14** and **15**, with the HLADH enzyme, in tandem with [Cp*Rh(bpy)H](OTf) co-factor regeneration. We were totally amazed that a 1,4-NADH dependent enzyme, horse liver alcohol dehydrogenase (HLADH), recognized the 1,4-NADH biomimics to convert achiral ketones to chiral alcohols; in control experiments, these biomimics provided similar results as the natural 1,4-NADH, both in yield and enantioselectivity, while a critical control experiment verified that in the absence of the biomimetic

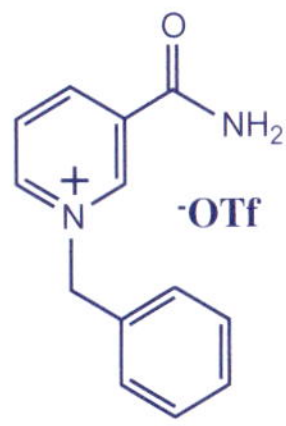

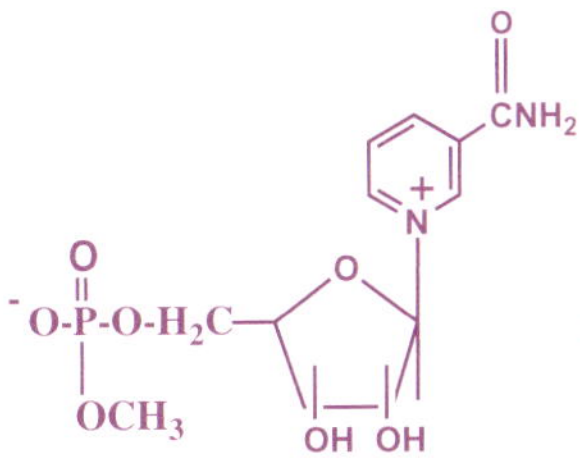

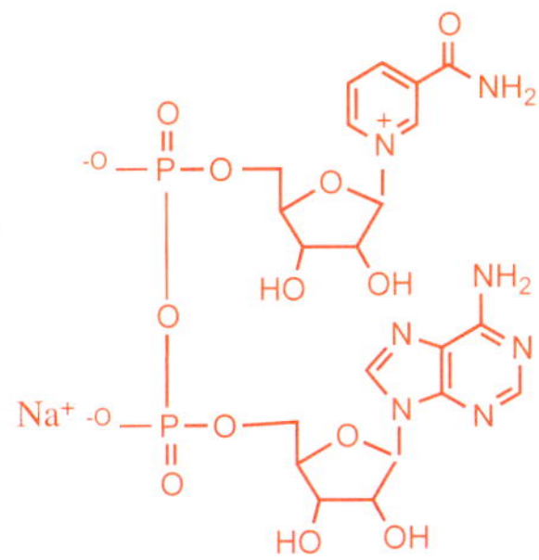

NAD+

14 **15**

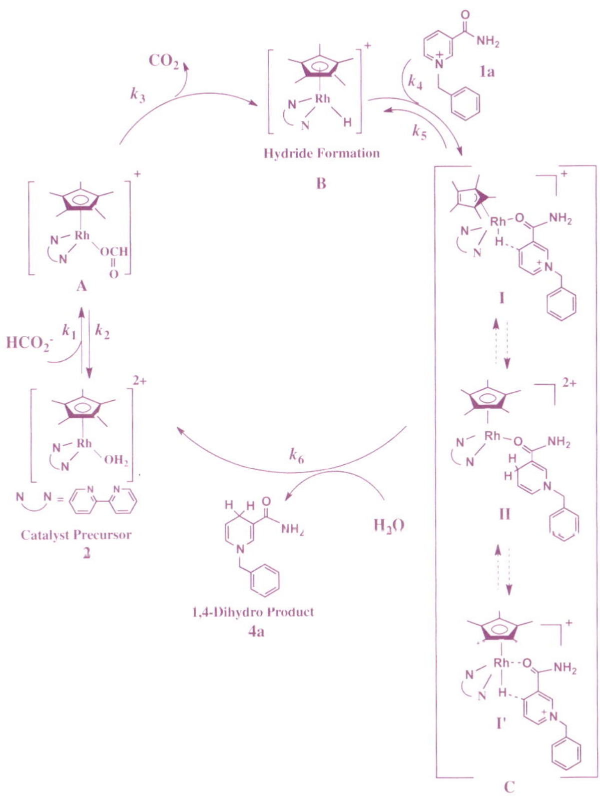

FIGURE 1.8 Plausible regioselective reduction mechanism of a NAD^+ biomimic with in situ formed [Cp*Rh(bpy)H](OTf).

co-factor, no chiral alcohols were formed; only racemic alcohols were formed from the direct reaction of [Cp*Rh(bpy)H](OTf) with the ketone (Scheme 1.3).[10a,b]

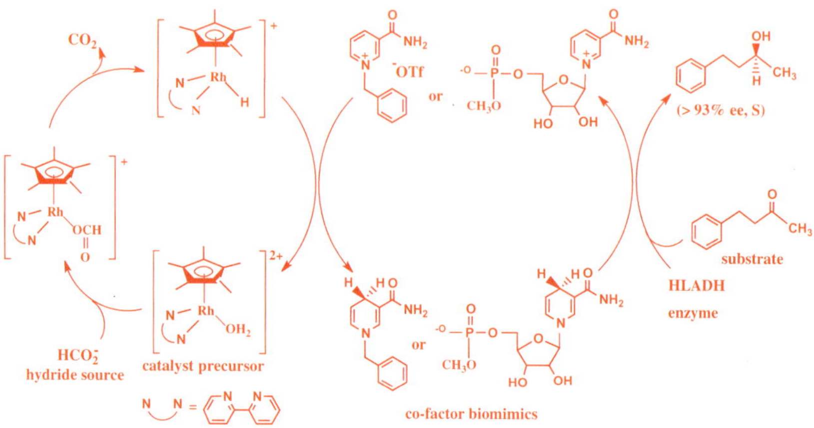

SCHEME 1.3 Tandem catalysis: Co-factor regeneration and chiral synthesis using HLADH with biomimetic NAD^+/1,4-NADH co-factors, **14** and **15**.

This unprecedented tandem biocatalysis result was also observed with monooxyenases, such as P450 BM-3 and P450 CAM, and was further enhanced by utilizing protein engineering (Scheme 1.4) to increase the molecular recognition process of the 1,4-NADH biomimetic co-factor and

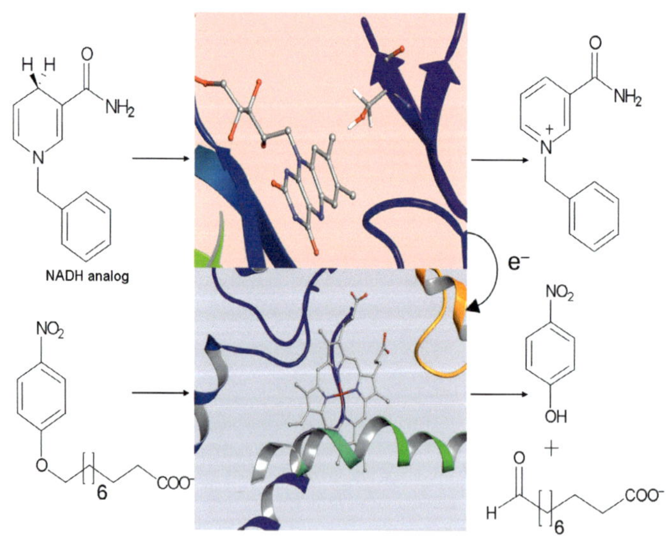

SCHEME 1.4 P450 BM3 Double Mutant: Trp to Ser, and Arg to Asp, via Site Directed Mutagenesis.

the FAD containing electron transfer proteins.[11] These results signified that organometallic chemistry and enzymology have been shown to be complementary, and that [Cp*Rh(bpy)H](OTf) co-factor regeneration was not compromised by enzyme interactions. This was also the case utilizing a monoxygenase, HbpA, a catecholase that converts phenols to catechols.[12]

1.7 Cp*Rh-HYDROXYTAMOXIFEN COMPLEXES: THEIR SYNTHESIS, BIOLOGICAL ACTIVITY, AND MOLECULAR DOCKING TO THE ISOFORMS OF THE ESTROGEN RECEPTOR, ERα AND ERβ

Another topic we have pursued was focused on the organometallic pharmaceutical discipline. The drug discovery paradigm was dramatically changed with the introduction of organometallic pharmaceuticals, as complexes that were more stable kinetically and thermodynamically, and whose properties could be adjusted to be hydrophobic for cell penetration, and hydrophilic inside the cell.[13a–q] Our approach was to study the chemistry of the *cis* and *trans* metabolites of tamoxifen, hydroxytamoxifens, **16** and **17**, with Cp*Rh solvato complexes, to identify novel chemistry with this multidentate ligand, and hopefully, provide evidence for biological activity.[14]

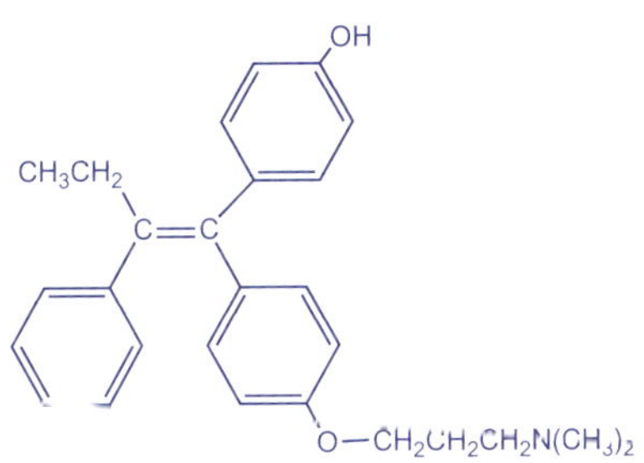

***cis*-16**

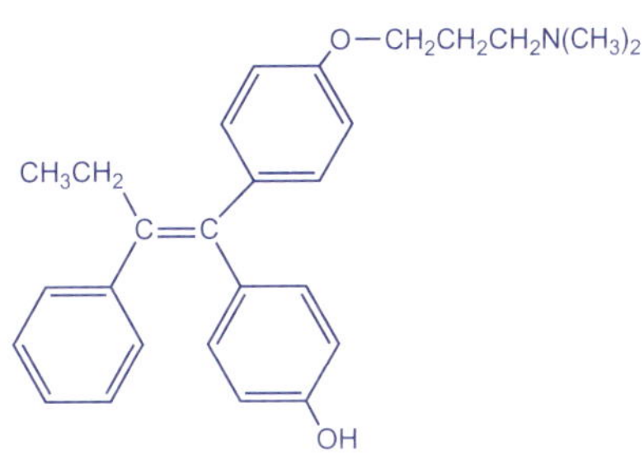

***trans*-17**

Our initial thought was to conduct all the hydroxytamoxifen ligand experiments in water; however, we quickly clarified an important point, that ligands, ***cis*-16** and ***trans*-17**, were not water soluble. The initial studies were accomplished via ^{1}H NMR in CH_3OH-d_4. This provided evidence that the kinetic products were η^1-N-Cp*Rh complexes, [1-butenyl-2-phenyl-1-(p-phenol)-1′-*p*-phenyl-(oxotrimethylene-3-dimethylamino)(η^1-N-η^5-pentamethylcyclopentadienylrhodium)(η^1-O-(CH_3OH)(OTf)](OTf), ***cis*-18(CH_3OH)** and ***trans*-19(CH_3OH)**.[14]

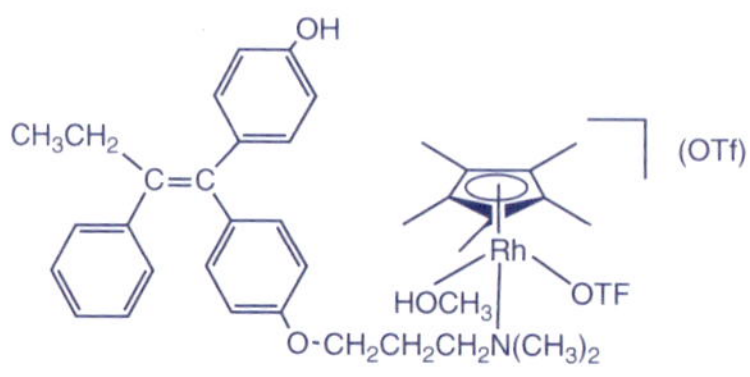

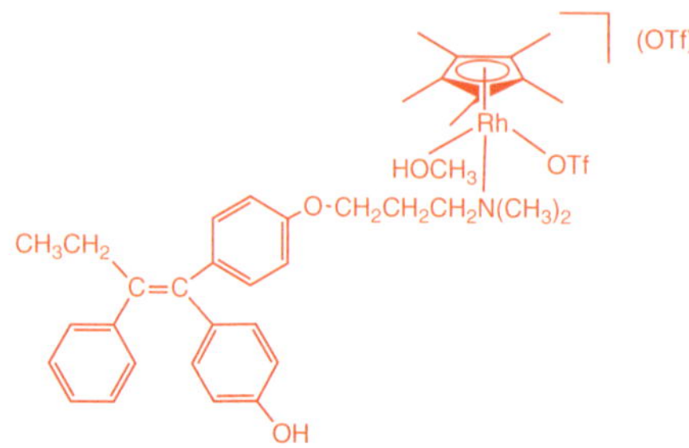

***cis*-18(CH$_3$OH)** ***trans*-19(CH$_3$OH)**

When we heated these η[1]-N-Cp*Rh complexes, ***cis*-18(CH$_3$OH)** and ***trans*-19(CH$_3$OH)**, individually to 60°C, we observed a novel and unprecedented N-π rearrangement that gave, after in-depth HMQC, HMBC, COSY, and NOESY correlation NMR studies, the *cis* or *trans*-[1-butenyl-2-phenyl-1-(η^5-pentamethylcyclopentadienyl- η^6-*p*-phenol)rhodium)-1′-*p*-phenyl-(oxo-trimethylene-3-dimethylamino)](OTf)$_2$, respectively, complex **[*cis*-20]$^{2+}$** or **[*trans*-21]$^{2+}$**.[14]

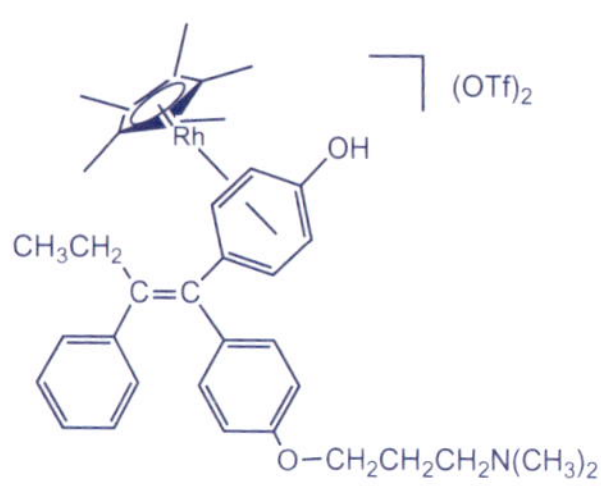

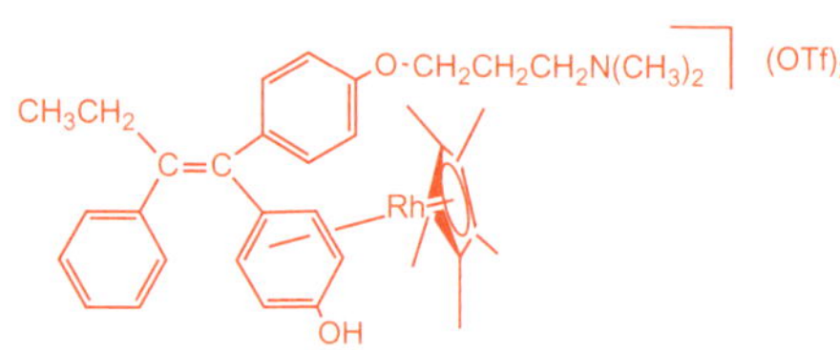

[*cis*-20]$^{2+}$ **[*trans*-21]$^{2+}$**

Moreover, we found that starting with ***cis*-18(CH$_3$OH)**, we observed the two η^6-Cp*Rh complexes, **[*cis*-20]$^{2+}$** and **[*trans*-21]$^{2+}$**, via NMR analysis of the new dimethylamino signals that appeared at 2.97 and 2.95 ppm; respectively, in the ratio of ~5:1, after 17 h of heating at 50°C in CD$_3$OD-d$_4$, reflecting the isomerization of ***cis*-18(CH$_3$OH)** to ***trans*-19(CH$_3$OH)** prior to the N-π rearrangement, as stated above.

Interestingly, there was a dramatic solvent effect on the rate of the N-π rearrangement.[14] We discovered that in CH$_2$Cl$_2$, the N-π rearrangement, starting with the *in situ* formed η[1]-N complex, was completed in 15 h at

22°C for both ***cis*-18(OTf)** and ***trans*-19(OTf)** to either [***cis*-20**]$^{2+}$ or [***trans*-21**]$^{2+}$ (Eq. 1.2), while in CH_3OH *no reaction* occurred at 22°C; *under these mild reaction conditions in CH_2Cl_2, neither* ***cis-18*(OTf)** *nor* ***trans-19*(OTf)** *was isomerized to a cis and trans mixture of both starting complexes.*

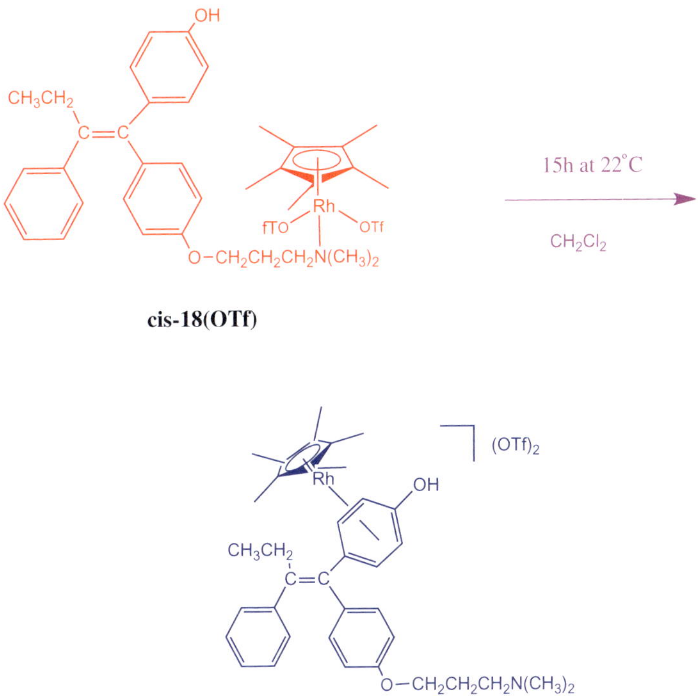

cis-18(OTf)

[cis-20(OTf)]$^{2+}$

(1.2)

1.7.1 The Mechanism of the N-π Rearrangement

The intramolecular and chemoselective N-π rearrangement involved both the monocationic Cp*Rh-N complex, and the inner sphere Cp*Rh-OTf bond dissociations, as highly important steps, in this reaction mechanism. The geometries of the complexes that existed in CH_2Cl_2 and CH_3OH solutions, including their recently derived DFT calculated energies, allowed us to determine the lowest energy pathway for this novel transformation.[15] Thus, the most plausible reaction mechanism for the N-π rearrangement is shown in Scheme 1.5.

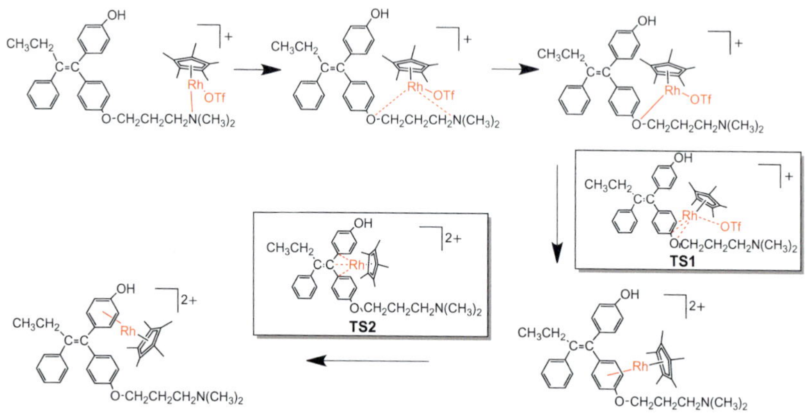

SCHEME 1.5 Plausible reaction mechanism for the N-π rearrangement. Outer sphere counter-ions are omitted for clarity. TS1 and TS2 are the DFT calculated transition states.[15]

1.7.2 Relative Binding Affinities of Complexes [cis-20]$^{2+}$ and [trans-21]$^{2+}$ for the ERα and ERβ Estrogen Receptors, and Growth Inhibition Properties of [cis-20]$^{2+}$ with Breast Cancer Cell Lines

Tamoxifen was found to be an antagonist breast cancer drug, that competitively inhibited the female hormone, estradiol, from binding to the hormone-dependent ERα estrogen receptor, as one important aspect for anticancer activity. Therefore, in order to determine the biological parameters of the water soluble **[*cis*-20]**$^{2+}$ and **[*trans*-21]**$^{2+}$ complexes, the relative binding affinities (RBA) of these complexes for the two isoforms of the estrogen receptors (ERα and ERβ, 4°C, 3 h) were determined using a competition experiment with [^{3}H]-estradiol, and provided values of 4.7% and 1%, respectively.[14] Thus, it clearly showed a significant difference between the two geometrical isomers, **[*cis*-20]**$^{2+}$ and **[*trans*-21]**$^{2+}$, which suggested differences in the binding conformation at the estrogen receptor. For comparison, Ferrocifen has an RBA of 11.5% (4°C, 3 h).[14] Clearly, **[*cis*-20]**$^{2+}$ was found to be moderately competitive with estradiol for binding to the ERα receptor. Both compounds were recognized by the two isoforms of the estrogen receptor; however, the RBA values for the *cis* isomer, **[*cis*-20]**$^{2+}$, on both isoforms of the receptor were significantly higher than those found for the *trans* isomer, **[*trans*-21]**$^{2+}$.[14]

The effect of the complex, **[*cis*-20]**$^{2+}$, on the growth of cancer cells has been tested on two breast cancer cell lines, MCF-7 and MDA-MB-231, which are respectively, hormone dependent and hormone independent. The **[*cis*-20]**$^{2+}$ complex showed a growth inhibition, on these two cell lines with IC_{50} values in the range of 1.25–11 μM. Thus, the **[*cis*-20]**$^{2+}$ isomer, which had a moderate RBA value, showed a antiestrogenic effect towards the MCF7 breast cancer cells, IC_{50} of 1.25 μM, and a modest anti-proliferative effect towards the hormone independent MDA-MB-231 breast cancer cells, with an IC_{50} of 11 μM.[14]

1.7.3 Computer Docking Studies of [cis-20]$^{2+}$ to the Estrogen Receptors, Hormone Dependent, ERα, and Hormone Independent, ERß, Including Thermodynamic Values of Binding

Since one important aspect of the bioactivity of all these potential organic, inorganic, and organometallic breast cancer drugs has been their action as an antagonist in competition with the female hormone, estradiol, we were interested in the binding modes of [***cis*-19**]$^{2+}$, the growth inhibitor that was analyzed against the MCF7 and MDA-MB-231 cell lines, ERα(+), IC_{50} = **1.25** μM, and ERα (−), IC_{50} = **11** μM, respectively. Cartoon representations of the predicted 3D structures of [***cis*-20**]$^{2+}$, were obtained using **3ERD** and **2FSZ** PDB structures as initial models of the ERα and ERβ receptor binding sites, respectively, and are shown in Fig. 1.9, along with the binding energies. The [***cis*-20**]$^{2+}$complex was bound in the pocket formed by helices H3, H4, H5, and H12, in a manner similar to that described before for other drugs.[14]

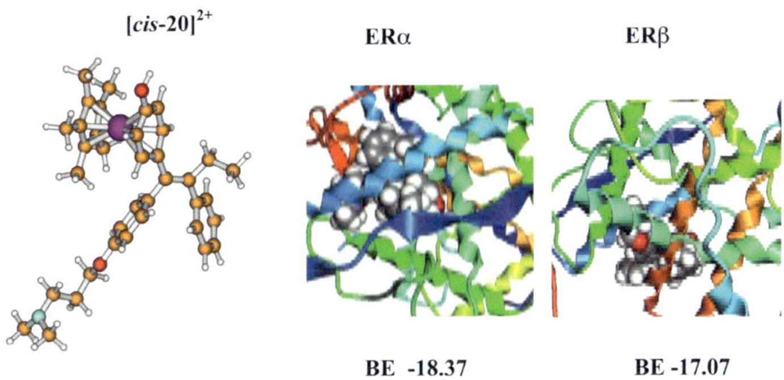

FIGURE 1.9 Cartoon representations of the ERα and ERβ receptor binding sites with docked [*cis*-20]$^{2+}$ (**left**), together with the corresponding calculated binding energies (kcal/mol). The docked guests are shown in the CPK rendering mode.

Analysis of the [***cis*-20**]$^{2+}$ interactions with the ERα and ERβ receptor binding sites demonstrated that the main contribution to the host-guest interaction energy came from van der Waal and electrostatic interactions of the dicationic complex with the surrounding amino acids. Interaction of the phenol hydroxyl group, and the ether oxygen groups of [***cis*-20**]$^{2+}$ with O atoms of the surrounding amino acids, has generally been shown to have repulsive characteristics. More importantly, it also could possibly be the compact configuration adopted by the guest, [***cis*-20**]$^{2+}$, within the ERα host binding site, such that the electrophilic H atom of the phenol group was not readily available for non-covalent bonding regimes.[14,15]

1.8 REACTIONS OF [Cp*Rh(H_2O)$_3$](OTf)$_2$ WITH G-PROTEIN-COUPLED RECEPTOR PEPTIDES: SYNTHESIS, DFT/2D NMR STRUCTURES, AND THE BIOLOGICAL CONSEQUENCES

The bioorganometallic chemistry discipline has continued to fully demonstrate that organometallic chemistry is compatible, at the interface, with biology. Another important current interest, the bioconjugation of peptides to organometallic complexes, has been studied extensively.[16] In general, the organometallic complexes have been bioconjugated terminally, to either the free amino or carboxyl groups of the designated peptide, which could potentially have deleterious effects on their bioactivity, while the majority of these bioconjugation reactions were conducted in organic solvents.

Moreover, the phenol side chain of tyrosine-containing G-protein-coupled receptors (GPCRs) peptides has been shown to be a potential component for molecular recognition, as well as biological activity. Furthermore, the GPCRs have been shown to influence the physiological responses to hormones, neurotransmitters, and environmental stimulants; and therefore, have importance as therapeutic targets for a wide spectrum of critical diseases.[17] For example, tyrosine kinases have been involved in autoimmune diseases and cancer, while tyrosine residues, in proteins, were shown to be crucial in some electron-transfer pathways, or at the active sites of some enzymes.

Surprisingly, the tyrosine residue of peptides had not been previously considered as a site of reactivity with organometallic reagents, in competition with other aromatic amino acid residues. Therefore, we discovered a facile, chemoselective bioconjugation reaction of important tyrosine amino acid components of GPCR peptides, in water, at room temperature, as a function of pH, by utilizing the air and water stable organometallic aqua complex, [Cp*Rh(H_2O)$_3$](OTf)$_2$.[18] The important tyrosine-containing GPCR peptides that were used to demonstrate this chemoselective bioconjugation technique are shown in the Chart 1.1, and they include, [Tyr^1]-leu-enkephalin, [Tyr^4]-neurotensin(8–13), and [Tyr^3]-octreotide, with the tyrosine residue being a potential molecular recognition component for binding to their respective receptor sites; the message-address paradigm.

A representative reaction, with 2D NMR structures of both the GPCR peptide, [Tyr^3]-octreotide, and its [(η6-Cp*Rh-Tyr^3)-octreotide]$^{2+}$ complex, are shown in Fig. 1.10, including the overlay of both 2D NMR structures. Furthermore, both adopted the backbone canonical β-turn structure at the well-defined ^{D}Trp^4-Lys^5-dipeptide, although their pharmacophore conformations were found to be dramatically different. In the 2D NMR structure of [(η^6-Cp*Rh-Tyr^3)-octreotide]$^{2+}$, the Cp*Rh group was coordinated to the side chain of Tyr^3 in an η^6 bonding mode, and consequently, pushed the side chain of the ^{D}Trp^4 residue away, and rotated toward the direction of the Lys^5 residue (Fig. 1.10). This resulted in the Lys^5 residue being flipped to the

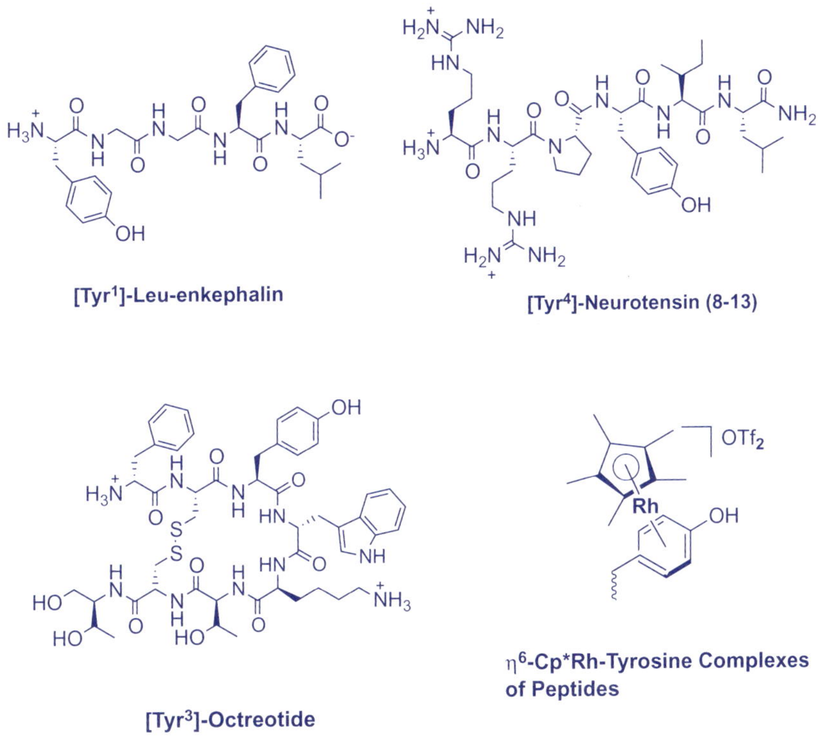

CHART 1.1 GPCR Peptides and their [(η^6-Cp*Rh-Tyr$^\#$)-GPCR Peptide]$^{2+}$ Complexes.

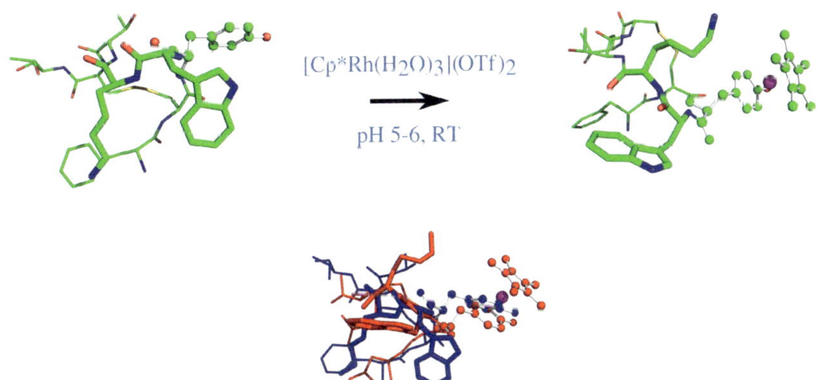

FIGURE 1.10 (**Top**) 2D NMR Structures of [Tyr^3]-octreotide (**left**), and the [(η^6-Cp*Rh-Tyr^3)-octreotide]$^{2+}$ complex (**right**), with reaction conditions, (**bottom**) overlay of [Tyr^3]-octreoide (**blue**) and [(η^6-Cp*Rh-Tyr^3)-octreotide]$^{2+}$(Red).[18]

other side of the pharmacophore (Fig. 1.10). The overall structural similarities of the peptide backbones, demonstrated by 2D NMR, were also confirmed by the CD spectra of [Tyr^3]-octreotide with minima at 196 and 218 nm, and a maximum at 230 nm, corresponding to an anti- parallel pleated β-sheet and type II′ β-turn, while [(η^6-Cp*Rh-Tyr^3)-octreotide]$^{2+}$ had two less intense, and slightly red-shifted minima at 198 and 221 nm, and no maximum around 230 nm.[18]

The plausible reason for the high chemoselectivity of the [η^6-Cp*Rh-$Tyr^{\#}$]$^{2+}$ product with all three peptides, at pH ~5 − 6, might be predicated on the dramatic lowering of the pK_a of the phenol hydrogen as the η^2/η^4-Cp*Rh-tyrosine intermediates transition to the final η^6-Cp*Rh-tyrosine complex; i.e., the back donation into the Rh d-orbitals from increased electron density into the phenol aromatic ring as the pH was raised from ~3 to ~5 − 6, provided a driving force for higher electrophilic reactivity. Moreover, the pH also appears

TABLE 1.2 Comparison of the GPCR Receptor Binding and Cancer Cell Growth Inhibition Activity of [(η^6-Cp*Rh-Tyr^1)-Leu-enkephalin]$(OTf)_2$ and [(η^6-Cp*Rh-Tyr^3)-Octreotide]$(OTf)_2$ With Their Peptide Precursors

Receptor	Substrates	
	[(η^6-Cp*Rh-Tyr^3)-Octreotide] $(OTf)_2$	[DTPA-^{D}Phe^1]-Octreotide
SST_2[a] (IC_{50}, nM)	15.8	13.8
IC_{50} Growth Inhibition (μM)		[Tyr^3]-Octreotide
MCF7	4.6	4.6
HT29	5.3	4.3
	[(η^6-Cp*Rh-Tyr^1)-Leu-enkephalin]$(OTf)_2$	[Tyr^1]-Leu-enkephalin
μ-OR[b] (EC_{50}, nM)	93.3	14.3
∂-OR[c]	15.6	4.4
μ−OR + ∂-OR	3.4	3.3
IC_{50} Growth Inhibition (μM)		
MCF7	4.1	2.8
HT29	4.6	3.8

[a]*Somatostatin receptor.*
[b]*μ-Opioid receptor.*
[c]*∂-Opioid receptor*

to have controlled the lack of reactivity of all the amino groups via protonation. The biological consequences of the η^6-Cp*Rh modification of tyrosine-containing peptides, were of interest; therefore, our initial bioassay studies were conducted with $[(\eta^6\text{-Cp*Rh-Tyr}^1)$-leu-enkephalin]$(OTf)_2$. The EC_{50} receptor binding value for the [Tyr^1]-leu-enkephalinan opioid receptor peptide, and that for $[(\eta^6\text{-Cp*Rh-Tyr}^1)$-leu-enkephalin]$(OTf)_2$, at the μ-opioid receptor (**μ-OR**), or ∂-opioid receptor (**∂-OR**), can be found in Table 1.2. It was shown that $[(\eta^6\text{-Cp*Rh-Tyr}^1)]$-leu-enkephalin]$(OTf)_2$, was an agonist, with nM potency, on cells expressing **μ-OR** **∂-OR** alone, as well as on cells co-expressing both μ-OR and ∂-OR (Table 1.2). Thus, η^6-coordination of Cp*Rh to the tyrosine residue of [Tyr^1]-leu-enkephalin lowered its potency toward cells expressing only **μ-OR** or **∂-OR**, but had a similar potency as [Tyr^1]-leu-enkephalin for cells co- expressing both **μ-OR** and **∂-OR**, which coincided with previous findings of distinctly different pharmacological profiles for co-expressed versus receptor cells alone, and complements the results with peptide, [Tyr^1]-leu-enkephalin. Moreover, a competitive binding experiment was performed on the GPCR somatostatin receptor, rat SST_2, with positive tumoral acinar pancreatic cells (AR42J), using [^{111}In-DTPA,^{D}Phe^1]-octreotide, as a radiotracer. Thus, from these results, $[(\eta^6\text{-Cp*Rh})\text{-Tyr}^3)$-octreotide]$(OTf)_2$ had a very similar affinity for the SST_2 receptor in comparison to [DTPA,^{D}Phe^1]-octreotide, a GPCR peptide with no tyrosine residue at position 3 (Table 1.2).[18]

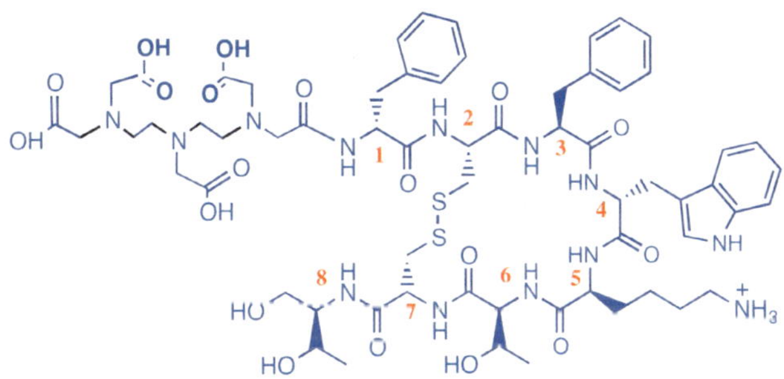

[DTPA,^{D}Phe^1]-Octreotide

Apparently, the Cp*Rh moiety η^6-bonded to the tyrosine residue, which was on the opposite side to the ^{D}Trp^4-Lys^5-pharmacophore (Fig. 1.10), had little effect on the binding of $[(\eta^6\text{-Cp*Rh})\text{-Tyr}^3)$-octreotide]$(OTf)_2$ to the GPCR SST_2 receptor. However, this result still suggested that the $[(\eta^6\text{-Cp*Rh})\text{-Tyr}^3)$-octreotide]$(OTf)_2$ residue was an important component in the molecular recognition process, at the SST_2 receptor site, and intimated that the flexibility of the receptor site can accommodate major changes in the pharmacophore conformation, but still retain bioactivity. Moreover, the affinity of [(η^6-Cp*Rh)-Tyr^3)-octreotide]$(OTf)_2$ for the SST_2 receptor, which has been shown to be the most abundant SST subtype in human tumors expressing the somatostatin receptors, was also found to be in the range of other somatostatin peptides.

In other bioassay experiments, the consequence of this Cp*Rh modification of peptides, [Tyr^1]-leu-enkephalin and [Tyr^3]-octreotide, on the in vitro growth inhibition of the breast adenocarcinoma cell line (MCF7) and the human colon carcinoma (HT29) cell line was determined. For these bioassays, both [(η^6-Cp*Rh-Tyr^1)-leu- enkephalin]$(OTf)_2$ and [(η^6-Cp*Rh-Tyr^3)-octreotide]$OTf)_2$ were tested and compared to peptides, [Tyr^1]-leu-enkephalin and [Tyr^3]-octreotide (Table 1.2). Interestingly, both [(η^6-Cp*Rh-Tyr^1)-leu-enkephalin]$(OTf)_2$ and [(η^6-Cp*Rh-Tyr^3)-octreotide]$OTf)_2$ had very similar IC_{50} values with those of peptides [Tyr^1]-leu-enkephalin and [Tyr^3]-octreotide, for both cell lines (Table 1.2). Since the MCF7 and HT29 cell lines were found to express opioid and somatostatin receptors, among others, this growth inhibition activity observed for both Cp*Rh-peptide complexes might be directly related to their GPCR binding regimes. [18] These latter bioassay studies also confirmed the findings of the GPCR binding experiments, in that, the antiproliferative activity of peptides [Tyr^1]-leu-enkephalin and [Tyr^3]-octreotide was still retained by the Cp*Rh modification of their tyrosine residues, complexes, and to reiterate, provided information on the potential role of the opioid and somatostatin receptors in the growth inhibition of MCF7 and HT29 cancer cells.

1.9 MOLECULAR DOCKING STUDIES OF A [(η^6-Cp*Rh-Tyr^1)-LEU-ENKEPHALIN]$^{2+}$ COMPLEX TO THE μ-, ∂-, AND κ – OPIOID RECEPTORS, IN COMPARISON TO THE PEPTIDE, LEU-ENKEPHALIN

G-Protein-Coupled Receptors (GPCRs), especially the μ-, ∂-, and κ-opioid receptors, have been some of the most extensively studied GPCRs, but what was missing was their unequivocal X-ray crystal structures, in order to better design non-addictive drug candidates for pain and addiction therapy. Recently, the X-ray structures of the μ-, ∂-, and κ-opioid receptors (OR) were published by two groups, with Kobilka et al. providing the μ- and ∂-opioid structures with their antagonist morphinan guest derivatives, while the κ-opioid receptor structure, also with an antagonist morphanin guest, was solved by Stevens et al (Scheme 1.6).[19a–c] The X-ray data showed a common seven transmembrane structure (Scheme 1.6, **left**, ∂-OR as an example) for the μ-, ∂-, and κ-opioid receptors, with conserved areas (Scheme 1.6, **center**), including conserved β-strand folds (**right**).

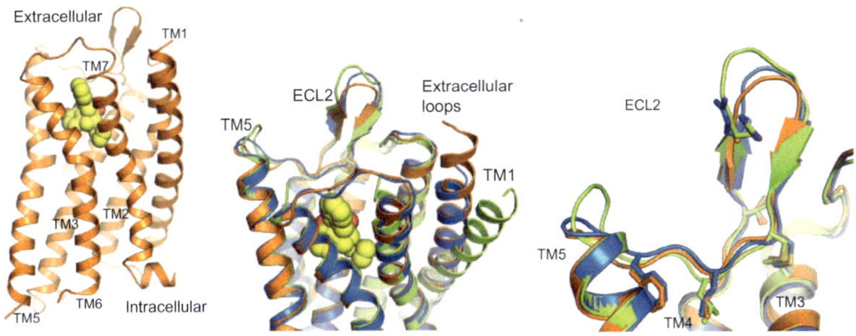

SCHEME 1.6 ∂-opioid receptor (**OR**) showing the typical seven-transmembrane structure (**left**); usual conserved fold structures for ∂-, μ-, κ-**ORs** (center, color coded); and the usual conserved β-strand folds, with an expansive, open binding pocket (**right**).[19b]

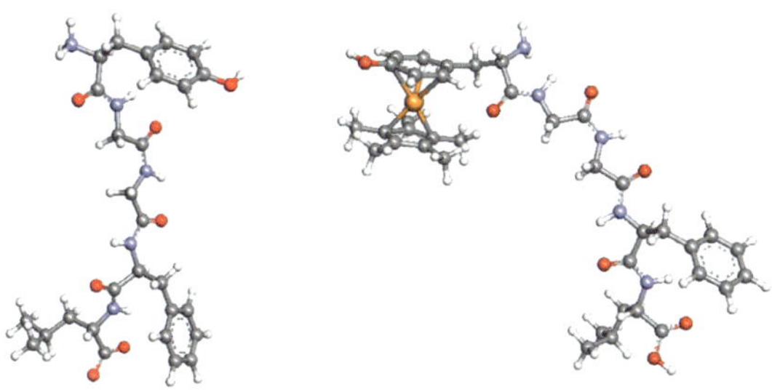

22 **23**

FIGURE 1.11 Optimized structures of [Tyr[1]]-Leu-enkephalin, as a zwitterion, **22**, and dication, $[(\eta^6\text{-Cp*Rh-Tyr}^1)\text{-Leu-enkephalin}]^{2+}$, a non-zwitterion, **23**.

1.9.1 Molecular Recognition Sequences of Peptide 22 and Complex 23 at the μ-, ∂-, and κ-Opioid Receptors

The important molecular recognition components of the opioid neuropeptide, [Tyr[1]]-Leu-enkephalin, **22** (Tyr[1]-Gly-Gly-Phe-Leu), and consequently complex **23**, was the terminal tyrosine residue with the phenol, terminal amino, and amide carbonyl groups, and the phenylalanine residue, which have been designated as prominent recognition points at the μ-, ∂-, and κ-opioid receptors, and labeled the message sequence (signal transduction) of the message-address paradigm for binding to the opioid receptors (Fig. 1.11).[20]

Moreover, the terminal leucine for **22**, and now for **23**, provided selectivity and/or reduced affinity for another opioid receptor, and has been designated the address sequence, while the spacers are the glycine, glycine amino acids between the message and address sequences (Fig. 1.12).[17] Therefore,

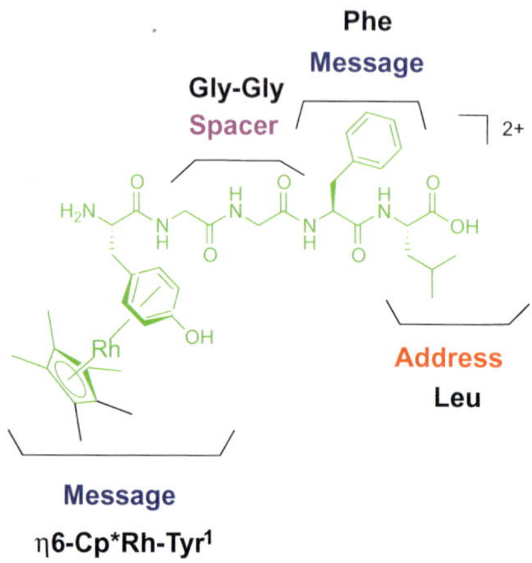

FIGURE 1.12 Message, spacer, and address sequences for molecular recognition at the μ-, ∂-, and κ-opioid GPCRs of complex **23**; the message-address paradigm.

from the Kobilka/Stevens et. al. X-ray studies of the μ-, ∂-, and κ- opioid receptors, it was found that the binding sites were separated into two regions, where the lower part of the binding pocket (message; signal transduction) was greatly conserved among all opioid receptors (Scheme 1.6), while the less conserved upper part of the binding pocket provided opioid receptor selectivity (address paradigm).[19a–c]

Thus, the message-address paradigm was validated for the pharmacological aspects with these opioid receptor X-ray structural results, in that any single drug provided both message and address recognition aspects, and were contained in an individual drug molecules that formed non-covalent H-bonds with receptor amino acid residues, and had conformational flexibility.[19a–c] However, it should be clearly pointed out that all three morphanin derivatives used in the X-ray structural determinations of the three opioid receptors were antagonists that do not elicit a biological response; and therefore, are not in biologically active conformations. This factor allowed us to verify with agonists, **22** and **23**, the differences in non-covalent binding to the opioid receptor amino acids, in comparison to the biologically inactive morphanin derivatives used for the X-ray structural analysis, and the possibility of using this information to better understand the non-covalent binding criteria that elicits bioactivity.[20]

We then evaluated the molecular recognition sequences, the message-address paradigm, that have been previously designated for neuropeptides, such as **22**, at GPCORs, and compared them to those found for complex **23** via molecular docking experimental calculations.[20] Furthermore, [Tyr^1]-Leu-enkephalin, **22**, an endogenous neuropeptide agonist, has been shown to bind to μ- and ∂-opioid receptor cells, but negligible/weak binding was found for the κ-opioid receptor.[20] However, we compared all three, μ-, ∂-, and κ-opioid receptors, in non-covalent binding with neuropeptide **22** and complex **23** (Fig. 1.11), to ascertain the differences in conformations and binding to all opioid receptor amino acid residues that were structurally characterized.[20]

The neuropeptide, **22**, docked at the μ-, ∂-, and κ-opioid receptors (Fig. 1.13), showed the conformation inside the opioid receptors (**left, center left**), the message sequences, tyrosine residue and phenylalanine carbonyl, along with the address sequence, leucine residue terminal carboxyl, as the major sites of non-covalent molecular recognition (**right**). Fig. 1.13 also shows (**center right**) that there are more than fifteen amino acid residues within 5 Å (**backside not shown**) of docked neuropeptide **22** at the μ-opioid receptor, more than fifteen residues at the κ-opioid receptor, and more than sixteen residues for the ∂-opioid receptor. The number of amino acid residue sites with non-covalent binding to neuropeptide **22**, within 3.5 Å (**right**), for the μ-opioid receptor, had four prominent H-bonding sites that included the message tyrosine phenol *O* and *H* atoms with the Ala 304 N-*H* atom, 2.84 Å,

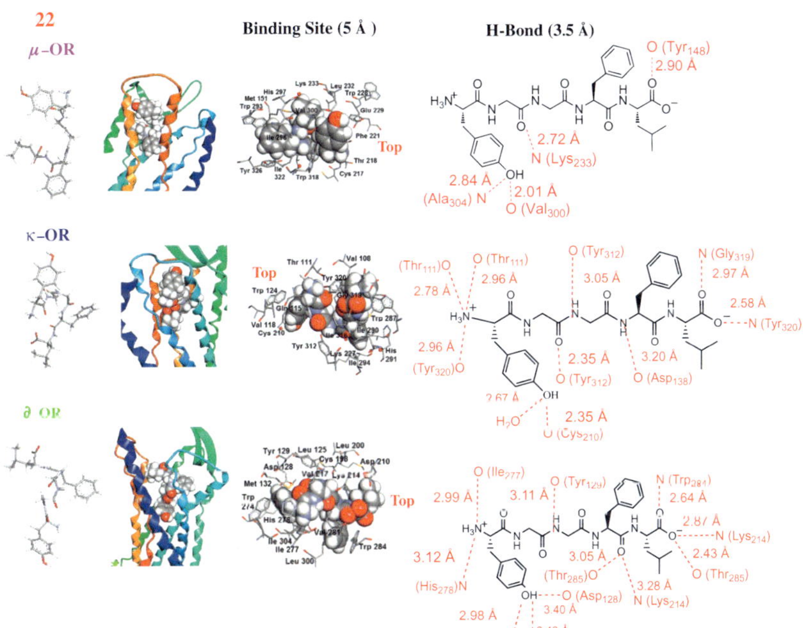

FIGURE 1.13 Conformation of **22** in the designated opioid receptor (**left, center left**); non-covalent binding regimes of [Tyr[1]]-Leu-enkephalin, **22**, to the μ-, κ-, and ∂- opioid receptors: the amino acid residues of the binding sites within 5 Å of docked peptide **22** (**center right**); and hydrogen bonds responsible for the non-covalent receptor interactions. H-bond lengths indicate the distance between the heavy atoms (**not hydrogen**) within 3.5 Å (**right**). **Top** refers to the binding pocket receptor opening.[20]

and the Val 300 carboxyl *O* atom, 2.01 Å; the spacer Gly carbonyl *O* atom with the N-*H* atom of Lys 233, 2.72 Å, while the address terminal Leu carboxyl *O* atom with the phenol *O* and *H* atoms of Tyr 148, 2.90 Å. The docked conformation of **22** in the μ-opioid receptor is shown on the left, to demonstrate the position in the receptors (**center left**).

The κ−opioid receptor with docked **22** had ten prominent H-bonding sites within 3.5 Å (Fig. 1.13) that included the message tyrosine phenol *O* and *H* atoms with the Cys 210 carboxyl *O* atom 2.35 Å, and a H_2O molecule *H* atom, 2.67 Å; the message NH_3^+ *H* atom to both the Thr 111 COO^- and *O*-H atoms, 2.96 Å and 2.78 Å; spacer Gly carbonyl *O* atom with the phenol O-*H* of Tyr 312, 2.35 Å; and the Gly N-*H* with the phenol *O* atom, Tyr 312, 3.05 Å; the message phenylalanine N-*H* bond with the amide carbonyl *O* atom of Asp 138, 3.20 Å; while the address terminal Leu COO^- atoms with the NH_3^+ *H* atom of Tyr 320, 2.58 Å; and the N-*H* of Gly 319, 2.97 Å, with the Leu carboxyl carbonyl.

The ∂ − opioid receptor with docked **22** had eleven prominent H-bonding sites within 3.5 Å that included the message tyrosine phenol O-*H* atoms with the amide carbonyl of Asp 128, 2.42 Å, message phenol *H*, to *O* Tyr 308, 2.98 Å, and Asp 128 carboxyl *H* with the message Tyr phenol *O*-H, 3.40 Å; the message NH_3^+ *H* atom to both the carboxyl of Ile 277, 2.99 Å, and *N* ring atom of His 278, 3.12 Å; message phenylalanine amide carbonyl with the O-*H* of Thr 285, 3.05 Å, and the N-*H* of Lys 214, 3.28 Å; address Leu carboxyl carbonyl with the ring N-*H*, Trp 284, 2.64 Å, and carboxyl *O* atom with the N-*H* of Lys 214, 2.87 Å, and the O-*H* of Thr 285, 2.43 Å.

Interestingly, Fig. 1.13 also demonstrated the differences in binding modes of the μ- and ∂-opioid receptors with **22**. For example, the message tyrosine phenol of **22** in the μ-opioid receptor was at the top binding, less conserved sites, while the message phenyl of phenylalanine and address leucine residue were at the inner conserved areas of the receptor. The ∂-opioid receptor had the reverse situation with guest, **22**, with the message tyrosine residue (signal transduction) in the more conserved inner areas, while the address (selectivity) leucine residue was in the less conserved top of the receptor.

It was clear that the ∂ − opioid receptor binding site of amino acids with guest **22** had the most critical message-address non-covalent H-bonding interactions (eleven for the ∂-OR versus four for the μ-OR) with neuropeptide, **22**. This was also reflected in the EC_{50} binding constants we found for **22** in our previous study, with values for the μ-opioid receptor of 14.3 nM versus the ∂-opioid receptor of 4.4 nM (vide infra); the ∂-opioid receptor having a potency/selectivity factor of 3.25X that of the μ-opioid receptor for **22**.[18]

While the κ-opioid receptor also had many non-covalent H-bonds (ten interactions) to the message-address structural aspects of **22**, it was stated in the literature that **22** had a weak binding value (EC_{50} > 1000 nM) to the κ-opioid receptor, [21] in contrast to the potency of the μ- and ∂-opioid receptors, which seems to indicate that selectivity to the κ-opioid receptor and its biological activity (EC_{50} value) are compatible; however, other factors not related to the non-covalent binding regimes must be responsible for the weak interactions of **22** and **23** to the κ-opioid receptor. Moreover, Asp 138 has been shown by Stevens et al to be important for selectivity to the κ-opioid receptor with the morphanin analog, **JDTic**, as the guest.[19c] Furthermore, we see a Asp 138 carboxyl carbonyl H-bond to the phenylalanine amide N-*H* of **22**, but this may not be enough stabilization to facilitate any selectivity to the κ-opioid receptor, or be important for biological activity, since **JDTic** is an antagonist, which does not elicit a biological response, while being conformationally stable.

The electronic effect of the [η^6-Cp*Rh-Tyr^1] moiety of **23** dramatically reduced the nucleophilicity of the phenol *O* atom, while making the O-*H* more acidic, and a *H*-donor. Further, the -NH_2 group was converted into a *H*-acceptor, as well as the amide C = *O*. Thus, the Mulliken charge differences between **22** and **23** were dramatic, and have changed the non-covalent bonding modes in all three opioid receptors in comparison to neuropeptide, **22**. Moreover, the *Cp**Rh moiety became a site for π−π and *C*H-π molecular recognition, while the *phenyl* group of the phenylalanine residue was also a site for π−π interactions; we found none of these π-π or CH-π interactions within 3.5 Å, for **22**. Thus, the important non-covalent *H*-bond, π−π, and C*H*-π interactions (within 3.5 Å) with selective amino acids in the conserved (bottom of the binding pocket) or less conserved (top of binding pocket) areas of the μ-, ∂-, and κ-opioid receptors for **23** are as follows. Fig. 1.14 shows the lowest energy conformation of **23** inside the μ-opioid receptor (left, center left), while more than seventeen amino acid residues (backside not shown) for the μ-opioid receptor that are within 5 Å (center right, **Top** refers to the receptor binding pocket opening), and eight H-bond, π−π, and CH-π non-covalent interactions (right) within 3.5 Å. Importantly, the [η^6-Cp*Rh-Tyr^1]$^{2+}$ residue of complex **23** was found to be in the more inner conserved area for the μ- and ∂ opioid receptors, and similar to the ∂-opioid receptor result with **22**, but opposite to the μ and κ-opioid receptor, with **1b**. Moreover, the conformation of **23** in the κ opioid receptor was found to have the [η^6-Cp*Rh-Tyr^1]$^{2+}$ residue in the somewhat less conserved area that might reflect its weak binding value (>1000 nM),[20] that was found for **22**, and also theoretically for this trend with **23**, since we found that **23** was less potent in binding to the μ- and ∂-opioid by a factor of ~2, in comparison to **22**.

The H-bond interactions for **23** in the μ-opioid receptor included a C−H bond of a Cp*CH_3- to the *N* of His 297, 2.9 Å; tyrosine amide carbonyl with the *H* of H_2O, 2.96 Å; O-*H* of Tyr 148, 2.19 Å and NH_3^+ of Leu 219, 2.74 Å, with both Gly amide carbonyls; NH_3^+ of Lys 233 with the phenylalanine amide carbonyl, 2.00 Å; and the NH_3^+ of Lys 233 with the Leu carboxyl carbonyl, 1.95 Å; no π−π interactions were found within 3.5 Å, but

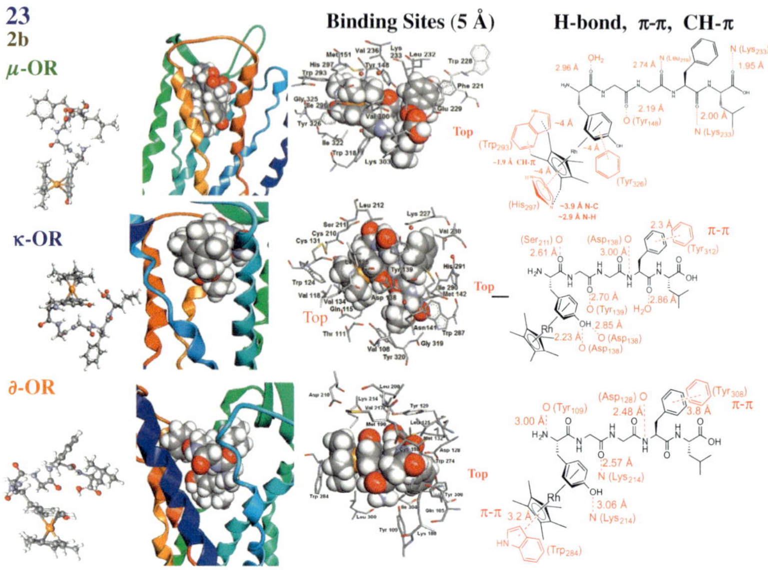

FIGURE 1.14 The lowest energy conformation of **23** inside the μ-, κ-, and ∂-opioid receptors (**left, center left**); the structural basis for non-covalent binding regimes of dication complex **23** with amino acid residues at the binding sites within 5 Å of the docked ligand (**center right**); hydrogen bonds, π−π, and CH-π non-covalent interactions responsible for the non-covalent receptor interactions, while H-bond lengths indicate the distance between the heavy atoms (not hydrogen) within 3.5 Å (**right**).[20]

His 297 was ~ 4 Å from the *Cp**Rh, as a hydrophobic amino acid residue, as well as Tyr 326 at ~4 Å from $[\eta^6\text{-}Cp^*\text{Rh-Tyr}^1]^{2+}$, while C*H*-π interactions with the Cp**CH*$_3$- residue, in **23** with Trp 293 at 1.9 Å (phenyl), and a possible weaker interaction at ~4 Å (indole). Tyr 326 also appeared to prevent H-bonding to the phenol *O*-H, and could be one reason the EC_{50} value was high at 93 nM, in comparison to the **22** value, of 14.3 nM.[18]

The interaction of **23** with the κ-opioid receptor shown in Fig. 1.14 provided twenty-two amino acid residues within 5 Å, six H-bonds, and one π−π interaction within 3.5 Å. Again, we see the important Asp 138 residue of the κ-opioid receptor, this time *H*-bonding to the $[\eta^6\text{-Cp}^*\text{Rh-}Tyr^1]^{2+}$ phenol *O*-H, carboxyl *H*, 2.23 Å, and *O*, 2.85 Å; Ser 211 terminal O-*H* bonding to the Tyr carbonyl, 2.61 Å; Tyr 139 O-*H* bonding to the Gly C = *O*; Asp 138 carboxyl O-*H* bonding to the *N*H of phenylalanine, 3.00 Å; H_2O molecule H-bonding to the phenylalanine carbonyl, 2.86 Å; phenyl group of phenylalanine, π−π interaction with the Tyr 312 phenyl group, 2.3 Å. Interestingly, no π−π interactions with aromatic amino acid residues were observed with the *Cp**Rh group of **23** at the less conserved areas of the κ-opioid receptor, but even with the phenol *O-H* H-bonding regimes, as noted above, the extrapolated EC_{50} value from **22** was speculated to be >1000 nM, indicating very weak binding to the κ-opioid receptor.

The interaction of **23** with the ∂-opioid receptor shown in Fig. 1.14 provided more than twenty amino acid residues within 5 Å, with four H-bonds and two π-π interactions within 3.5 Å. The message portion, for signal transduction, of **23** with a $[\eta^6\text{-Cp*Rh-}Tyr^1]^{2+}$ group clearly shows the *Cp** group's π−π interaction with the Trp 284 residue indole ring, 3.2 Å. This is an important find, since Trp amino acid residues are highly hydrophobic/lipophilic; therefore, the hydrophobic *Cp** group would stabilize its conformation with this π−π interaction in the inner part of the ∂-opioid receptor (Fig. 1.14). Furthermore, the other message portion of **23**, the phenyl group of the phenylalanine residue, revealed another critical π−π interaction with the Tyr 308 phenol, 3.8 Å. Moreover, H-bond, non-covalent interactions were also important for conformational stability of **23** inside the ∂-opioid receptor. For example, the terminal Tyr NH_2 group of **23**, an H-bond acceptor, had a H-bond from the Tyr 109 phenol O-*H*, 3.00 Å, while the Lys 214 NH_2, H-bonds with the **23** Tyr phenol O-*H*, 3.06 Å; both part of the message paradigm. The Lys 214 NH_2 also H-bonds with an Gly amide carbonyl, 2.57 Å. Interestingly, no other amino acid residues from the μ- and κ-opioid receptors H-bonds to the terminal Tyr NH_2 group of **23**. This was also reflected in the EC_{50} value of **23** at the ∂-opioid (15.6 nM) versus the μ-opioid receptor (93 nM), with a ∂/μ ratio of 0.17, the ∂-opioid receptor being more selective to **23**, without having any H-bonds to the address portion, the protonated, terminal Leu carboxylic acid.

When we compared **22** to **23** in each intra and inter series of μ-, κ-, and ∂-opioid receptors, we saw that no similarities of their respective non-covalent binding regimes, within the message-address paradigm, were apparent (Figs. 1.13 and 1.14). This clearly demonstrates the varied binding regimes for **22** to **23** in each opioid receptor, as epitomized in the lowest energy conformations (Figs. 1.13 and 1.14), including the message and address aspects, and the degree of H-bond, π−π, and C*H*-π interactions, all of which dictates signal transduction (message region) and opioid receptor selectivity (address region), and therefore, their overall EC_{50} binding values. Thus, **22** binds both to the μ- and the ∂-opioid receptors, with selectivity to the ∂-opioid, by a factor, ∂/μ, of 0.31.[18]

The anomaly was the κ-opioid receptor, which has many non-covalent interactions with the message tyrosine residue, and the address terminal Leu carboxylic acid (Fig. 1.13); however, its EC_{50} binding value was found to be >1000 nM;[21] a weak binding to **22**, and we postulate a similar situation for **23**, from the published results of the μ- and ∂-opioid receptors, where **22** was more potent in comparison to **23**, by factors of 0.15 (**22** /**23**, μ-OR) and 0.28 (**22** /**23**, ∂-OR); thus, **22** was more selective to the **∂-OR** in comparison to **23**, by a factor, ∂/μ = 1.86.[18] The significant receptor binding differences revolved around the total lack of π−π or C*H*-π non-covalent interactions at the μ-, κ-, and ∂-opioid receptors for **22** in comparison to **23**. We presume that when you bind a $[(\eta^6\text{-Cp*Rh-Tyr}^1)]^{2+}$ moiety to **23**, you create a more lipophilic/hydrophobic molecule, **23**, which when docked to an opioid receptor will seek receptor aromatic amino acid residues for π-π and *CH*-π non-covalent interactions, that are weak; however, multiple weak, non-covalent interactions provide an overall fundamental stabilization in binding guest compounds to all GPCRs.

Furthermore, a detailed analysis of the conformations of **22** and **23** inside the binding sites of the opioid receptors showed that their weak binding to the κ-opioid receptor might be due to the high internal energy of the guests, **22** and **23**. Indeed, torsional rotation was considered in the docking calculations as being free, while the distortion effect of the scoring function accounts only for the entropy effects, due to the constraints in conformations of both **22** and **23**, and the receptor amino acid residues.[20] However, accurate calculations showed that conformations of sufficiently large organic molecules, such as biological and pharmacological compounds, could differ in energy by 10 kcal/mol, and were more dependent on the presence of internal hydrogen bonds and dipole and quadrupole contacts of the Ar···O and Ar···N variety.[20] Our DFT calculations showed that the **22** conformation, inside the κ-OR binding pocket, was 18.5 and 12.7 kcal/mol higher in energy than the corresponding conformations inside the μ- and δ-opioid receptors, respectively. For complex **23**, the corresponding differences were 5.9 and 9.6 kcal/mol. By comparing these values, along with the estimated docking energies, this demonstrated that bonding in the κ-opioid receptor pocket was found to be energetically much less favorable in comparison with the μ-and ∂-opioid receptors, especially for **22**, and could be a major factor in **22** and **23** having weak binding values in the κ-opioid receptor.

1.10 OVERVIEW OF THE BIOORGANOMETALLIC CHEMISTRY DISCIPLINE

Since the X-ray crystal structure of the NiFe hydrogenase was determined in 1995 by Fontecilla-Camps et al., *Nature*, **1995**, *373*, 580, and found to be an ancient organometallic complex with the NiFe metal centers being bound to CN and CO groups, i.e., C-M bonds, as well as the X-ray crystal structure of CH_3-CoB_{12}, also a natural product, by Marzilli et al. *J. Am. Chem. Soc.*, **1985**, *107*,1729, and our own discovery in 1983, that methyl and phenylarsonic acids were natural products in ancient, algae kerogen formations of oil shale,[22] the relevance of the bioorganometallic chemistry discipline has risen dramatically. We hope we have demonstrated that organonmetallic chemistry, at the interface with biology, has a unique place in the scientific lexicon, and that the diversity of opportunities for this exciting field has no limits.

ACKNOWLEDGMENTS

I wish to thank all the undergraduate students and postdoctoral fellows, that worked with me at LBNL/UC Berkeley, as well as the senior colleagues, that are named in the publications, for the important contributions they made to the Bioorganometallic Chemistry program we developed together. Special thanks to Irena Efremenko of the Weizmann Institute, Rehovot, Israel, who provided all the quantum chemical calculations, as well as, Gérard Jaouen and Paul Dyson, who helped me initiate an organometallic pharmaceutical program, and Nils Metzler-Nolte, who facilitated my research ideas on Cp*Rh-GPCR peptide complexes during visiting professorships in their respective laboratories in Paris, Lausanne, and Bochum. Support by the US Department of Energy and LBNL Director's Funds are gratefully acknowledged under Contract No. DE AC02-05CH11231, while National Institute of Environmental Health Sciences is also acknowledged for support of the Bioorganotin metabolism studies.

*To whom all correspondence should be sent: rhfish@lbl.gov

REFERENCES

1. (a) Fish RH, Kimmel EC, Casida JE. *J. Organometal. Chem* 1975;**93**:C1. (b) Fish RH, Kimmel EC, Casida JE. *J. Organometal. Chem* 1976;**118**:41. (c) Fish RH, Casida JE, Kimmel EC. *ACS Symposium Series* 1978;**82**:82. (d) Aldridge WN, Casida JE, Fish RH, Kimmel EC, Street BW. *Biochem. Pharmacol* 1977;**26**:1997.
2. (a) Fish RH, Casida JE, Kimmel EC. *Tetrahedron Lett* 1978;3515. (b) Fish RH, Broline BM. *J. Organometal. Chem* 1977;**136**:C41. (c) Fish RH, Broline BM. *J. Organometal. Chem* 1978;**159**:255.
3. Fish RH, Price RT. *Organometallics* 1989;**8**:225.
4. de Vos D, Willem R, Gielen M, van Wingerden KE, Nooter K. *Metal-Based Drugs* 1998;**5**:179.
5. Parvin B, Maestre MF, Fish RH, Johnston WE, Method and Apparatus for Accurately Manipulating an Object during Microelectrophoresis Patent issued, September, 23, 1997. U.S. Patent 5,671,086.
6. Smith DP, Chen H, Ogo S, Elduque AI, Eisenstein M, Olmstead MM, Fish RH. *Organometallics* 2014;**33**:2389 and references therein.
7. Ogo S, Buriez O, Kerr JB, Fish RH. *J. Organomet. Chem* 1999;**589**:66.
8. (a) Chen H, Ogo S, Fish RH. *J. Am. Chem. Soc* 1996;**118**:4993. (b) Ogo S, Nakamura S, Chen H, Isobe K, Watanabe Y, Fish RH. *J. Org. Chem* 1998;**63**:7151. (c) Bakhtiar R, Chen H, Ogo S, Fish RH. *Chem. Commun* 1997;2135. (d) Elduque A, Carmona D, Oro LA, Eisenstein M, Fish RH. *J. Organometal. Chem* 2003;**668**:123.
9. Lo HC, Leiva C, Buriez O, Kerr JB, Olmstead MM, Fish RH. *Inorg. Chem* 2001;**40**:6705.
10. (a) Lo HC, Fish RH. *Angew. Chem. Int. Ed* 2002;**41**:478. (b) Lo HC, Ryan JD, Kerr JB, Clark DS, Fish RH. *J. Organometal. Chem* 2017;**839**:38.
11. Ryan JD, Fish RH, Clark DS. *ChemBioChem* 2008;**9**:2579.
12. Lutz J, Hollmann F, Ho TV, Schnyder A, Fish RH, Schmid A. *J. Organometal. Chem* 2004;**689**:4783.
13. (a) Fish RH, Jaouen G. *Organometallics* 2003;**22**:2166 and references therein. (b) Allardyce CS, Dyson PJ, Ellis DJ, Heath SL. *Chem. Commun* 2001;1396. (c) Allardyce CS, Dyson PJ, Ellis DJ, Salter PA, Scopelliti R. *J. Organometal. Chem* 2003;**668**:35. (d) Morris RE, Aird RE, Murdoch PD, Chen HM, Cummings J, Hughes ND, et al. *J. Med.*

Chem 2001;**44**:3621. (e) Chen H, Parkinson JA, Parsons S, Coxall RA, Gould RO, Sadler PJ. *J. Am. Chem. Soc* 2002;**124**:3064. (f) Wang F, Chen H, Parkinson JA, Murdoch PD, Sadler PJ. *Inorg. Chem* 2002;**41**:4509. (g) Jaouen G, Top S, Vessières A, Alberto R. *J. Organomet. Chem* 2000;**600**:25 and references therein. (h) Jaouen G. *Chemistry in Britain* 2001;**36**. (i) Köpf-Maier P. *Eur. J. Chim. Pharmacol* 1994;**47**:1. (j) Melchart M, Sadler PJ. In: Jaouen G, editor. *Bioorganometallics*. Wiley-VCH; 2005. Chap 2, p 39. (k) Jaouen G, Top S, Vessières A. In: Jaouen G, editor. *Bioorganometallics*. Wiley-VCH; 2005. Chap 3, p 65. (l) Alberto R. In: Jaouen G, editor. *Bioorganometallics*. Wiley-VCH; 2005. Chap 4, p 97. (m) Dagani R. *Chemical and Engineering News* 2002;**80**:23 September 16, 2002 issue. "The Bio Side of Organometallics." (n) Dorcier A, Hartinger CG, Scopelliti R, Fish RH, Keppler BK, Dyson PJ. *J. Inorg. Biochem* 2008;**102**:1066. (o) Peacock AFA, Sadler PJ. *Chem. Asian J* 2008;**11**:1886. (p) Hartinger CG, Dyson PJ. *Chem. Soc. Rev.* 2009;**38**:391. (q) Strohfeldt K, Tacke M. *Chem. Soc. Rev* 2008;(37):1174.

14. Top S, Efremenko I, Rager N-M, Vessieres A, Jaouen G, Fish RH. *Inorg. Chem* 2011;**50**:271.
15. Efremenko I, Top S, Martin JML, Fish RH. *Dalton Trans* 2009;4334.
16. (a) Metzler-Nolte N. Labelling of peptides and PNA with organometallics. In: Jaouen G, editor. *In Bioorganometallics*, 179. Weinheim: Wiley-VCH; 2005. p. 125. (b) Albada B, Metzler-Nolte N. *Chem. Rev* 2016;**116**:11797 and references therein.
17. GPCRs reviews: (a) Rosenbaum DM, Rasmussen SGF, Kobilka B. *Nature* 2009;**459**:356. (b) Gentilucci L. *Curr. Top. Med. Chem* 2004;**4**:19. (c) Lappano R, Maggiolini M. *Nat. Rev. Drug Discovery* 2011;**10**:47. (d) Baselga J. *Science* 2006;**312**:1175. (e) Klabunde T, Hessler G. *ChemBioChem* 2002;**3**:928.
18. Albada HB, Wieberneit F, Dijkgraaf I, Harvey JH, Whistler JL, Stoll R, Metzler-Nolte N, Fish RH. *J. Am. Chem. Soc* 2012;**134**:10321 and references therein.
19. (a) Manglik A, Kruse AC, Kobilka TS, Thian FS, Mathiesen JM, Sunahara RK, Pardo L, Weis WI, Kobilka BK, Granier S. The X-ray crystal structure of the μ-opioid receptor bound to a morphinan antagonist, **β-FNA**. *Nature* 2012;**485**(321):326. (b) Granier S, Manglik A, Kruse AC, Kobilka TS, Thian FS, Weis WI, et al. Structure of the ∂-opioid receptor bound to **naltrindole**. *Nature* 2012;**485**:400–4. (c) Wu H, Wacker D, Mileni M, Katritch V, Han GW, Vardy E, Liu W, Thompson AA, Huang X-P, Carroll FI, Mascarella SW, Westkaemper RB, Mosier PD, Roth BL, Cherezov V, Stevens RC. Structure of the human κ-opioid receptor with **JDTic**. *Nature* 2012;**485**:327–32.
20. Efremenko I, Fish RH. *Organometallics* 2015;**34**:4117 and reference therein.
21. Raynor K, Kong H, Chen Y, Yasuda K, Yu L, Bell GI, et al. *J. Pharmacol. Exp. Ther.* 1993;**45**:330.
22. (a) Fish RH, Walker W, Tannous RS. *Energy & Fuels* 1987;**1**:243. (b) Fish RH, Tannous RS, Walker W, Weiss CS, Brinckman FE. *J. Chem. Soc., Chem. Commun* 1983;490.

Chapter 2

Design of Bioorganometallic Conjugates Composed of Dipeptides or Nucleobases to Induce Chirality-Organized Structures

Toshiyuki Moriuchi[1] and Toshikazu Hirao[2]

[1]*Division of Molecular Materials Science, Graduate School of Science, Osaka City University, Osaka, Japan,* [2]*The Institute of Scientific and Industrial Research, Osaka University, Osaka, Japan*

2.1 INTRODUCTION

Biomolecules such as peptides and nucleobases play an important role in the formation of well-organized structures like proteins, enzymes, and DNA to exhibit the specific functions. Hydrogen bonding[1] plays a vital in regulating and modulating the three-dimensional structure and function of biological systems by virtue of its directionality and specificity.[2–4] The tuneability and reversibility of hydrogen bonding are also key factors in the physical properties of molecular assemblies. Protein secondary structures such as α-helices, β-sheets, and β-turns are formed to serve as an important role in protein folding, which is mostly driven and stabilized by complementary hydrogen bonding and hydrophobic interaction of side chains.[5,6] The most common secondary structures are α-helices and β-sheets. In contrast to α-helices, the structure and stability of β-sheets are less understood. Hydrogen bonds are formed between the CO and NH of two different strands in β-sheets, wherein a series of 12-membered hydrogen-bonded rings are involved in parallel β-sheets and alternating series of 10-membered and 14-membered hydrogen-bonded rings are formed in antiparallel β-sheets. A turn is also an element of secondary structure in proteins. The most common form of a turn is β-turn,

Advances in Bioorganometallic Chemistry. DOI: https://doi.org/10.1016/B978-0-12-814197-7.00002-9

wherein hydrogen bond between the CO of residue i and the NH of residue i + 3 is formed to give a 10-membered hydrogen-bonded ring. γ-Turn is constituted by a 7-membered hydrogen-bonded ring through the formation of hydrogen bond between the CO of residue i and the NH of residue i + 2. It is difficult to predict the geometry of structural motifs of protein folding from the sequence of amino acids. Considerable effort has been devoted to the design of protein secondary structure mimics by using short peptides for gaining fundamental insight into the factors affecting the protein structure and stability, and also for designing pharmacologically useful compounds. Control of hydrogen bonding has drawn great attention in the construction of functional molecular assemblies. Utilization of the chirality-organization properties of short peptides based on their hydrogen bonding sites and chiral centers is regarded as a convenient strategy to chirality-organized molecular assemblies. On the other hand, the double helical DNA is created by adenine-thymine and guanine-cytosine complementary base pairing, which are mainly stabilized by hydrogen bonds and $\pi-\pi$ interactions.[7] The complementary hydrogen bonding of nucleobases is considered to be a powerful tool for constructing various well-organized molecular assemblies based on the directionality and specificity of hydrogen bonds. The conjugation of biomolecules with organometallic compounds has attracted much attention to afford novel bioinspired systems depending on both properties. In the past two decades, a variety of bioorganometallic conjugates have been designed to demonstrate the formation of the chirality-organized molecular assemblies from the view point of bioorganometallic chemistry,[8–12] which is a hybrid area between biology and organometallic chemistry. Ferrocene is one of the most stable organometallic compounds with a sandwich-shaped structure and the most useful one among metallocenes due to a reversible redox properties and two rotatory coplanar cyclopentadienyl (Cp) rings.[13] The inter-ring spacing of ferrocene is about 3.3 Å, which is appropriate for hydrogen bonding between attached peptide strands on the two Cp rings as observed in β-sheets. Employment of a ferrocene unit as an organometallic scaffold with a central reverse-turn unit is envisioned to be a convenient approach to explore the hydrogen bonding ability of introduced peptide strands. The inter-base-pair spacing of about 3.3 Å in the B-form of DNA is suited to induce metal–metal interaction. Controlled arrangement of organometallic groups by using self-assembling properties of nucleobases is thought to be a relevant approach to induce metal–metal interaction between conjugated organometallic groups.

In this chapter, we focus on advances in the design of chirality-organized ferrocene-dipeptide conjugates to guide controlling formation of protein secondary structures for the peptidomimetic basis. Our ongoing research on the controlled arrangement of the bioorganometallic conjugates composed of nucleobases to induce metal-metal interaction is also described.

2.2 CHIRALITY ORGANIZATION OF FERROCENE-DIPEPTIDE CONJUGATES

The first example of a ferrocene-amino acid conjugate, ferrocenylalanine, was synthesized in 1957.[14–16] After its synthesis, a variety of ferrocene-amino acid or peptide conjugates have been investigated to gain insights into a peptidomimetic basis for protein folding and to create well-organized molecular assemblies.[17–25] The utility of ferrocene as a molecular scaffold for leading intramolecular hydrogen bonds has been demonstrated in the ferrocene-amino acid conjugate using a valine unit by Herrick and co-workers.[26] A series of ferrocene-dipeptide conjugates is designed by the introduction of dipeptide chains into the ferrocene scaffold to induce chirality-organized structures based on the formation of protein secondary structures through intramolecular hydrogen bonding of the dipeptide chains.[27–43] Our design of ferrocene-dipeptide conjugates is based on symmetrically introducing two dipeptide chains of alanyl-proline or prolyl-alanine sequence into the ferrocene scaffold as a central reverse-turn unit (Fig. 2.1). An advantage of the use of alanine and proline depends on alanine being the smallest chiral amino acid and a sterically constrained proline is

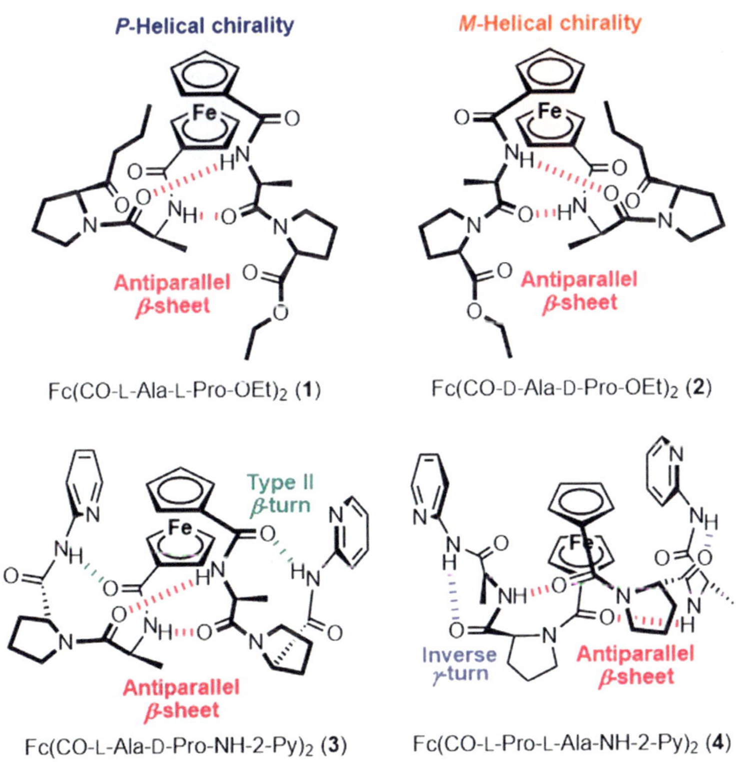

FIGURE 2.1 Ferrocene-dipeptide conjugates **1-4**.

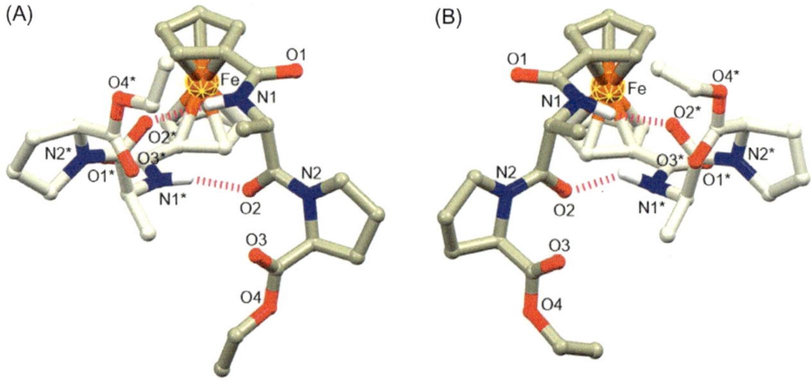

FIGURE 2.2 Molecular structures of (A) **1** and (B) **2**.

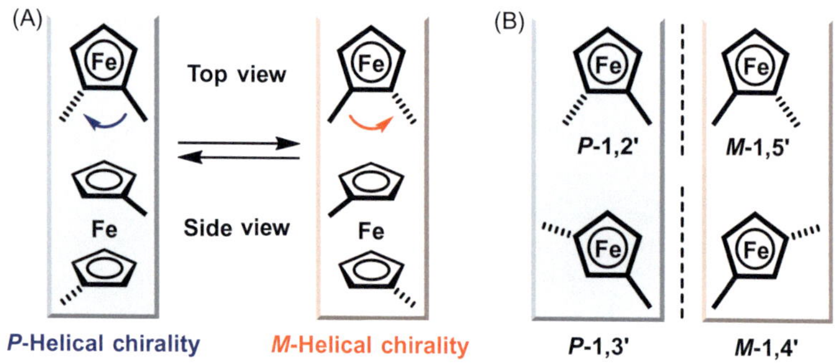

FIGURE 2.3 Schematic representation of 1,*n*′-disubstituted ferrocenes. (A) Top and side views of *P*-helical chirality and *M*-helical chirality, respectively, and (B) stereoisomers of the axial chirality.

expected to serve as a turn inducer. The "Herrick" pattern of interchain intramolecular antiparallel β-sheet-like hydrogen bonding between CO (Ala) and NH (Ala of another chain) of each dipeptide chain is formed to induce the chirality-organized structure in the crystal structure of the ferrocene-dipeptide conjugate **1** bearing the L-dipeptide chains of -L-Ala-L-Pro-OEt (Fig. 2.2A) although a wide range of relative orientations of the introduced dipeptide chains are possible depending on two rotatory Cp rings.[27–29] There are interconvertible conformational enantiomers, *P*- and *M*-helical conformations, based on the presence of a torsional twist about the Cp(centroid)-Fe-Cp(centroid) axis in the 1,*n*′-disubstituted ferrocene as shown in Fig. 2.3.[19,20,22,23] The ferrocene-dipeptide conjugate **1** shows *P*-helical chirality with *P*-1,2′ helical conformation of the ferrocenoyl moiety. The formation of interchain intramolecular antiparallel β-sheet-like hydrogen bonds induces *P*-helical chirality based on the chirality-organized structure.

M-Helical chirality with *M*-1,5′ helical conformation of the ferrocenoyl moiety is driven by the D-dipeptide chains of -D-Ala-D-Pro-OEt (Fig. 2.2B). Consequently, the introduction of the chiral dipeptide chains into the ferrocene scaffold leads the chirality organization by restriction of the torsional twist through the formation of interchain intramolecular antiparallel β-sheet-like hydrogen bonds.[29]

A positive Cotton effect at the absorbance region of the ferrocenoyl moiety, based on *P*-helical chirality of the ferrocenoyl moiety, is observed in the CD spectrum of the ferrocene-dipeptide conjugate **1**, indicating that the chirality-organized structure as observed in the crystal structure is preserved even in solution. As expected, the ferrocene-dipeptide conjugate **2** with *M*-helical chirality of the ferrocenoyl moiety shows a negative Cotton effect.

Controlling the formation of protein secondary structures is possible by changing chirality and sequence of amino acids. The combination of the ferrocene scaffold as a central reverse-turn unit with the L-Ala-D-Pro heterochiral dipeptide sequence induces the simultaneous formation of antiparallel β-sheet-like structure through intramolecular hydrogen bonding between CO (Ala) and NH (Ala of another chain) and type II β-turn-like structure through intramolecular hydrogen bonding between the CO of the ferrocenoyl moiety and the NH adjacent to the pyridyl moiety of the same dipeptide chain, as confirmed by the molecular structure of the ferrocene-dipeptide conjugate **3**, wherein *P*-helical chirality of the ferrocenoyl moiety is formed (Fig. 2.4A).[35] The helical chirality of the ferrocenoyl moiety is likely to be controlled by the absolute configuration of the alanyl α-carbon atom,[44,45] because *P*-helical chirality of the ferrocenoyl moiety is observed with **1** bearing -L-Ala-L-Pro-OEt homochiral dipeptide chains. Furthermore, the molecular structure of the ferrocene-dipeptide conjugate **4**, in which the L-Pro-L-Ala homochiral dipeptide chains are introduced into the ferrocene scaffold, reveals the formation of antiparallel β-sheet-like structure through intramolecular hydrogen bonding between the CO of the ferrocenoyl moiety and NH (Ala of another

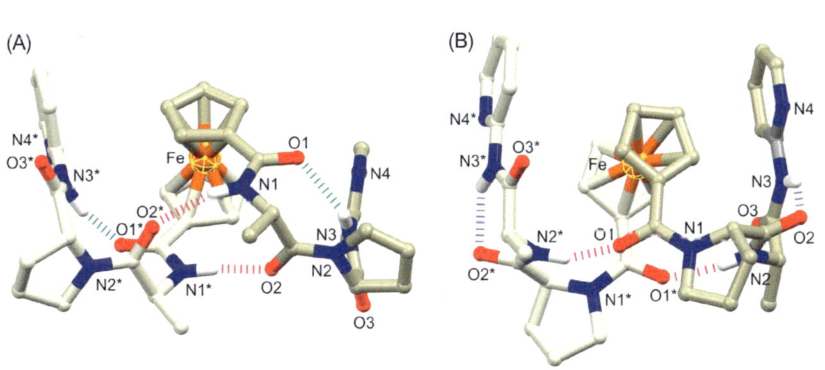

FIGURE 2.4 Molecular structures of (A) **3** and (B) **4**.

chain) and an inverse γ-turn-like structure through intramolecular hydrogen bonding between CO (Pro) and the NH adjacent to the pyridyl moiety of the same dipeptide chain simultaneously, as shown in Fig. 2.4B.[36]

As mentioned above, the helical chirality of the ferrocenoyl moiety in the ferrocene-dipeptide conjugates derived from 1,1′-ferrocenedicarboxylic acid is induced by the intramolecular hydrogen bonding of the dipeptide chains.[23,27–31,33–36,38,41] In these systems, the helical chirality of the ferrocenoyl moiety depends solely on the absolute configuration of the α-carbon atom of an amino acid adjacent to the ferrocene scaffold, irrespective of the absolute configuration of the second amino acid. The regulation of the conformational flexibility of the dipeptide chains can be served as a controlling factor for inducing the helical chirality and creating protein secondary structures in ferrocene-dipeptide conjugates. Cyclization of the introduced dipeptide chains is considered to be one simple strategy to restrict the conformational flexibility of the dipeptide chains, triggering the reorientation of the hydrogen bonding interactions. Cyclic ferrocene-dipeptide conjugates are designed as a means of restriction of flexibility of the dipeptide chains (Fig. 2.5).[42] Cystamine ($H_2NCH_2CH_2S\text{-}SCH_2CH_2NH_2$) is employed as a C-terminal linker to cyclize two dipeptide chains introduced into the ferrocene scaffold, wherein the corresponding ring-opened derivatives can be readily

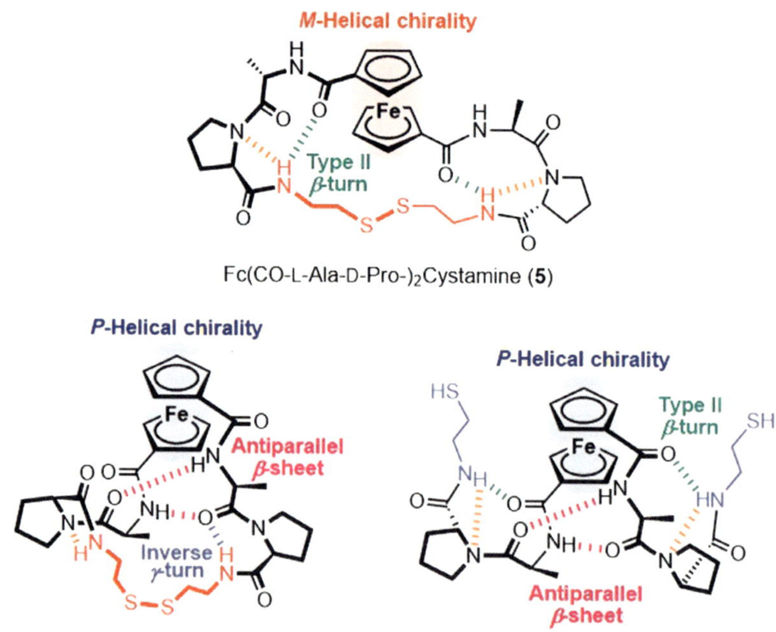

Fc(CO-L-Ala-D-Pro-)$_2$Cystamine (**5**)

Fc(CO-L-Ala-L-Pro-)$_2$Cystamine (**6**)

Fc(CO-L-Ala-D-Pro-NHCH$_2$CH$_2$SH)$_2$ (**7**)

FIGURE 2.5 Ferrocene-dipeptide conjugates **5-7**.

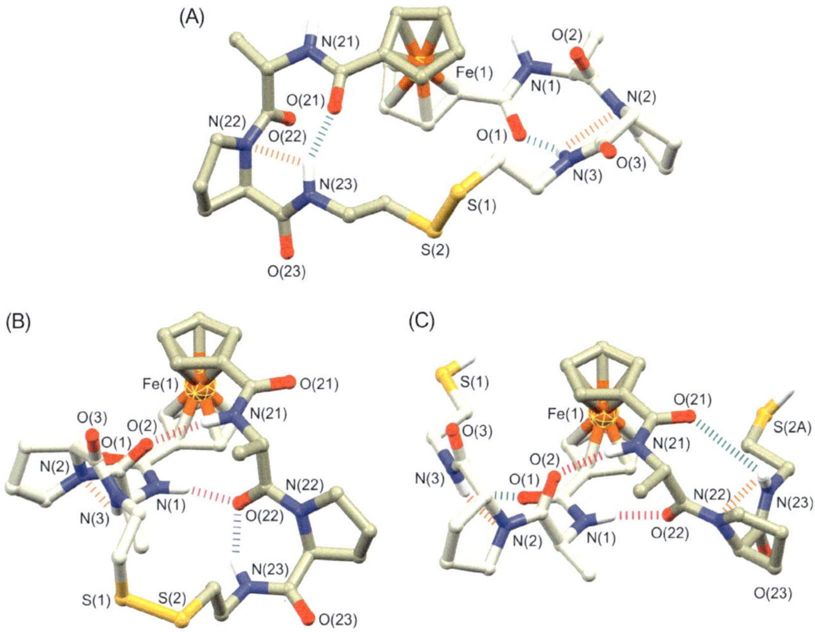

FIGURE 2.6 Molecular structures of (A) **5**, (B) **6**, and (C) **7**.

generated by the reductive cleavage of the disulfide bond to the thiols. The molecular structure of the cyclic ferrocene-dipeptide conjugate **5** is characterized by the creation of a type II β-turn-like structure through the formation of intramolecular hydrogen bonds between the CO of the ferrocenoyl moiety and the NH of cystamine in each dipeptide moiety, inducing an *M*-helical chirality with an *M*-1,4′ helical conformation of the ferrocenoyl moiety as depicted in Fig. 2.6A.[42] The intramolecular N-H•••N hydrogen bonds between the NH of cystamine and the nitrogen of Pro is formed to stabilize a type II β-turn-like structure. On the contrary, *P*-helical chirality with *P*-1,2′ helical conformation of the ferrocenoyl moiety is induced through the formation of intramolecular antiparallel β-sheet-like hydrogen bonds between the CO of Ala and the NH of Ala (another dipeptide moiety) in the cyclic ferrocene-dipeptide conjugate **6** with the homochiral dipeptide (-L-Ala-L-Pro-cystamine-L-Pro-L-Ala-) (Fig. 2.6B).[42] A conformational conversion from a type II β-turn-like structure to an antiparallel β-sheet-like structure by changing the chirality of the second amino acid prolyl moiety is demonstrated to control the induced chirality of the ferrocenoyl moiety. An inverse γ-turn-like structure through the formation of intramolecular hydrogen bond between one of the CO of the Ala and the NH of cystamine of same dipeptide moiety is also observed in **6**. The helical chirality of the ferrocenoyl moiety can be controlled by cyclic/acyclic conversion. The acyclic ferrocene-dipeptide conjugate **7** bearing the dipeptide chains of -L-Ala-L-Pro-$NHCH_2CH_2SH$, which is synthesized by the reductive cleavage of the disulfide bond in cyclic conjugate **1**, reveals *P*-helical chirality with a *P*-1,2′ helical conformation of the ferrocenoyl moiety based on the chirality organization through intramolecular antiparallel β-sheet-like hydrogen bonding between CO (Ala) and NH (Ala of another chain) of each dipeptide chain as shown in Fig. 2.6C.[42]

Secondary structures of peptide chains are expected to be stabilized and/or regulated by the complexation with metal ions. The chirality-organized ferrocene-dipeptide conjugate **8** bearing dipeptide chains of -L-Ala-L-Pro-NH-2-Py through two intramolecular interchain hydrogen bondings affords the 1:1 *trans*-palladium complex **9** by the reaction with $PdCl_2(MeCN)_2$, wherein the conformational regulation is stabilized in both solution and solid states (Fig. 2.7).[31] The single-crystal X-ray structure determination of **9** reveals the pseudo-helical conformation by the complexation and chirality organization based on the preservation of the intramolecular interchain hydrogen bonds as shown in Fig. 2.7.[31]

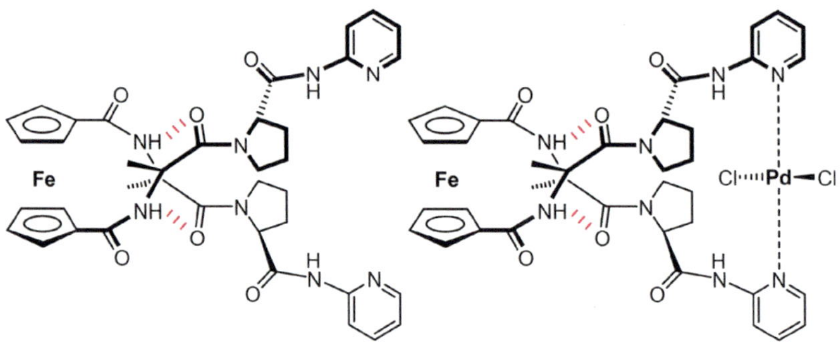

Fc(CO-L-Ala-L-Pro-NH-2-Py)$_2$ (**8**) Fc(CO-L-Ala-L-Pro-NH-2-Py)$_2$–$PdCl_2$ (**9**)

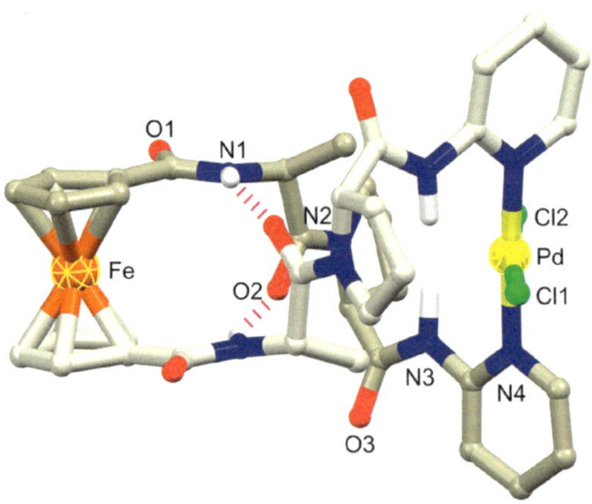

FIGURE 2.7 Molecular structure of **9**.

The organization of host molecules by self-assembling is a useful strategy to construct active receptors.[46] Ferrocene-peptide conjugates have been demonstrated to act as receptors for anions[47,48] and biomolecules.[49–53] The two amido pyridyl moieties of the ferrocene-dipeptide conjugate **10** bearing dipeptide chains of -L-Ala-L-Pro-NH-2-PyMe are well organized for binding of dicarboxylic acids by the chirality organization through two intramolecular hydrogen bondings (Fig. 2.8).[34] The size-selective and chiral recognition of dicarboxylic acids can be performed by multipoint hydrogen bondings of the amido pyridyl moieties of **10**. The ferrocene-dipeptide conjugate **10** affords the 1:1 complex with a series of dicarboxylic acids **11**, wherein the chirality-organized binding sites of **10** shows the size-selective binding of dicarboxylic acids. The highest association constant is obtained with adipic acid (**11c**, $K_a = 2.1 \times 10^4\ M^{-1}$). The chirality-organized binding sites of **10** are able to discriminate the chirality of guest molecules. In fact, the chirality-organized ferrocene-dipeptide conjugate **10** binds benzoyl-L-glutamic acid (**11 f**, $K_a = 5.5 \times 10^3\ M^{-1}$) approximately fifteen times more tightly than benzoyl-D-glutamic acid (**11 g**, $K_a = 3.7 \times 10^2\ M^{-1}$).

A chirality-organized structure through intramolecular interchain hydrogen bondings is not formed in the case of the ferrocene-dipeptide conjugate **12** bearing only one -L-Ala-D-Pro-NH-2-Py heterochiral dipeptide chain, and the left-handed helically ordered molecular arrangement through a network of intermolecular hydrogen bonds is created in a solid state as depicted in Fig. 2.9.[35]

The position of the pyridyl nitrogen of the C-terminal pyridyl moiety is found to control the chirality organization through the self-organization. In contrast to the structure of **12**, the ferrocene-dipeptide conjugate **13** bearing only one -L-Ala-D-Pro-NH-4-Py heterochiral dipeptide chain forms the chirality-organized structure through intramolecular type II β-turn-like hydrogen bonding between the CO of the ferrocenoyl moiety and the NH adjacent to the pyridyl moiety (Fig. 2.10A),[39] in which the β-turn mimic is

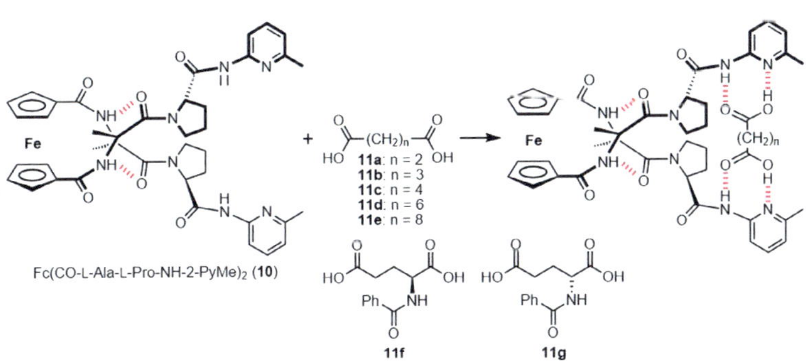

FIGURE 2.8 Selective recognition of dicarboxylic acid in the ferrocene-dipeptide conjugate **10**.

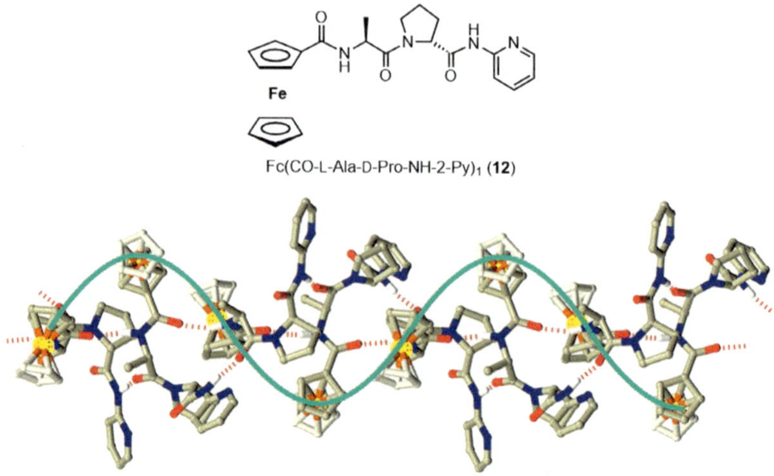

FIGURE 2.9 A portion of a layer containing the helical molecular assembly in the crystal packing of **12**.

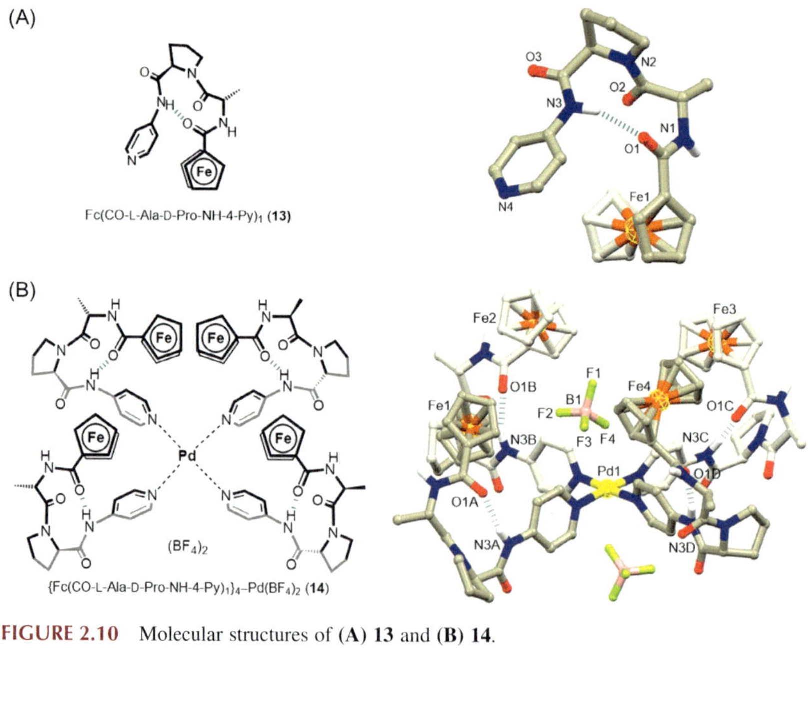

FIGURE 2.10 Molecular structures of **(A) 13** and **(B) 14**.

created by utilizing the minimum-sized peptide chain. The 4:1 palladium complex **14** is obtained by the reaction of the ferrocene-dipeptide conjugate **13** with 0.25 molar equivalents of $[Pd(MeCN)_4](BF_4)_2$ (Fig. 2.10B).[39] The single-crystal X-ray structure determination of **14** reveals that the four chirality-organized ferrocene-dipeptide conjugates are assembled around a palladium center in the same direction to form a chiral pocket, wherein one BF_4^- counter anion is accommodated (Fig. 2.10B). In the palladium complex **14**, each ferrocene-dipeptide conjugate preserves a β-turn-like structure through the intramolecular hydrogen bonding despite of complexation, wherein a type II β-turn-like structure of one ferrocene-dipeptide conjugate transforms into an intermediate between a type I′ and type III′ β-turn-like structures.

2.3 BIOORGANOMETALLIC CONJUGATES COMPOSED OF NUCLEOBASES TO INDUCE METAL–METAL INTERACTION

Platinum(II) compounds are known to show metallophilic Pt(II)–Pt(II) interaction through d_{Z^2}•••d_{Z^2} and/or π–π interactions, inducing the intriguing photophysical and photochemical properties.[54–60] Controlled arrangement of platinum(II) centers is of great importance to bring out the hidden talent in platinum(II) compounds. The d^{10}•••d^{10} closed shell aurophilic bonding interaction in gold(I) compounds[61–65] have been also demonstrated to play a key role in determining aggregated structures and luminescence properties.[61,66,67] The distance between two base pairs of DNA is about 3.3 Å, which is appropriate for the metallophilic interaction. The utilization of self-assembling properties of nucleobases is allowed to induce the controlled arrangement of metal centers. The introduction of the uracil group as a nucleobase into the luminescent organoplatinum(II) compound [4-octyloxy-(C^N^N)PtCl] affords the bioorganometallic platinum(II) compounds **U6Pt** and **U5Pt** (Fig. 2.11).[68] Molecular scaffolds **ND** and **AD**, having two 2,6-dihexamidopyridyl moieties as a complementary hydrogen bonding site for the uracil moiety, are focused on to control the molecular arrangement of the bioorganometallic platinum(II) compounds **U6Pt** and **U5Pt**. Naphthalene and anthracene are employed as an aromatic rigid framework in molecular scaffolds **ND** and **AD**, respectively.

The bioorganometallic platinum(II) compound **PtU6** shows the emission band around 730 nm, derived from the triplet metal-metal-to-ligand charge transfer (^{3}MMLCT) excited state resulting from Pt(II)-Pt(II) interaction with a concomitant decrease of an emission band at 590 nm based on the triplet metal-to-ligand charge transfer (^{3}MLCT) and/or ligand-to-ligand charge transfer (^{3}LLCT) excited state in the presence of the molecular scaffold **ND** (Fig. 2.12A).[68] The molecular scaffold **ND** induces the aggregation of **PtU6** through the complementary hydrogen bonding and π–π interaction between the uracil moieties as well as [4-octyloxy-(C^N^N)Pt] ligands in a solution

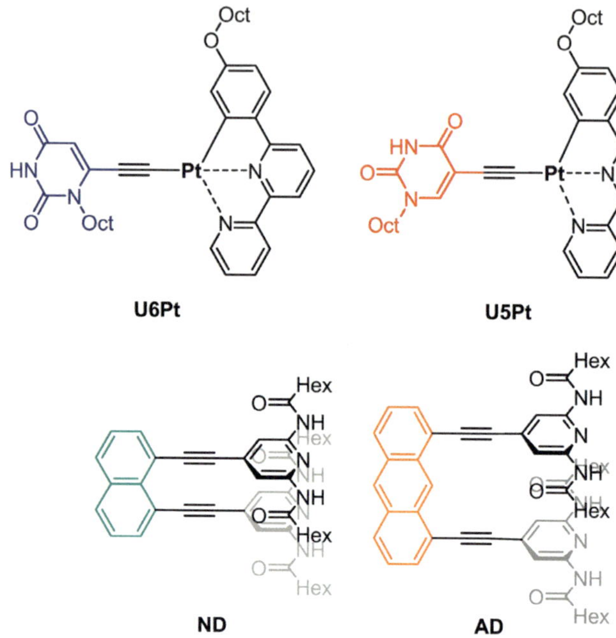

FIGURE 2.11 Bioorganometallic platinum (II) compounds **PtU6** and **PtU5**, molecular scaffolds **ND** and **AD**.

state. The direction of hydrogen bonding sites is a key factor to arrange the [4-octyloxy-(C^N^N)Pt] moieties regularly. Such ^{3}MMLCT emission resulting from Pt(II)–Pt(II) interaction is hardly observed in the case of the bioorganometallic platinum(II) compound **PtU5** wherein the direction of hydrogen bonding sites of the uracil moieties is different compared to **PtU6** (Fig. 2.12B).[68] The Pt(II)–Pt(II) interaction in aggregated compounds is considered to depend on an interbase distance. By the addition of 0.5 molar equivalents of molecular scaffold **AD** to a dichloromethane solution of **PtU6** results in the increase of the ^{3}MMLCT emission at around 730 nm based on Pt(II)–Pt(II) interaction and the decrease of the ^{3}MLCT/^{3}LLCT emission in the emission spectrum as shown in Fig. 2.12C.[68] The low intensity of the ^{3}MMLCT emission and the high intensity of the ^{3}MLCT emission are observed with the **AD** composed of anthracene in comparison to the emission with the molecular scaffold **ND** composed of naphthalene. The longer interbase distance is likely to cause the weak Pt(II)–Pt(II) interaction. The tuning of the emission properties of the organoplatinum(II) compounds is achieved by changing the direction of hydrogen bonding sites and the molecular scaffold size, which depend on the aggregated structures.

Guanine (G) derivatives have been heavily used in supramolecular chemistry because they self-assemble into G-ribbons, G-quartets, G-octamers, and columnar aggregates.[69–72] The most common form of guanine aggregates is

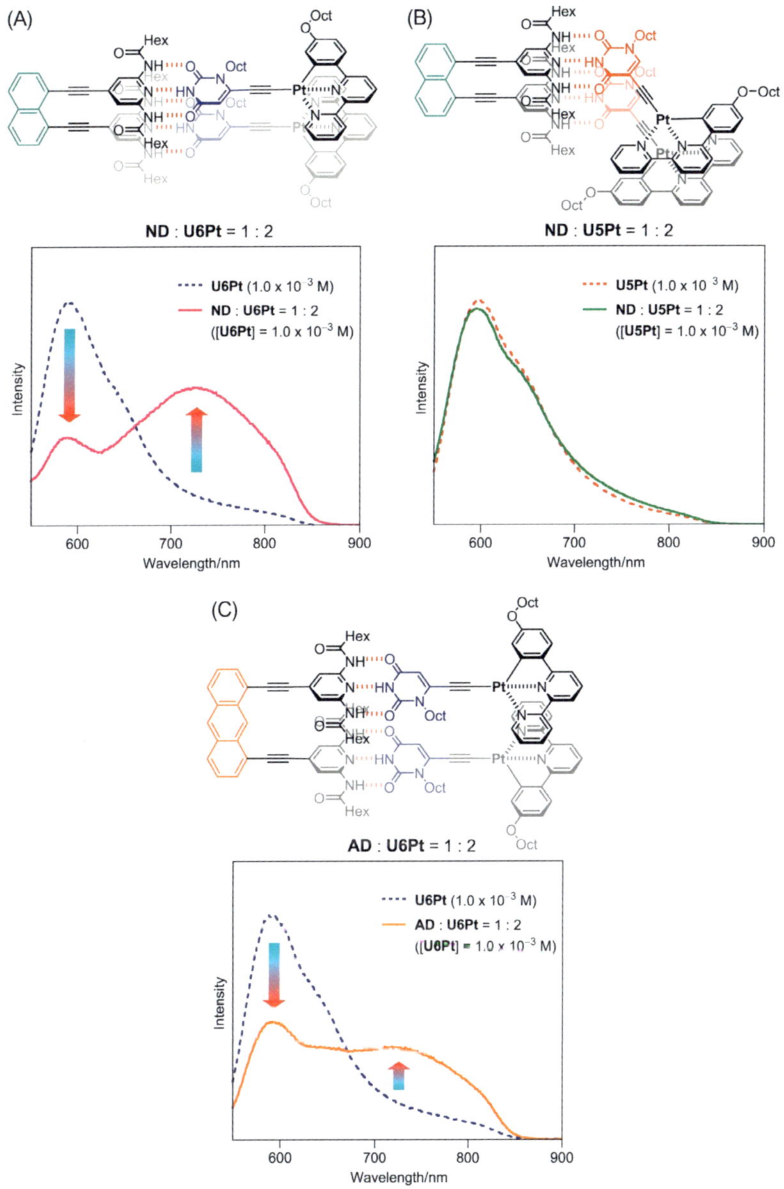

FIGURE 2.12 (A) Emission spectra of **U6Pt** and **ND-U6Pt**, (B) **U5Pt** and **ND-U5Pt**, and (C) **U6Pt** and **AD-U6Pt** in CH_2Cl_2 ($\lambda_{ex} = 530$ nm).

the G-octamer, constructed by two G-quartets and a cation. The distance between the two quartets is about 3.3 Å, which is a suitable distance for metal–metal interaction. The G-octamer is regarded as an ideal and promising scaffold to arrange the introduced functional units in preprogrammed positions. The bioorganometallic gold(I) compound **GAu(I)CNPh** composed of guanosine and the Au(I) isonitrile moiety is demonstrated to form the G-octamer in the presence of potassium ion, as depicted in Fig. 2.13, exhibiting a switchable emission based on Au(I)–Au(I) interaction, wherein the formation of the G-octamer serves as a reliable scaffold for the arrangement of the Au(I) isonitrile moieties to induce Au(I)–Au(I) interaction.[73]

Control of aggregation-induced emission based on Au(I)–Au(I) interaction is possible by using the G-octamer formation. The emission based on Au(I)–Au(I) interaction is not observed in the case of the bioorganometallic gold(I) compound **GAu(I)PPh$_3$** composed of guanosine and the Au(I)PPh$_3$ moiety although the expected octamer is formed by the addition of 0.25 molar equivalents of KPF$_6$ (Fig. 2.14).[74] This is probably due to the bulky triphenylphosphine ligand to make the Au(I) moiety far away each other. The G-octamers formed by trivalent cations have been reported to be more closely stacked than those formed by mono- and divalent cations.[75,76] The La(OTf)$_3$-mediated G-octamer exhibits the emission based on Au(I)–Au(I) interaction due to close stacking (Fig. 2.14).[74]

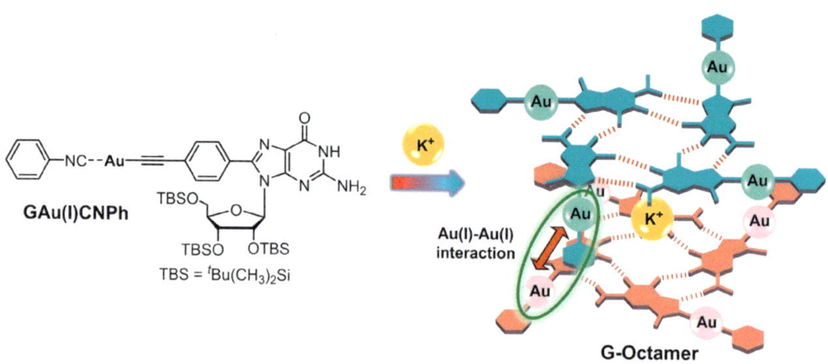

FIGURE 2.13 Formation of G-octamer of **GAu(I)CNPh** upon addition of KPF$_6$ to induce Au(I)-Au(I) interaction.

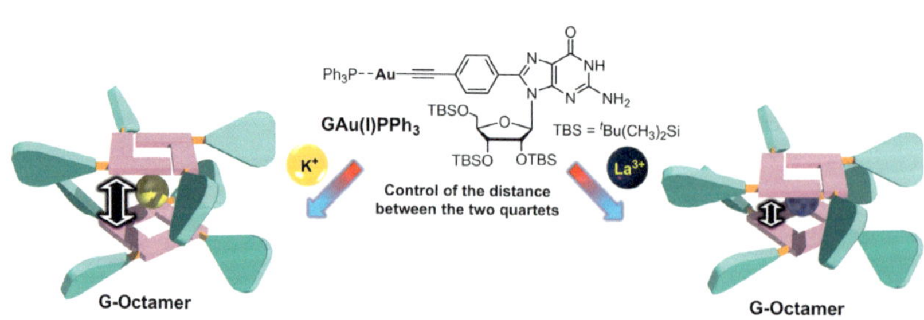

FIGURE 2.14 Formation of G-octamer of **GAu(I)PPh$_3$** upon addition of KPF$_6$ or La(OTf)$_3$.

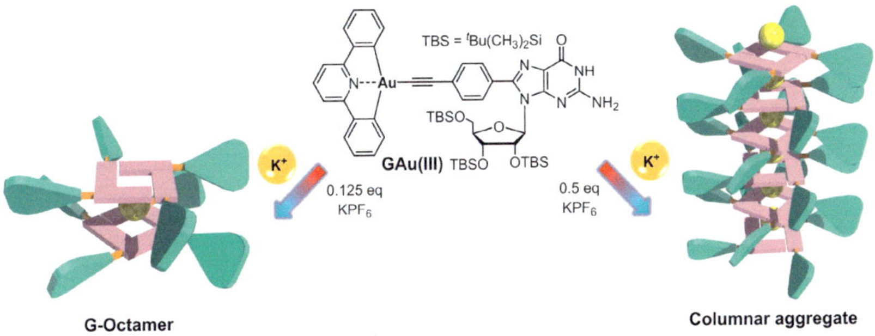

FIGURE 2.15 Formation of G-octamer or columnar aggregate of **GAu(III)** depending on the amount of KPF_6.

The bioorganometallic gold(III) compound **GAu(III)** composed of guanosine and the organogold(III) [(C^N^C)Au] moiety is designed to demonstrate the controlled formation of G-octamer and columnar aggregate. The complexation of **GAu(III)** with 0.125 molar equivalents of KPF_6 affords the G-octamer. In contrast, the addition of 0.5 molar equivalents of KPF_6 induces the formation of the columnar aggregate as shown in Fig. 2.15.[77]

The control of the aurophilic interaction is important for the design of aggregated gold(I) compounds. The utilization of a bridging ligand is considered to be a convenient strategy for the controlled arrangement of the Au(I) centers. The semirigid bridging diphosphine ligand is envisioned to arrange Au(I) centers in the same side of the ligand and promote induction of the aurophilic interaction. The dinuclear organogold(I)-uracil conjugates with the bridging diphosphine ligands and is designed to control the arrangement of Au(I) centers and self-assembly properties of the uracil moieties.

The single-crystal X-ray structure determination of the dinuclear organogold(I)-uracil conjugate **(U6Au)$_2$(μ-Xantphos)** with Xantphos as a bridging diphosphine ligand reveals an intramolecular aurophilic Au(I)–Au(I) interaction, wherein the semirigid xanthene backbone plays an important role in the arrangement of the phosphorus atoms as a coordination site on the same side to induce intramolecular Au(I)–Au(I) interaction (Fig. 2.16A).[78] *R*- and *S*-Enantiomers based on torsional twist about Au(I)–Au(I) axis are present in the crystal structure of **(U6Au)$_2$(μ-Xantphos)** (Figs. 2.16A and 2.17). Each enantiomer is connected alternately through intermolecular hydrogen bonding between the uracil moieties, affording the hydrogen-bonded assembly as depicted in Fig. 2.16B.

The chirality induction in Au(I)-Au(I) axis is possible by using the bridging diphosphine ligand with axial chirality. An intramolecular Au(I)–Au(I) interaction based on the aurophilic interaction is also observed in the crystal structure of the dinuclear organogold(I)-uracil conjugate **(U6Au)$_2$(μ-*R*-BINAP)**, wherein the chirality of Au(I)–Au(I) axis is induced by the

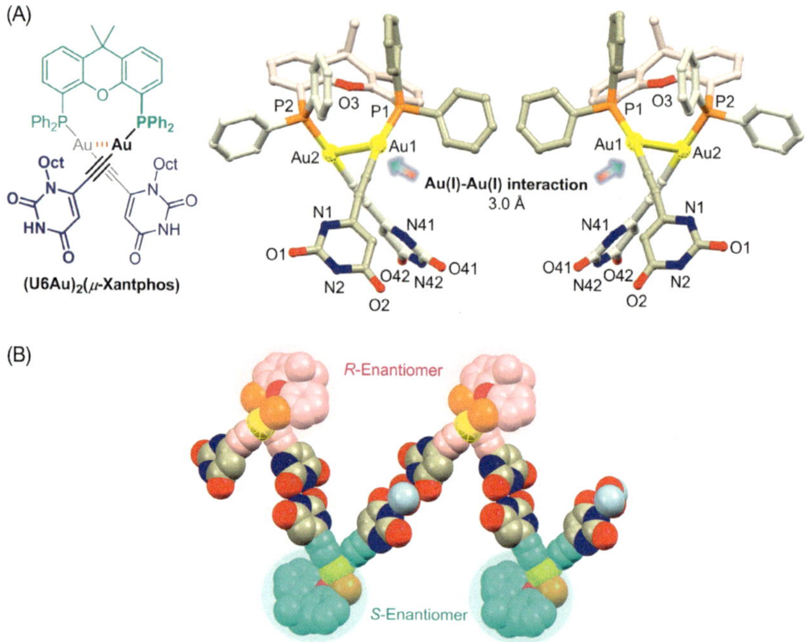

FIGURE 2.16 (A) Crystal structures of the *R*- and *S*-enantiomers of the dinuclear organogold (I)-uracil conjugate **$(U6Au)_2(\mu$-Xantphos)** and (B) the heterochiral hydrogen-bonded assembly through intermolecular hydrogen bonds between the uracil moieties of **$(U6Au)_2(\mu$-Xantphos)** (hydrogen atoms and octyl moieties are omitted for clarity).

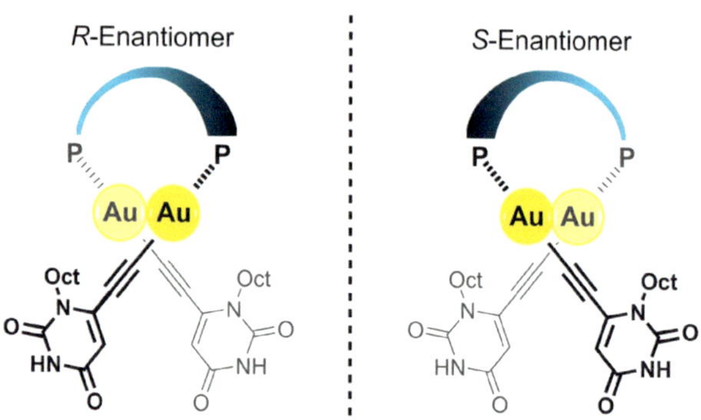

FIGURE 2.17 *R*- and *S*-Enantiomers of the dinuclear organogold(I)-uracil conjugate with the bridging diphosphine ligands. The broken line is the mirror plane.

utilization of (*R*)-BINAP as a bridging diphosphine ligand with axial chirality to adopt *R,R*-configuration as shown in Fig. 2.18A.[78] A helical molecular assembly is created by self-assembling through intermolecular hydrogen bonds between the uracil moieties (Fig. 2.18B).

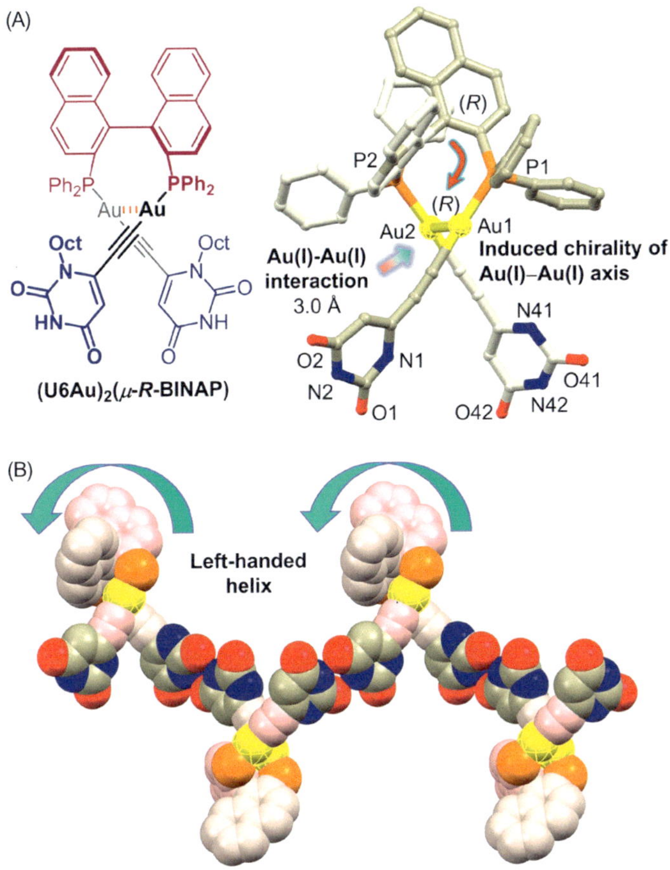

FIGURE 2.18 (A) Crystal structure of the dinuclear organogold(I)-uracil conjugate **(U6Au)$_2$(μ-R-BINAP)** and (B) the hydrogen-bonded assembly through intermolecular hydrogen bonds between the uracil moieties (hydrogen atoms and octyl moieties are omitted for clarity).

2.4 CONCLUSION

A variety of bioorganometallic conjugates are designed by the conjugation of organometallic compounds and biomolecules such as dipeptides and nucleobases. Ferrocenes are demonstrated to serve as a reliable organometallic scaffold for the construction of protein secondary structures such as β-sheets, β-turns, and γ-turns through intramolecular hydrogen bondings, wherein the introduced dipeptide strands are regulated in the appropriate positions, affording chirality-organized structures. In the ferrocene-dipeptide conjugates, the introduction of dipeptide chains into the ferrocene scaffold permits conformational enantiomerization of the ferrocenoyl moiety by restricting the torsional twist around the Cp(centroid)-Fe-Cp(centroid) axis and the C (ipso)-CO bond through the formation of intramolecular hydrogen bonds between the dipeptide chains. Controlled formation of protein secondary structures is achieved by changing chirality and sequence of amino acids. Stabilization of the chirality conformational regulation and control of the assembled structure of the dipeptide chains of the ferrocene-dipeptide conjugates are demonstrated by complexation with palladium(II) compounds. These bioorganometallic chemical models of protein secondary structures provide not only fundamental insight into understanding protein folding but also design basis for pharmacologically useful compounds and artificial biomaterials. The complementary hydrogen bonding of nucleobases is also demonstrated to serve as a powerful tool for controlled arrangement of bioorganometallic conjugates based on the directionality and specificity of hydrogen bonds, wherein tuning of the emission properties of the bioorganometallic conjugates is achieved through control of metal–metal interaction by changing the direction of hydrogen bonding sites and the molecular scaffold size. The controlled formation of G-octamer, in which the distance between two quartets is easily regulated by changing the introduced metal ion, is utilized to perform switchable emission properties based on Au(I)–Au(I) interaction. A helical molecular assembly is able to be designed by the chirality induction of Au(I)–Au(I) axis in the dinuclear organogold(I)-uracil conjugate.

The utilization of self-assembling properties of biomolecules such as peptides and nucleobases is envisioned to be a powerful strategy for the construction of the architectural well-organized molecular assemblies without chemical synthesis. The field of bioorganometallic chemistry has been attracting a great deal of attention and developing rapidly. The design of innovative bioorganometallic systems might be achieved as conjugation between organometallic compounds and biomolecules.

ACKNOWLEDGMENTS

We express special thanks all of our talented co-workers for their contribution to this chapter. Financial support from the Japan Science and Technology Agency and the Japan Society for the Promotion of Science is gratefully appreciated.

REFERENCES

1. Jeffrey GA. *An introduction to hydrogen bonding*. 1st ed New York: Oxford University Press; 1997.
2. Conn MM, Rebek Jr. J. Self-assembling capsules. *Chem Rev* 1997;**97**:1647–68.
3. Archer EA, Gong H, Krische MJ. Hydrogen bonding in noncovalent synthesis: Selectivity and the directed organization of molecular strands. *Tetrahedron* 2001;**57**:1139–59.
4. Prins LJ, Reinhoudt DN, Timmerman P. Noncovalent synthesis using hydrogen bonding. *Angew Chem Int Ed* 2001;**40**:2382–426.
5. Kyte J. *Structure in protein chemistry*. New York: Garland; 1995.
6. Branden C, Tooze J. *Introduction to protein structure*. 2nd ed New York: Garland; 1998.

7. Saenger W. *Principles of nucleic acid structure*. New York: Springer-Verlag; 1984.
8. Jaouen G, Vessiéres A, Butler IS. Bioorganometallic chemistry: A future direction for transition metal organometallic chemistry? *Acc Chem Res* 1993;**26**:361–9.
9. Severin R, Bergs R, Beck W. Bioorganometallic chemistry - transition metal complexes with α-amino acids and peptides. *Angew Chem Int Ed* 1998;**37**:1634–54.
10. Fish RH, Jaouen G. Bioorganometallic chemistry: Structural diversity of organometallic complexes with bioligands and molecular recognition studies of several supramolecular hosts with biomolecules, alkali-metal ions, and organometallic pharmaceuticals. *Organometallics* 2003;**22**:2166–77.
11. Jaouen G. *Bioorganometallics; biomolecules, labeling, medicine*. Weinheim: WilC-VCH; 2006.
12. Hartinger CG, Dyson PJ. Bioorganometallic chemistry—from teaching paradigms to medicinal applications. *Chem Soc Rev* 2009;**38**:391–401.
13. Togni A, Hayashi T. *Ferrocenes*. Weinheim: Wiley-VCH; 1995.
14. Schlögl K. Über Ferrocen-Aminosäuren und verwandte Verbindungen. *Monatsh Chem* 1957;**88**:601–21.
15. Hauser CR, Lindsay JK. Certain alkylations with the methiodide of *N,N*-dimethylaminomethylferrocene. Synthesis of an α-amino acid having the ferrocene group. *J Org Chem* 1957;**22**:1246–7.
16. Osgerby JM, Pauson PL. Ferrocene derivatives. Part VI. DL-ferrocenylalanine. *J Chem Soc* 1958;656–60.
17. Moriuchi T, Hirao T. Highly ordered structures of peptides by using molecular scaffolds. *Chem Soc Rev* 2004;**33**:294–301.
18. van Staveren DR, Metzler-Nolte N. Bioorganometallic chemistry of ferrocene. *Chem Rev* 2004;**104**:5931–86.
19. Kirin SI, Kraatz H-B, Metzler-Nolte N. Systematizing structural motifs and nomenclature in 1,n′-disubstituted ferrocene peptides. *Chem Soc Rev* 2006;**35**:348–54.
20. Moriuchi T, Hirao T. Ferrocene-peptide bioconjugates. In: Simonneaux G, editor. *Bioorganometallic chemistry*, vol. 17. Berlin: Springer-Verlag; 2006. p. 143–75.
21. Salmain M, Metzler-Nolte N. Bioorganometallic chemistry of ferrocene. In: Stepnicka P, editor. *Ferrocenes*. Chichester: John Wiley & Sons; 2008. p. 499–639.
22. Lataifeh A, Kraatz S, Kraatz H-B. Designer peptides: Attempt to control peptide structure by exploiting ferrocene as a scaffold. *Eur J Inorg Chem* 2009;3205–18.
23. Moriuchi T, Hirao T. Design of ferrocene-dipeptide bioorganometallic conjugates to induce chirality-organized structures. *Acc Chem Res* 2010;**43**:1040–51.
24. Moriuchi T, Hirao T. Dipeptide-induced chirality organization. *J Incl Phenom Macrocycl Chem* 2012;**74**:23–40.
25. Hirao T, Moriuchi T, Groβ A. Bioconjugates to induce chirality organization. In: Hirao T, editor. *Functuionalized redox systems*. Japan: Springer; 2015. p. 111–50.
26. Herrick RS, Jarret RM, Curran TP, Dragoli DR, Flaherty MB, Lindyberg SE, Slate RA, Thornton LC. Ordered conformations in bis(amino acid) derivatives of 1,10-ferrocenedicarboxylic acid, *Tetrahedron Lett* 1996;37:5289–92.
27. Nomoto A, Moriuchi T, Yamazaki S, Ogawa A, Hirao T. A highly ordered ferrocene system regulated by podand peptide chains. *Chem Commun* 1998;1963–4.
28. Moriuchi T, Nomoto A, Yoshida K, Hirao T. Characterization of ferrocene derivatives bearing podand dipeptide chains (-L-Ala-L-Pro-OR). *J Organomet Chem* 1999;**589**:50–8.
29. Moriuchi T, Nomoto A, Yoshida K, Ogawa A, Hirao T. Chirality organization of ferrocenes bearing podand dipeptide chains: Synthesis and structural characterization. *J Am Chem Soc* 2001;**123**:68–75.

30. Moriuchi T, Nomoto A, Yoshida K, Hirao T. Intramolecular conformational control in ferrocenes bearing podand dipeptide chains. *Organometallics* 2001;**20**:1008–13.
31. Moriuchi T, Yoshida K, Hirao T. Complexation stabilized conformational regulation of ferrocene bearing podand dipeptide chains (-L-Ala-L-Pro-NHPy). *Organometallics* 2001;**20**:3101–5.
32. Moriuchi T, Yoshida K, Hirao T. Structural characterization and complexation behavior of ferrocene bearing dipeptide chain (-L-Ala-L-Pro-NHPy). *J Organomet Chem* 2001;**637–639**:75–9.
33. Moriuchi T, Yoshida K, Hirao T. Hydrogen-bonding-directed molecular assembly of ferrocene bearing dipeptide chains (-L-Ala-L-Pro-NHPyMe) as an organometallic crystal architecture. *J Organomet Chem* 2003;**668**:31–4.
34. Moriuchi T, Yoshida K, Hirao T. Chirality-organized ferrocene receptor bearing podand dipeptide chains (-L-Ala-L-Pro-NHPyMe) for the selective recognition of dicarboxylic acids. *Org Lett* 2003;**5**:4285–8.
35. Moriuchi T, Nagai T, Hirao T. Chirality organization of ferrocenes bearing dipeptide chains of heterochiral sequence. *Org Lett* 2005;**7**:5265–8.
36. Moriuchi T, Nagai T, Hirao T. Induction of γ-turn-like structure in ferrocene bearing dipeptide chains via conformational control. *Org Lett* 2006;**8**:31–4.
37. Moriuchi T, Fujiwara T, Hirao T. Complexation-induced conformational regulation of ferrocene-dipeptide conjugates to nucleate γ-turn-like structure. *J Organomet Chem* 2007;**692**:1353–7.
38. Moriuchi T, Nagai T, Fujiwara T, Honda N, Hirao T. Simultaneous formation of antiparallel β-sheet-like and type II β-turn-like structures based on introduction of dipeptide chains with heterochiral sequence into ferrocene scaffold. *Heterocycles* 2008;**76**:595–603.
39. Moriuchi T, Fujiwara T, Hirao T. β-Turn-structure-assembled palladium complexes by complexation-induced self-organization of ferrocene-dipeptide conjugates. *Dalton Trans* 2009;**14**:4286–8.
40. Moriuchi T, Nishiyama T, Tayano Y, Tohnai N, Hirao T. Self-assembled structures of ferrocene-L-carnosine conjugates. *J Organomet Chem* 2017;**839**:78–82.
41. Moriuchi T, Wu H, Tayano Y, Hirao T. Structural characterization of chirality-organized ferrocene-dipeptide conjugates having pyridine *N*-oxide moieties. *Asian J Org Chem* 2017;**6**:1250–6.
42. Moriuchi T, Nishiyama T, Nobu M, Hirao T. Control of helical chirality of ferrocene-dipeptide conjugates by the secondary structure of dipeptide chains. *Chem Eur J* 2017;**23**:12704–8.
43. Moriuchi T, Nishiyama T, Tayano Y, Hirao T. Controlled self-assembling structures of ferrocene-dipeptide conjugates composed of Ala-Pro-$NHCH_2CH_2SH$ dipeptide chain. *J Inorg Biochem* 2017;**177**:259–65.
44. Kirin SI, Wissenbach D, Metzler-Nolte N. Unsymmetrical 1,*n*′-disubstituted ferrocenoyl peptides: Convenient one pot synthesis and solution structures by CD and NMR spectroscopy. *New J Chem* 2005;**29**:1168–73.
45. Heinze K, Beckmann M. Conformational analysis of chiral ferrocene-peptides. *Eur J Inorg Chem* 2005;3450–7.
46. Rebek Jr. J. Simultaneous encapsulation: Molecules held at close range. *Angew Chem Int Ed* 2005;**44**:2068–78.
47. Gallagher JF, Kenny PTM, Sheehy MJ. Electrochemical anion recognition by redox-responsive unnatural amino acid derivatives; the X-ray crystal structure of *N*-ferrocenoyl

glycine methyl ester, (η^5-C_5H_5)Fe(η^5-C_5H_4)CONHCH$_2$COOMe. *Inorg Chem Commun* 1999;**2**:200–2.

48. Gallagher JF, Kenny PTM, Sheehy MJ. Synthesis and electrochemical anion recognition by novel redox-responsive ferrocenoyl dipeptide ester derivatives; ^{1}H NMR anion complexation studies. *Inorg Chem Commun* 1999;**2**:327–30.
49. Georgopoulou AS, Mingos DMP, White AJP, Williams DJ, Horrocks BR, Houlton A. Bifunctional ferrocene derivatives for molecular recognition of DNA duplexes. *J Chem Soc, Dalton Trans* 2000;2969–74.
50. Plumb K, Kraatz H-B. Interaction of a ferrocenoyl-modified peptide with papain: Toward protein-sensitive electrochemical probes. *Bioconjugate Chem* 2003;**14**:601–6.
51. Miklán Z, Szabó R, Zsoldos-Mády V, Reményi J, Bánóczi Z, Hudecz F. New ferrocene containing peptide conjugates: Synthesis and effect on human leukemia (HL-60) cells. *Biopolymers* 2007;**88**:108–14.
52. Mahmoud KA, Kraatz H-B. A bioorganometallic approach for the electrochemical detection of proteins: A study on the interaction of ferrocene-peptide conjugates with papain in solution and on Au surfaces. *Chem Eur J* 2007;**13**:5885–95.
53. Kerman K, Kraatz H-B. Electrochemical probing of HIV enzymes using ferrocene-conjugated peptides on surfaces. *Analyst* 2009;**134**:2400–4.
54. Roundhill DM, Gray HB, Che C-M. Pyrophosphito-bridged diplatinum chemistry. *Acc Chem Res* 1989;**22**:55–61.
55. Houlding VH, Miskowski VM. The effect of linear chain structure on the electronic structure of Pt(II) diimine complexes. *Coord Chem Rev* 1991;**111**:145–52.
56. Kato M. Luminescent platinum complexes having sensing functionalities. *Bull Chem Soc Jpn* 2007;**80**:287–94.
57. Eryazici I, Moorefield CN, Newkome GR. Square-planar Pd(II), Pt(II), and Au(III) terpyridine complexes: Their syntheses, physical properties, supramolecular constructs, and biomedical activities. *Chem Rev* 2008;**108**:1834–95.
58. Develay JAG, Develay S, Rochester DL, Murphy L. Optimising the luminescence of platinum(II) complexes and their application in organic light emitting devices (OLEDs). *Coord Chem Rev* 2008;**252**:2596–611.
59. McGuire Jr R, McGuire MC, McMillin DR. Platinum(II) polypyridines: A tale of two axes. *Coord Chem Rev* 2010;**254**:2574–83.
60. Wong KM-C, Yam VW-W. Self-assembly of luminescent alkynylplatinum(II) terpyridyl complexes: Modulation of photophysical properties through aggregation behavior. *Acc Chem Res* 2011;**44**:424–34.
61. Pyykkö P. Theoretical chemistry of gold. *Angew Chem, Int Ed* 2004;**43**:4412–56.
62. Schmidbaur H. Ludwig mond lecture. High-carat gold compounds. *Chem Soc Rev* 1995;**24**:391–400.
63. Pyykkö P. Strong closed-shell interactions in inorganic chemistry. *Chem Rev* 1997;**97**:597–636.
64. Katz MJ, Sakai K, Leznoff DB. The use of aurophilic and other metal–metal interactions as crystal engineering design elements to increase structural dimensionality. *Chem Soc Rev* 2008;**37**:1884–95.
65. Schmidbaur H, Schier A. A briefing on aurophilicity. *Chem Soc Rev* 2008;**37**:1931–51.
66. Yam VW-W, Lo KK-W. Luminescent polynuclear d^{10} metal complexes. *Chem Soc Rev* 1999;**28**:323–34.
67. Yam VW-W, Cheng EC-C. Highlights on the recent advances in gold chemistry-A photophysical perspective. *Chem Soc Rev* 2008;**37**:1806–13.

68. Moriuchi T, Sakamoto Y, Noguchi S, Fujiwara T, Akine S, Nabeshima T, Hirao T. Design and controlled emission properties of bioorganometallic compounds composed of uracils and organoplatinum(II) moieties Design. *Dalton Trans* 2012;**41**:8524–31.
69. Davis JT. G-Quartets 40 years later: From 5′-GMP to molecular biology and supramolecular chemistry. *Angew Chem Int Ed* 2004;**43**:668–98.
70. Davis JT, Spada GP. Supramolecular architectures generated by self-assembly of guanosine derivatives. *Chem Soc Rev* 2007;**36**:296–313.
71. Lena S, Masiero S, Pieraccini S, Spada GP. The supramolecular organization of guanosine derivatives. *Mini-Rev Org Chem* 2008;**5**:262–73.
72. Aimé C, Nishiyabu R, Gondo R, Kaneko K, Kimizuka N. Controlled self-assembly of nucleotide-lanthanide complexes: Specific formation of nanofibers from dimeric guanine nucleotides. *Chem Commun* 2008;6534–6.
73. Meng X, Moriuchi T, Kawahata M, Yamaguchi K, Hirao T. A G-octamer scaffold via self-assembly of a guanosine-based Au(I) isonitrile complex for Au(I)–Au(I) interaction A G-octamer scaffold. *Chem Commun* 2011;**47**:4682–4.
74. Meng X, Moriuchi T, Sakamoto Y, Kawahata M, Yamaguchi K, Hirao T. La(OTf)$_3$-mediated self-organization of the guanosine having alkynyl-Au(I)PPh$_3$ moiety to induce Au(I)–Au(I) interaction. *RSC Adv* 2012;**2**:4359–63.
75. Kwan ICM, She YM, Wu G. Trivalent lanthanide metal ions promote formation of stacking G-quartets. *Chem Comm* 2007;4286–8.
76. Shinoda S, Noguchi T, Ikeda M, Habata Y, Tsukube H. Luminescent double-decker type guanine octets with trivalent lanthanide cations: In situ self-assembling and stability evaluation in homogeneous organic media. *J Incl Phenom Macrocycl Chem* 2011;**71**:523–7.
77. Meng X, Moriuchi T, Tohnai N, Miyata M, Kawahata M, Yamaguchi K, Hirao T. Synthesis and assembling properties of bioorganometallic cyclometalated Au(III) alkynyls bearing guanosine moieties. *Synthesis Org Biomol Chem* 2011;**9**:5633–6.
78. Sakamoto Y, Moriuchi T, Hirao T. Dinuclear organogold(I) complexes bearing uracil moieties: Chirality of Au(I)–Au(I) axis and self-assembling. *CrystEngComm* 2015;**17**:3460–7.

Chapter 3

Ferrocene Peptide-based Supramolecular Gels: Current Trends and Applications

Natashya Falcone[1,2] and Heinz-Bernhard Kraatz[1,2,3]
[1]*Department of Physical and Environmental Sciences, University of Toronto Scarborough, Toronto, ON, Canada,* [2]*Department of Chemical Engineering and Applied Chemistry, University of Toronto, Toronto, ON, Canada,* [3]*Department of Chemistry, University of Toronto, Toronto, ON, Canada*

3.1 INTRODUCTION

Gels are well-known materials commonly found in everyday application use including hair gels and soft contact lenses; they are also heavily used in scientific research applications as agar gel plates for bacteria culture growth and polyacrylamide gels in electrophoresis. A gel is defined as a semisolid material made up of a solid gelator and a liquid solvent that exhibits no flow in steady state.[1,2] In contrast to macromolecular gel materials, supramolecular gels consist of low-molecular weight gelators that self-assemble into a 3D network, encapsulating solvent molecules.[2] They can also be classified into categories such as physical gels, where the assembly results from physical bonds, or chemical gels, such as conventional polymer gels where the assembly is a result from chemical bonds.[1,3] While conventional chemical polymer gels are interconnected through covalent bonds forming strong rigid crosslinks, these gels cannot be re-dissolved and are thermally irreversible.[1,3] Supramolecular gels formed by physical interactions are formed through non-covalent interactions including hydrogen bonding, metal coordination, van der Waals interactions, $\pi - \pi$ stacking, and hydrophobic interactions.[4–6] These types of intermolecular forces are known to have an effect on gel volume phase transitions which play a role in the formation of a supramolecular gel. Phase transitions of a gel are a reversible and discontinuous folding and unfolding of the gel network. Many supramolecular gels undergo a sol-gel phase transition, which in some cases can be fully reversible, going from a flowing liquid, the sol phase, to a non-flowing gel phase.[7] Phase

Advances in Bioorganometallic Chemistry. DOI: https://doi.org/10.1016/B978-0-12-814197-7.00003-0

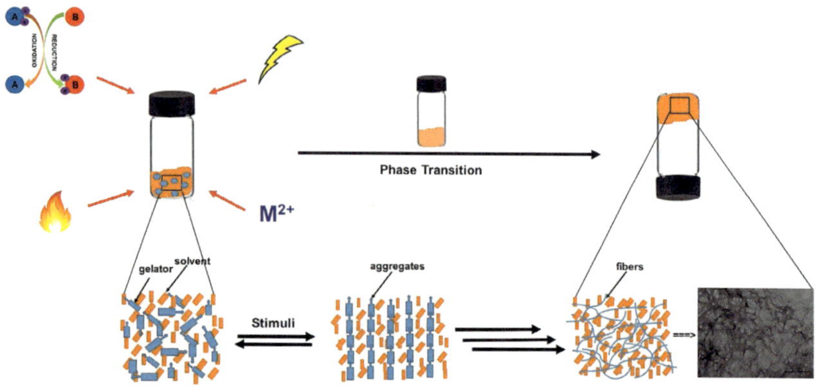

FIGURE 3.1 Scheme representing phase changes of a solution into a supramolecular gel. The first vial is a solution showing the gelator and solvent components separately. As the solution is mixed or exposed to an additional source, such as the ones represented in the figure—redox agents, light, heat, or metal ions—the solution then aggregates together and eventually forms a supramolecular gel with a 3D nanofiber matrix.

transitions in gels provides a means of studying the four interactions, ionic, hydrophobic, van der Waals, and hydrogen bonding, that are responsible for gelation. In this chapter, we focus on, and explore how, different stimuli affect the compounds' phase transition from solution into gel (Fig. 3.1).

Among various gelator compounds capable of forming supramolecular gels, amino acids and peptides have by far been the most optimal gelators for self-assembly.[4] Peptides and amino acids as organic building blocks possess a number of interesting features including hydrogen bonding, chirality, and metal binding, making them attractive and versatile components.[6] They also possess significantly strong and stable structures, they have excellent biocompatibility and biodegradability, and possess self-assembling behavior in responses to slight environmental changes. Due to the dynamic and reversible nature of the non-covalent interactions mentioned, supramolecular peptide gels may undergo phase transitions in response to different stimuli such as temperature, pH, solvent, light, and redox reactions.[8] Das et al. reported a selenoester mediated chemical ligation to explore redox active peptide self-assembly.[9] They showed that cysteine and glycine peptides have provided insight into the ligation-driven, self-assembly process and how the sol-gel transition was highly dependent on oxidation and reduction of cysteine and cysteine-based peptides.[9] Diaz et al. synthesized tetrapeptide gelators with azobenzene groups as a photochromic group to use light irritation as a stimulus to trigger the sol-to-gel phase transition[10] (Fig. 3.2).

They found the ability of these small peptides bearing the azobenzene side chain to form stable supramolecular gels that are responsive to several stimuli including light.[10] The stimuli-responsiveness of these materials makes them attractive materials in all types of applications including medicinal, drug

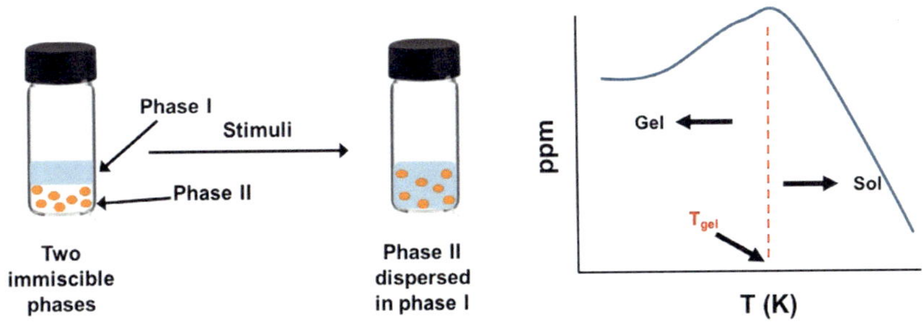

FIGURE 3.2 Scheme depicting the phase transition of a supramolecular gel and a representative graph of the temperature-induced chemical shift, the inflection point corresponds to the T_{gel}. Graph redrawn from Fatás, P.; Bachl, J.; Oehm, S.; Jiménez, A. I.; Cativiela, C.; Díaz Díaz, D. Chem. Eur. J. 2013, 19 (27), 8861–8874.

delivery, and tissue engineering. Nagasaki et al. designed a new local protein delivery system by using a protein-loaded, redox-active, injectable gel (RIG) and showed that this gel has the potential to be an injectable sustained-release carrier.[11,12] They also showed that the protein (IL-12) loaded in the RIG gel showed tumor growth inhibition in tumor-bearing mice compared to the IL-12 protein alone, indicating the potential as a platform technology for sustained release to provide high antitumor efficacy.[11] Regan et al. designed a 34 amino acid tetratricopeptide which adopts a helix-turn-helix structure that forms a super-helix with eight repeats per helical turn and was reported to form a hydrogel. The rheological properties of this gel show that this compound exhibited elasticity with good mechanical integrity and the elasticity was found to be optimal to encapsulate and release both large and small molecules for the potential use in controlled drug delivery applications.[12] Chauhan et al. also designed a dipeptide-based hydrogel with high mechanical strength and stability, responsive to temperature, pH, and salt perturbations, for applications in drug delievery.[13] They showed that the gel could entrap and release bioactive molecules in a sustained manner with the drug release rate being modulated by pH and salt concentration with no cytotoxic effects to mammalian cell cultures. As mentioned, some peptide-based supramolecular hydrogels possess medicinal properties. Banerjee et al. designed thermos-responsive amphiphilic peptide hydrogels with a new structural motif that possesses thixotropic properties and demonstrates antibacterial activity.[14] Two of the synthesized hydrogels exhibit remarkable antibacterial properties against both Gram-positive and Gram-negative bacteria with no cytotoxicity, holding promise for the development of soft materials as antibiotics.

In this chapter, we highlight the assembly of peptide conjugates into supramolecular gels in response to different environmental factors, with a focus on ferrocene (Fc)-peptide conjugates. We discuss how different stimuli and interactions—such as redox, temperature, H-bonding, and chirality—in these conjugates, play a role in the phase transitions affecting gelation.

3.2 STIMULI-RESPONSIVE PEPTIDES

Peptide gelators possess advantages in forming reversible supramolecular assemblies through intermolecular hydrogen-bonding interactions in response to various stimuli. Stimuli-responsiveness is an attractive property and is based on the alteration of the secondary structure, solubility, and/or intermolecular interactions. Peptide-based stimuli responsive biomaterials have exciting prospects in areas including controlled drug delivery, tissue engineering and regenerative medicine, bio-sensing, and bio-diagnostics.[3] Examples of biomedical applications of peptide gels include Akiyama's peptide hydrogel consisting of 16 amino acids, which was assessed for transplantation of mucosal cells to repair a damaged ear.[15] Zheng et al. synthesized a KLD-12 peptide gelator that includes lysine, leucine, and aspartic acid, and studied its biocompatibility with a host rabbit in the hope of repairing the degenerated nucleus in intervertebral disks.[16] The same team explored a different peptide hydrogel for facilitating sustained release of cytokines with applications in regenerative medicine.[17]

Any given stimulus can be classified as having a physical, chemical, or biological origin. Physical stimuli include light, sonication, and temperature. Stimuli of a chemical nature include pH, ionic strength, solvent, and changes in redox state, while biological stimuli include reactions triggered by enzymes, binding of antigens, and receptors.[3] In the following section, a number of stimuli in respect to their effect on gelation will be discussed.

3.2.1 Temperature and pH Dependent Gelation

Thermo-responsive peptides are one of the most commonly studied peptide gelators. Here, temperature is the primary stimulus responsive for the formation of a supramolecular gel, and the strengths of the interactions are temperature dependent.[3] Cooling from an elevated temperature is a very common approach for forming supramolecular gels. Xu et al. reported a thermo-responsive gel of the dipeptide conjugate Fmoc-Ala-Ala, where the increase of temperature induced a hydrogel.[18] Adjusting the pH of a compound in solvent is another widely used stimulus for obtaining a supramolecular gel. It has been reported to be the most effective and simplest chemical method to trigger gelation because a small amount of acid or base can lead to a large shift.[3] A change of pH in amino acids can affect the charges on peptide gelators leading to compounds being able to gel in various pH's. Ulijin's group reported that for Fmoc-Phe-Phe, a decrease in pH induces self-assembly to form a fibrous network, which is shown in Fig. 3.3.[19] Furthermore, it was demonstrated that Fmoc-Phe-Phe undergoes multiple phase transitions leading into gelation and the dissociation of the gel. As the compound went through the first transition in response to the pH change, the solution became slightly cloudy and with time formed a supramolecular gel. As the samples went through the second

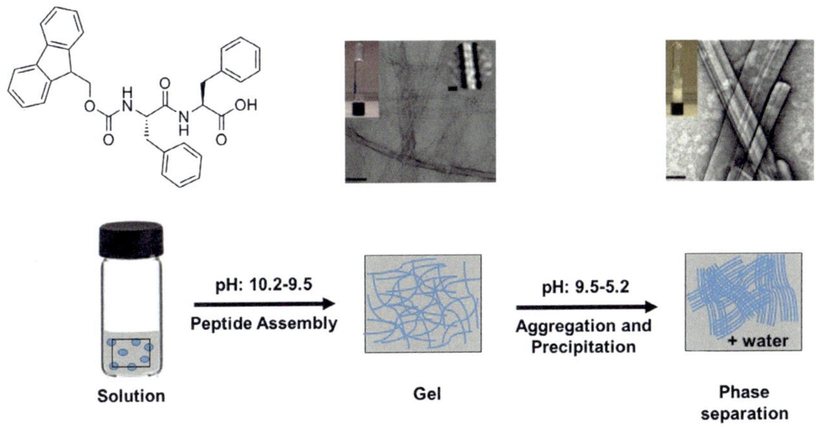

FIGURE 3.3 Scheme showing the proposed self-assembly mechanism of Fmoc-FF from high to low pH above the critical gelation concentration as a function of the peptide degree of ionization. *Adapted with permission from Tang, C.; Smith, A. M.; Collins, R. F.; Ulijn, R. V.; Saiani, A. Langmuir 2009, 25 (16), 9447–9453. Copyright 2009 American Chemical Society.*

transition, a phase separation occurs where there was a clear liquid phase at the top and precipitate at the bottom. This shows that the behavior is dependent on pH as the first transition corresponds with self-assembly into fibrils while the second transition induced aggregation and dissociation of the gel resulting in an aggregated phase and water phase.

Afrasiabi and Kraatz also demonstrated thixotropic properties of a peptide gel, where under normal conditions the property of the gel is viscous and under stressed conditions, such as heat or sonication, become less viscous; when the stress is removed the gel is able to return back to normal.[20] The thixotropy of the gel was followed by temperature-dependent circular dichroism spectroscopy, where the CD signal at 500 nm gradually vanishes as the supramolecular assembly is being destabilized at elevated temperatures as shown in Fig. 3.4.

3.2.2 Enzyme-Dependent Gelation

Enzymes are increasingly attracting attention in the generation and applications of supramolecular gels. In enzymatic gelation, a precursor molecule undergoes an enzymatic conversion into a gelator molecule, normally through bond cleavage or bond formation.[3] This is an attractive method because it uses natural biological agents in the supramolecular self-assembly. It was reported that enzymes as biological catalysts can be used to mediate hydrogel formation with the advantages of cross-linking under non-toxic, mild conditions and having high chemo- and regio-selectivity.[21] A number of enzyme-responsive materials have been described based on

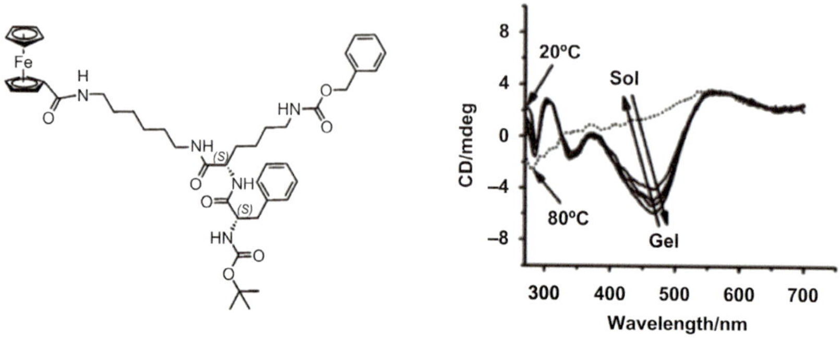

FIGURE 3.4 Structure of Fc-peptide conjugate which exhibits thixotropic effects, followed by the CD spectrum exhibiting the effect by showing the prominent peak at 500 nm reduces to zero as the temperature increases, showing the decrease in supramolecular assembly. *Adapted with permission from, R.; Kraatz, H-B. Chemistry: A European Journal 2013, 19 (51), 17296–17300. Copyright 2013 ChemPubSoc Europe.*

supramolecular assemblies that respond to a range of different enzymes including lipases, proteases, phosphatases, kinases, acyl transferases, and redox enzymes. The use of enzymes in supramolecular materials opens up a number of possible applications in regenerative medicine and drug delivery. A non-gelling phosphorylated Fmoc-tyrosine precursor can be dephosphorylated by a phosphatase to induce gelation. In addition, there are number of studies that demonstrate the use of enzyme-triggered intramolecular acyl migration in modified peptides to induce gel formation.[22,23] Xu et al used an acid phosphatase enzyme to catalyze the molecular hydrogelation of a precursor to a peptide derivative gelator.[24]

This is shown in Fig. 3.5 where a precursor molecule, such as the phenylalanine derivative shown, was converted into a gelator compound and formed a hydrogel after reacting with an acid phosphatase enzyme and thus entrapping the enzyme into the gel matrix. Groll et al. also reported a HRP-mediated preparation of a redox-sensitive disulfide-cross-linked hydrogel through the gelation of thiomers without the requirement of hydrogen peroxide.[21]

In the next section, metal-mediated gel formation will be discussed, in which metal coordination to peptides is key in forming a stable gel material.

3.2.3 Metal Ion Induced Self-Assembly

Metal ions have also been reported to increase stabilities of supramolecular gels. Protein-metal ion interactions are ubiquitous in nature and can be utilized for controlling the self-assembly of supramolecular materials. The use of different metal ions was reported to provide wide control over the mechanical strength as well as self-healing properties. Ali et al. described a

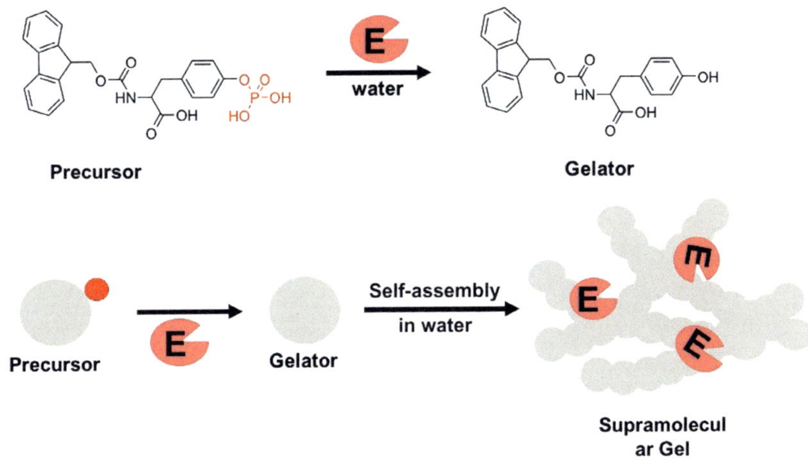

FIGURE 3.5 Illustration of the acid phosphatase enzyme in the enzymatic hydrogelation by converting the precursor molecules (7 or 9) into the hydrogelators (8 or 10). *Adapted from Wang, Q.; Yang, Z.; Ge, W.; Wang, L.; Xu, B. Soft. Matt. 2008, 4, 550–553. Royal Society of Chemistry.*

tunable supramolecular hydrogel that was obtained through self-assembly of a Zn^{2+} peptide-hyaluronic acid. It was found that the addition of Zn^{2+} triggers folding of the peptides into four helix bundles resulting in a hydrogel, and the chelation of the Zn^{2+} ions resulted in disassembly of the hydrogel.[25] The Kraatz group reported the ability of a peptide-based gelator to form self-healing hydrogels after stimulation by metal coordination and specifically explored the self-assembly of a phenylalanine derivative in a phosphate buffer in the presence of a range of divalent metal ions.[26] The experimental results demonstrate that the metal ions provide significant control over the mechanical and thixotropic properties of the hydrogels. The stability of the hydrogel increases with a decrease in ionic radius, and the strength of metallo-hydrogels increased with increasing stability of the complex formed. Specifically, the Cd^{2+} hydrogel was found to have the strongest mechanical strength, while the Mn^{2+} hydrogel was found to exhibit the weakest mechanical strength.[26]

Along the same line of argument, the mechanical and thermal properties of a C12-phenylalanine hydrogel in phosphate buffer is modulated by the presence of Ca^{2+}, Co^{2+}, and Ni^{2+}. The hydrogel displayed high thermal and mechanical stability.[27] Fig. 3.6 shows a scheme of the self-assembly of the gelator with the presence of the metal ions Ca^{2+}, Co^{2+}, and Ni^{2+}. It was reported that the role of metal binding is binding the two gelator molecules through the complex formation which self-assembled by the help of non-covalent interactions.[27]

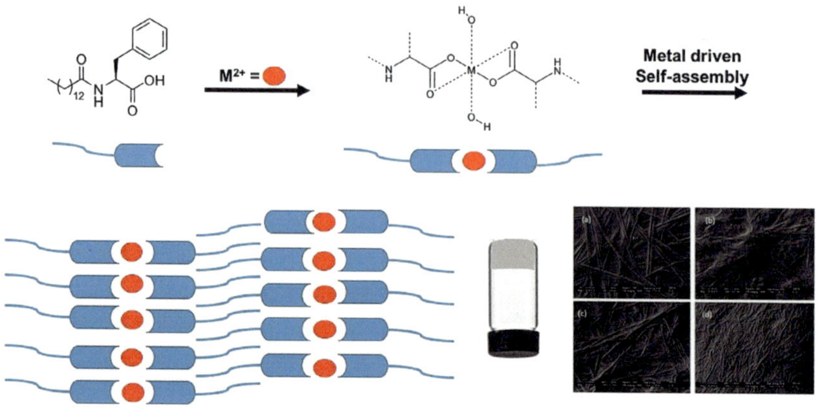

FIGURE 3.6 This scheme highlights the importance of metal ions as a stimulus in self-assembly. The blue structure represents the phenylalanine derivative (structure shown) while the metal ion is represented by the red circle. In the presence of the metal ions (Ca^{2+}, Co^{2+}, Ni^{2}) additional non-covalent interactions are present between the two gelator molecules influencing the self-assembly into gelation. Upon metal ions, the morphology changes which is shown in the SEM images of the gel (a) and with the addition of the metal ions (b-d). *Adapted with permission from Basak, S.; Singh, I.; Banerjee, A.; Kraatz, H.-B. RSC Adv. 2017, 7 (24), 14461–14465. Royal Society of Chemistry.*

3.2.4 Redox-Dependent Gelation

Redox agents as a stimulus provides another useful method for controlling the self-assembly of small molecules. This is where oxidation or reduction of the gelator promotes phase transitions by inducing gelation or reversing the gel back into solution state with the opposite agent. Nilsson et al. demonstrated that the reduction of a disulfide bond in a cyclic peptide was an optimal strategy for controlling the peptide's self-assembly into a hydrogel.[28] We will continue to discuss and emphasize redox-triggered gelation later on in this chapter when discussing Fc-peptide conjugates (Fig. 3.7).

3.3 FERROCENE-PEPTIDE CONJUGATES

3.3.1 Introduction of Ferrocene-Peptide Conjugates

Bio-organometallic Fc amino acid and peptide conjugates have been explored extensively, in particular the Fc group has been exploited largely as a scaffold to support peptide secondary structural motifs. In this section, Fc peptides are discussed in respect to their interesting structural properties and response to various stimuli. Since the report of these systems by Herrick and coworkers, a range of interesting systems have been reported that support turn-like structures and are even suitable for metallo-ligands.[29] Early structural property reports of these compounds include when Kraatz et al.

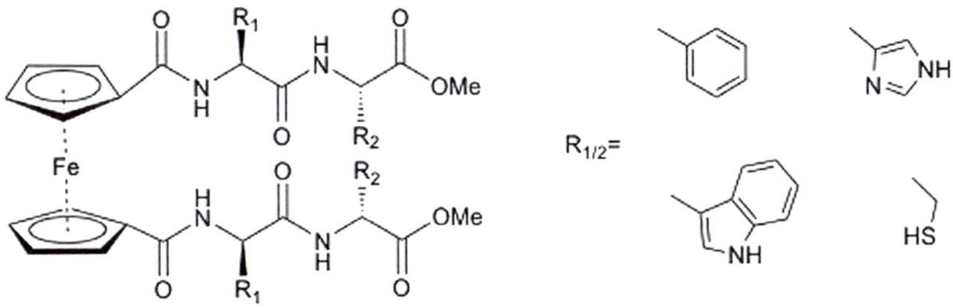

FIGURE 3.7 The general structure of a ferrocene dicarboxylic acid dipeptide conjugate. R groups can include amino acid side chains such as phenylalanine, tryptophan, histidine, and serine.

reported synthetic results of Fc-carboxylic acid/amino acid couple reactions and the structural properties of two ferrocenoyl-amino acid ester compounds, while Hirao et al. incorporated a dipeptide (Ala-Pro-OEt) into ferrocene gaining insight into the structural factors for the molecular arrangement of these compounds.[30,31] It was also stressed in both cases that strong intramolecular hydrogen bonds form between the amide NH and carbonyl oxygen atoms.[31,32] These systems were then explored further in terms of their response to different stimuli including metal binding and redox which eventually led to stimuli-responsive gelators. Saweczko et al. looked at the system of Fc-Gly_c-OEt and showed that the redox potential of the Fc attached to helical and non-helical olgiopeptides is influenced by the peptides secondary structure as the diglycine substituents established a H-bonding pattern unlike that in free diglycine.[33] Appoh et al. evaluated the interaction of six Fc-amino acid conjugates with metal ions Mg^{2+}, Ca^{2+}, Zn^{2+}, La^{3+}, Tb^{3+} and an interesting change was observed in the selectivity for metal binding as a function of amino acid residue.[34] These interesting stimuli-responsive properties are further discussed with additional stimuli in terms of their abilities to induce or reverse gelation in these fc-peptide compounds.

Zhang et al. explored and reported the first Fc-peptide conjugate, Fc CO-Phe-OH capable of forming a hydrogel with multi-stimuli responsive properties.[35] Images of the morphologies of these compounds were obtained using atomic force microscopy, scanning electron microscopy, and transmission electron microscopy, and these images show the scaffolds of the Fc-phenylalanine compound to show well-organized homogenous fibers. This gelator was also found to respond to changes in pH and to chemical redox, giving it interesting and attractive reversible properties. From this, Fc-peptide conjugates have attracted a lot of attention due to their interesting properties. Our group has reported extensively on Fc-peptide gelators including Afrasiabi et al. who reported responsiveness of Fc-dipeptide gelators to various external stimuli including thermal, sound, redox, and mechanical stress.[20,36,37] The remainder of this chapter will highlight some examples.

3.3.2 Ferrocene-Peptide Gelators

Peptide conjugates of Fc have been explored as mimics for secondary structural elements commonly found in proteins and display a range of interesting supramolecular structural features as peptides engage in inter- and intramolecular H-bonding interactions. Typical arrangements include the formation of β-sheet-like structures, turns, and β-helical arrangements. As mentioned, amino acids possess a number of interesting features, among them is their ability to engage in hydrogen bonding and metal bonding. Since the first example of a fc-peptide gelator reported previously, the Kraatz group explored the ability of various Fc-peptide conjugates to self-assemble into supramolecular structures and the ability to be triggered by different stimuli. A list of the Fc peptide gel compounds synthesized can be found in Table 3.1. This section will discuss some of the Fc peptide gels synthesized and highlight some of the interesting properties that were reported for these conjugates.

It was discovered that the interplay between intra- and intermolecular interactions make Fc-peptides an interesting group of conjugates to be studies as a small structural and/or environmental change can influence gelation properties dramatically.[6] In one of our first gelation studies, we reported two Fc-peptide conjugates that were capable of forming gels in response to various external signals including thermal, redox, and mechanical stress, since then we went on to report many more examples. We explored the properties of the Fc dicarboxylic acid derivative of L- and D-tryptophan Fc[CO-D-Trp-OMe]$_2$ and Fc[CO-L-Trp-OMe]$_2$, these compounds were found to self-assemble in toluene to form a supramolecular nanofibrillar network structure, However, it was not reported to form a supramolecular gel but instead formed a viscous solution.[38] We've shown that this tryptophan derivative displayed intermolecular hydrogen bonding involving the indole nitrogen which is shown in Fig. 3.8.[39]

In 2017, a Fc conjugate of Trp-Trp-OMe was reported to form a supramolecular gel with more complexity compared to the mono-Trp conjugate, as it combines the features of a semi-Herrick structure with the ability to interact on the supramolecular level.[6] It was discovered that multiple indolic nitrogens of the tryptophan moiety were responsible for gel formation, while exhibiting the additional tryptophan moieties increase the ability for the supramolecular interactions.[6] This compound was shown to possess many interesting properties including supramolecular gel formation, self-healing abilities, helicity, and electrochemical redox reversibility. We further demonstrated the importance of intermolecular hydrogen bonding in the role of gelation through the compound's molecular structure obtained by X-ray crystallography as shown in Fig. 3.9. These examples highlight the effects of hydrogen bonding in gelation. The rest of the chapter will discuss chirality and redox as stimuli for the gelation of these compounds.

TABLE 3.1 List of Supramolecular Fc-Peptide Gels and Their Corresponding Properties

Compound	Minimum Gelation Concentration (MGC) w/v %	T_{gel} (°C)	Morphology	Stimuli
Fc(CO-Trp-Trp-OMe)$_2$	2.7	54	fibers	Redox, temperature,
Fc-CO-((S)-Phe-(S)-Phe-(R)-Leu-OMe)	0.1	52	fibers	Chirality, temperature
Fc-CO-((S)-Phe-(R)-Phe-(R)-Leu-OMe)	0.1	48	fibers	Chirality, temperature
Fc-CO-((R)-Phe-(R)-Phe-(R)-Leu-OMe)	0.1	55	fibers	Chirality, temperature
Fc-CO-((R)-Phe-(S)-Phe-(R)-Leu-OMe)	0.1	47	fibers	Chirality, temperature
Fc-(CO-Phe-Phe-Ala-OMe)	0.12	N/A	fibers	Chirality, temperature
Fc-(CO-C$_{10}$-CO-Phe-Phe-Ala-OMe	0.5	N/A	fibers	Chirality, temperature
Fc-(CO-C$_3$-Phe-Phe-Ala-OMe)$_2$	0.38	N/A	fibers	Chirality, temperature
Fc-(CO-OMe) CO-Phe-Phe-Ala-OMe	0.4	N/A	fibers	Chirality, temperature
Fc-(CO-Trp-OMe)$_2$	0.15 (self-assembly)	60	fibers	Redox, temperature
Fc-CO-Val-Phe-Phe-OMe	1.4	56	fibers	Redox, temperature, sonication
LPA-CO-Phe-Phe-OH	15 mM	51–90	fibers	Temperature, sonication, metal-ion, pH
C$_8$-CO-Phe-Phe-OH	5 mM	42	fibers	Temperature, sonication, pH
C$_{12}$-CO-Phe-OH	16 mM	41	fibers	Metal-ion, temperature, pH

Source: LPA = lipoic acid, C_8 = 8 carbon chain, C_{12} = 12 carbon chain

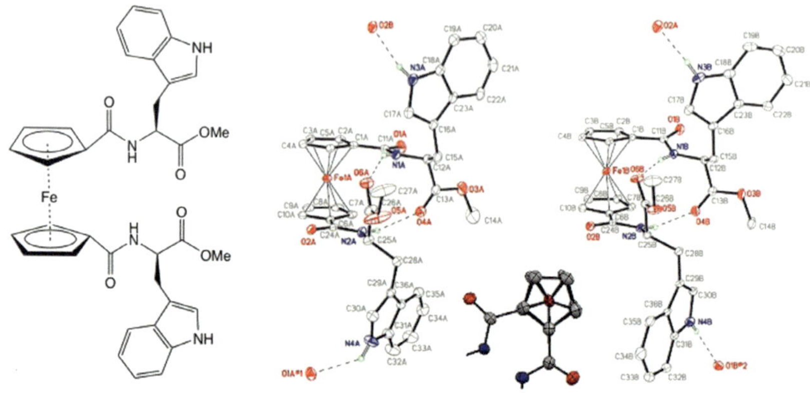

FIGURE 3.8 Chemical structure of Fc-(CO-Trp-OMe)$_2$ ORTEP diagram showing hydrogen bonding interactions of two independent molecules A and B obtained from single-crystal X-ray diffraction patterns of Fc[CO-L-Trp-OMe]$_2$, (d[N(1A)···O(6A)] = 3.06 Å and d[N(2A)···O (4A)] = 3.02 Å for molecule A). (c) shows the conformation of the Fc core showing the dihedral angle between two ring-bound substituents.

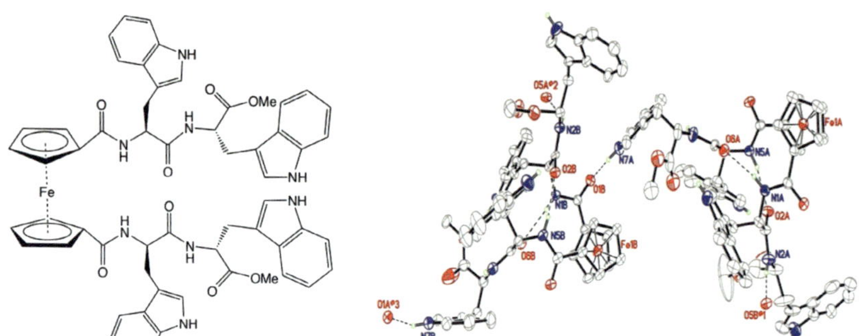

FIGURE 3.9 Chemical structure of Fc[CO-Trp-Trp-OMe]$_2$. OTREP diagrams showing hydrogen-bonding interactions between two independent molecules A and B as obtained from single-crystal X-ray diffraction studies of Fc-conjugate 1. The intermolecular hydrogen bonding interactions between two independent molecules are shown on the right (d[N(7A)—O(1B)] = 2.764A).

3.3.3 Chirality Guiding Self-Assembly

Chirality has been reported to play an important role in controlling and mediating the self-assembly of gels. There have been a few reports on molecular chirality–regulated tuning of gelation and nanostructures. Wu et al. designed two types of chiral gelators that were responsive to enantiomeric purity and they found that chiral gelators normally self-assemble into a helical conformation in a single direction and that a small amount of enantiomer with opposite chirality will mismatch the self-assembly direction.[40] It was revealed that the mismatch of chirality destroyed the one-dimensional self-assembly and thus disrupted the gel network. Smith et al. have shown that altering the chirality of one amino acid can lead to changes in the gelation behavior in terms of fibrous nature.[41] However, the current understanding about the impact of chirality on peptide gelation is limited. It was discovered that by altering the chirality of amino acids on the Fc scaffold allowed tuning of nanoscale morphology.[4] And while enantiomers showed no impact on gelation, the two diasteromeric gelators showed cross-linked nanofibrillar networks while the non-gelators produced nanorods and small irregular fibers.[4] Fig. 3.10 shows that the compounds exhibiting homochirality formed a supramolecular gel while the others did not.

Homochirality plays a critical role for the structure and function of systems including self-replication cycles, however chirality-guided interactions between peptide molecules in the context of functionality is rarely explored. The role of chirality in peptides and how chirality affects the recognition of peptides and their self-assembly into larger aggregates, were explored.[5]

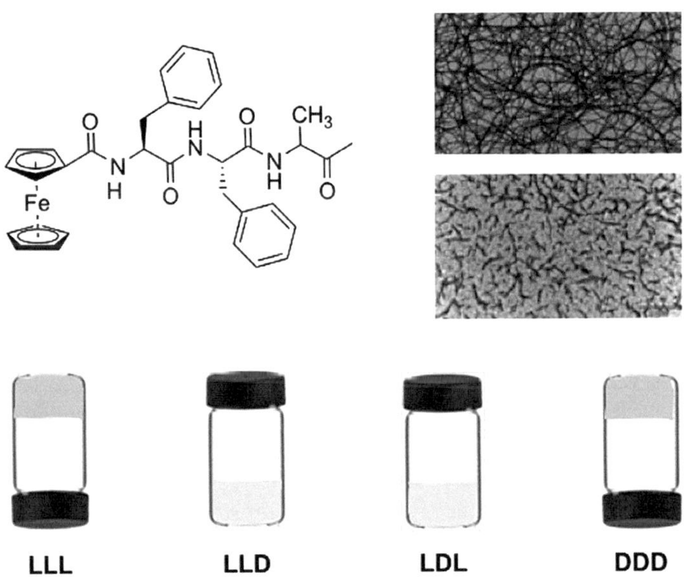

FIGURE 3.10 The top image shows the chemical structure of mono substituted Fc peptide conjugate. TEM images are shown where the top image shows the nanofiber matrix of homochiral pepide gels while the heterochiral peptides have smaller needle like morphologies. The bottom image is a scheme of the chirality-guided gelation test for the Fc-peptide conjugates in toluene at 5 mM concentration. Inverted vials correspond to gels where non-inverted vials are in non-gel state. *Adapted with permission from Adhikari, B.; Singh, C.; Shah, A.; Lough, A.; Kraatz, H-B. Chemistry: A European Journal 2015, 32, 11560–11572. Copyright 2015 ChemPubSoc Europe.*

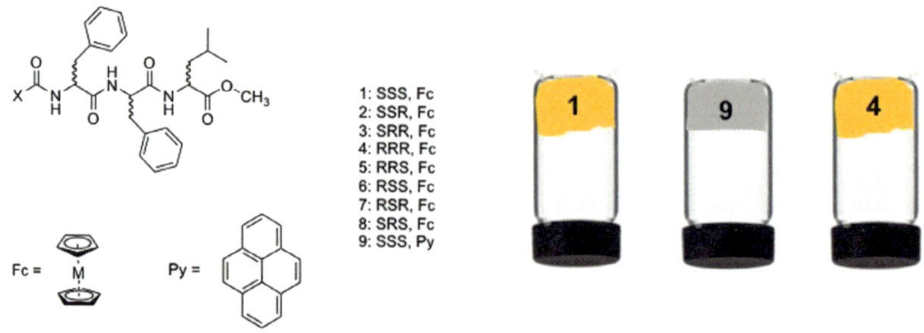

FIGURE 3.11 Chemical structures of stereoisomers of Fc- and pyrene-appended tri-peptides (1–8 and 9, respectively). Self-assembly of homochiral organogelators formed into gel nanofibers. Homochiral Fc-based gelator 1 and 4 form non-fluorescent fibers, whereas 9 forms fluorescent fibers. Homochiral peptides 1, 9, and 4 were reported to form a gelator.

Specifically, they studied the self-assembly of eight isomers of a tripeptide Phe-Phe-Leu, and labeled them with a Fc group and a pyrene (Py) group which were explored due to the fact that Py is highly fluorescent and the Fc tag serves as a fluorescence quencher, as shown in Fig. 3.11. We clearly outlined the critical role that peptide stereochemistry plays in the self-assembly of peptides as homochiral self-assemblies were shown to be stronger and more stable than those of heterochiral gels self-assembly. Through these examples, we show how chirality of the peptides can have an impact in gelation and how homochiral peptides were found to form stronger aggregates and a uniform nano-fibrous structure. Frequency sweep rheology studies were performed to examine the mechanical strength of these homo/heterochiral organogels and it was found that the storage modulus value of homochiral gels (4338 and 3363 Pa) were higher than those of heterochirality (114 and 188 Pa) which indicate the higher strength of homochiral gels.

3.3.4 Redox-Active Gelators

Here, we continue discussing redox agents as stimuli-influencing phase transitions leading to gelation. The construction of redox-active nanomaterials that efficiently produce redox-induced morphological features is an attractive goal in the design of functional stimuli-responsive materials. Redox-responsive materials that are constructed from small molecules possessing a supramolecular nano-architecture are rare. As was previously mentioned, Fc imparts redox responsiveness and it was observed that a Fc-tryptophan conjugate was found to self-assemble in toluene to form a supramolecular nanofibrillar network structure.[38] The nanofibers were responsive toward oxidation/reduction and show thermo and redox reversibility. The redox state phase change of the Fc bio-conjugate lead to structural changes in the

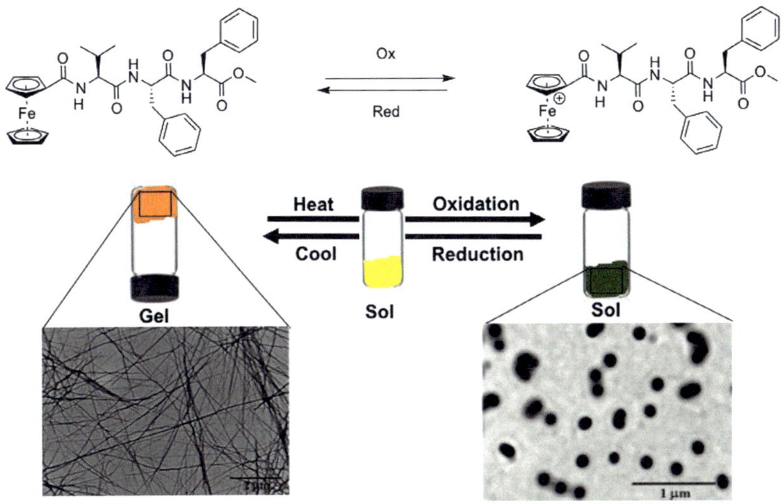

FIGURE 3.12 A schematic representation of redox-induced morphological transformations of the Fc-peptide conjugates (chemical structure shown at the top). The conjugate self-assembles into a B-sheet-like conformation to form a nanofibrous gel with a fiber morphology with an average width of 45 nm and a length of a few micrometers, which upon oxidation produced micelle-like spherical structure morphologies with an average diameter of 155 nm. This change in morphology is reversible and reduction of the spherical micelles with ascorbic acid, followed by sonication, results in the formation of fibrillary morphologies. *Adapted from Adhikari, B.; Kraatz, H.-B. Chem. Commun. 2014, 50 (42), 5551–5553. Royal Society of Chemistry.*

assembly. The redox-induced reversible morphological transformations were observed between nanofiber and spheroid morphologies.

These redox-induced morphological changes are also seen in another example shown in Fig. 3.12 where the addition of the oxidizing agent Fe $(ClO_4)_3$ resulted in a transition of the gel to a solution state, a color change was also observed—going from orange into green—which attributed to the oxidation of Fc to a ferrocenium cation.[42] After the addition of a reducing agent such as ascorbic acid, reversal of the changes induced by $Fe(ClO_4)_3$ was observed.

The Fc tryptophan-tryptophan dicarboxylic acid gelator mentioned previously was also shown to exhibit dramatic redox-triggered structural changes from a fibrillary network to spheres.[6] These examples further support redox agents acting as a stimulus that influence phase transitions in gelation and the molecular morphologies of the compounds.

3.4 CONCLUSION AND FUTURE PERSPECTIVES

This chapter summarizes the interesting properties of stimuli-responsive peptide gelators with a focus on Fc-peptides conjugates. In this overview, the role of intra- and intermolecular hydrogen bonding in guiding self-assembly has been highlighted. In some cases, even metal coordination and the abilities of metal ions and chirality induced gelation. We also explored Fc as a redox stimulus for self-assembly. We were able to exhibit interesting properties of these systems that has allowed us to study the process of gel formation in greater detail and to allow the study of fundamental gel properties. The application of the redox chemistry of Fc in electrochemical sensor devices is an area that is expected to grow. For example, applications in areas related to the detection of biochemical molecules, environmental pollutants, drugs, or other small molecules are within reach. Fc-peptide conjugates can also be explored when applied in biomedical applications, as natural fibrillary proteins of the extracellular matrix can be mimicked by the self-assembly of peptides. This allows for molecules that can be used for tissue engineering, drug delivery, and reduce cytotoxic effects of peptide antimicrobial agents. We have just scratched the surface on these interesting systems and this is just the beginning of exploring them and their potential applications.

REFERENCES

1. Yu G, Yan X, Han C, Huang F. Characterization of supramolecular gels. *Chem. Soc. Rev* 2013;**42**(16):6697–722.
2. Banerjee S, Das RK, Maitra U. J. Supramolecular Gels 'in action. *Mater. Chem* 2009;**19** (37):6649.
3. Du X, Zhou J, Shi J, Xu B. Supramolecular Hydrogelators and Hydrogels: From soft matter to molecular biomaterials. *Chem. Rev* 2015;**115**(24):13165–307.
4. Adhikari B, Singh C, Shah A, Lough AJ, Kraatz HB. Amino acid chirality and ferrocene conformation guided self-assembly and Gelation of Ferrocene-peptide conjugates. *Chem. Eur. J* 2015;**21**(32):11560–72.
5. Basak S, Singh I, Ferranco A, Syed J, Kraatz H-B. On the role of chirality guiding the self-assembly of peptides. *Angew. Chem. Int. Ed.* 2017;**56**:13288–92.
6. Falcone N, Basak S, Dong B, Syed J, Ferranco A, Lough A, et al. A Ferrocene-Tryptophan Conjugate: The Role of the Indolic Nitrogen in Supramolecular Assembly. *Chempluschem* 2017;**82**(10):1282–9.
7. Ye E, Chee PL, Prasad A, Fang X, Owh C, Yeo VJJ, et al. Supramolecular soft biomaterials for biomedical applications. *Mater. Today* 2014;**17**(4):194–202.
8. Ilmain F, Tanaka T, Kokufuta E. Volume transition in a gel driven by hydrogen bonding. *Nature* 1991;**349**(6308):400–1.
9. Rasale DB, Maity I, Das AK. *In situ* generation of redox active peptides driven by selenoester mediated native chemical ligation. *Chem. Commun* 2014;**50**:11397–400.
10. Fatás P, Bachl J, Oehm S, Jiménez AI, Cativiela C, Díaz Díaz D. Multistimuli-responsive supramolecular organogels formed by low-molecular-weight peptides bearing side-chain azobenzene moieties. *Chem. Eur. J.* 2013;**19**(27):8861–74.

11. Ishii S, Kaneko J, Nagasaki Y. Development of a long-acting, protein-loaded, redox-active, injectable gel formed by a polyion complex for local protein therapeutics. *Biomaterials* 2016;**84**:210–18.
12. Grove TZ, Osuji CO, Forster JD, Dufresne ER, Regan L. Stimuli-responsive smart gels realized via modular protein design. *J. Am. Chem. Soc* 2010;**132**(40):14024–6.
13. Panda JJ, Mishra A, Basu A, Chauhan VS. Stimuli responsive self-assembled hydrogel of a low molecular weight free dipeptide with potential for tunable drug delivery. *Biomacromolecules* 2008;**9**:2244–50.
14. Nandi N, Gayen K, Ghosh S, Bhunia D, Kirkham S, Sen SK, et al. Amphiphilic Peptide-Based Supramolecular, Noncytotoxic, Stimuli-Responsive Hydrogels with Antibacterial Activity. *Biomacromolecules* 2017;**18**:3621–9.
15. Akiyama N, Yamamoto-Fukuda T, Takahashi H, Koji T. In situ tissue engineering with synthetic self-assembling peptide nanofiber scaffolds, PuraMatrix, for mucosal regeneration in the rat middle-ear. *Int. J. Nanomedicine* 2013;**8**:2629–40.
16. Sun J, Zheng Q, Wu Y, Liu Y, Guo X, Wu W. Biocompatibility of KLD-12 peptide hydrogel as a scaffold in tissue engineering of intervertebral discs in rabbits. *J. Huazhong Univ. Sci. Technol. - Med. Sci* 2010;**30**(2):173–7.
17. Gelain F, Unsworth LD, Zhang S. Slow and sustained release of active cytokines from self-assembling peptide scaffolds. *J. Control. Release* 2010;**145**(3):231–9.
18. Zhang Y, Gu H, Yang Z, Xu B. Supramolecular hydrogels respond to ligand-receptor interaction. *J. Am. Chem. Soc* 2003;**125**:13680–1.
19. Tang C, Smith AM, Collins RF, Ulijn RV, Saiani A. Fmoc-diphenylalanine self-assembly mechanism induces apparent pKa shifts. *Langmuir* 2009;**25**(16):9447–53.
20. Afrasiabi R, Kraatz HB. Stimuli-responsive supramolecular gelation in ferrocene-peptide conjugates. *Chem. Eur. J.* 2013;**19**(51):17296–300.
21. Singh S, Topuz F, Hahn K, Albrecht K, Groll J. Embedding of active proteins and living cells in redox-sensitive hydrogels and nanogels through enzymatic cross-linking. *Angew. Chem. Int. Ed* 2013;**52**(10):3000–3.
22. Ulijn RV. Enzyme-responsive materials: a new class of smart biomaterials. *J. Mater. Chem* 2006;**16**(23):2217.
23. Yang Z, Xu B. A simple visual assay based on small molecule hydrogels for detecting inhibitors of enzymes. *Chem. Commun.* 2004;**1**(21):2424–5.
24. Gao Y, Zhao F, Wang Q, Zhang Y, Xu B. Small peptide nanofibers as the matrices of molecular hydrogels for mimicking enzymes and enhancing the activity of enzymes. *Chem. Soc. Rev* 2010;**39**(9):3425.
25. Selegård R, Aronsson C, Brommesson C, Dånmark S, Aili D. Folding driven self-assembly of a stimuli-responsive peptide-hyaluronan hybrid hydrogel. *Sci. Rep* 2017;**7**(1):1–9.
26. Basak S, Singh I, Kraatz H-B. Ion-Dependent Modulation of Self-Healing Hydrogels. *Chemistry Select* 2017;**2**(1):451–7.
27. Basak S, Singh I, Banerjee A, Kraatz H-B. Amino acid-based amphiphilic hydrogels: metal ion induced tuning of mechanical and thermal stability. *RSC Adv* 2017;**7**(24):14461–5.
28. Bowerman CJ, Nilsson BL. A reductive trigger for peptide self-assembly and hydrogelation. *J. Am. Chem. Soc* 2010;**132**(28):9526–7.
29. Herrick RS, Jarret RM, Curran TP, Dragoli DR, Flaherty MB, Lindyberg SE, et al. Ordered conformations in bis (amino acid) derivatives of 1, 1′-ferrocenedicarboxylic acid. *Tetrahedron Lett* 1996;**37**(30):5289–92.
30. Kraatz H, Lusztyk J, Enright GD. Ferrocenoyl amino acids: A synthetic and structural study. *Inorg. Chem* 1997;**1669**(4):2400–5.

31. Moriuchi T, Nomoto A, Yoshida K, Hirao T. J. Characterization of ferrocene derivatives bearing podand dipeptide chains (-L-Ala-L-Pro-OR). *Organomet. Chem* 1999;**589**(1):50–8.
32. Kirin SI, Kraatz H-B, Metzler-Nolte N. Systematizing structural motifs and nomenclature in 1,n'-disubstituted ferrocene peptides. *Chem. Soc. Rev* 2006;**35**(4):348–54.
33. Saweczko P, Enright GD, Kraatz HB. Interaction of ferrocenoyl-dipeptides with 3-aminopyrazole derivatives: beta-sheet models? A synthetic, spectroscopic, structural, and electrochemical study. *Inorg. Chem* 2001;**40**(17):4409–19.
34. Appoh FE, Sutherland TC, Kraatz HB. Ferrocenoyl-amino acids: redox response towards di- and trivalent metal ions. *J. Organomet. Chem.* 2005;**690**:1209–17.
35. Sun Z, Li Z, He Y, Shen R, Deng L, Yang M, et al. Ferrocenoyl phenylalanine: A new strategy toward supramolecular hydrogels with multistimuli responsive properties. *J. Am. Chem. Soc* 2013;**135**(36):13379–86.
36. Afrasiabi R, Kraatz HB. Rational design and application of a redox-active, photoresponsive, discrete metallogelator. *Chem. Eur. J.* 2015;**21**(21):7695–700.
37. Afrasiabi R, Kraatz HB. Small-peptide-based organogel kit: Towards the development of multicomponent self-sorting organogels. *Chem. Eur. J.* 2013;**19**(47):15862–71.
38. Adhikari B, Afrasiabi R, Kraatz H. Ferrocene-tryptophan conjugate: An example of a redox-controlled reversible supramolecular nanofiber network. *Organometallics* 2013;**32**:5899–905.
39. Adhikari B, Lough AJ, Barker B, Shah A, Xiang C, Kraatz H-B. Bis-amino Acid Derivatives of 1,1'-Ferrocenedicarboxylic Acid. Structural, Electrochemical, and Metal Ion Binding Studies. *Organometallics* 2014;**33**:4873–87.
40. Liu C-W, Su M, Li X-L, Xue T, Liu N, Yin J, et al. Multi-stimuli-responsive chiral organogels based on peptide derivatives. *Soft Matter* 2015;**11**(28):5727–37.
41. Hirst AR, Smith DK, Feiters MC, Geurts HPM. Two-component dendritic gel: Effect of stereochemistry on the supramolecular chiral assembly. *Chem. Eur. J.* 2004;**10**(23):5901–10.
42. Adhikari B, Kraatz H-B. Redox-triggered changes in the self-assembly of a ferrocene-peptide conjugate. *Chem. Commun* 2014;**50**(42):5551–3.

Chapter 4

Metalated Amino Acids and Peptides: Programmable Metal Array Fabrication and Application to Supramolecular Catalyst

Hikaru Takaya[1], Katsuhiro Isozaki[1], Ryota Yoshida[1,2], Takafumi Shanoh[1,2], Nobuhiro Yasuda[3] and Masaharu Nakamura[1]

[1]*Institute for Chemical Research, Kyoto University, Kyoto, Japan,* [2]*Department of Energy and Hydrocarbon Chemistry, Graduate School of Engineering, Kyoto University, Kyoto, Japan,* [3]*Japan Synchrotron Radiation Research Institute (JASRI), Kouto, Japan*

4.1 INTRODUCTION

Metalated amino acids and peptides are one of the most widely used and developed bioorganometallic compounds.[1] There has been numerous research in bioanalytical and biomedical applications, where various biologically active, radioactive metal complexes are conjugated into amino acids and peptides to provide biomarkers, contrast agents, and drugs.[1b–3] However, there are non-biological applications of metalated amino acids and peptides to functional materials, such as supramolecular architectures and catalysts, based on their inherent properties of self-assembly and molecular recognition. We have successfully formulated programmable fabrications of sequence-controlled multi-metalated peptides and their supramolecular architectures using α-side chain-metal-conjugated amino acids and peptides. Synthesis and the precise structure determination of these novel bioorganometallic molecules and the application to supramolecular catalyst are also described in this chapter.

Advances in Bioorganometallic Chemistry. DOI: https://doi.org/10.1016/B978-0-12-814197-7.00004-2

4.2 CLASSIFICATION OF METALATED AMINO ACIDS AND PEPTIDES

Reported metalated amino acids and peptides are roughly classified into three types as shown in Fig. 4.1.[1,3–7] The *N*- and *C*-terminus types of amino acids, where the metals or metal complexes conjugated to the *N*-/*C*-terminus of amino acid/peptide, are the most commonly reported forms especially in bioanalytical and biomedical research. The advantage of these two types is that the metal moiety can be readily incorporated by the simple condensation reaction of a metal complex or ligand for the coordination of metals at a late stage. This merit was fully demonstrated in the synthesis of SPECT and MRI contrasting agents using short life time and labile metals, such as Tc and Gd, and protein-binding peptides, where a minimized step synthesis is required to avoid leaching metal during the synthesis.[2] The third type of side-chain-conjugated metalated amino acids and peptides are generally synthesized through a cumbersome multi-step process to furnish the metal conjugation-site involving protection/deprotection processes of the *N*-/*C*-terminus. However, the *N*-/*C*-terminus can be fully available for further conjugation of various amino acids, peptides, and proteins. Such diversely-oriented molecular design is suitable for the programmable construction of highly-ordered hierarchical structure and installation of various functional groups. For this reason, we choose the side chain-metalated type as the platform of our research aiming to produce metalated-amino acids/peptides-based highly-functional materials.

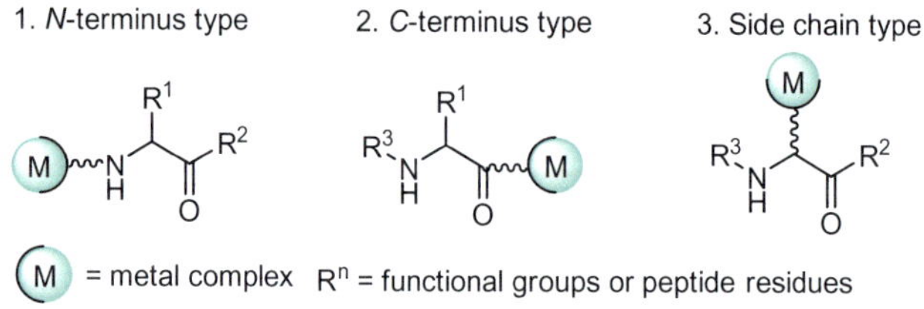

FIGURE 4.1 Classification of metalated amino acids.

4.3 SYNTHESIS OF α-SIDE CHAIN-METALATED AMINO ACIDS AND PEPTIDES

Synthesis of side chain-metalated amino-acids and peptides is commonly carried out by following two routes, as shown in Fig. 4.2. The first route is the post-complexation or post-metalation route: Metalation is performed by the reaction of metal salt with amino acid/peptide-bearing coordination site conjugated into the α-side chain. The other is the metal complex conjugation route where the metal complex is conjugated by appropriate coupling chemistry, such as through a cross-coupling reaction and condensation reaction.

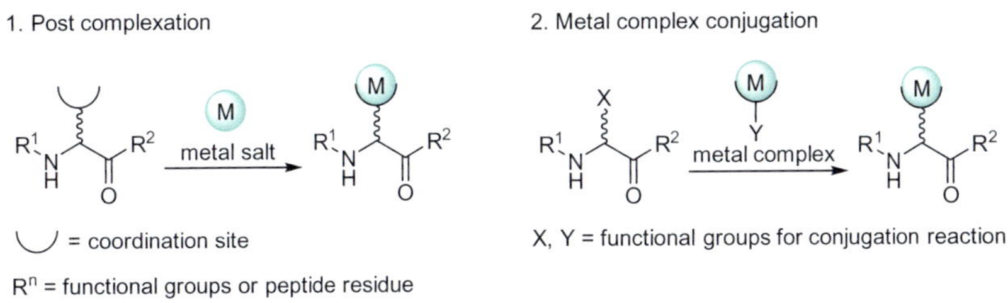

FIGURE 4.2 Synthesis of side chain-metalated amino acids and peptides.

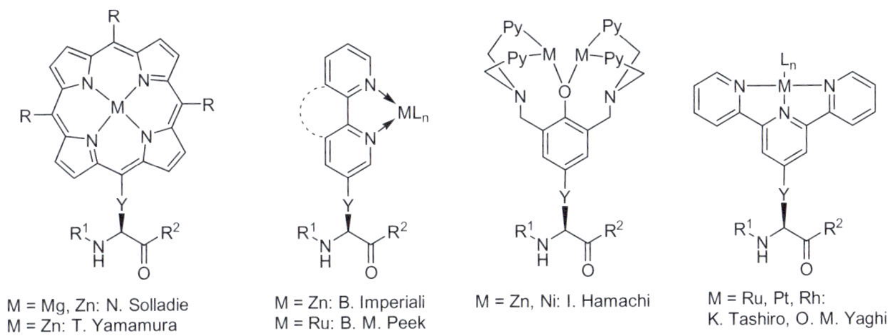

FIGURE 4.3 Selected examples of metalated amino acids and peptides synthesized by post complexation route.

The post-complexation route has the merit of late-stage metalation and one can readily introduce a variety of metals into the desired amino acids and peptides through simple and easy synthetic operation. Therefore, most side-chain metalated amino acids/peptides have been synthesized by this route and widely applied to bioanalytical chemistry.[1a,2,4] *N*-Heterocyclic scaffolds such as pyridines[8] and porphyrins[9] are commonly used for the coordination site (Fig. 4.3) while their rich coordination chemistry provides significant contributions to expand synthetic diversity of metalated amino acids/peptides.

In comparison with the post-complexation route, the second metal complex conjugation route is generally troublesome and time-consuming because of the necessity to carefully consider choosing the molecular design of metal complexes to avoid undesirable side reactions, such as metal- and ligand leaching during the coupling reaction. However, if one can find substantially robust metal complexes under coupling conditions, the merit of higher metal unit convertibility enables us to establish a programmable synthesis of sequence-controlled metal arrays, as described in the next section.

4.4 SYNTHESIS OF α-SIDE CHAIN-METALATED PEPTIDES

Synthesis routes of α-side chain-metalated peptides are summarized in Fig. 4.4. The post-metalation route is the rewriting of the routes depicted in

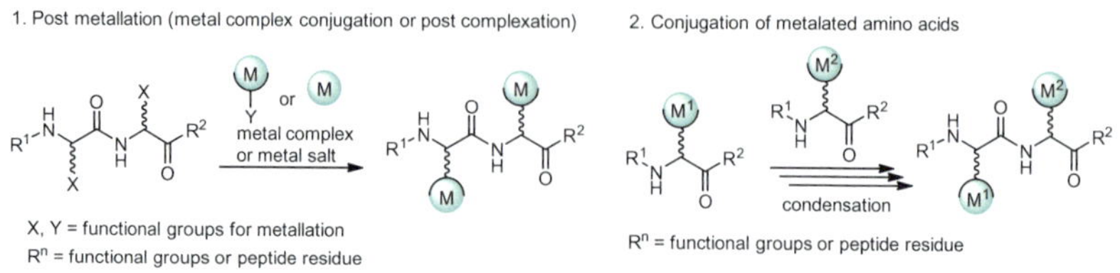

FIGURE 4.4 Synthesis of α-side chain-metalated peptides.

Fig. 4.2, where the metalation of peptides is simply performed in one-step by conjugation of the metal complex to the α-side chain by appropriate coupling reactions or by incorporation of metal salt into the coordination site conjugated to the α-side chains prior to metalation.[4] The second route of conjugation of metalated amino acids is carried out by the condensation of metalated amino acids or peptides to afford the corresponding multi-metalated peptides.

This route requires multi-step synthesis including *N-/C*-terminus protection/deprotection processes to obtain desired peptides while being aware of undesirable metal- or ligand-detaching; however, there is a crucial advantage that metal sequence can be absolutely defined as the condensation sequence of metalated amino acids. For this reason, we chose this route to synthesize multi-metalated peptides to construct dimensionally- or hierarchically-controlled metal arrays, as described in the following section.

4.5 PD- AND PT-COMPLEX-CONJUGATED GLUTAMIC ACID

We designed a glutamic acid-based metalated amino acid as a platform of sequence-controlled metal array fabrication (Fig. 4.5). The significantly robust *N*-benzylidenalkylamine complexes of palladium and platinum

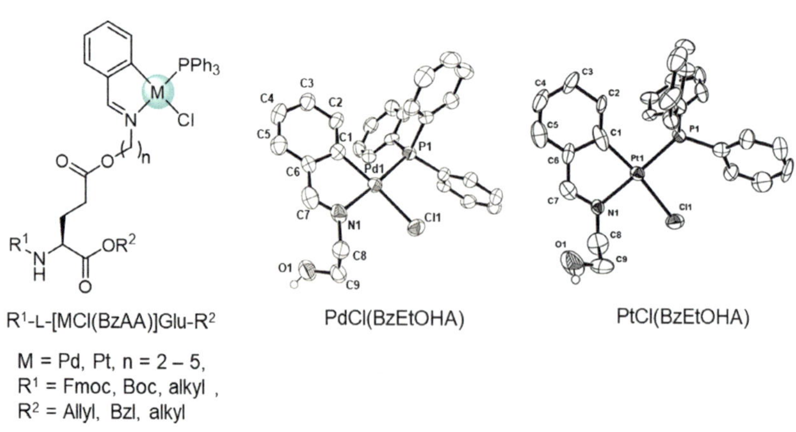

FIGURE 4.5 Pd- and Pt-conjugated glutamic acid (R^1-L-[MCl(BzAA)]Glu-R^2) and X-ray structures of the corresponding benzilydenealkylamine complexes (M = Pd and Pt, $n = 2$).

(Pd(BzAA)) enabled us to perform conventional solution-phase peptide synthesis involving *N-/C*-terminus protection/deprotection processes without metal detaching, affording the desired multi-metalated glutamic acids peptides. By using these metalated-peptides, dimensionally-controlled metallo-supramolecular architecture was successfully fabricated through an ultrasound-induced, self-assembly process.[6,10–12]

4.6 SYNTHESIS OF PD- AND PT-COMPLEX-CONJUGATED GLUTAMIC ACIDS/PEPTIDES

The palladium and platinum-conjugated glutamic acid R^1-L-[MCl(BzAA)] Glu-R^2 can be synthesized in three steps including syntheses of BZAA ligands from benzaldehyde and its complexes, as shown in Scheme 4.1.[6,10,12]

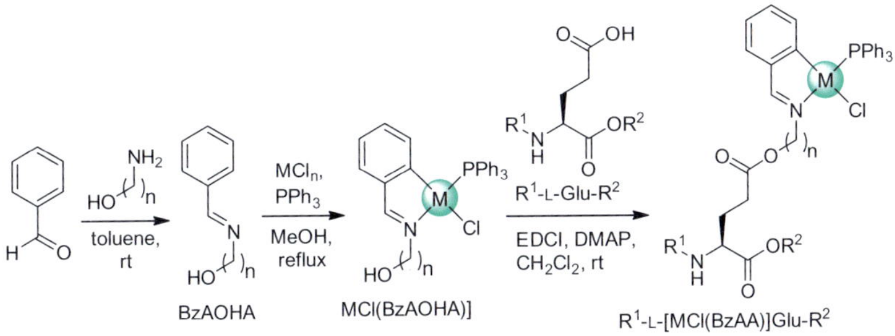

SCHEME 4.1 Synthesis of Pd- and Pt-conjugated glutamic acid (R^1-L-[MCl(BzAA)]Glu-R^2).

The simple and practical solution-phase reactions allowed large-scale preparation of 10 g per batch using a 3 L flask. The resulting metalated glutamic acids show sufficient stability under conventional conditions of peptide synthesis for acidic *N*/*C*-deprotection with elevated temperature (ca. 100 °C). Any decompositions including metal leaching could not be clearly observed under these conditions by ^{1}H NMR monitoring.

We examined the condensation of metalated-amino acids by the reaction of in situ prepared *C*-terminus free Pd-conjugated glutamic acids Fmoc-L-[PdCl(BzAA)]Glu-OAllyl **1**[13] and *N*-terminus free Pd-conjugated glutamic acid H-L-[PdCl(BzAA)]Glu-NHBu **2** using EDCI as a condensation reagent. As a result, the corresponding Pd-conjugated glutamic acid dipeptides Fmoc-L-{[PdCl(BzAA)]Glu}$_2$-NHBu **3–6** were formed in 50%–62% yields as shown in Scheme 4.2.[6,10] Although the product yields stayed moderate due to the steric hindrance of bulky Pd-complex units, no metal leaching byproducts were observed in the reaction mixture. The resulting selective formation of dipeptides **3–6** with recovery of unreacted *C*/*N*-terminus free Pd-glutamic acids indicates that this motif and Pd moiety is stable and well-tolerated under the peptide condensation conditions.

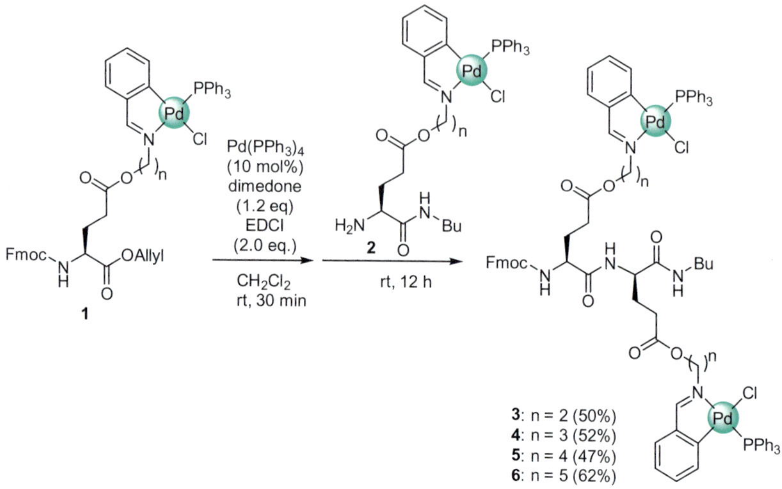

SCHEME 4.2 Synthesis of Pd-conjugated glutamic acid dipeptides Fmoc-{L-[PdCl(BzAA)]Glu}$_2$-NH-Bu.

Programmable fabrication of the heterometallic array was successfully accomplished through peptide synthesis using Pd- and Pt-conjugated glutamic acids as metal units.[6] We can synthesize two possible sequence isomers of Pd-Pt and Pt-Pd heterometallic dipeptides **9** and **10** selectively, as well as Pd-Pd and Pt-Pt homometallic dipeptides **3** and **11**, by the condensation of Fmoc-L-[MCl(BzAA)]Glu-H (M = Pd or Pt) prepared from Fmoc-L-[MCl(BzAA)]Glu-OAllyl (**1**: M = Pd, **7**: M = Pt) and H-L-[MCl(BzAA)]Glu-NHBu (**2**: M = Pd or **8**: M = Pt) as shown in Scheme 4.3. Notably, this is the first example of comprehensive metal sequence fabrication using metalated amino acid as a metal unit. Construction of metallo-supramolecular frameworks based on the ultrasound-induced self-assembly of metalated dipeptides **3** and **9–11** will be discussed later.

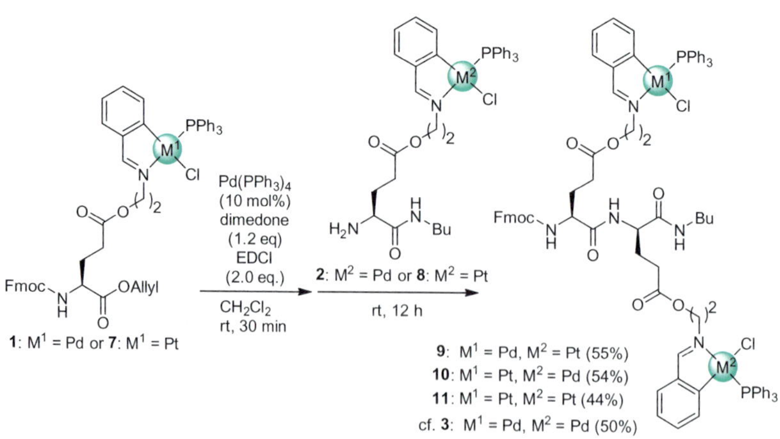

SCHEME 4.3 Synthesis of controlled metal arrays using Pd- and Pt-conjugated glutamic acids.

The advantage of the molecular design of our Pd- and Pt-glutamic acids was demonstrated in the elongation of the peptide chain to synthesize tri- and tetra-metalated peptides. Generally, solubility of metalated peptides in organic solvent drastically decreases with an increasing number of amino acid residues. However, our Pd- and Pt-conjugated amino acids and peptides show excellent solubility in various organic solvents such as benzene, chlorobenzene, toluene, CH_2Cl_2, $CHCl_3$, EtOAc, THF, and Et_2O by the effect of the highly soluble PPh_3 ligand. Therefore, solution-phase synthesis of tripeptide Fmoc-L-{[MCl(BzAA)]Glu}$_3$-NHBu (**12:** M = Pd, **13**: M = Pt) and tetrapeptide Fmoc-L-{[MCl(BzAA)]Glu}$_4$-NHBu (**14:** M = Pd, **15**: M = Pt) can be performed by the same procedure of dipeptide synthesis with similar product yields (Fig. 4.6).[10,11]

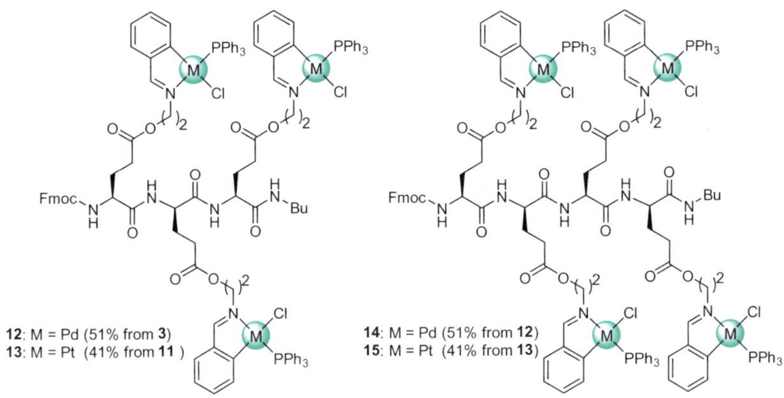

FIGURE 4.6 Tri- and tetrametalated-peptides derived from Pd- and Pt-conjugated glutamic acids.

4.7 ULTRASOUND INDUCED SELF-ASSEMBLY OF PD- AND PT-COMPLEX-CONJUGATED GLUTAMIC PEPTIDES

Peptide self-assembly has been recognized as a key platform to construct precisely structure-controlled materials gained intense interest from interdisciplinary research areas.[14] In the self-assembling process of peptides, programmed formation of amide-amide hydrogen bonds ($C=O\cdots H-N$) at the desired point, direction, and timing is an essential factor to obtain dimensionally- and hierarchically-controlled architectures. During the course of research on metalated amino acids/peptide-based supramolecular gel formation, we found an interesting ultrasound-induced hydrogen-bond switching mechanism as shown in Scheme 4.4.[6,10] In the metalated glutamic acids/peptides dissolved into aprotic organic solvents, a self-locked structure **16** is predominantly formed by the intramolecular hydrogen bonding of

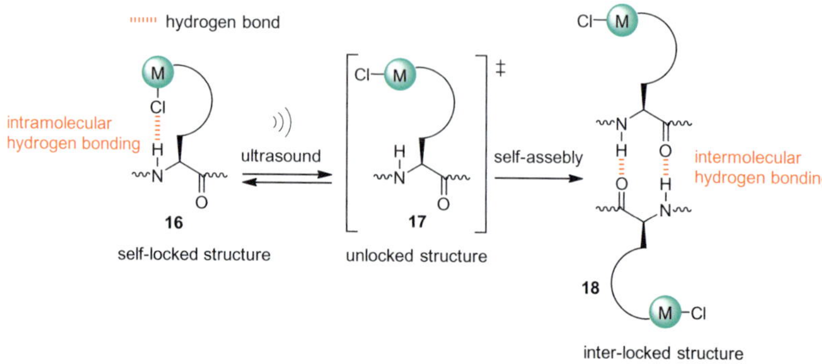

SCHEME 4.4 Ultrasound-induced switching of intra- and intermolecular hydrogen bonds of metalated amino acids/peptides.

M–Cl···H–N between the chloride ligand and amide N-H.[15] Irradiation of ultrasound to the solution can break the intramolecular hydrogen bond to give hydrogen a bond-free unlocked structure **17** as a short-lived unstable intermediate, followed by self-assembly of amino acids/peptides which proceeds along with the formation of an inter-locked structure **18** through intermolecular amide-amide hydrogen bonds to give the corresponding supramolecular assembly possessing a β-sheet-type structural topology.[16,17]

Representative results of the ultrasound-induced self-assembly of Pd- and Pt-conjugated glutamic dipeptide are illustrated in Fig. 4.7. Upon 60 s irradiation of ultrasound (0.45 W/cm^2, 40 kHz) to EtOAc solutions of dipeptides **3** and **9–11**, the peptides undergo self-assembly with the spontaneous formation of opaque gels. The observed reversible sol-gel transformation under a sonication-heating cycle exhibits that the gelation is driven by non-covalent interaction of hydrogen-bonds. This gelation specifically occurs under sonication and usual heat–cool procedures gives an amorphous precipitate without the formation of gel.[18] SEM images of xerogel of **3** show an entangled fiber structure being typical of supramolecular peptide assemblies and the lamellar pattern with 3 nm pitch found on the surface of belt-like fibers coincides with the formation of β-sheet type aggregate. Interestingly, the minimum gelation concentration increased with the increase of Pt-conjugated glutamic acids, such as Pd-Pd $<$ Pd-Pt $\approx$ Pt-Pd $<$ Pt-Pt. This phenomenon can be explained by the hydrogen bond strength of M–Cl···H–N where the hydrogen bond of Pd–Cl···H–N is more stable than Pt–Cl···H–N because of the larger electric dipole moment of Pd-Cl compared with Pt-Cl[6,15] and higher stabilization of inter-β-sheet interaction would contribute to reducing the minimum gelation concentrations of Pd-conjugated dipeptides.

To determine the precise self-assembled structures of Pd- and Pt-conjugated glutamic acid dipeptides, we carried out X-ray scattering analysis (WAX and SAXS) and IR analysis of the obtained gels. Structural

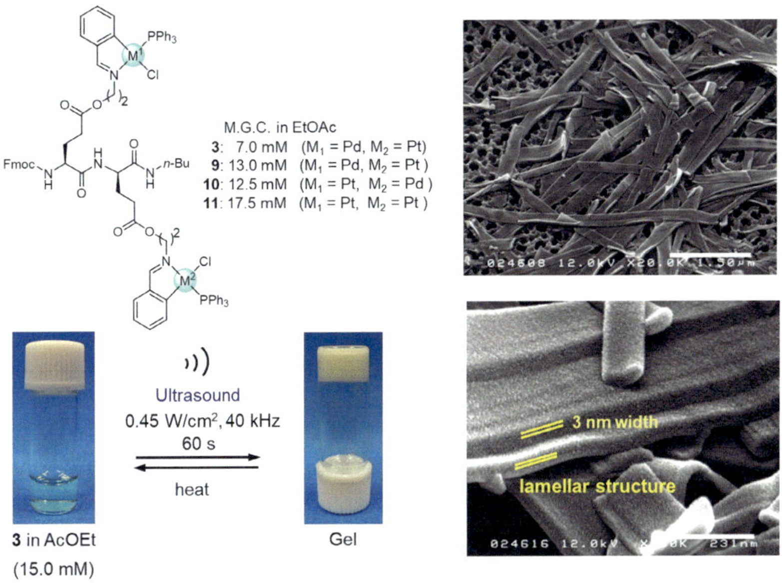

FIGURE 4.7 Ultrasound-induced supramolecular gel formation of Pd- and Pt-bound glutamic acid dipeptides **3** and the minimum gelation concentration of dipeptides **3** and **9**–**11** in EtOAc with a SEM image of supramolecular xerogel of **3**.

characterization revealed that the structure of dipeptide supramolecular assembly is influenced by the solvent used for the sonication gelation. The supramolecular gels of Pd-Pt dipeptide **9** and Pt-Pd dipeptide **10** prepared from a chlorobenzene solution consist of parallel β-sheet assemblies with the formation of intermolecular π–π stacking of *N*-terminus fluorenyl moieties, as illustrated in Fig. 4.8. In the parallel β-sheet aggregates, the Pd- and Pt complex moieties are aligned into the same side and the same orientations along the *N*/*C*-direction of the peptide backbone to afford a dimensionally controlled heterometal array through the supramolecular gel formation, as we expected for this molecular design (Fig. 4.9).

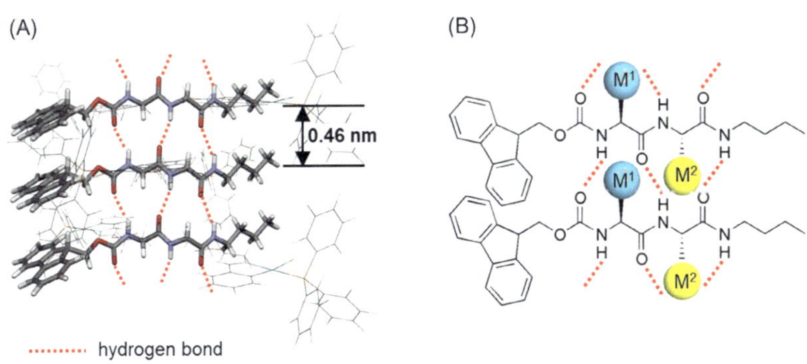

FIGURE 4.8 Parallel β-sheet type self-assembled structures of Pd- and Pt-conjugated glutamic acid dipeptides.

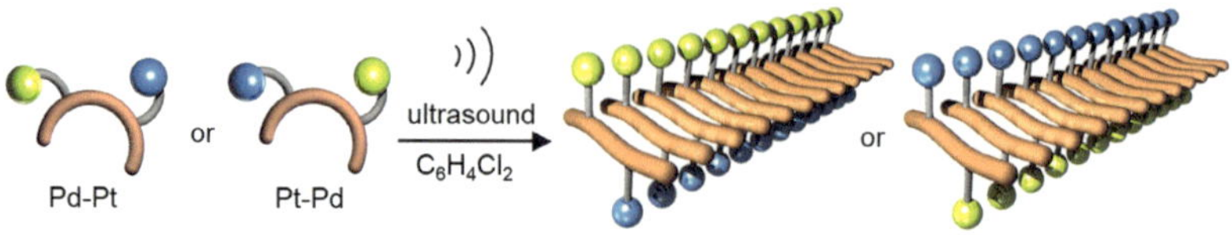

FIGURE 4.9 Controlled heterometal assembling in parallel β-sheet type self-assemblies.

To gain insight into ultrasound-induced supramolecular gelation, we conducted a measure of the kinetics of gelation to determine the rate constant under varied sonication times and frequencies. A kinetic experiment using 7.0 mM EtOAc solution of Pd-Pd dipeptide **3** revealed that the gelation rate of dipeptide obeys a first-order dependence on the sonication time of over 80 s irradiation, but an induction period was observed for short-time irradiation of less than 60 s without the formation of gel. This result suggested that the ultrasound-induced gelation process involves a sonication-induced initiation step followed by a spontaneous propagation process. The gelation rate also depends on the frequency of ultrasound: The rate exponentially increases with decreasing frequencies and the curve profile showed agreement with the relationship of the frequency and energy of micro-cavitation which was formed by ultrasound irradiation.[19] These results strongly indicate that micro-cavitation is a key factor to promote the self-assembly of metalated peptides.

Considering the results, we proposed a mechanism of ultrasound-induced gelation as depicted in Fig. 4.10. The intramolecular hydrogen bond of self-locked metalated dipeptides is cleaved under extremely high temperature and

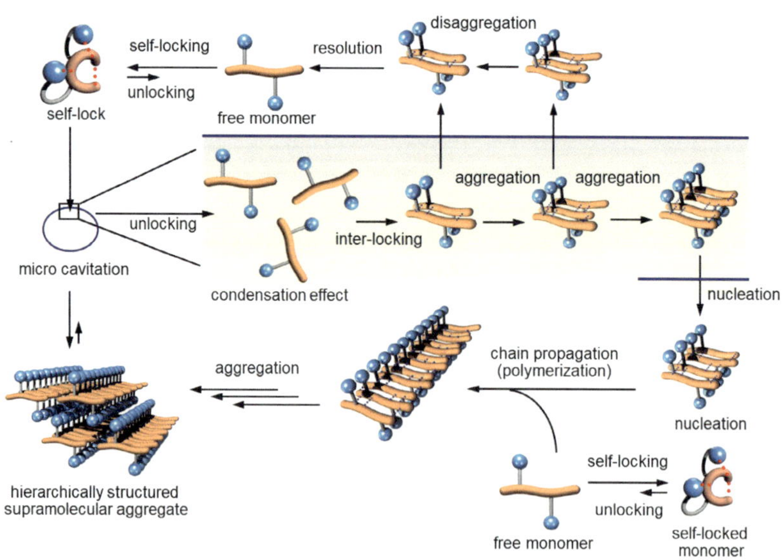

FIGURE 4.10 Proposed mechanism of ultrasound-induced gelation of Pd- and Pt-bound glutamic acid dipeptides.

pressure conditions arisen from ultrasound micro-cavitation to give the corresponding unlocked metastable intermediate.

Supersaturation at the interfacial surface of the cavitation bubble induces an intermolecular association of the unlocked intermediate to give metastable, tiny oligomers. Notably, such extremely high energy and high concentration conditions cannot be provided by thermal processes. After collapsing the micro-cavitation, the resulting well-grown insoluble oligomers act as nuclei for the spontaneous propagation of self-assembling peptides to provide β-sheet supramolecular aggregates. Such supramolecular polymerization have gained an upsurge of interest as a living supramolecular polymerization and variety of monomers bearing both π-stacking and hydrogen bonding parts as key factors have been developed.[20]

4.8 SYNTHESIS OF NCN-PINCER PD- AND RU-COMPLEX-CONJUGATED NORVALINES

In this decade, transition-metal complexes on a pincer scaffold have emerged with an upsurge of interest as a privileged class of organometallic compounds, owing to their extremely robust nature, high catalytic activity, excellent photo-/electric properties, and their extensive versatility in interdisciplinary research areas.[21] We consider that these properties of pincer complexes will open up new horizons for bioorganometallic chemistry especially for the platform of advanced catalysts. In contrast to supramolecular gel preparation, catalytic applications of metalated amino acids and peptides are required to be more stable and robust under more severe and varied conditions in the presence of highly reactive reagents, bases, oxidizing agents, and so on. We have developed a variety of pincer complex-conjugated norvalines in which stable and catalytically active NCN-pincer palladium- and ONO-pincer ruthenium-complexes are covalently conjugated into α-side chain of norvaline derivatives, as depicted in Fig. 4.11.[7,22,23] The molecular design strategy based on pincer scaffold conjugated through robust carbon—carbon bond linkage enabled us to perform various catalytic reactions, such as nucleophilic cycloaddition, cross-coupling, and oxidation reactions.

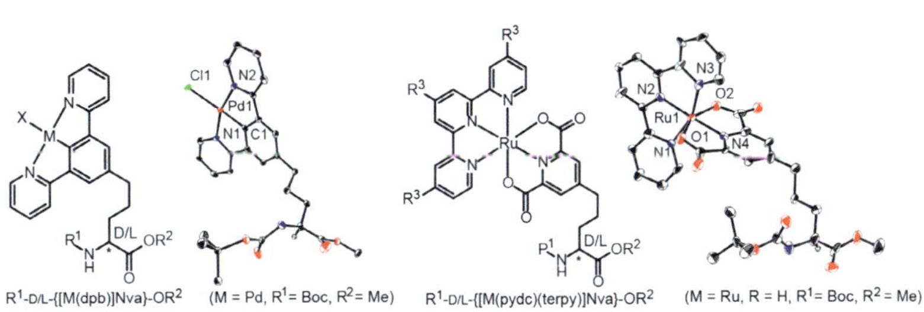

FIGURE 4.11 NCN-pincer and ONO-pincer complex-conjugated norvalines.

NCN-pincer palladium complex-conjugated norvalines can be synthesized by using Suzuki-Miyaura cross-coupling reactions originally developed by Taylor[24] and von Koten[25] for functionalization of amino acids side chains as shown in Scheme 4.5.[7,26] The cross-coupling reaction was performed by the reaction of 9-BBN adduct L/D-**19** in situ prepared from commercially available protected allylglycines Boc-L/D -Allygly-OMe with NCN-pincer palladium complex PdCl(dpb-Br). After treatment of the resulting reaction mixture with KCl for conversion of Pd-Br to Pd-Cl of the conjugated pincer complex gave the desired Pd-conjugated norvalines L/D-**20** through a one-pot, three-step process.

SCHEME 4.5 Synthesis of NCN-pincer Pd-complex-bound norvalines.

Large-scale synthesis of L/D-**20** in a 5 g per batch using a 3 L flask can be performed without loss of chiro-optical purity which was unequivocally confirmed by chiral HPLC and single X-ray structure analysis. Notably, only a few examples have been reported on the determination of the absolute configuration of metalated amino acids, including our works,[7,23,24,26,27] due to its poor crystallinity caused by inherent fluxionality of the molecules.

As we expected, the NCN-pincer Pd-conjugated norvaline **20** showed sufficient stability under acidic and basic conditions for *C*- and *N*-terminus conversions (Schemes 4.6 and 4.7). Deprotections of *C*-OMe and *N*-Boc were efficiently carried out by treatment of **20** with LiOH or HCl to provide the corresponding *C*/*N*-free Pd-conjugated norvalines **21** and **23** almost quantitatively without leaching palladium. The resulting *C*/*N*-free intermediates **21** and **23** underwent condensation with various amines and carboxylic acids

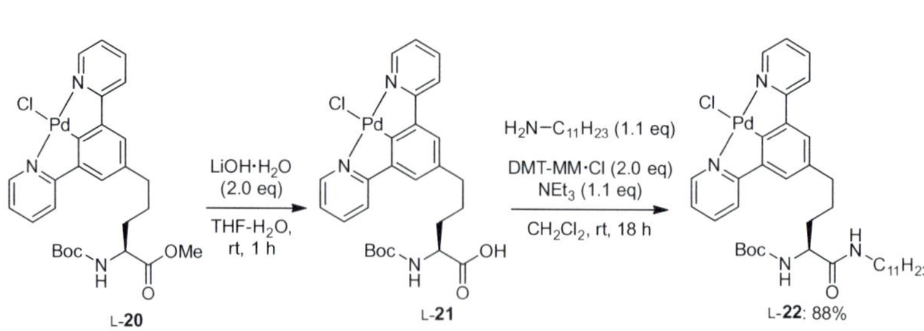

SCHEME 4.6 *C*-terminus conversion of Pd-conjugated norvaline.

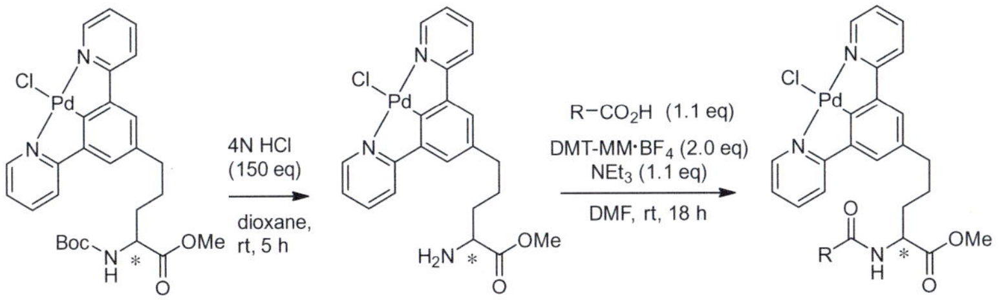

L/D-**20**

L/D-**23**

L-**24**: R = $C_{11}H_{23}$ (88%)
D-**25**: R = $C_{11}H_{23}$ (88%)
L-**26**: R = CH_3 (75%)
L-**27**: R = 2,4-NO_2-C_6H_3 (44%)

SCHEME 4.7 *N*-terminus conversion of Pd-conjugated norvaline.

using a DMT-MM condensing agent[28] to provide *C*/*N*-functionalized metalated norvalines **22** and **25–28** in good yields without the formation of metal leaching by-products and loss of optical purity. Notably, these *C*/*N*-conversions can be performed in a one-batch, one-pot process without isolation and purification of **21** and **23**.

The DMT-MM-mediated condensation proved highly effective for the coupling of extremely sterically bulky metalated norvalines to afford the corresponding metalated norvaline peptides, as shown in Scheme 4.8. The coupling of *N*/*C*-free Pd- or Pt-conjugated norvalines, which can be used as synthesized without purification after the deprotection process, proceeded efficiently by DMT-MM-mediated condensation to provide the corresponding Pd-/Pt-conjugated norvaline dipeptides **31–34**. No metal leaching or scrambling was observed with recovery of starting *N*-/*C*-free metalated norvalines. Up to tetra- to hexa-metalated peptides were successfully synthesized both for NCN-pincer

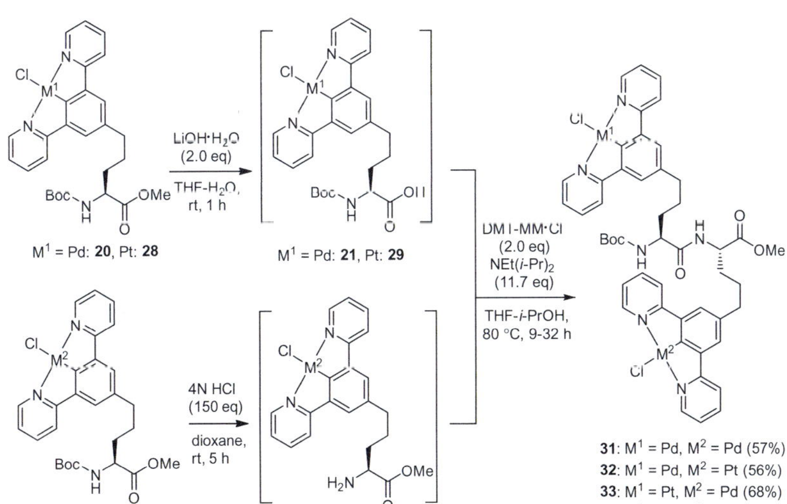

M^1 = Pd: **20**, Pt: **28**

M^1 = Pd: **21**, Pt: **29**

M^2 = Pd: **20**, Pt: **28**

M^2 = Pd: **23**, Pt: **30**

31: M^1 = Pd, M^2 = Pd (57%)
32: M^1 = Pd, M^2 = Pt (56%)
33: M^1 = Pt, M^2 = Pd (68%)
34: M^1 = Pt, M^2 = Pt (73%)

SCHEME 4.8 Controlled metal array fabrication based on Pd- and Pt-conjugated norvalines.

Pd/Pt-conjugated norvalines and ONO-pincer Ru-conjugated norvalines on a gram scale without loss of optical purity.[7,29]

4.9 SELF-ASSEMBLY OF NCN-PINCER PD- AND ONO-PINCER RU-COMPLEX-CONJUGATED NORVALINES AND APPLICATIONS TO SUPRAMOLECULAR CATALYSTS

Supramolecular self-assembly has been established as a powerful technique for the fabrication of well-defined nano-/micro-scale molecular architectures. The embedded side-chain functional groups in the resulting hierarchically structured porous supramolecular architecture can provide an efficient catalytic site as well as an enzyme-active site to promote chemo- and stereoselective organic reactions.[30] We envision that self-assemblies of metalated amino acids and peptides act as a supramolecular catalyst possessing metal-accumulated nano-size pores for substrate binding and catalytic sites.[6,10]

Supramolecular gel of NCN-pincer Pd-conjugated norvaline L-**22** was prepared through the conventional heat−cool process from the toluene solution of process L-**22**. Microscopic observation of the resulting gel by means of cryo-TEM demonstrated the formation of a typical fibrous network structure with well-regulated nanoscale patterns, as illustrated in Fig. 4.12. The

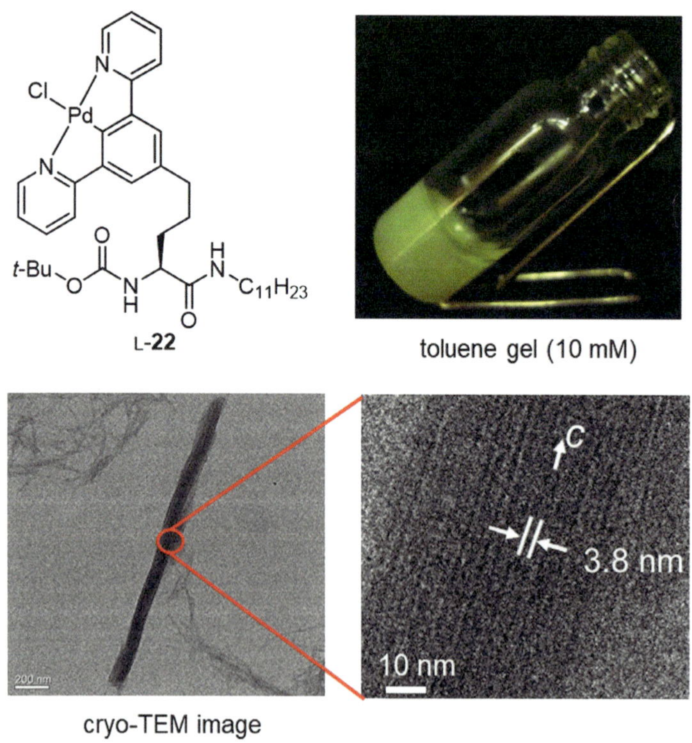

FIGURE 4.12 Supramolecular gel of L-**22** and cryo-TEM images.

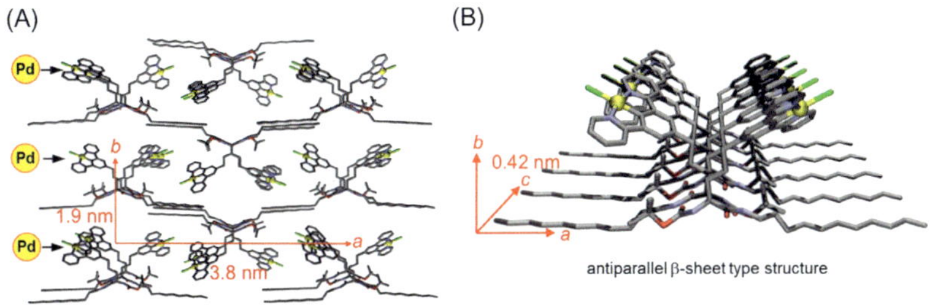

FIGURE 4.13 Supramolecular structure of L-**22** in gel state.

precise supramolecular structure was successfully solved through a computer-aided molecular modeling study using all the data obtained from X-ray scattering analysis (WAX, SAXS), IR, and UV-vis spectroscopy, as shown in Fig. 4.13.[7,26]

The observed 3.8 nm-spaced periodic pattern on the fiber surface is well-agreed with the alternating arrangement of Pd-containing layers and alkyl chain layers. Inside the supramolecular assembly, formation antiparallel β-sheet type association through intermolecular hydrogen-bonding is clearly confirmed with accumulation of PdCl(dpb) moieties stacking up along with the hydrogen-bonding axis (Fig. 4.13).

Firstly, we conducted a catalytic application of the supramolecular self-assembly of a NCN-pincer Pd-complex conjugated norvaline L-**22** as a heterogeneous catalyst in aqueous reaction conditions, where hydrophobic organic substrate is readily adsorbed and condensed into supramolecular gel to accelerate the reaction. In fact, the cyclization of 4-alkynoic acid proceeded efficiently in water to afford the corresponding ene-lactone in the presence of a xerogel particle prepared from toluene gel of L-**22** (Fig. 4.14).[30]

Notably, higher catalytic efficiency of gel state catalyst was clearly demonstrated by lower yields of the product under homogeneous condition in CH_2Cl_2. When substrate 4-alkynoic acid added to the water in the presence of a xerogel particle, the lemon yellow xerogel promptly swelled through the absorption of the substrate to afford an orange gel. This result adequately corroborates the proposed absorption/condensation effects of supramolecular gel catalyst.

Supramolecular catalysis was also investigated in ONO-pincer Ru-conjugated norvaline-catalyzed oxidation.[23,31] The inherent oxidation resistance of amino acids and peptides often observed in enzymes and proteins is suitable as a scaffold catalyst under oxidation conditions using hydrogen peroxide. We designed and synthesized Ru-conjugated norvaline L-**35**[23,27] In this amino acid, the α-side chain is covalently bound to ONO-pincer ruthenium complex of Ru(pydc)(terpy) which shows excellent catalytic activity toward oxidation of various organic compounds.[32] In the presence of very

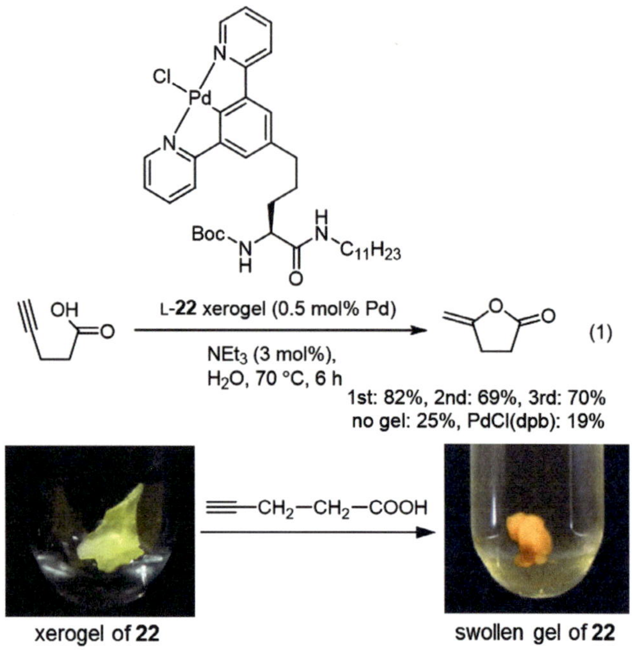

FIGURE 4.14 Supramolecular gel-catalyzed cycloaddition of alkynoinc acid. Pictures are before/after substrate addition to water in the presence of a xerogel particle of L-**22**.

small amounts of L-**35** (0.01 mol%), oxidation of alcohols provides the corresponding aldehydes and ketones highly efficiently using hydrogen peroxide as a terminal oxidant. As shown in Fig. 4.15, the higher catalytic activity of Ru-conjugated norvaline L-**35** compared with the parent Ru(pydc)(terpy) complex was found out through the higher yield of cyclohexanone. To our best knowledge, this is the first example of enhanced catalytic activity among Ru-conjugated amino acids ever reported. Gilbertson, Robinson, and Albrecht independently developed NHC–Ru complex-conjugated amino acids for catalytic olefin metathesis and transfer hydrogenation reactions; however, no enhancement of catalytic activities was found to be comparable or less than the parent ruthenium complexes.[33] In situ SAXS measurement showed formations of micellar nanoparticles in both the aqueous and organic phases of the reaction mixture. We consider that these micelle act as a phase-transfer catalyst to increase the catalytic efficiency. Actually, the addition of surfactant such as sodium dodecyl sulfate to the Ru(pydc)(terpy)-catalyzed oxidation of cyclohexane is effective to improve the yield of cyclohexanone drastically. IPC analysis also supports the proposed phase-transfer ability of L-**35**; the content of ruthenium in L-**35**-catalyed oxidation mixture were almost equal, despite a small amount of ruthenium detected in organic phase of Ru(pydc)(terpy)-catalyzed oxidation mixture.

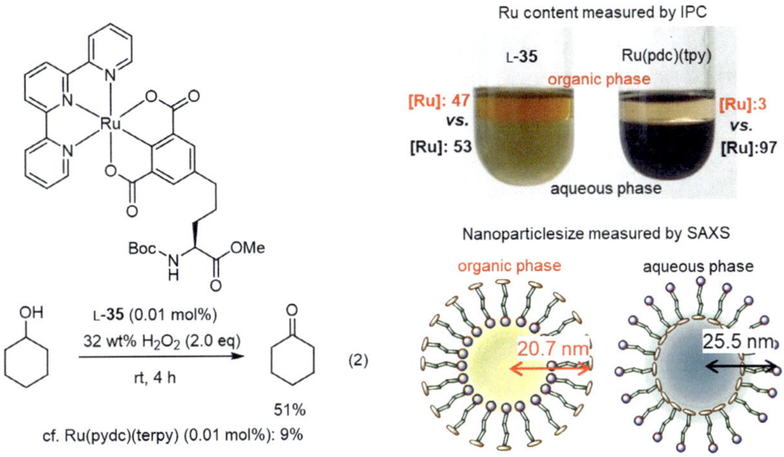

FIGURE 4.15 ONO-pincer Ru-conjugated norvaline-catalyzed oxidation of alcohol. The pictures and images show the biphasic reaction mixtures of L-**35**- and Ru(pydc)(terpy)-catalyzed oxidation of cyclohexanol and the Ru contents measured by ICP, and the schematic illustrations of micellar particles found in the reaction mixture by SAXS measurement.

The role of norvaline moiety of L-**35**-catalyzed oxidation investigated on the catalytic oxidation of 1,3-dimethoxybenzene to 2-methoxybenzophenone as depicted in Fig. 4.16. A series of ruthenium complexes **36** and **37** bearing norvaline analogous parts was prepared to compare the catalytic activity with Ru-conjugated norvaline L-**35** and the parent Ru(pydc)(terpy). Among

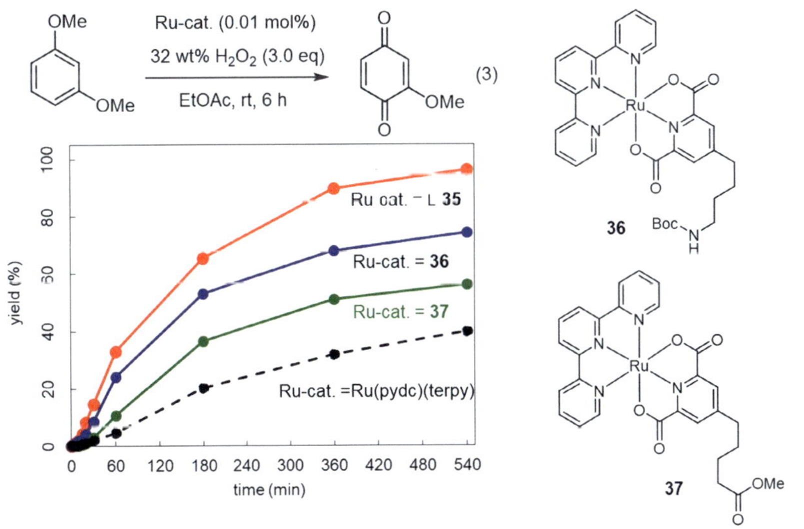

FIGURE 4.16 Structure-catalytic activity relationship for a series of analogous ruthenium catalyst based on Ru-conjugated norvaline L-**35**.

all the catalysts examined, L-**35** showed the highest catalytic activity toward the oxidation of methoxybenzene. Interestingly, the initial reaction revealed a second-order dependence on the concentration of ruthenium catalyst in L-**35**-catalyzed oxidation despite the first-order dependence in Ru(pydc)(terpy)-catalyzed oxidation. These results strongly support the hypothesis that the norvaline moiety is an essential part for enhanced catalysis. This unique catalytic property is induced by the formation of micellar supramolecular self-assembly acting as a phase-transfer catalyst in this biphasic oxidation system.

4.10 CONCLUSION

We synthesized a series of α-side chain-conjugated metalated amino acids based on glutamic acid and norvaline in which stable and robust half-pincer- and pincer-complexes of palladium, platinum, and ruthenium were covalently conjugated through appropriate coupling reactions. The robust nature of these metalated amino acids enabled us to synthesize multi-metalated oligopeptides by the condensation of metalated amino acids based on conventional peptides chemistry involving deprotection processes under acidic, basic, or high-temperature conditions. Notably, the molecular design of these metalated amino acids and peptides allowed various *N*-/*C*-functionalization without any detectable metal leaching and loss of optical purity. The scalability of the synthesis of metalated amino acids/peptides in up to 10 g makes it possible to supply adequate quantities of metalated amino acids and peptides, being important material and catalyst research for screening a huge number of reaction parameters. Because of these advantages of our metalated amino acids and peptides, we succeeded to achieve sequence-controlled metal array fabrication and dimensionally-controlled supramolecular self-assembly possessing multi-lamellar β-sheet architectures. Applications of these amino acids and peptides self-assemblies to a supramolecular catalyst were also accomplished together with a substantially large enhancement of catalytic activity compared to the parent complexes. The results clearly demonstrate that the emergent effect on metalated amino acids and peptides can be achieved by the conjugation of biomolecules and organometallic complexes. We believe that α-side chain-metalated amino acids and peptides will open up new horizons for the further application of bioorganometallic compounds in multidisciplinary areas and contribute to the design of innovative bio-inspired organometallic catalysts, such as peptide-based artificial metalloenzymes.

REFERENCES

1. (a) Albad B, Metzler-Nolte N. *Chem. Rev.* 2016;**116**:11797. (b) Metzler-Nolte N, Salmain M. In: Stepnicka P, editor. *Ferrocenes: Ligand, Materials and Biomolecules*. West Sussex: Wiley; 2008. p. 499. (c) Metzler-Nolte N, Jaouen G, editors. *Bioorganometallics*. Weinheim: Wiley-VCH; 2006.

2. (a) Correia JDG, Paulo A, Raposinho PD, Santos I. *Dalton Trans.* 2011;**40**:6144. (b) Lee S, Xie J, Chen X. *Chem. Rev.* 2011;**110**:3087. (c) Liu S. *Bioconjugate Chem.* 2009;**20**:2199. d) Liu S, Edwards DS. *Chem. Rev.* 1999;**99**:2235.
3. Severin K, Bergs R, Beck W. *Angew. Chem. Int. Ed.* 1998;**37**:1634.
4. Dirscherl G, König B. *Eur. J. Org. Chem.* 2008;**597**.
5. Takaya H. In: Matsuo Y, Higuchi M, Negishi Y, Yoshizawa M, Uemura T, Takaya H, et al., editors. *Briefs in Molecular Science Series*, Vol. 26. New York: Springer; 2013. p. 49. Chapter 6.
6. Isozaki K, Haga Y, Ogata K, Naota T, Takaya H. *Dalton Trans* 2013;**42**:15953.
7. Ogata K, Sasano D, Yokoi T, Isozaki K, Yoshida R, Takenaka T, et al. *Chem. Eur. J.* 2013;**19**:12356.
8. (a) Imperiali B, Fisher SL. *J. Am. Chem. Soc.* 1991;**113**:8527. (b) Mecklenburg SL, Peek BM, Erickson BW, Meyer TJ. *J. Am. Chem. Soc.* 1991;**113**:8540. (c) Hamachi I, Ojida A, Fujishima S, Honda K, Nonaka H, Uchinomiya S. *Chem. Asian J.* 2010;**5**:877. (d) Vairaprakash P, Ueki H, Tashiro K, Yaghi OM. *J. Am. Chem. Soc.* 2011;**133**:759.
9. (a) Aubert N, Troiani V, Gross M, Solladié N. *Tetrahedron Lett.* 2002;**43**:8405. (b) Hasobe T, Saito K, Kamat PV, Troiani V, Qiu H, Solladié N, et al. *J. Mater. Chem.* 2007;**17**:4160. (c) Yamamura T, Suzuki S, Taguchi T, Onoda A, Kamachi T, Okura I. *J. Am. Chem. Soc.* 2009;**131**:11719.
10. Isozaki K, Takaya H, Naota T. *Angew. Chem. Int. Ed.* 2007;**46**:2855.
11. Takaya H, Haga Y, Isozaki K, Ogata K, Naota T. *Chem. Lett.* 2014;**43**:1167.
12. Isozaki K, Ogata K, Haga Y, Sasano D, Ogawa T, Kurata H, et al. *Chem. Commun.* 2012;**48**:3936.
13. Pd(0)-catalyzed. deprotection of allyl ester of 1 proceeded through the formation of π-allyl Pd complex to give C-terminus free Pd-glutamic acid Fmoc-L-[PdCl(BzAA)]Glu]-H and 2-allyldimedone.
14. a) Löwik DWPM, van Hest JCM. *Chem. Soc. Rev* 2004;**33**:234. (b) Frauenrath H, Jahnke E. *Chem. Eur. J* 2008;**14**:2942. (c) Das A, Ghosh Suhrit. *Chem. Commun* 2016;**52**:6860.
15. (a) Kojima T, Noguchi D, Nakayama T, Inagaki Y, Shiota Y, Yoshizawa K, et al. *Inorg. Chem.* 2008;**47**:886. (b) Park YJ, Kim J-S, Youm K-T, Lee N-K, Ko J, Park H, et al. *Angew. Chem., Int. Ed* 2006;**45**:4290 2006. (c) Kuwabara J, Takeuchi D, Osakada K. *Bull. Chem. Soc. Jpn.* 2005;**78**:668. (d) Rivas JCM, de Rosales RTM, Parsons S. *Dalton Trans.* 2003;**2156**.
16. For the pioneering research on ultrasound-induced structure controll and its application to supramolecular gellation, see; Naota, K. Koori, *J. Am. Chem. Soc. 127*, 9324 (2005).
17. (a) Bardelang D, Camerel F, Margeson JC, Leek DM, Schmutz M, Zaman MB, et al. *J. Am. Chem. Soc.* 2008;**131**:3313. (b) Ke D, Zhan C, Li ADQ, Yao J. *Angew. Chem., Int. Ed.* 2006;**50**(3715). (2011). (c) Maity S, Kumar P, Haldar D. *Soft Matter* 2011;**7**:5239. (d) Ye F, Chen S, Tang GD, Wang X. *Colloids and Surfaces A: Physicochem. Eng. Aspects* 2014;**452**:165. (e) Bardelang D, Zaman Md B, Moudrakovski IL, Pawsey S, Margeson CJ, Wang D, et al. *Adv. Mater.* 2008;**20**:4517.
18 Supramolecular gelation by heat-cool process, see Fracaroli AM, Tashiro K, Yaghi OM. *Inorg. Chem* 2012;**51**:6437.
19. The ultrasound micro-cavitation can generate high-energy-density sites with extremely high temperature, pressure, and supersaturation around the local environment after collapsing. The cavitation energy is strongly correlated with the size of micro bubble in which lower frequency of ultrasound generates larger bubbles with higher collapsing energy; see also ref. 6 and references cited therein.

20. (a) Robinson ME, Nazemi A, Lunn DJ, Hayward DW, Boott CE, Hsiao MS, et al. *ACS Nano* 2017;**11**:9162. (b) Fukui T, Kawai S, Fujinuma S, Matsushita Y, Yasuda T, Sakurai T, et al. *Nature Chemistry* 2017;**9**:493.
21. (a) Albrecht M, van Koten G. *Angew. Chem. Int. Ed.* 2001;**40**:3750. (b) van Koten G. *J. Organometallic. Chem* 2013;**730**:156. (c) Chase PA, Gossage RA, van Koten G. Modern organometallic multidentate ligand design strategies: The birth of the privileged "pincer" ligand platform. In: van Koten G, Gossage RA, editors. *The Privileged Pincer-Metal Platform: Coordination Chemistry & Applications, Top. Organometallic. Chem*, Vol. 54. Springer; 2016. p. 1–15. (d) Murugesan S, Kirchner K. *Dalton Trans.* 2016;**45**:416.
22. Takaya H, Iwaya T, Ogata K, Isozaki K, Yokoi T, Yoshida R, et al. *Synlett* 2013;**24**:1910.
23. Isozaki K, Yokoi T, Yoshida R, Ogata K, Hashizume D, Yasuda N, et al. *Chem. Asian J.* 2016;**11**:1076.
24. (a) Collier PN, Campbell AD, Patel I, Taylor RJK. *Tetrahedron Lett.* 2000;**41**:7115. (b) Collier PN, Campbell AD, Patel I, Raynham TM, Taylor RJK. *J. Org. Chem.* 2002;**67**:1802.
25. Guillena G, Rodríguez G, Albrecht M, van Koten G. *Chem. Eur. J.* 2002;**8**:5368.
26. Ogata K, Sasano D, Yokoi T, Isozaki K, Yasuda N, Ogawa T, et al. *Chem. Lett.* 2012;**41**:194.
27. (a) Rose-Munch F, Aniss K, Rose E, Vaisserman J. *J. Organomet. Chem.* 1991;**415**:223. (b) C. Sergheraert, A. Tartar, J. Organomet. Chem. 240, 163 (1082); (c) Wolff JM, Sheldrick WS. *J. Organomet. Chem.* 1997;**531**:141. (d) Brunner H, Kçnig W, Nuber B. *Tetrahedron: Asymmetry* 1993;**4**:699. (e) Jackson RFW, Tuner D, Block MH. *Synlett* 1996;**862**. (f) Dialer H, Polborn K, Ponikwar W, Sünkel K, Beck W. *Chem. Eur. J.* 2002;**8**:691.
28. (a) Kamiński ZJ, Kolesińska B, Kolesińska J, Sabatino G, Chelli M, Rovero P, et al. *J. Am. Chem. Soc.* 2005;**127**:16912. (b) Kunishima M, Ujigawa T, Nagaoka Y, Kawachi C, Hioki K, Shiro M. *Chem. Eur. J.* 2012;**18**:15856.
29. Takaya H, Yokoi T, Yoshida R, Isozaki K, Kawakami T, Takenaka T, et al. *Chem. Lett.* 2017;**46**:665.
30. (a) Rodríguez-Llansola F, Miravet JF, Escuder B. *Chem. Eur. J* 2016;**16**:8480. (b) Rodríguez-Llansola F, Escuder B, Miravet JF. *J. Am. Chem. Soc* 2009;**131**:11478. (c) Rodríguez-Llansola F, scuder BE, Miravet JF. *Org. Biomol. Chem* 2009;**7**:3091. (d) Guler MO, Stupp SI. *J. Am. Chem. Soc* 2007;**129**:12082.
31. Ogata K, Sasano D, Yokoi T, Isozaki K, Seike H, Takaya H, et al. *Chem. Lett.* 2012;**41**:498.
32. Yoshida R, Isozaki K, Yokoi T, Yasuda N, Sadakane K, Iwamoto T, et al. *Org. Biomol. Chem.* 2016;**14**:7468.
33. (a) Nishiyama H, Shimada T, Itoh H, Sugiyama H, Motoyama Y. *Chem. Commun.* 1863. (1997)). (b) Shi F, Tse MK, Beller M. *Chem. Asian J.* 2007;**2**:411.

FURTHER READING

Xu et al., 2005 2011 2015(a) Xu G, Gilbertson SR. *Org. Lett.* 2005;**7**:4605. (b) Monney A, Venkatachalam G, Albrecht M. *Dalton Trans.* 2011;**40**:2716. (c) Gleeson EC, Wang ZJ, Jackson WR, Robinson AJ. *J. Org. Chem.* 2015;**80**:7205.

Chapter 5

Controlling Peptide Conformation Using Tungsten Alkyne Coordination

Timothy P. Curran
Department of Chemistry, Trinity College, Hartford, CT, United States

5.1 INTRODUCTION

Proteins possess three major secondary structures: The α-helix, the β-turn, and the β-sheet in both the antiparallel and parallel forms (Fig. 5.1). The combination and arrangement of these secondary structures provides the basis for the overall three-dimensional conformation, or tertiary structure, of most proteins.[1] Further, it has been noted that a large number of biological processes are mediated by protein–protein binding,[2] which involves interactions between one or more secondary structures on one protein with one or more secondary structures on the second protein. In cases where protein–protein interactions have an adverse effect on human health it has been proposed that peptides constrained to particular secondary structures, or non-peptides whose 3D shapes mimic that of a given secondary structure, may prove to be useful as drug molecules.[3]

The most notable example of a peptide being used in this manner is the anti-HIV medicine enfuvirtide.[4] This peptide inhibits HIV entry into host cells, and it accomplishes this by binding to the envelope protein gp41 found on the surfaces of T-cells.[5] The active conformation of enfuvirtide is an α-helix, and this α-helix binds to a helix bundle on gp41.[6] Once enfuvirtide is bound to gp41, HIV is unable to bind to T-cells.[5]

Most peptides, when removed, and isolated from the native protein, do not maintain their secondary structure. Instead, the isolated peptides display an ability to adopt multiple solution conformations. This likely happens because the secondary structure was enforced by interactions with the rest of the protein. Once those constraints are removed, the peptide displays increased conformational mobility. To hold the isolated peptide in the desired conformation requires the introduction of different conformational

Advances in Bioorganometallic Chemistry. DOI: https://doi.org/10.1016/B978-0-12-814197-7.00005-4

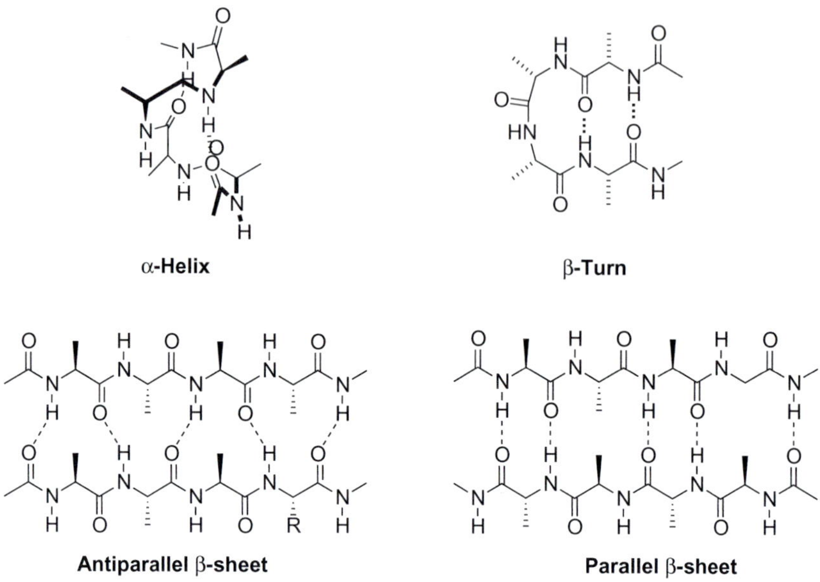

FIGURE 5.1 The major secondary structures found in proteins: the α-helix, the β-turn, and antiparallel and parallel β-sheets.

constraints. To date, the majority of these constraints have involved the introduction of covalent linkages of organic moieties between two or more loci on the peptide chain, most notably for α-helices[7] and β-sheets.[8] That organic linkages have been used is not surprising since most peptide chemists are trained as organic chemists.

Until 2000, using organometallic moieties for constraining peptides to a particular secondary structure had not been extensively explored. Organometallic constraints offer some key advantages over their organic counterparts. First, the formation of a metal-ligand complex can often be achieved in fewer steps than the lengthy and complicated syntheses required for some covalently constrained peptides. Second, there is a wide array of transition metal-ligand combinations that might be used. Third, the presence of the transition metal in these organometallicpeptides provides a unique spectroscopic chromophore that may be used to assess the behavior and biological activity of the constrained peptide. Fourth, the presence of the transition metal might help facilitate the solution of an X-ray crystal structure. Fifth, it is possible to change the oxidation states of many transition metals, providing a way to change the charge-state of the organometallicpeptide.

In seeking organometallic systems capable of controlling peptide conformation, the properties of the ligands are important. First, it should be possible to easily locate the ligand on either the amino-acid side chain, or at the N- or C-termini of the peptide. Second, the ligand should be unreactive during the usual course of peptide synthesis, which would facilitate construction of the desired peptide. Third, the ligand should readily react with its transition metal partner and reliably form the desired complex. Fourth, the conditions under which the ligand reacts with the transition metal should not alter the chemical identity of the peptide. Finally, the transition metal should be able to coordinate two or more ligands, allowing for the placement of multiple peptides around the metal center.

For nearly 20 years, research in the Curran group has explored the possibility of using coordination of alkynes to tungsten as a vehicle for controlling peptide conformation. The coordination of alkynes to tungsten was first observed by Tate and Augl in 1963 by their synthesis and characterization of **1**, W(CO)(3-hexyne)$_3$.[9] Subsequent work explored other alkyne ligands for this complex, and this was followed by the preparation and characterization of a number of mono- and bis-alkyne complexes. A group of researchers, led notably by McDonald and Templeton, made and examined a large number of mono- and bis-alkyne complexes and studied the nature of the bonding between the metal and the alkyne ligands. This work was summarized in a review paper by Templeton from 1989.[10]

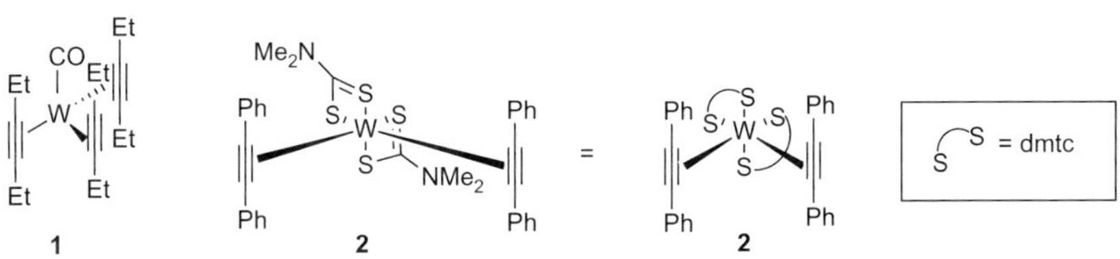

Of the many complexes examined, those that also possess dithiocarbamate (R_2NCS_2) ligands are particularly attractive, since the bis-alkyne complexes of these species were found to be stable in air (for example **2**).[11] The long-term goal of constraining peptides to a particular secondary structure would be to use them as probes for exploring the connection between structure and biological activity. Studies of this nature presumably would be done in conditions that mimic the natural states of cells and living organisms, which in most cases requires an oxygenated milieu. Having an air stable complex that would constrain a peptide to a particular secondary structure would be very useful for elucidating these structure–function relationships.

This chapter summarizes the studies that have examined the conformational behavior of tungsten alkynylpeptide complexes, and details that this chemistry has been successful in generating sheet and turn structures.

5.2 INITIAL STUDIES

The first work in this area focused on coordinating two alkynylpeptides to a single tungsten center. The working hypothesis for these studies was that the tungsten would hold the two amino acids or peptides in close proximity, and hydrogen bonding, like that seen in β-sheets, would develop. These alkynylamino acids were reacted under an inert atmosphere with $W(CO)_3(dmtc)_2$[12] in refluxing methanol to yield the bis-alkyne complexes **3, 4a-b** and **5a-b**. These bis-alkyne complexes were readily purified by flash chromatography, and their identities established by elemental analysis, mass spectrometry and ^1H NMR spectroscopy.[13]

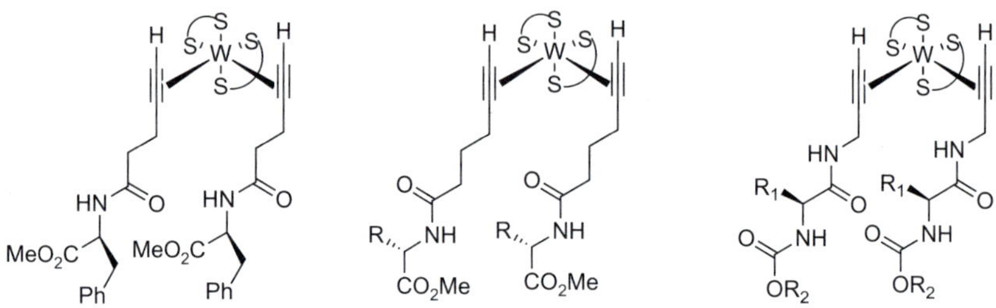

3

4a R = CH_2Ph
4b R = $(CH_2)_4NHBoc$

5a R = CH_2Ph
5b R = $(CH_2)_4NHBoc$

The use of mass spectrometry for establishing the identity of these and other tungsten-alkynylpeptide complexes has proved to be of great value. Elemental analyses of these complexes proved to be somewhat challenging. On the other hand, the use of electrospray mass spectrometry (ESMS) was a reliable and easy method for establishing the molecular identities of the complexes. Methanol solutions of the tungsten-alkynylpeptide complexes made up in borosilicate glass test tubes reliably produce the M + Na cation during ESMS analysis. This M + Na cation has a distinct and unique pattern for the molecular ion, due to the makeup of tungsten, which has four major isotopes. Matching the molecular ion pattern of the M + Na cation in the ESMS to the theoretical isotope pattern[13] is sufficient for establishing the molecular identity of the complex.

The key question with complexes **3**, **4a-b**, and **5a-b** was whether intramolecular hydrogen bonds developed between the two alkynylamino acids linked to tungsten. From the prior work of Herrick and Templeton done on molybdenum bis-alkyne complexes it was known that the two alkyne ligands are arranged parallel to each other in these complexes,[14] making it possible for the NH on one of the amino acids to hydrogen bond with a C = O on the other amino acid. Intramolecular hydrogen bonds can be discovered using a ^1H NMR experiment, a DMSO titration.[15–20] In this experiment the ^1H NMR spectrum of the compound under study is recorded in $CDCl_3$ and the chemical shifts of the NH protons recorded. Then, small amounts of d_6-DMSO are added to the $CDCl_3$ solution and the ^1H NMR spectrum taken again and the chemical shifts of the NH protons recorded. This process is repeated for solutions with a percentage DMSO composition between

0% and 25%. If an NH proton is involved in an intramolecular hydrogen bond, its chemical shift will be insensitive to the addition of the DMSO, which is an aggressive hydrogen bond acceptor. On the other hand, if an NH proton is only exposed to solvent, its chemical shift will undergo a dramatic change, typically about 1 ppm, over the course of the experiment.

The ^{1}H NMR spectra of **3**, **4a-b**, and **5a-b** were recorded in $CDCl_3$. A notable feature of these spectra was the appearance of the resonances for the two alkyne hydrogens. As noted by earlier studies, these protons undergo a large change in chemical shift upon coordination of the tungsten, moving from the region between 2–3 ppm to the region above 11 ppm. Since there are no other resonances in these complexes that would appear near 11 ppm, the resonances of the alkyne hydrogens provide a clear window into the conformational behavior of these species.

Shown in Fig. 5.2 is the alkyne hydrogen region for **4b**. The alkyne hydrogen should appear as a singlet in these spectra, and from a simple view of **4b** one would expect to see two singlets since the magnetic environments for each alkyne hydrogen is slightly different. As shown in Fig. 5.2 there are at least five singlets visible in the spectrum of **4b**. Multiple singlets arise for two reasons. First, the alkyne ligands in these complexes can equilibrate between syn- and anti-orientations (Fig. 5.3). Second, when the two amino acid derivatives coordinate to tungsten they generate a chiral center at tungsten, and both sets of ligand orientations are obtained. Thus, **4b** is really a mixture of the two diastereomers **4b**′ and **4b**″ (Fig. 5.4), and both **4b**′ and **4b**″ can equilibrate between the syn- and anti-orientations (Fig. 5.3).

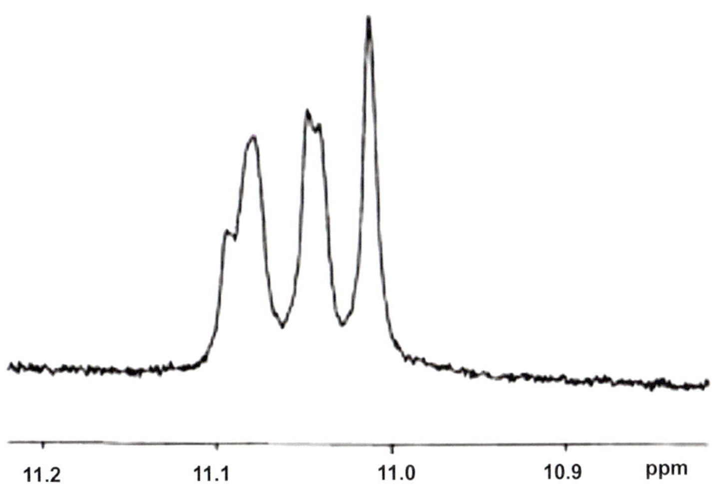

FIGURE 5.2 The alkyne hydrogen region in the ^{1}H NMR spectrum for a $CDCl_3$ solution of complex **4b**. At least five different singlets are visible in this spectrum, which shows that the alkynes in **4b** adopt both the syn- and anti-orientations relative to the tungsten center.

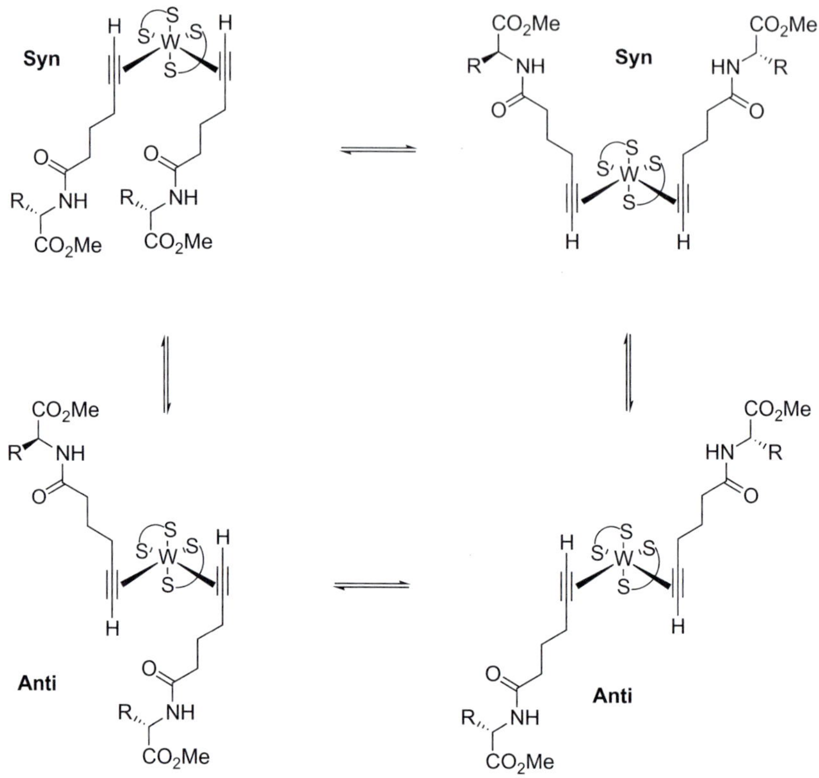

FIGURE 5.3 Syn- and anti-isomerism in tungsten bis-alkyne complexes as illustrated for **4a-b**. At room temperature the alkynes slowly rotate about the metal center, and that slow rotation gives rise to separate NMR spectra for the syn, and anti conformers. This isomerism is responsible for the large number of resonances seen in the NMR spectra of **3**, **4a-b** and **5a-b**, as illustrated for the alkyne hydrogen resonances for **4b** shown in Fig. 5.2.

For intramolecular hydrogen bonding to occur between the two amino acid derivatives, the alkyne ligands need to adopt and maintain the syn-orientation. The spectra for **3**, **4a-b,** and **5a-b** show that these complexes adopt both the syn and anti orientations. Variable temperature NMR of **3**, **4a-b** and **5a-b** showed that the alkyne hydrogen resonances will coalesce to a single peak in around 60°C, but will revert to their original pattern of multiple singlets when cooled back to 23°C.[13] This behavior shows that in these tungsten-bisalkynylpeptides the syn- and anti-orientations are in slow equilibrium at 23°C, behavior that matches the dynamic behavior seen in simple tungsten bis-alkyne complexes.[14]

Finally, the chemical shifts of the NH protons were examined in a DMSO titration. Owing to the presence of the two diastereomers and the conformational isomerism associated with the alkyne ligands, there were

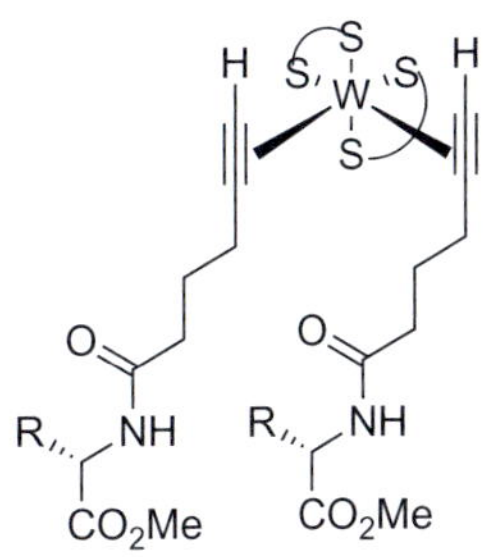

4a′ R = CH_2Ph
4b′ R = $(CH_2)_4NHBoc$

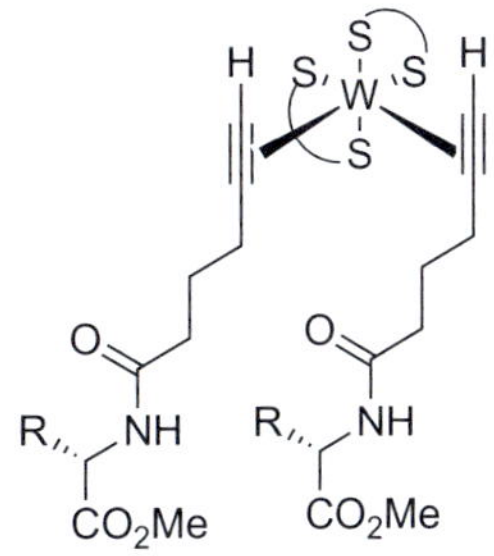

4a″ R = CH_2Ph
4b″ R = $(CH_2)_4NHBoc$

FIGURE 5.4 Depiction of the two diastereomers obtained when **4a-b** are formed. The tungsten center becomes chiral upon formation of the complex, with the dmtc ligands assuming two possible orientations. Since the amino acids on the alkyne ligands possess chiral centers, the resulting product mixture contains the two diastereomers, **4a**′ and **4a**″, and **4b**′ and **4b**″.

multiple peaks for each NH in **3**, **4a-b** and **5a-b**. All of the NH peaks were followed in the DMSO titration; all of their chemical shifts underwent a dramatic change in chemical shift, indicating that there were no intramolecular hydrogen bonds present in these complexes. Subsequent studies with alkynylpeptide complexes **6** and **7**, where a dipeptide and a tripeptide were coordinated to tungsten, gave similar results.[21] The conclusion made from this data was that the two peptides in these complexes are simply too far apart to form intramolecular hydrogen bonds.

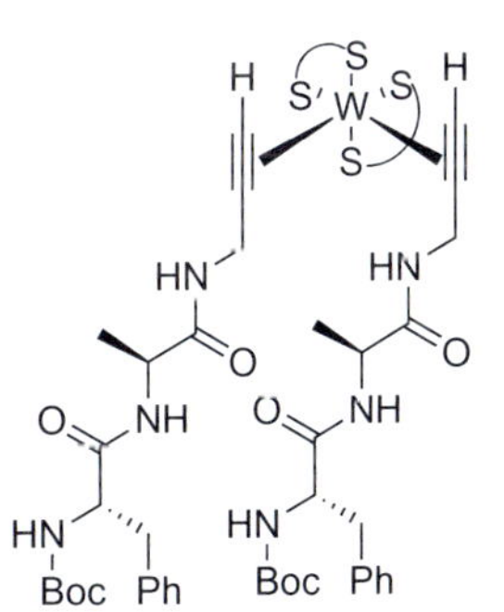

6

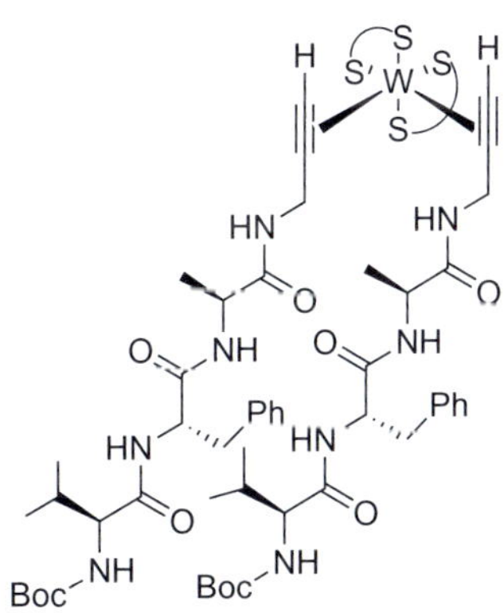

7

5.3 MONOALKYNYLPEPTIDE COMPLEXES

Given that two peptides attached to tungsten did not hydrogen bond with each other to form a β-sheet structure, examined next were some monoalkyne complexes that might accomplish the same goal, generation of a constrained β-sheet, using tungsten-alkyne coordination.

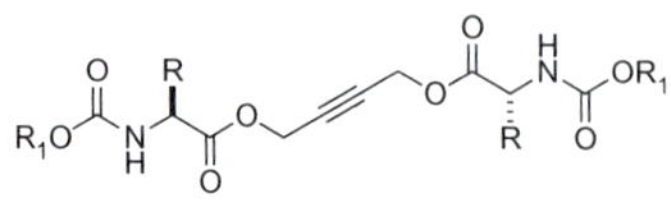

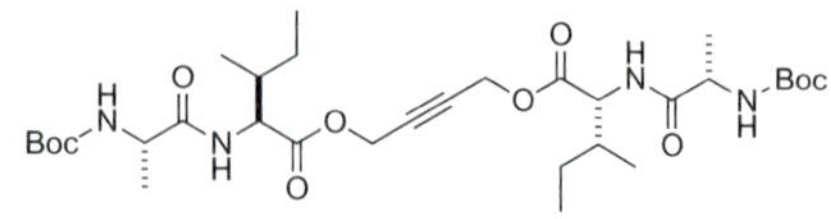

9

8a R1 = $C(CH_3)_3$, R = $(CH_2)_4NHCbz$
8b R1 = CH_2Ph, R = CH_3
8c R1 = CH_2Ph, R = $CH(CH_3)CH_2CH_3$
8d R1 = $C(CH_3)_3$, R = $CH(CH_3)CH_2CH_3$

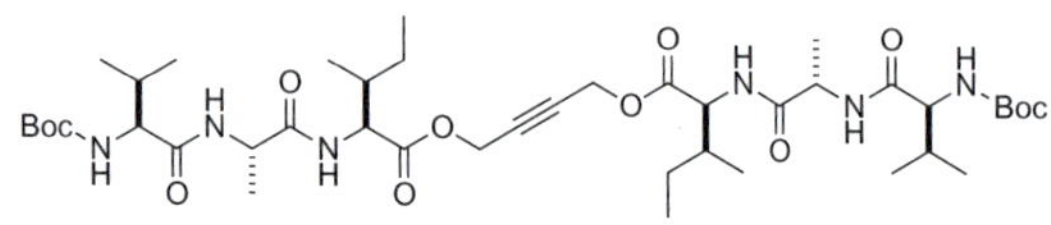

10

In the first approach, two identical peptides were linked to 1,4-dihydroxy-2-butyne to generate the symmetrical alkynylpeptides **8–10**. These alkynes were then reacted with $W(CO)_3(dmtc)_2$ in CH_2Cl_2 to generate the monoalkyne complexes **11–13**.[22] These complexes, which were moderately unstable in air, were purified by flash chromatography and stored under vacuum. They were characterized by ESMS and ^1H NMR spectroscopy.

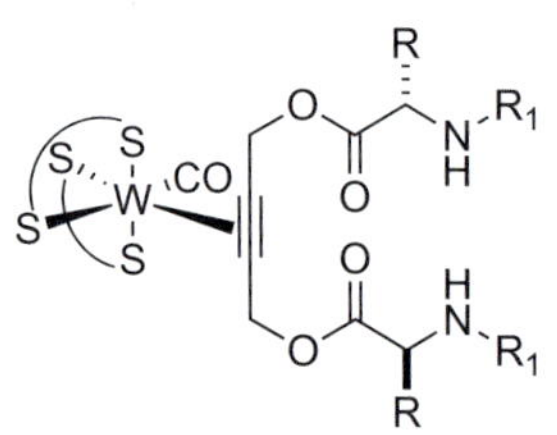

11a R_1 = Boc, R = $(CH_2)_4NHCbz$
11b R_1 = Cbz, R = CH_3
11c R_1 = Cbz, R = $CH(CH_3)CH_2CH_3$

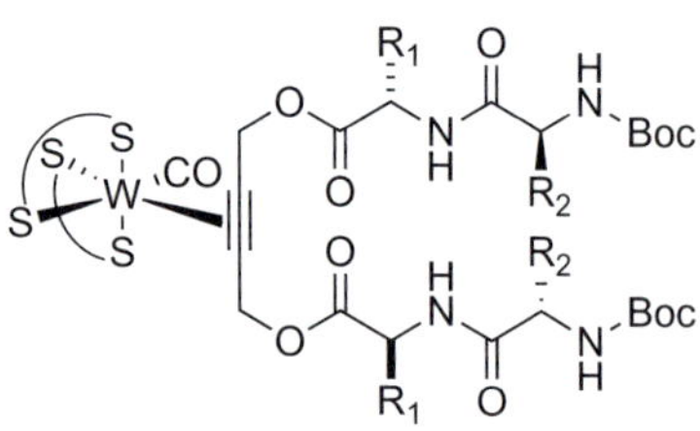

12 R_1 = $CH(CH_3)CH_2CH_3$, R_2 = CH_3

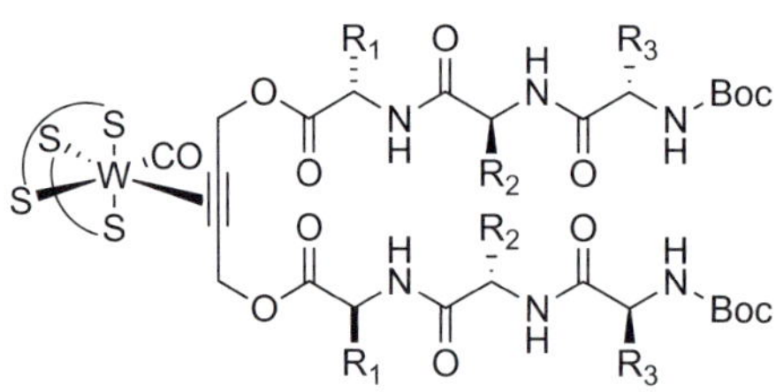

13 R_1 = $CH(CH_3)CH_2CH_3$, R_2 = CH_3, R_3= $CH(CH_3)_2$

The idea for using the symmetrical alkynylpeptides **8–10** as the ligands for tungsten was to remove alkyne rotation as a source of conformational isomerism in these complexes, since no change in the molecule would occur upon alkyne rotation. The other reason for using **8–10** was the potential for the two peptides emanating from the alkyne to hydrogen bond with each other, as is suggested by the structural drawings for **11–13**. There was reason to believe that this would occur based on the behavior of similar peptide

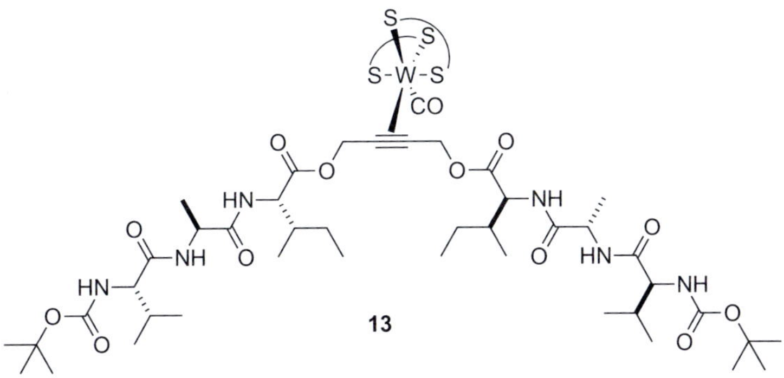

FIGURE 5.5 The extended conformation for monoalkyne complex **13** that is consistent with the finding that none of the NH protons are engaged in intramolecular hydrogen bonds.

derivatives made from 1,4-diamino-2-butyne.[20] If the two peptide components in **11–13** were to hydrogen bond with each other then the tungsten and the alkyne would act as a β-turn mimetic, holding the two peptides side-by-side.

Unfortunately, examination of these species using the DMSO titration did not reveal any intramolecular hydrogen bonding. All of the NH protons were exposed to solvent. It was concluded that the peptide portions of **11–13** do not curl back on each other, but rather extend away from each other as shown in Fig. 5.5 for **13**.[22]

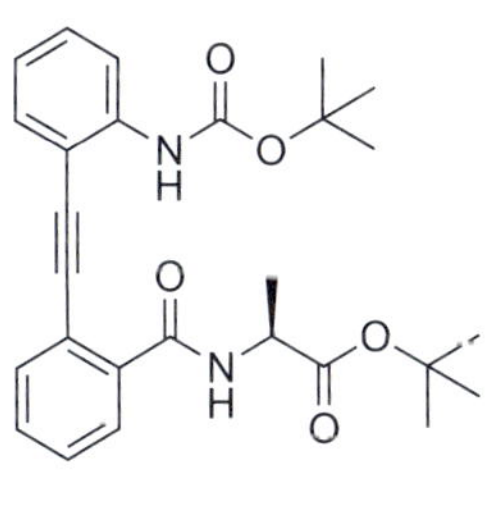

14

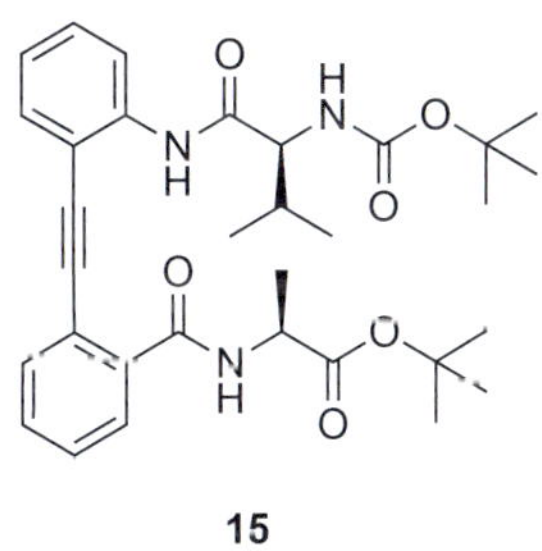

15

A second avenue of inquiry for creating organometallic β-sheet complexes was an examination of an alkynylpeptide system known to adopt a β-sheet. As delineated by Kemp and Li,[23,24] peptidyldiphenylacetylenes **14–15** adopt a β-sheet conformation. Would **14–15** retain their β-sheet conformation when the alkyne is coordinated to tungsten? To answer this question, **14–15** were prepared and then reacted, under inert atmosphere, with $W(CO)_3(dmtc)_2$ in CH_2Cl_2 to afford the mono-alkyne complexes **16–17**, which were obtained as an inseparable mixture of the two diastereomers **16a-b** and **17a-b**.[25] As with **11–13**, **16–17** were moderately air stable. They could be purified by flash chromatography and characterized by ESMS and ^{1}H NMR spectroscopy.

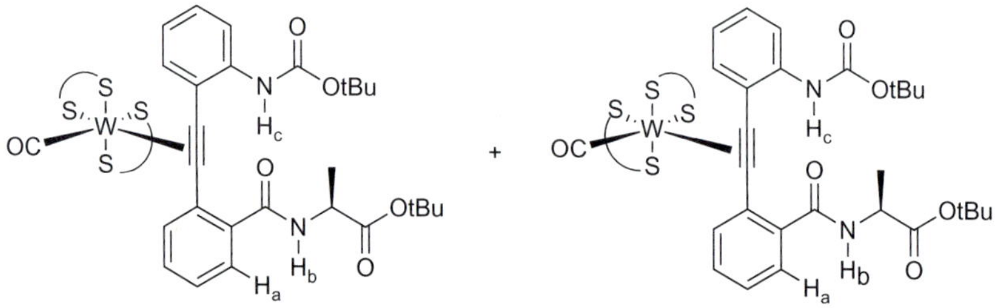

16a **16b**

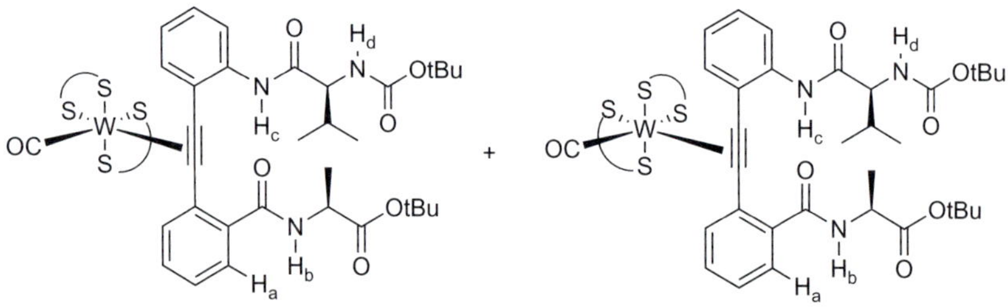

17a **17b**

Complexes **16**–**17** were subjected to the DMSO titration procedure, and the data showed that an NH proton in **16** and **17** was involved in an intramolecular hydrogen bond. Using COSY data this proton was identified as H_c, the NH proton expected to be in the intramolecular hydrogen bond if the peptide retained its β-sheet structure. This indicated that the β-sheet structure was retained upon coordination of the tungsten. Further evidence supporting this conclusion came from the NOESY spectra of **16** and **17**, which showed a crosspeak between H_a and H_b. These two protons are only in close proximity if **16** and **17** adopt the β-sheet conformation.[25]

Since the β-sheet was retained in the mono-alkyne complex, experiments were undertaken to see if the β-sheet was retained in the bis-alkyne complexes. Much to our surprise, the bis-alkyne complexes proved exceedingly difficult to obtain. After some careful experimentation a small amount of the bis-alkyne complex of **14** was obtained, but it rapidly falls apart,[26] and this behavior has stymied experiments to determine whether the β-sheet structure is retained.

5.4 METALLACYCLICPEPTIDES

Given that the initial studies had not produced intramolecular hydrogen-bonding between two peptides, except in the case of a peptidomimetic already in a β-sheet conformation, the next experiments focused on whether tungsten-alkyne coordination could generate β-turn structures. Two approaches were explored for generating the turn structure. In one case alkynes were appended to the N- and C-termini of peptides, and the dialkynylpeptides were coordinated to tungsten to form a cyclic structure that includes the tungsten center. These novel molecules were termed metallacyclicpeptides. It was hypothesized that tying the N- and C-termini to the tungsten center would force the peptide to adopt a turn structure. In the second approach, dipeptides bearing an alkyne on the side chain of each amino acid were prepared, and the dialkynyldipeptides were then coordinated to tungsten to form a different set of metallacyclicpeptides. This second set of metallacyclicpeptides was intended to study whether tying the side chains of two consecutive amino acids in a peptide would enforce a turn structure.

The metallacyclicpeptides **18–21**, which link the N- and C-termini of the peptide via a tungsten bis-alkyne complex, were prepared and characterized in the usual manner.[27,28] As with the previous tungsten bis-alkynylpeptide complexes that had been studied, **18–21** were also obtained as an inseparable mixture of diastereomers. ESMS here was very important for the characterization of **18–21**, as cyclization involving reaction of $W(CO)_3(dmtc)_2$ with the dialkynylpeptide could potentially lead to the formation of oligomeric species. The M + Na cation in the ESMS gave definitive proof that the desired monomeric complex had been obtained.

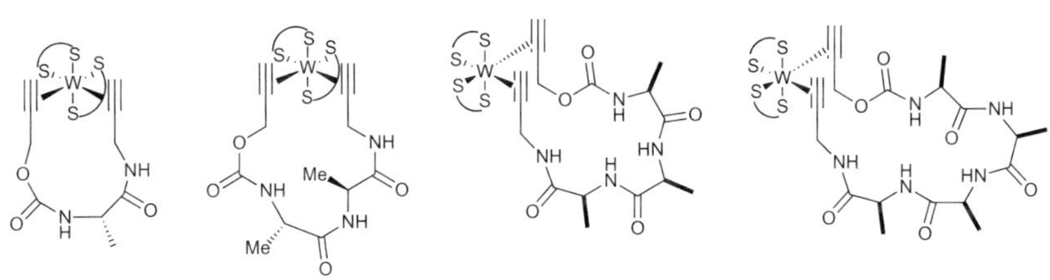

18 **19** **20** **21**

At the outset it had been thought that formation of the metallacycle would force the tungsten bis-alkyne complex to assume only the syn orientation. This was disproved by the ^{1}H NMR spectra of **18–21**, which displayed multiple singlets in the region around 11 ppm for the alkyne hydrogens, indicating that these metallacyclicpeptides do adopt both the syn- and anti-orientation of the alkyne about the metal center. This conclusion was enforced by the behavior of **18–21** in variable temperature experiments, where the alkyne hydrogen singlets coalesced to one peak as the temperature was raised. The temperature of coalescence was measured for **18–21**, and it was found that the temperature *increased* as the peptide chain became longer, with the metallacyclictetrapeptide **21** having the highest temperature of coalescence. At first thought this trend appeared counterintuitive; it had been expected that the ring should have more, not less flexibility, as its size increased. On further thought, however, if the ring system possesses intramolecular hydrogen bonds, then these would have to be broken to allow for interconversion between the syn- and anti-alkyne orientations; the presence of intramolecular hydrogen bonds and the extra energy needed to break them would explain the increase in the temperature of coalescence as the ring system grows larger. To see whether there were any intramolecular hydrogen bonds present, DMSO titrations were performed, and NH protons involved in intramolecular hydrogen bonds were discovered. From this data it was concluded that **20** was likely in a γ-turn conformation, and **21** was likely in a β-turn conformation.[28]

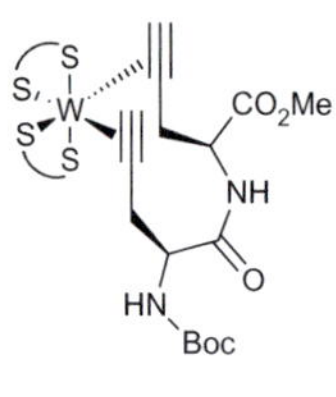

22

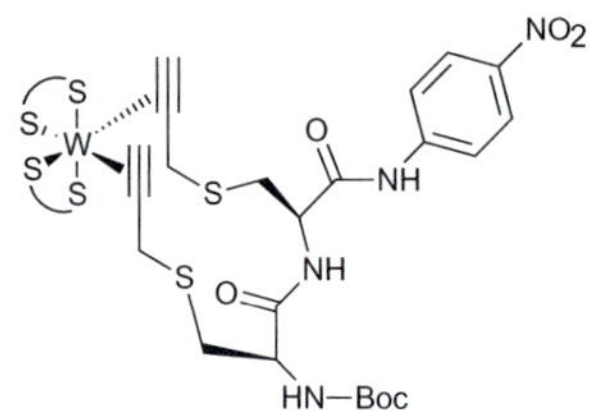

23

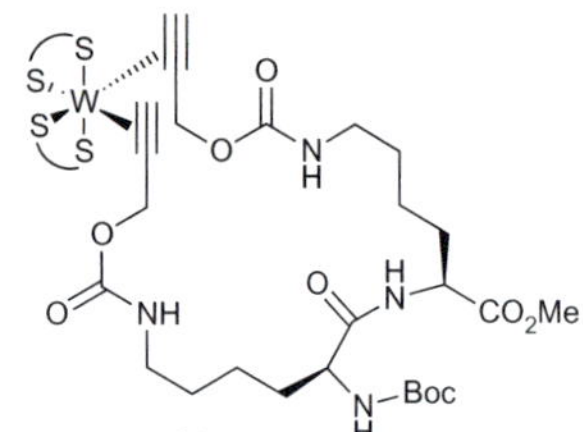

24

In the second study, a series of three dialkynyldipeptides were prepared and coordinated to tungsten to yield the complexes **22–24**.[29] Coordination of the tungsten formed metallacycles of varying ring sizes, and as with prior tungsten bis-alkynylpeptides they were obtained as an inseparable mixture of two diastereomers. The smallest complex was derived from cyclization of dipropargylglycine (**22**), the second from cyclization of a dialkynyl derivative of cysteine (**23**), and the largest complex from cyclization of a dialkynyl derivative of lysine (**24**). The ^{1}H NMR spectra of all three metallacyclicpeptides showed multiple peaks for the alkyne hydrogens, indicating again that the syn- and anti-orientations of the alkyne ligands had been adopted. Variable NMR studies on the complexes showed that the alkyne hydrogen resonances of the two largest complexes, **23** and **24**, coalesced as the temperature was raised, with the larger ring having the lowest temperature of coalescence. This result, which is opposite from the results obtained with metallacyclicpeptides **18–21**, indicates that there is no intramolecular hydrogen bonding involving the NH protons in either **23** or **24**. In contrast, with the smallest complex **22**, the alkyne hydrogen resonances did not undergo any change as the temperature was raised. This indicated that while **22** adopts both the syn and anti orientations, the ring system is too small and rigid to allow for equilibration. Thus, whatever orientation is formed when the ring system is closed by coordination of the alkynes to tungsten, that orientation cannot be changed.

In summary, with metallacyclicpeptides it was found that cyclization of side-chain alkynes failed to generate turn structures,[29] but N- to C-termini cyclization in tri- and tetrapeptides generated structures that had intramolecular hydrogen bonds and were likely in γ- or β-turn structures.[28]

5.5 FERROCENE TUNGSTEN BIS-ALKYNE RING SYSTEM

One common finding from the exploration of tungsten alkynylpeptide complexes has been the conformational flexibility of the alkyne ligand about the metal center. In all cases examined, whether the complex was acyclic or cyclic, both the syn- and anti-orientations of the alkyne ligand have been observed. In no case had the alkyne ligands been limited to only one of the two possible orientations, even in cases (like **22**), where a rigid ring system had been produced. This posed the question as to whether any tungsten bis-alkyne complex could be limited to only the syn or anti orientation. To answer this question a number of dialkynes were cyclized by reaction with $W(CO)_3(dmtc)_2$ and the resulting complexes examined for their conformational behavior.

From this work it was found that the answer to the question is yes, there are tungsten bis-alkyne complexes that are limited to only the syn alkyne orientation. To date, these are the cyclic complexes **25** and **26**, which were obtained as a mixture of two enantiomers.[30]

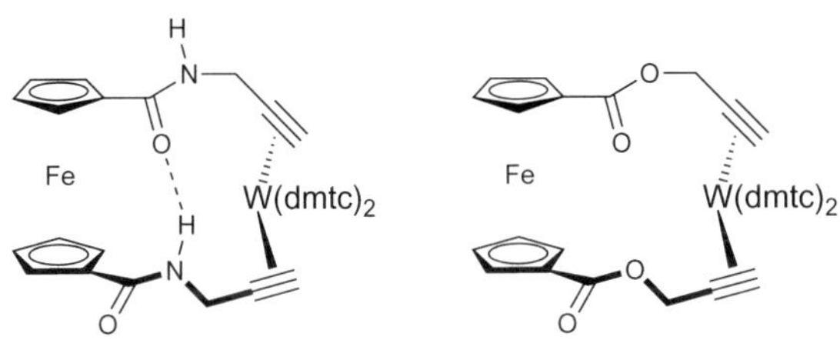

25 **26**

Both **25** and **26** incorporate both a ferrocene moiety and a tungsten bis-alkyne complex in the ring system. Both the amide (**25**) and the ester (**26**) assume only one solution conformation, and the complex is locked into that one conformation, with the two alkynes likely held to the syn orientation. In the case of **25**, one of the amide NH protons is hydrogen bonded to the carbonyl across the ring system, while the other NH is not. The similar conformational behavior of **25** and **26** (which cannot form an intramolecular hydrogen bond), however, shows that this intramolecular hydrogen bond is not responsible for limiting the conformational mobility of **25**. Rather, it is the unique shape and size of the ring system that holds the alkynes to the syn-orientation; small increments in the size of the ring result in the adoption of both the syn- and anti-orientations.

In analyzing the structure of **25**, it is thought that the two alkynes are spaced about 3.3 Å apart, the same spacing as the two Cp rings in ferrocene.

This is very close to the 3.5 Å spacing seen between the two peptide strands in a β-sheet.[1] This led to the hypothesis that peptides appended to the two alkynes in **25** might form intramolecular hydrogen bonds and adopt a β-sheet conformation.

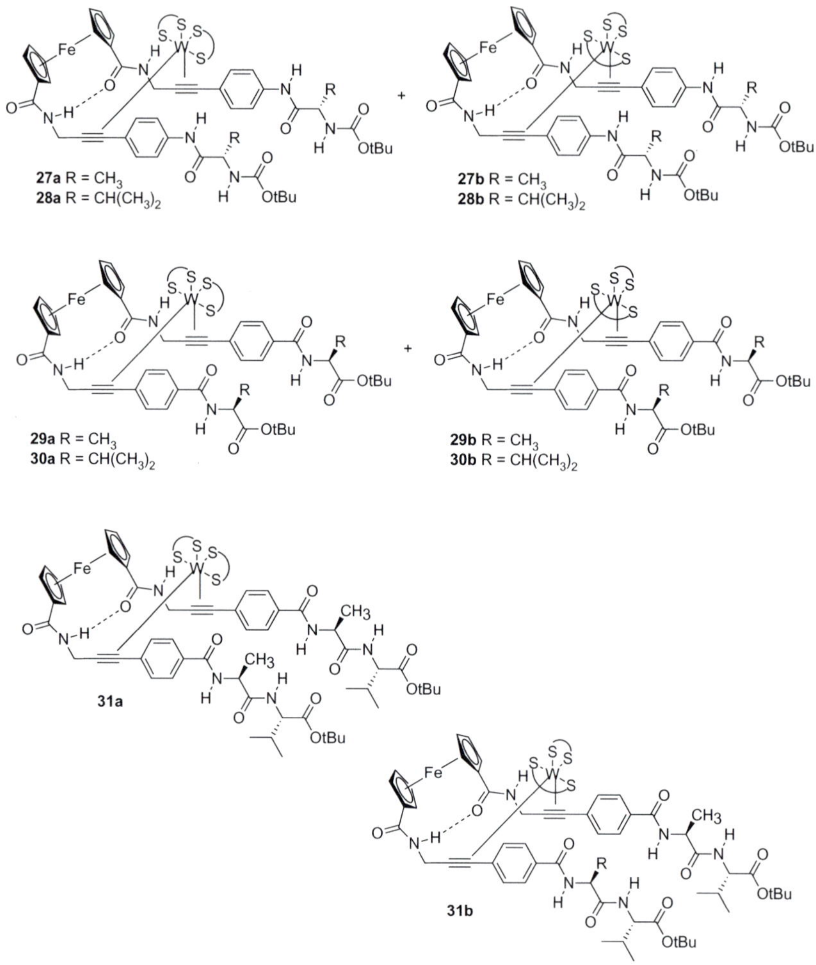

To test this hypothesis, the amino acid, and peptide derivatives **27–31** were prepared and examined.[31] These molecules were readily assembled using the Sonogashira reaction to attach peptide-derived iodoaniline or iodobenzoic acids to the parent ferrocenyldialkyne, followed by coordination of the tungsten to produce an inseparable mixture of two diastereomers (**a-b**). As usual the complexes were characterized by ESMS and NMR spectroscopy.

One question about these derivatives was whether appending groups to the two alkynes in **25** would disrupt the rigid ring system. It was discovered that the ring systems in **27–31** were identical to the ring system in **25**. The chemical shifts and coupling constants of the Cp, NH and methylene protons in these ring systems were nearly identical to those found in **25**. One notable difference, however, between **25** and **27–31** were the appearance of the dmtc methyl groups. With **25** the ^{1}H NMR spectrum shows four singlets for the four dmtc methyls as expected for the two enantiomers. In **27–31** there are eight singlets for the dmtc methyls, and this is consistent with these species being a mixture of two diastereomers. Although it is often true that diastereomers can be separated, this has not been the case so far with the diastereomers of **27–31**, which have co-eluted in thin layer chromatography, flash chromatography and analytical HPLC using either normal or reverse phase media. But the presence of the two diastereomers is discernable in the doubling of resonances in the NMR spectra.

To ascertain whether the appended amino acids and peptides in **27–31** form intramolecular hydrogen bonds, they were subjected to DMSO titrations.[31] In all cases, only one intramolecular hydrogen bond was detected, but it was from the NH embedded in the tungsten-ferrocene ring system. None of the amino acid or peptidyl NH protons in **27–31** were found to be involved in intramolecular hydrogen bonds. This suggests that the benzene rings situated between the tungsten-ferrocene ring system and the peptide might be aligned parallel to each other, and that parallel alignment keeps the two peptides stacked on top of each other, as opposed to being pointed at each other which is the orientation that would allow for intramolecular hydrogen-bonding between the two peptide strands. Efforts to confirm this using X-ray crystallography are being pursued.

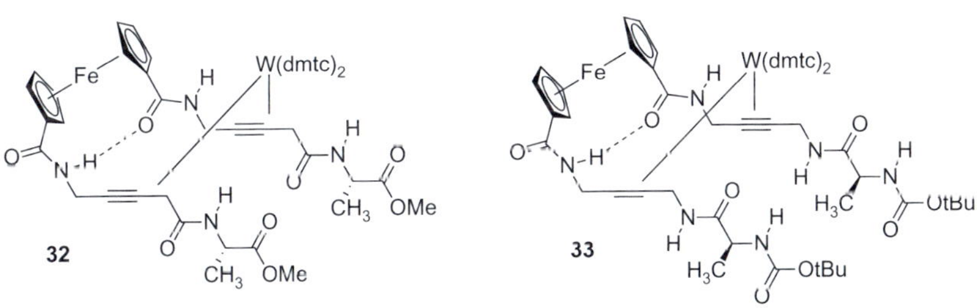

32 **33**

To give this system more conformational flexibility in order to allow for intramolecular hydrogen-bonding the benzene ring has been replaced by methylene groups. Currently under study are **32** and **33**.[32] In **32** the link between the peptide and the bimetallic ring system is a methylenecarboxyl, while in **33** the link is a methyleneamine. The carboxyl and amine groups provide a functional group that can be acylated by amino acids or peptides. Work to prepare these novel bioorganometallics and study their conformational behavior is in progress.

ACKNOWLEDGMENTS

This research was supported by grants from the United States National Science Foundation (CHE94-17783, CHE-0305325, CHE-0619275, CHE-0963165 and CHE-1464761), a Henry Dreyfus Teacher-Scholar Award, a Summer Undergraduate Research Fellowship from the Division of Organic Chemistry of the American Chemical Society, and the Trinity College Faculty Research Committee. The author is indebted to the undergraduate students from the College of the Holy Cross and Trinity College who have performed the majority of this work, and to Professors Richard Herrick and David Henderson for their assistance.

REFERENCES

1. Creighton TE. *Proteins: structures and molecular principles*. New York: W H. Freeman; 1984.
2. Wendt MD, editor. *Protein-protein interactions*. Berlin Heidelberg: Springer-Verlag; 2012.
3. Dömling A, editor. *Protein-protein interactions in drug discovery*. Wiley-VCH Verlag GmbH & Co. KGaA; 2013.
4. Manfredi R, Sabbatani S. A novel antiretroviral class (fusion inhibitors) in the management of HIV infection. Present features and future perspectives of enfuvirtide (T-20). *Curr. Med. Chem.* 2006;**13**:2369–84.
5. Yi HA, Fochtman BC, Rizzo RC, Jacobs A. Inhibition of HIV entry by targeting the envelope transmembrane subunitgp41. *Curr HIV Res* . 2016;**14**:283–94.
6. Biron Z, Khare S, Quadt SR, Hayek Y, Naider F, Anglister J. The 2F5 epitope Is helical in the HIV-1 entry inhibitor T-20. *Biochem* 2005;**44**:13602–11. Available from: https://doi.org/10.1021/bi0509245.
7. Walensky LD, Bird GH. Hydrocarbon-stapled peptides principles, practice, and progress. *J. Med. Chem.* 2014;**57**:6275–88. Available from: https://doi.org/10.1021/jm4011675.
8. Nowick JS. Exploring β-sheet structure and interactions with chemical model systems. *Acc. Chem. Res.* 2008;**41**:1319–30. Available from: https://doi.org/10.1021/ar800064f.
9. Tate DP, Augl JM. A Novel acetylenic. Complex of tungsten(0) carbonyl. *J Am Chem Soc* 1963;**85**:2174–5. Available from: https://doi.org/10.1021/ja00897a036.
10. Templeton JL. Four-electron alkyne ligands in molybdenum(II) and tungsten(II) complexes. In: Stone FGA, West R, editors. *Advances in Organometallic Chemistry*. Academic Press; 1989. p. 1–100. <https://doi.org/10.1016/S0065-3055(08)60352-4>.
11. Templeton JL, Herrick RS, Morrow JR. Electronic spectroscopy and electrochemistry of alkyne dithiocarbamate complexes of molybdenum(II) and tungsten(II). *Organome* 1984;**3**:535–41. Available from: https://doi.org/10.1021/om00082a004.
12. Burgmayer SJN, Templeton JL. Synthesis and structure of a seven-coordinate molybdenum carbonyl fluoride derivative: [Et4N][Mo(CO)2(S2CNEt2)2F]. *Inorg Chem* 1985;**24**:2224–30. Available from: https://doi.org/10.1021/ic00208a022.
13. Curran TP, Grant AL, Lucht RA, Carter JC, Affonso J. π-Ligands for generating transition metal – peptide complexes: coordination of amino acid derivatives to tungsten utilizing alkyne ligands. *Org Lett* 2002;**4**:2917–20. Available from: https://doi.org/10.1021/ol026298a.

14. Herrick RS, Templeton JL. Syntheses, spectral properties, and dynamic solution behavior of bis(alkyne)bis(dithiocarbamato)molybdenum(II) complexes. *Organome* 1982;**1**:842–51. Available from: https://doi.org/10.1021/om00066a015.
15. Hanessian S, Papeo G, Fettis K, Therrien E, Viet MTP. Synthesis of 310-Helix-inducing constrained analogues of l-proline. *J Org Chem* 2004;**69**:4891–9. Available from: https://doi.org/10.1021/jo0401422.
16. Maji SK, Haldar D, Bhattacharyya D, Banerjee A. Conformational heterogeneity of a turn mimetic pseudo-peptide: comparison of crystal state, solution and theoretically derived structures. *J Molecul Structure* 2003;**646**:111–23. Available from: https://doi.org/10.1016/S0022-2860(02)00619-1.
17. Vijayalakshmi S, Rao RB, Karle IL, Balaram P. Comparison of helix-stabilizing effects of alpha,alpha-dialkyl glycines with linear and cycloalkyl side chains. *Biopolymers* 2000;**53**:84–98 <https://doi.org/10.1002/(SICI)1097-0282(200001)53:1 < 84::AID-BIP8 > 3.0.CO;2-W>.
18. Jain RM, Rajashankar KR, Ramakumar S, Chauhan VS. First observation of left-handed helical conformation in a dehydro peptide containing two l-Val residues. Crystal and solution structure of Boc-l-Val-ΔPhe-ΔPhe-ΔPhe-l-Val-Ome. *J Am Chem Soc* 1997;**119**:3205–11. Available from: https://doi.org/10.1021/ja961460o.
19. Karle IL, Pramanik A, Banerjee A, Bhattacharjya S, Balaram P. ω-Amino acids in peptide design. Crystal structures and solution conformations of peptide helices containing a β-Alanyl-γ-Aminobutyryl Segment. *J Am Chem Soc* 1997;**119**:9087–95. Available from: https://doi.org/10.1021/ja970566w.
20. Curran TP, Marques KA, Silva MV. Bis(amino acid) derivatives of 1,4-diamino-2-butyne that adopt a C2-symmetric turn conformation. *Org Biomol Chem* 2005;**3**:4134–8. Available from: https://doi.org/10.1039/b508608f.
21. T.P. Curran, C. M. Warner, unpublished work.
22. Curran TP, Smith WE, Hendrickson PC. Conformational behavior of symmetrical and unsymmetrical mono(alkynylpeptide)-tungsten complexes. *J Organom Chem* 2012;**711**:15–24. Available from: https://doi.org/10.1016/j.jorganchem.2012.03.021.
23. Kemp DS, Li ZQ. 2-Amino-2′-carboxydiphenylacetylenes as β-turn mimetics. synthesis and conformational properties. *Tetrahedron Lett* 1995;**36**:4175–8. Available from: https://doi.org/10.1016/0040-4039(95)00775-8.
24. Kemp DS, Li ZQ. A short β-sheet containing proline nucleated by a 2,2′-substituted tolan β-turn mimetic. *Tetrahedron Lett* 1995;**36**:4179–80. Available from: https://doi.org/10.1016/0040-4039(95)00776-9.
25. Curran TP, Boynton AN, Berk SM, Pedro E-MC. Alkynyl β-sheet peptidomimetics retain their anti-parallel sheet conformation when coordinated to tungsten. *J Organome Chem* 2015;**782**:31–6. Available from: https://doi.org/10.1016/j.jorganchem.2014.08.004.
26. T.P. Curran, E.-M. Pedro, A. N. Boynton, unpublished work.
27. Curran TP, Yoon RSH, Volk BR. N-terminus to C-terminus metallacyclicpeptides employing tungsten–alkyne coordination. *J Organome Chem* 2004;**689**:4837–47. Available from: https://doi.org/10.1016/j.jorganchem.2004.09.062.
28. Curran TP, Lesser AB, Yoon RSH. Turn conformations in a metallacyclictripeptide and a metallacyclictetrapeptide induced by tungsten–alkyne coordination. *J Organome Chem* 2007;**692**:1243–54. Available from: https://doi.org/10.1016/j.jorganchem.2006.08.093.
29. Curran TP, McTeague TA, Nguyen VD, Yennie CJ, Handali PR, Sanderson-Brown JP, et al. Synthesis and conformational behavior of metallacyclicdipeptides derived from

coordination of side chain alkynylamino acids to tungsten. *J Organomet Chem* 2016;**806**:12–21. Available from: https://doi.org/10.1016/j.jorganchem.2016.01.023.

30. Curran TP, Lawrence AP, Murtaugh TS, Ji W, Pokharel N, Gober CB, et al. Conformationally rigid cyclic tungsten bis-alkyne complexes derived from 1,1′-dialkynylferrocenes. *J Organomet Chem* 2017;**846**:24–32. Available from: https://doi.org/10.1016/j.jorganchem.2017.05.048.
31. T.P. Curran, L. Davidson, N. Pokharel, C. B. Gober, M. Phillip, J. Frempong, unpublished work.
32. T.P. Curran, P. Handali, J. Sanderson-Brown, T. Nguyen, unpublished work.

Chapter 6

NCN-Pincer Organometallics in Bioorganometallic Chemistry

Gerard van Koten
Organic Chemistry and Catalysis, Faculty of Science, Utrecht University, Utrecht, The Netherlands

6.1 INTRODUCTION

The privileged ligand platform known as the "pincer"[1,2] has been the subject of active research in a variety of disciplines: these include, homogeneous catalysis (in particular in connection with (in)organic synthesis), the catalytic conversion of CO_2, and in the production of H_2, in materials science and biochemistry.[2–8] Pincer metal complexes are considered attractive because of their surprising thermal stability, ease of construction through their modular composition and the ample possibilities to tune their reactivity. In Fig. 6.1, a commonly used representation of pincer-metal complexes is shown with indications as to where changes in its basic pincer-platform structure have been made to allow for the tuning of their properties as (immobilized, molecularly enlarged, or dendritic) homogeneous catalysts,[9] sensor materials, or compounds with special photochemical or photophysical characteristics.[5]

The application of pincer-metal complexes as building blocks, active site or sensor materials, or as part of a larger biomolecule requires the availability of highly selective synthetic routes for both their synthesis and incorporation.[9] In practice, the introduction of the metal site in a pincer building block is often planned as the final stage of a complicated, multistep synthetic protocol. An alternate approach is, however, the pre-synthesis of the pincer-metal building block followed by a postmodification reaction that allows its incorporation, with the correct topology, into a larger carrier molecule, namely a large biomolecule. As the formation of the metal-carbon bond is usually the most difficult step, the availability of the postmodification route can be of paramount importance.[9,10]

The present contribution focusses on a description of research that was carried out in Van Koten's laboratory in the field of bioorganometallic

Advances in Bioorganometallic Chemistry. DOI: https://doi.org/10.1016/B978-0-12-814197-7.00006-6

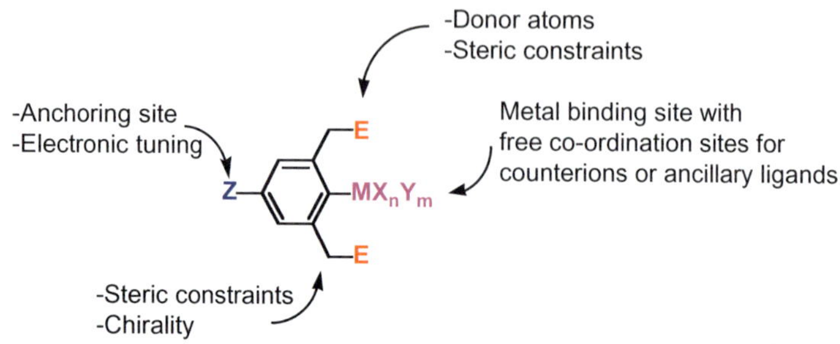

FIGURE 6.1 Showing the ECE-pincer aryl metal complexes and some of the possibilities to change the platform. It must be noted that only one of the many possibilities that compose a "pincer platform" is shown.[1,7] In the case of the NCN-pincer metal complexes, E = NMe_2. In this report M is a formal d^8-metal atom, mostly Pt^{II}, Pd^{II} or Ni^{II}, *cf.* Ref. 5.

chemistry with the aim to explore the use of pincer organometallics as functional building blocks. This chapter is an adaptation and overview of various compilations on this subject from his group and is, thus, not meant as a comprehensive review of this rapidly expanding area of pincer chemistry.[11]

The pincer platform of choice in this research has been the NCN-one, comprised of the central bis *ortho*-substituted aryl ring that is bonded via C_{ipso} to a metal center. The binding of both *ortho* amine substituents complements the potential κ^3-*N, C, N* tridentate-coordination motif by the pincer ligand and provides enhanced stability to the central C_{ipso}-metal bond. Our initial choice for the use of *ortho*-CH_2NMe_2 substituents emerged from our previous and ongoing organo-lithium, -copper, and -tin chemistry.[12–15] The N-atom is relatively small and the NMe_2 grouping is exclusively a σ-donor one which could be advantageous for the binding of a metal in redox processes involving higher formal oxidation states, e.g., Pt^{II} to Pt^{IV} or Pd^{II} to Pd^{IV}. The binding of the resulting NCN-pincer platform to the metal center was anticipated to result in complexes with a rigid and essentially planar *mer*-coordination geometry.[1] In reality, however, binding modes were found to be surprisingly diverse and versatile. Whereas the *mer*-coordination appeared as a predominant property of, for example, related PCP-pincer platforms, the NCN-pincer ligand showed, in addition, the existence of a variety of other, different bonding motifs. These include *C*-monodentate, *N, C*-bidentate, either *mer*- or *fac-N, C, N*-tridentate to even 2e–3c bonding of C_{ipso} to two metal centers, depending on, or more correctly, in concert with, the nature of the metal site with its ancillary ligands.[16] Later research by Roddick showed, however, that with proper choice of PR_2 donor sites (e.g., PMe_2, or $P(CF_3)_2$) that the PC(aryl)P-pincer ligand also displays some of the "alternate" coordination geometries encountered with the NCN pincer analogue.[17]

In the following section, the synthesis of an NCN pincer-metal building block as well as its functionalization, by either prior- or postmodification of the prefabricated NCN-pincer complex will be discussed. The examples will be mostly taken from our findings with NCN-pincer metal compounds in which the metal is a formal d^8 one and X is an anion or halido ligand. In particular, the numerous bio-organometallic applications involving $Pt^{II}X(NCN)$ are in essence only possible because of the remarkable stability of this arylplatinum(II) platform. This inertness to reactivity is displayed over a wide pH range, in air, and at elevated temperatures.

6.2 SYNTHESIS OF FUNCTIONAL NCN-PINCER METAL COMPLEXES AS BUILDING BLOCKS

6.2.1 Installing the C_{ipso}-Metal Bond

In the 1970s, our initial studies on the organometallic aspects of the NCN-pincer platform were primarily concentrated on the development of different routes for the synthesis of NCN-pincer aryl metal complexes. Direct metalation reactions involving C-H activation of pincer H-$C_{sp}{}^2$ arene or H-$C_{sp}{}^2$ alkene bonds with a suitable metal salt, as used by Shaw in his seminal studies on corresponding PCP-pincer ligands,[18] did not work out for the synthesis of our NCN-pincer aryl metal compounds. As we and others established in later studies, regio-selective C-H activation appeared only successful when both flanking substituents contain strongly coordinating (i.e., in both a σ- and a π-sense: e.g., P or S) donor atoms. In the case of bis *ortho*-amine or -imine groupings, this N-metal coordination is clearly too weak and led to the formation of mixtures of bis *ortho* (2,6-) and mono *ortho* (2,4-) regio-isomers. Thus, applicable routes for the required selective metalation of NCN-pincer arene starting materials had to be developed. These involved; (i) converse oxidative addition of a C_{ipso}-Br bond, (ii) prior bis *ortho* directed lithiation of a C_{ipso}-H bond in *apolar* solvents[19,20] followed by trans-metalation of the pure lithium reagent with a suitable metal salt, (iii) C_{ipso}-H/C'_{ipso}-M trans-cyclometallation,[21] and (iv) *ortho* directed electrophilic palladation of a C_{ipso}-$SiMe_3$ grouping.[22,23] For the synthesis of NCN-pincer ETM (early transition metal) complexes, transmetalation of the corresponding pincer gold(I) complexes with a suitable ETM salt was used in lieu of the corresponding lithium starting materials.[24] This was due to the fact that organo-Li reagents often react as reductants and, moreover, produce difficult-to-remove lithium salt side-products and/or early TM complexes incorporating Li materials. An obvious advantage of the pincer-gold/ETM-salt transmetallation route is the quantitative formation of the (pure) NCN-pincer ETM complex while the concomitantly-formed gold(I) metal salt could be easily recovered and thereafter re-used.[24]

6.2.2 PostModification of Pincer Platinum Compounds[9,25,26]

Use of bis(*p*-tolyl)bis(diethylsulfide)platinum(II) (the *Canty* reagent) allowed for the chemo-selective conversion of *para*-iodo-NCN-pincer bromide

1 into corresponding *para*-iodo-NCN-pincer platinum bromide **2**. It is interesting to note that the reaction with $Pt(PPh_3)_4$ leads to the chemo-selective formation of the *para*-$PtI(PPh_3)_2$-substituted product **3 only**, *i.e.*, the product arising from the alternate, regio-selective attack of the *para*-C-I bond (Scheme 6.1).

The *para*-iodo-NCN-pincer platinum iodo compound **4** (Scheme 6.2), analogue of **2** prepared by a selective Br/I-exchange reaction, appeared to be a crucial and versatile starting material for a range of postmodification reactions following straightforward organic synthetic routes (Scheme 6.2: for

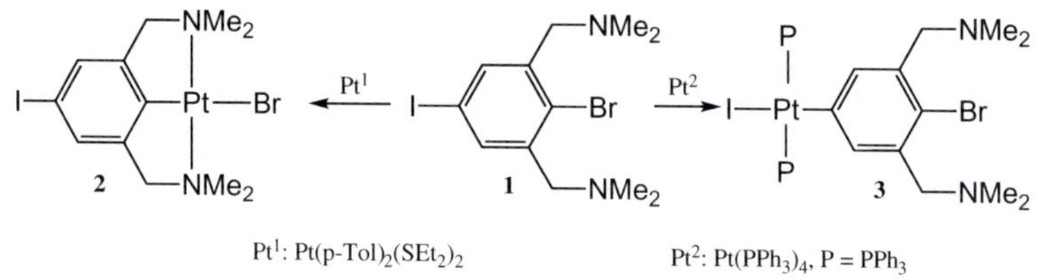

SCHEME 6.1 Chemo-selective postmodification reactions addressing either I-C_{para} or Br-C_{ipso} bond, respectively.[9]

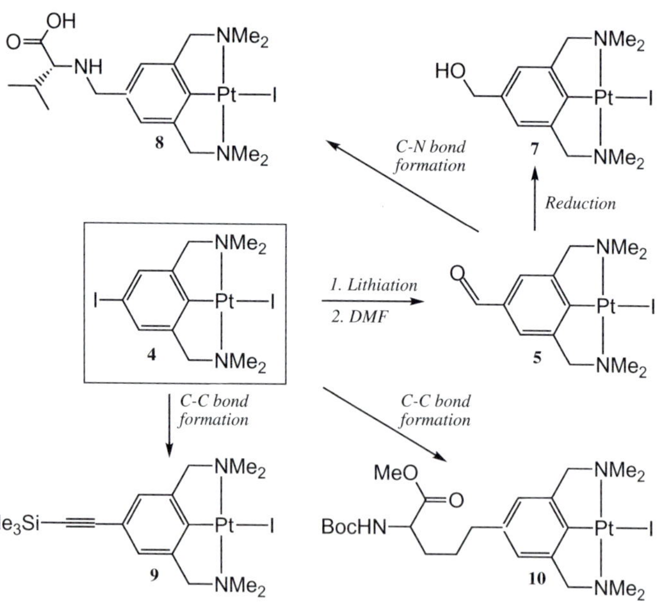

SCHEME 6.2 Examples of postmodification of the organometallic diiodide NCN-pincer compound **4** using common conversions in organic synthesis. Key is the selective conversion of **4** into the lithio-platinum intermediate **4-Li** with retention of the C_{ipso}-Pt bond. Quantitative formation of the *para*-trimethylsilyl derivative **6** (not shown) is direct evidence for the existence of the **4-Li** intermediate.[9]

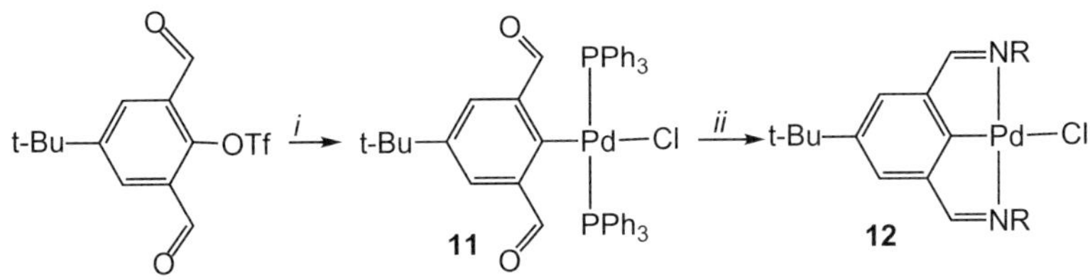

SCHEME 6.3 An early example of the synthesis of a NCN-pincer metal complex **12** via the "ligand-introduction" strategy; *i*, $Pd_2(dba)_3CHCl_3$, PPh_3, LiCl and *ii*, RNH_2, MeCN, reflux, under O_2 atm.[27]

details see Ref. 9). It was found that many of these reactions could be extended to the corresponding nickel and palladium complexes.

A key-step in this scheme is the quantitative formation of the *para*-lithio-NCN-pincer platinum iodido intermediate **4-Li** with retention of the C_{ipso}-Pt bond. Quenching reaction of the lithio-platinum intermediate with DMF or Me_3SiCl afforded the corresponding *para*-substituted derivatives such as aldehyde **5** and trimethylsilyl-containing **6** (not shown).

An alternate approach is the so-called "ligand-introduction strategy" which was first presented in work by Uozumi.[27] In this case, synthesis of the NCN-metal complex involves prior palladation of a functionalized aromatic ring that already contains two substituents in the *meta* position. In a subsequent step, both *meta* groups are converted into the projected bis *ortho* coordinating groups. A seminal example is shown in Scheme 6.3, in which the two formyl groups *ortho* to the C_{ipso}-Pd bond in **11** are converted into the *N*-coordinating imino groups in **12**.

6.2.3 Reactions of NCN-Pincer Metal Compounds with Electrophiles

The outcome of early work involving reactions of NCN-Pincer platinum complexes with electrophiles had already pointed to the surprising stability of the C_{ipso}-Metal bond. Reactions of parent NCN-pincer platinum chloride **13** with Cl_2 (or $Cu^{II}Cl_2$) afforded the corresponding platinum(IV) trichlorido compound rather than products originating from oxidative cleavage of the C_{ipso}-platinum(II) bond.[28] The same type of selectivity is observed in the reaction of the NCN-pincer platinum bromide with Br_2 producing the corresponding platinum(IV) tribromido. The reaction with I_2, however, afforded rather than the oxidative addition product, very stable, violet colored **14**, PtI (NCN)(η^1-I_2).[29–31] The square pyramidal structure of **14** features an end-on bonded I_2-molecule to the platinum(II) center; this (Lewis base-Lewis acid) donor-acceptor Pt-to-I_2 type of interaction involves binding of the formal (filled and hence donor) Pt d_{z^2} orbital to an empty (acceptor) σ^* orbital of I_2. For an extensive discussion of this first example of stable Pt(II)-electrophile

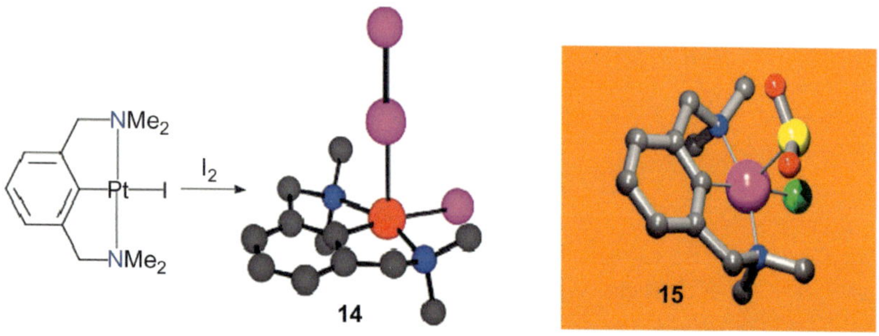

SCHEME 6.4 See also Refs. 29–31 Formation of **14**, PtI(NCN)(η^1-I_2) (reverse reaction occurs in the presence of Lewis base, e.g. NEt_3) and molecular structure in the solid state of **14** (lengthening of the I–I distance to 2.822(1) Å, *cf.* 2.715(6) Å in solid I_2, confirms donation of charge from the filled donor Pt $d_z{}^2$ into the empty σ^* of the end-on bonded I_2. Binding of SO_2 to PtI (NCN) yielding **15**, Pt^{II}I (NCN)(η^1-SO_2), is highly reversible (molecular structure of **15** in the solid state is shown; Pt-S 2.479(1) Å, Pt-S-O (av.) 104°).[32–34]

complexes, see Ref. 31. Further studies of reactions of PtI(NCN)(η^1-I_2) with Lewis bases revealed that the binding of I_2 is indeed reversible; on reaction of **14** with Lewis bases, e.g., trialkylamines, parent PtI(NCN) was reformed quantitatively (*vide infra*).[30] Applications of such chemistry in a bio-organometallic sense will be detailed below.

$$Pt^{II}X(NCN) + X_2 \rightarrow Pt^{IV}X_3(NCN)\ (X = Cl \text{ or } Br)$$

Parent NCN-pincer platinum iodide reacted both in solution and as a solid reversibly with (gaseous) SO_2 to yield Pt^{II}I(NCN)(η^1-SO_2), **15**. NMR spectroscopic and computational studies showed that, also in this case, that $Pt^{II} \rightarrow SO_2$ binding occurs via interaction of the formally filled donor Pt $d_z{}^2$ into the empty σ^* orbital on sulfur see Scheme 6.4.[32–34]

Notably, the formation of both **14** and **15** is highly selective, reversible and thus leaves the parent compound unaffected. Together with the fact that both complexes display deep, characteristic colors (violet and orange, respectively) made them ideal candidates to be used as sensory building blocks in bio-materials (*vide infra* and Refs. 32–34 for further details).

6.3 BIOMOLECULAR APPLICATIONS OF NCN-PINCER PLATINUM(II) BUILDING BLOCKS

Organometallic complexes attached to biomolecules have been successfully used in bio-assays, e.g., in immunoassays,[35] for studies of their antibacterial and antiproliferative properties[36] or as structural probes, cf. the use ferrocene amino acids as “turn inducers”.[37–40] We aimed to study and apply the Pt (NCN) group in various avenues of investigation within the sphere of bio-organometallic chemistry as the platinum fragment is small in size and contains Pt as a detectable heavy atom that, moreover, is also sufficiently NMR active (^{195}Pt: $I = ½$; natural abundance 33.8%).

6.3.1 An Example of Postmodification; Synthesis of "*Pincercifen*", an Organometallic Analogue of *Tamoxifen*

We were inspired by the pioneering work of Jaouen et al., who developed organometallic *Tamoxifen* derivatives that displayed a good cytotoxic and anti-estrogenic profile.[41] These include *Ferrocifen* **16a** and *hydroxy-Ferrocifen* **16b**. We thus set out to prepare organometallic analogues of **16** in which the ferrocene moiety had been replaced by a PtX(NCN)-type functional unit. It is of interest to note that this organometallic building block replaces the phenyl group that originally is present in *hydroxy-Tamoxifen*. This latter compound is a well-known anti-tumor agent used in the treatment of breast cancer. Whereas the cytotoxicity of *Ferrocifen* had been attributed to the redox properties of the ferrocenyl group, researchers also found that the binding of **16b** to the estrogen receptor protein was slightly lower when compared with *Tamoxifen* itself. It was inferred that the decreased binding might be due to the increased steric bulk of the Cp_2Fe unit when compared to the aforementioned phenyl ring. The group of Jaouen also developed *Tamoxifen* derivatives in which the phenyl group was replaced by various oxaliplatin-groupings that, however, being simple coordination compounds, are more prone to non-selective decomposition under physiological conditions.[41] We anticipated that using the organometallic PtX(NCN)-grouping with its proven kinetic stability under physiological conditions, would provide an interesting, alternate opportunity and perhaps displaying less toxic side effects during transport and delivery towards the target cells.

Synthesis of *pincercifen* **17**, having the *Z*-phenyl ring of *Tamoxifen* replaced by an NCN-pincer platinum entity was realized via a postmodification reaction of [PtCl(NCN-C(=O)Et-4)] with the appropriate diaryl ketone using a McMurry hetero-coupling reaction.[42] A series of *para*-substituted NCN-pincer platinum complexes [PtX(NCN-R-4)] (R = C(=O)H, C(=O)Me and C(=O)Et), X = Cl or Br) was synthesized, *cf.* Scheme 6.2. Two *pincercifen* derivatives, **17** and pivaloyl protected *hydroxy-pincercifen* **18** were prepared, see Scheme 6.5. The latter complex is the analogue of *hydroxy-Tamoxifen* functionalized with a pivaloyl protecting group. Notably *hydroxy-Tamoxifen* is the metabolite of *Tamoxifen* and is considered actually to be responsible for the observed anti-cancer activity. For *pincercifen* **17**, it was possible to separate the geometric *E/Z*-isomers of which the *Z*-isomer was analyzed using single crystal X-ray diffraction and revealed similar structural properties as tamoxifen analogues, *e.g.* **16** and *Tamoxifen* itself.

Up to now, the binding affinity of **17**-*Z* and **18**-*E* to the target molecules has not been tested. The steric differences of between **17** and **18**, on the one hand, and *hydroxy-Tamoxifen*, on the other hand, can be expected to be small

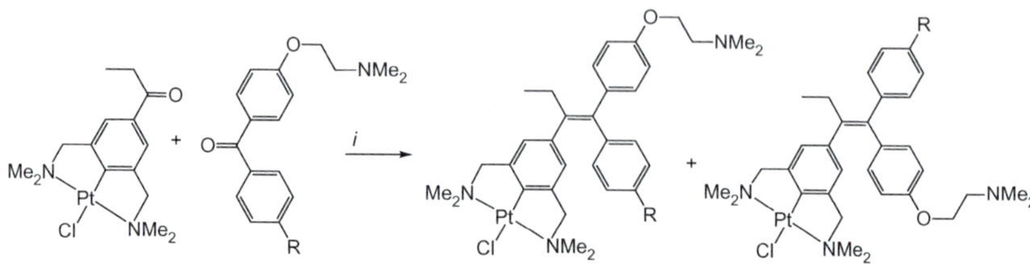

SCHEME 6.5 Synthesis of *E/Z* complexes **17**-*E* and -*Z*; (i) TiCl4(THF)2, Zn(0), THF, reflux, 16 h; (ii) H2O, K2CO3 (for **17** (*E/Z*) 36%).[42]

(when compared with *Ferrocifen*) because of the similar size and the planarity of the NCN-pincer platinum head group. Earlier studies clearly demonstrated that the Pt-halido head group is particularly well-suited for non-covalent hydrogen bonding with a variety of X-H functionalities, e.g., X is RO- and RCC-.[43] Note that a cationic NCN-pincer platinum head group is an excellent binding site itself for biomolecules containing nitrogen or sulfur donor groupings, e.g., peptides (vide infra).[44,45]

6.3.2 NCN-Pincer Platinum(II) Groupings as Organometallic Peptide Labels and Sensors

Using the postmodification approach (*vide supra*, see Scheme 6.2), NCN-pincer platinum units covalently bonded to the N-, and C-termini as well as to the α-C atom of different amino acids were prepared.[9,46] During these preparations, the covalent C_{ipso}-Pt bond remained, as predicted, unaffected. Subsequently, the reversible complexation formation of the various compounds in a SO_2 atmosphere was also tested.[5,30] It was found that instantaneous, strong, diagnostic coloration even for the de-protected amino acid groupings, was observed (see Scheme 6.6), i.e. the formation of the 5-coordinate $Pt^{II}Br(NCN)(\eta^1\text{-}SO_2)$ complex is immediate and completely reversible upon removal of the SO_2 atmosphere. This system affords detection limits in the ppm range. Further studies revealed that the SO_2 complexation process appeared insensitive to other atmospheric gases (e.g., O_2, CO, HCl, and humidity).[32] As SO_2 is a physiologically important gas,[50] the use of these very stable NCN-pincer platinum(II) halido labels, having a small size, as sensitive sensors in biochemical and medicinal applications is open for further studies, both in vitro and in vivo.

The abovementioned experimental observations pointed to the strong potential for the NCN-pincer platinum unit as a biomarker for peptides.[51] The feasibility of this approach was tested for the diagnostic coloration of

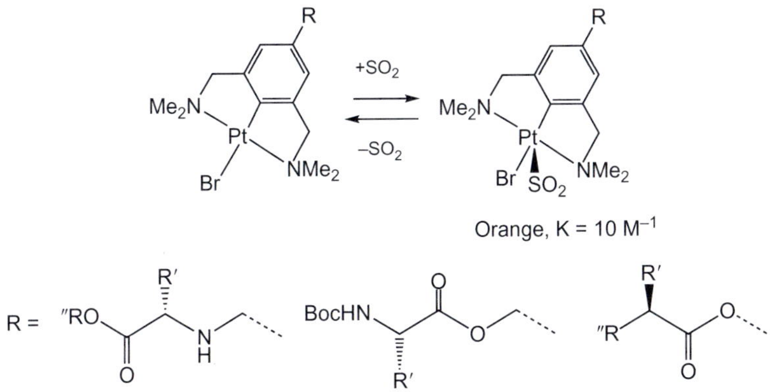

R′ (R″) =

19 iPr (Me)
20 iPr (H)
21 Bz (Me)
22 Bz (H)
23 CH_2COOH (Me)
24 Bz ($NHCiPrCO_2Me$)

25 iPr
26 Bz
27 H
28 CH_2COOH

29 iPr (NHBoc)
30 Bz (NHBoc)
31 iPr (NH_3Br)

SCHEME 6.6 Some examples of enantio-pure NCN-pincer platinum bromide complexes each having an amino acid (ester) *para*-substituent; **19–24** at the N-terminus,[47,48] **25–31** at the ester-O or CH_2O-terminus.[48,49] See also Scheme 6.2 for another example, **10**, in which the NCN pincer platinum unit is bonded at the α-C-terminus of the amino acid. Characteristic ^{195}Pt NMR data for both the pincer (di)protected amino acids and their corresponding SO_2 complexes were measured (see data listed in Table 1 of Ref. 11).

the marked peptide material with both gaseous SO_2 and with either I_2-vapor or with a solution of KI_3.

Solid-phase peptide synthesis on resin of the oligopeptides **32**-G-P-P-F-P and **33**-X-G-P-P-F-P-F (Scheme 6.7) was achieved with the de-protected NCN-pincer labels **32** and **33**, respectively. In both cases the characteristic, reversible orange coloration upon exposure of the loaded resins with an SO_2-containing atmosphere was observed. An alternate, mild coloration method involves the treatment of the loaded resins **32**-G-P-P-F-P and **33**-X-G-P-P-F-P-F with aqueous solutions of KI_3. A persistent violet/blackish coloration of the beads due to the formation of the Pt-η^1-I_2 binding of the label was observed that disappeared upon treatment of the beads with a solution of NEt_3 in DMF. Marker **33**, which lacks the labile sec-amine grouping present in **32**, appeared more stable towards acidic de-protection protocols and was preferred in subsequent studies. Consequently, the whole cycle of removal of the side chain protecting groups, photo-cleavage and analysis of the pincer-metal labeled peptide could be studied in detail. Investigation of the sensitivity of I_2-marking process showed that a low 6%-capping degree of the amine-termini with **33** was sufficient to induce visible coloration of the loaded resin (as observed visually with the human eye).[51]

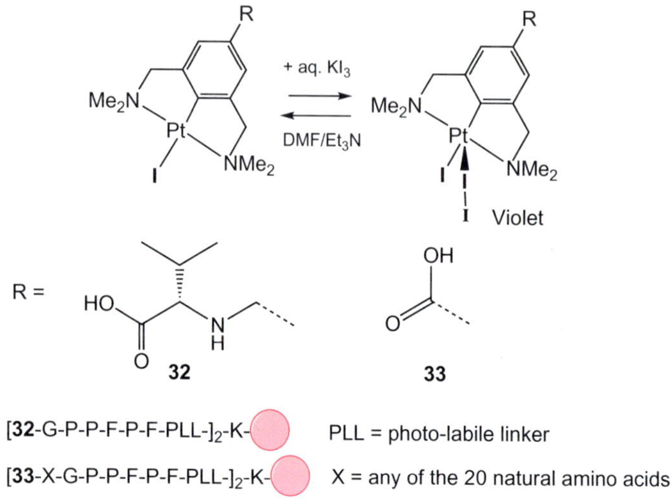

SCHEME 6.7 Reversible coloration of NCN-pincer platinum labels by KI_3. Proof-of-principle of the use of the solid-phase biomarkers **32** and **33** attached onto oligopeptides.[51]

6.3.3 Study of Carbohydrate-Protein Interactions using NCN-pincer Platinum Complexes as SPR Enhancers

Surface Plasmon Resonance (SPR) has proven to be an excellent technique for both the qualitative and quantitative study of, among others, carbohydrate-carbohydrate and carbohydrate-protein interactions.[52,53] As the SPR response is proportional to the accumulation of mass on the sensor surface, a serious constraint of SPR deals with the dimensions of the analyte-molecules. This situation is particularly true for the study of weak carbohydrate-protein interactions. At the time that we started our collaborative study with the group of Kamerling,[54] the poor availability of high-molecular mass oligosaccharides hampered the wide applicability of the SPR technique. To this end, a series of low molecular mass mono- and disaccharides carbohydrates labeled with the NCN-pincer platinum chlorido grouping (combining a small molecular size with the presence of a heavy Pt atom) were synthesized using our postmodification technique, see Scheme 6.8. These materials were used to assay the carbohydrate–protein interactions with the immobilized lectins RCA_{120} and ConA, being standard proteins for the study of carbohydrate-protein interactions.[54]

These studies revealed that the specific binding of the NCN-pincer platinum unit labeled saccharides **34–38** led to a strong, detectable SPR signal enhancement as compared to the pincer-free saccharides, even at low analyte concentrations (9 μM). Control studies confirmed that signal enhancement occurred exclusively upon incorporation of the NCN-pincer platinum(II) label which proves that the sensitivity increase is only due to the heavy atom

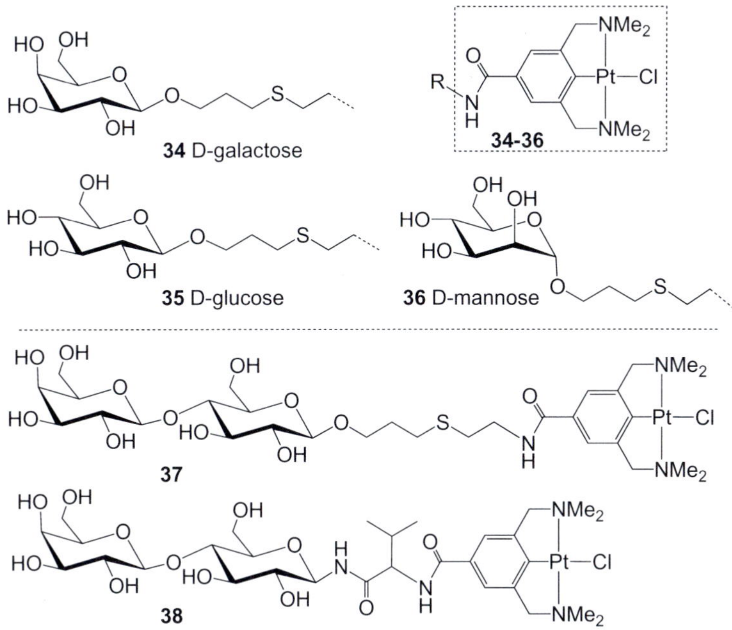

SCHEME 6.8 Various NCN-pincer unit labeled carbohydrates that were synthesized via post-modification of parent NCN-pincer platinum chloride. These were tested as SPR sensitizers in carbohydrate-protein interaction studies.[54]

effect of the platinum(II) ion. It is believed that, in addition to the mass increase effect by the Pt(II) ion, that a significant additional interaction between the platinum electrons and the evanescent wave, produced in proximity of the sensor-chip surface, is responsible for the observed phenomenon.[54] Moreover, variation of the linker length between the NCN-pincer platinum label and the carbohydrate, as for labeled **37** and **38** (Fig. 4), did not influence the response significantly.

6.4 ECE-PINCER METAL COMPLEXES AND PROTEINS

6.4.1 NCN-Pincer Palladium Catalyst Connected to Mono- and Di-Peptides

Transition metal complexes comprised of peptides with attached phosphine ligands have been successfully explored for applications as homogeneous catalysts in asymmetric hydrogenation, alkylation and allylation.[56–58] It was anticipated that a peptide chain connected to a catalytically active ECE-pincer metal grouping could generated the proper chiral environment to the metal site thus leading to stereoselective metal-catalyzed conversions between substrate and reagent. As a proof-of-principle, a small series of *para*-mono- and di-peptide substituted NCN-pincer palladium bromide compounds were synthesized using the alternate approach, i.e., first the labeled NCN-pincer ligand was synthesized followed by bis *ortho* palladation with $Pd_2(dba)_3(CHCl_3)$. The resulting peptide palladium bromido compounds were converted into the corresponding cationic NCN-pincer palladium pre-catalysts which then were tested in the aldol condensation reaction of methyl cyano acetate and benzyl aldehyde to give *cis*/*trans* mixtures of the resulting bis-substituted oxazolines (*i.e.*, 4,5-dihydro-2-oxazoles) see Scheme 6.9.[55]

Like the postmodified platinum derivative, these Pd-compounds appeared to have surprising stability. For example, deprotection of **39** and **43** with LiOH could be achieved without affecting the C_{ipso}-Pd bond. The catalytic results revealed that for each of the pre-catalysts **39–42** better than 94% conversion to the oxazoline products was observed with a *trans*/*cis* ratio (about 65:35) that was similar to that of the parent NCN-pincer palladium cationic catalyst. It is obvious that given the short lengths of the peptide substituent, comprising just monopeptides or dipeptides, the creation of a chiral micro-environment (i.e., second coordination sphere) around the metal site could not be achieved. However, on the other hand, these experiments demonstrated that even the presence of a free carboxylic acid group in the catalyst's structure, as in **40**, did not affected the outcome of the catalysis.

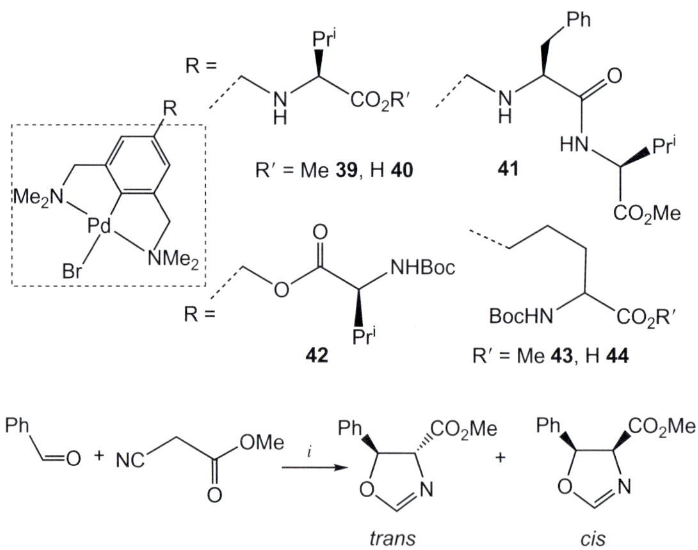

SCHEME 6.9 NCN-pincer palladium bromides with different monopeptides and dipeptides which were used as pre-catalysts in the aldol condensation reaction of benzaldehyde with methyl cyano acetate: reaction conditions, *i*, **39–42** (1 mol%)/$AgBF_4$, i-Pr_2EtN (10 mol%) in CH_2Cl_2 at room temperature (RT). AgBr was removed from the reaction solution prior to catalysis.[55]

6.4.2 Hybrid Catalysts: ECE-Pincer Metal Catalysts Covalently Bonded to Proteins

Rather than continue our research with smaller peptides and carbohydrates, we started to study the modification of ECE-pincer metal complexes with larger proteins. This project was connected to our interest, at that time, with the development of molecularly enlarged homogeneous catalysts for organic synthesis that after use could be readily separated from the product solution by filtration using nano-membrane technologies.[59] Eventually, we could demonstrate that these catalysts could be isolated and reused or even applied, for example, in a membrane reactor under continuous flow conditions. To this end, we designed and synthesized various nano-sized carrier molecules decorated with a number of ECE-pincer metal catalysts: two examples are show in Fig. 6.2.[61] They have sizes differences: About 2.5 nm for the carbosilane CS-$(NCNNiCl)_{12}$ species (**45**),[60] carrying twelve NCNNiCl catalysts, and about 1.7 nm for the cartwheel type molecule, **46**.[61] The latter species, C_6-$(SCSPdCl)_6$, has a central arene ring with six SCSPdCl catalysts. An obvious difference between these two types of species is their molecular

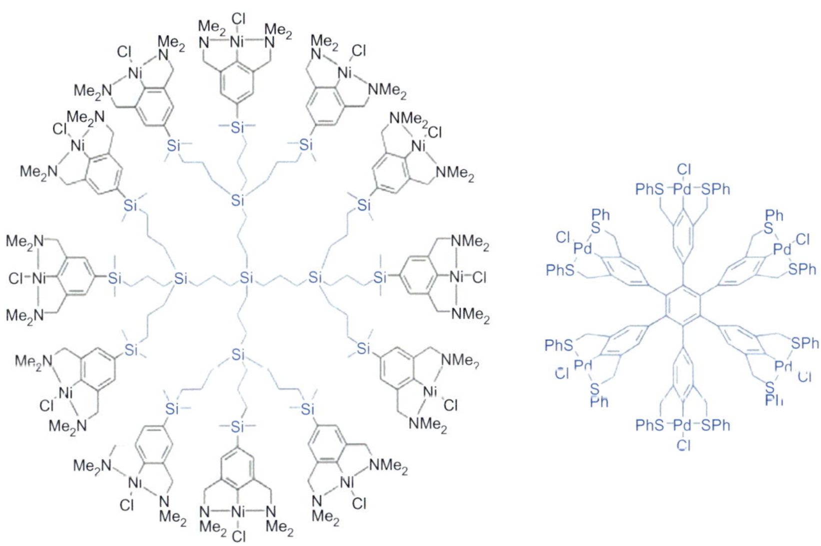

FIGURE 6.2 Two examples of molecularly enlarged catalytic species. A pronounced molecular flexibility of the nano-sized catalyst (*cf.* **45**, left hand structure, with that of the structure of **46** at the right hand) will negatively affect its retention by the membrane in a membrane reactor, by enhanced leakage of the polymeric catalyst species through membrane pores and defaults. Note that narrowing the pores of the membrane would lower the throughput of the membrane reactor. Moreover, insufficient separation of the catalytic sites could lead to mutual deactivation of neighboring sites resulting in a lower overall catalytic efficiency of the nano-sized species, a so-called *negative* dendritic effect.[60] The cartwheel type of species, *cf.* **46** and **47**, has a rigid structure resulting in significantly more isolated catalytic sites.[61]

flexibility. The polymer-type structure of CS-$(NCNNiCl)_{12}$ is extremely flexible with a hydrodynamic volume that will depend on the polarity of the solvent. The structure of the cartwheel species **46** was designed to be rigid so that its size would be independent of the nature of the solvent. The latter two features appeared favorable for attaining the required, high retention numbers ($\sim$100%) for reactions performed in a membrane reactor. That kind of retention could be achieved, for example, by extending the C_6-arene core of **46** with another layer of six benzene rings with each benzene ring itself being meta-substituted with two mono-cationic SCS-pincer palladium catalyst, i.e., $[C_6[C_6H_3(\text{-SCSPd(solvent)})_2\text{-}3,5]_6]^{12+}$, **47**. The nature of the disk-type appearance of the dodeca-cationic species **47**, amounting to a 3.2 nm diameter and 2.4 nm thickness, resulted in continuous retention of $>$99.9% using a MPF-60 nano-filter in a membrane flow reactor. The double Michael addition (Eqn. 6.1) could be run for more than 25 hours under continuous catalysis conditions using **47** as catalyst showing what is typically termed a *positive* "dendritic" effect (for details see Ref. 60).

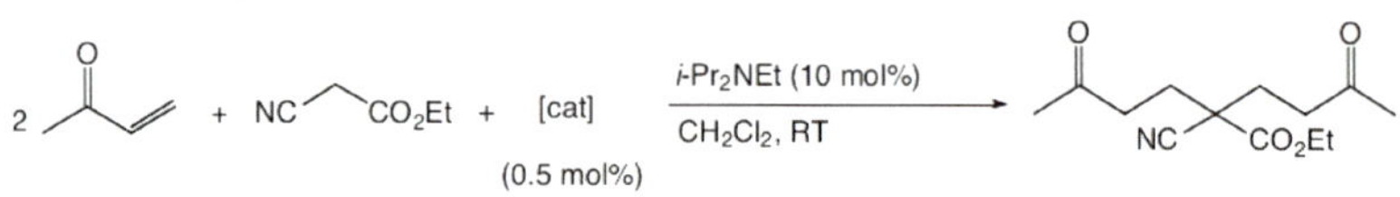

(6.1)

An alternate approach for the molecularly enlarged catalyst species discussed above, is to use a protein as the nano-sized carrier and to use its active site for the directed anchoring of a man-made homogeneous metal catalyst. In a seminal report, Wilson and Whitesides demonstrated Rh-mediated hydrogenation catalysis using a biotin-derived diphosphine ligand (i.e., N, N-bis(2-diphenylphosphinoethyl)biotinamide) which is anchored, via the biotin unit, to the protein *Avidin* by a strong non-covalent interaction.[62] The resulting hybrid catalyst could be then used for H_2 addition to α-acetamidoacrylic acid. Full conversion to N-acetylalanine was obtained displaying modest stereo-selectivity. Since then modification of proteins with transition metal complexes have become an important subject of study. Further examples of the chemogenetically improved biotin-(strept)avindin system were successfully explored by Ward[63–66] and co-workers while the group of Watanabe studied *apo*-myoglobin that was modified through coordination by, e.g., Salen complexes.[67] Roelfes and Feringa developed a very efficient system based on double-stranded DNA which was used as asymmetric scaffold for transition metal catalyzed Diels-Alder and Michael addition reactions in a stereo-selective manner (Fig 6.3).[68]

In a collaborative project with Egmond and co-workers,[69–73] we decided to address a whole protein class, i.e., the serine hydrolases, which are a large and well-studied class of proteins. Serine hydrolases have as a common feature consisting of their irreversible and covalent inhibition by phosphonate

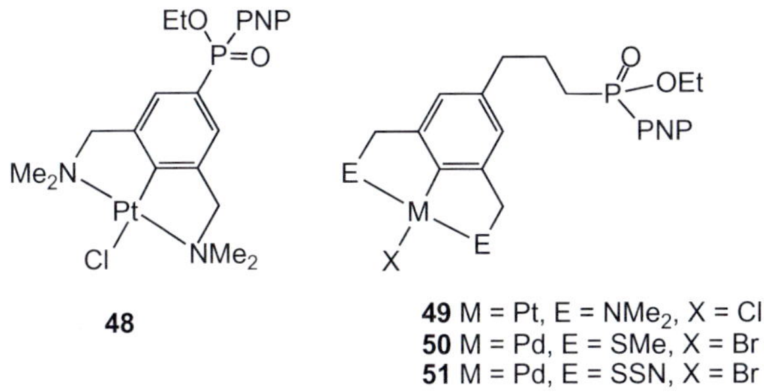

FIGURE 6.3 Various phosphinate inhibitors with different tether lengths (none: **48** and C_3H_6: **49–51**). PNP is *para*-nitrophenolate and $[PNP]^-$ is yellow colored *para*-nitrophenolate anion; its formation during the inhibition (see Scheme 6.10) can be monitored by UV-vis spectroscopy). Note that the inhibitors **48–51** are racemic, *i.e.*, 1:1 mixtures of the R_P- and S_P-enantiomers.[69]

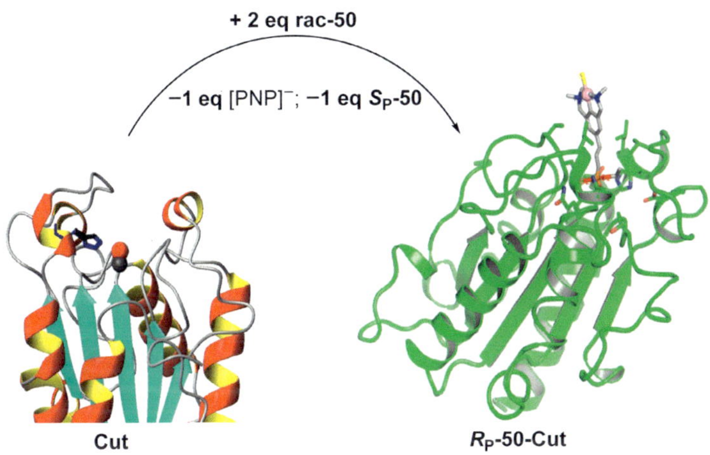

SCHEME 6.10 Irreversible inhibition of a serine hydrolase (molecular structure around the serine active site is shown) by phosphonate inhibitor **rac-50** with a C_3 tether leading to covalently anchored SCS-pincer palladium bromide species ***R*****P-50-Cut**. Structure of the SCS-pincer metal protein hybrid is shown; inhibited P center has the R_P-configuration.[73]

inhibitors. The mechanism of the resulting covalent anchoring of the phosphorus atom of the phosphonate by reaction of the catalytically active serine residue and a phosphonate, e.g., $RP(O)(OEt)_2$ is well-studied.[70] We argued that by using phosphonates with R containing a catalytically active unit, in our case an ECE-pincer metal moiety, a novel type of molecularly enlarged ECE-pincer metal protein hybrid catalyst, possibly with special properties would become available. This active site directed approach is attractive for a number of reasons. Firstly, the covalent binding of the man-made catalyst to the protein provides the stability to the hybrid catalyst assembly that would be required for its application in a membrane reactor. This is particularly true under continuous flow conditions provided that, in combination, robust catalytic units are used. Secondly, as the serine hydrolases are water soluble the hybrid catalyst will also be aqueous soluble, allowing catalysis in a benign, and environmentally friendly reaction medium. Thirdly, through the covalent anchoring, the protein gets, in a sense, a new catalytic site. Depending on the length of the tether connecting the inhibited phosphorus atom and the metal site the catalytically active grouping will be drawn near the surface of, or perhaps even becoming encapsulated within, the chiral protein matrix (e.g., the former active site). This situation could create an asymmetric, second coordination sphere around the anchored catalyst. Ideally, the asymmetric environment would transfer its chiral information to processes that take place at the embedded catalyst.

As a proof-of-concept, a series of ECE-pincer metal phosphonate inhibitors, examples are **48–51**, were synthesized partly by applying the postmodification approach, see Figure 6.3.[69]

For our study, we selected *Cutinase* (Cut) which is a relatively small lipase (21 kDa) and one that is stable and highly active and lacks the "lid," present in other lipases, which covers the hydrophobic active site. These characteristics make Cut an ideal model enzyme for inhibition studies. The inhibition reactions were carried out with a 1:2 molar ratio of Cut to phosphonate inhibitor (racemic) and was monitored by UV-vis spectroscopy.[70] In Scheme 6.10, this is shown for the enantioselective reaction of Cut with **rac-50**: during the inhibition process water-insoluble **rac-50** was converted into the water-soluble R_P-**50-Cut**. By dialysis, the latter product was purified and separated from its, non-inhibited, *S*-**50** enantiomer. R_P-**50-Cut** was identified by mass spectrometry (m/z 21128.1 vs. wild type cutinase: m/z 20603.9 $\pm$ 0.2, calc. 20604) while its molecular structure and the R_P-configuration in the solid state[73] could be unambiguously be established by an X-ray structure determination. It must be noted that the presence of the NCN-pincer Pt group further functions as a site-selective tool for the phasing of the raw protein diffraction data. This aspect of protein crystallography is of obvious future releveance.

Similarly, the inhibition reaction of phosphonate **49** with Cut was studied. Elucidation of the molecular structure of the inhibited material in the solid state surprisingly revealed that the isolated hybrid was the S_P-**49-Cut** one, i.e., in this case it was the S_P-phosphonate that was the faster reacting enantiomer.[73] An overlay of the two structures showed that the pincer head groups which have different sizes and polarities, i.e., NCN-PtCl, **49** versus SCS-PdBr, **50**, occupy different pockets in the "mouth area" of Cut. Preliminary kinetic studies (solution: ^{31}P NMR) suggested that in spite of the relatively long C_3-tether, the subtle differences of the SCS-pincer and NCN-pincer units affect the inhibition reaction of the respective P centers of the phosphonate units. However, these results did not necessarily provide information to help explain the different stereo-chemical outcome of the aforementioned inhibition reactions. However, these studies did reveal also that the length of the tether between the pincer unit and the P(phosphonate) center was of importance for the rate of the inhibition process. Full inhibition of the **48** when compared with **49**, which have the NCN-pincer head group in common, but differ with respect to the length of the tether, was about 500 times slower.[74] This likely reflects the fact that, for inhibition of **48**, the NCN-pincer head group has to be drawn more deeply within the molecular structure of the protein.

6.4.3 Application of ECE-Pincer-Cutinase Hybrid Systems in Coordination Chemistry, Catalysis and in Sensoring

Having well-defined hybrid species like S_P-**49-Cut** in hand, we set out to study the coordination properties of the embedded pincer metal halido species. An interesting example is shown in Scheme 6.11. It was known that monocationic, dimeric $[(\text{NCN-pincerPd})_2\text{-}\mu\text{-Cl}]^+$ $[52]^+$, (counter-anion has to be non-nucleophilic, *e.g.*, BF_4^-), is formed by reaction of NCN-pincer palladium with the corresponding NCN-pincer palladium cation: i.e., it is a single μ-Cl^- bridge which binds between two NCN-pincer palladium cations that is responsible for the formation of $[52]^+$.[75]

As in the case of S_P-**49-Cut**, the NCN-pincer platinum head group clearly resides at the periphery of the **Cut** enzyme. We studied whether the type of dimerization present in $[52]^+$could also be realised with S_P-**49-Cut**. To our surprise (and delight), the performed phosphonate inhibition reaction under *chloride poor* reaction conditions afforded, after purification by dialysis, crystals that appeared to be **[Cut-NCN-Pt-Cl-Pt-NCN-Cut]$^+$** (i.e., $[53]^+$). Obviously, this case reveals that coordination chemistry is possible with protein-modified complexes.

In a separate study, we explored coordination to the platinum center of S_P-**49-Cut** by various anionic, neutral and cationic phosphines with different

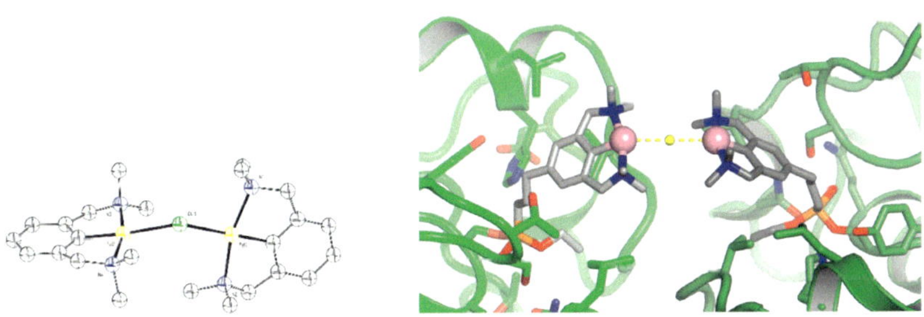

$[52]^+$ $[53]^+$ is [Cut-NCN-Pt-Cl-Pt-NCN-Cut]$^+$

SCHEME 6.11 Structure of $[52]^+$ in the solid state: two (NCN-pincer Pd)$^+$ cations brought together by a μ-Cl anion. Crystalizing S_P-**49-Cut** under Cl-poor conditions afforded $[53]^+$, **[Cut-NCN-Pt-Cl-Pt-NCN-Cut]$^+$**, in which a similar μ-Cl anion bridging occurs.[73]

steric bulk and charge (in D_2O and tris(hydroxylmethyl)aminomethane, *i.e.*, Tris, as buffer). ^{31}P NMR and ESI-MS studies indicated that the coordination of the various phosphines to the cationic Pt center is affected by the surrounding protein surface. This is particularly evident in cases when reactions of the phosphine $P(C_6H_4(SO_3Na)\text{-}3)_3$ are carried out in the presence of the Tris buffer. Under these physiological conditions, a change of the *N, C, N*-tridentate coordination mode of the NCN-pincer head unit, effective in S_P-**49-Cut**, into a *C*-monodentate one is observed (monitored by ^{31}P NMR). Whereas reaction of parent $[Pt(NCN)(H_2O)]^+$-cation with $P(C_6H_4(SO_3Na)\text{-}3)_3$ afforded the expected *trans*-phosphine complex, *mer-N, C, N*-$[Pt(NCN)(P(C_6H_4(SO_3Na)\text{-}3)_3)]$, ^{31}P NMR data of the S_P-**49-Cut**-phosphine complex in solution pointed to *cis*-coordination of $P(C_6H_4(SO_3Na)\text{-}3)_3$ to C_{ipso} instead.[76]

Inspired by the aforementioned results, we performed site-selective serine-hydrolase labeling studies with a luminescent organometallic NCN-pincer type **54**, Scheme 6.12. The linking of the luminescent label to the serine reactive phosphonate group was achieved by means of "click" chemistry.

The NCN-platinum label **PNP-P-54** is a robust dye with absorbance and emission maxima between 374–380 and 482–493 nm, respectively, and quantum yields between 0.04 and 0.15. Label **PNP-P-54** was tested in serine hydrolase specific binding studies using gel electrophoresis.[77]

The first example of a heterogeneous bifunctional catalytic system was thereafter designed; a situation inspired by the aforementioned results with the bio-organometallic chemistry of NCN-pincer metal complexes. In a collaboration with Bäckvall and co-workers, a catalytic system was synthesized comprised of an organometallic racemization catalyst, namely **55**, combined with an enzymatic acylation catalyst, namely CALB (Novozym 435) consisting of lipases immobilized on beads.[78] First, a calibration study was performed to get a reliable estimate of the amount of accessible CALB active sites on the beads. These beads were treated with the (racemic) phosphonate **PNP-P-55** (only one enantiomer reacted to **CALB-P-55**) until the residual lipase activity of the beads remained relatively constant, as monitored by

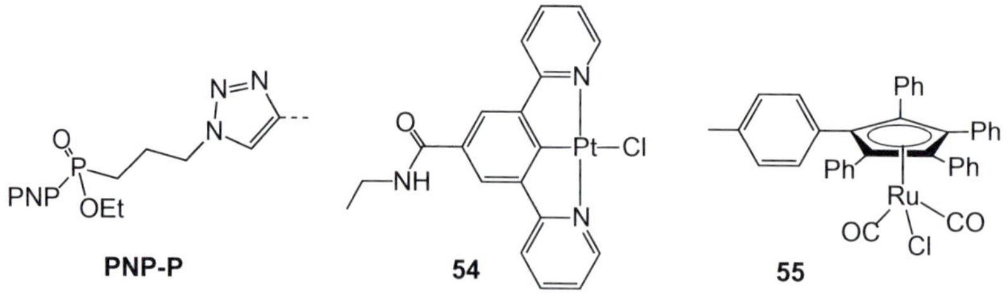

PNP-P 54 55

SCHEME 6.12 This scheme shows the *para*-nitrophenolate phosphonate scaffold **PNP-P** that combined with either **54** (luminescent NCN-pincer platinum chloro head group) affords the phosphonate **PNP-P-54** or with **55** (racemization catalyst), yields the phosphonate **PNP-P-55**. The preparation of these phosphonates has been described in Refs. 77 and 78.

back titration with carboxylic ester (PNP-)C(=O)C_3H_7 (Scheme 13). The surface loading reached was rather low (0.0125 mmol [Ru]g^{-1}) but through this inhibition approach the amount of available (non-inhibited) catalytic sites of CALB relative to the number of Ru sites can be controlled. This can be used in order to fine-tune the composition of the catalytic material.

With the available catalytic material, **CALB-P-55**, a proof-of-concept study of this novel bifunctional catalytic system for the Dynamic Kinetic Resolution (DKR) reaction of *rac*-1-phenylethanol (*rac*-**56**) with isopropenyl acetate to *R*-1-phenylethyl acetate (*R*-**57**) was carried out. Separate studies involved, *i.* the racemization of (*S*-**56**) to *rac*-**56** (Eqn. 6.2) and *ii.* the selective acetylation of *rac*-**56** to *R*-1-phenylethyl acetate catalyzed by either untreated CALB beads (Eqn. 6.3) or by Ru loaded **CALB-P-55** ones (Eqn. 6.4). Under the chosen reaction conditions (K_3PO_4 as base, toluene, RT), the **CALB-P-55** beads catalyzed the complete racemization of *S*-**56** in 24 h which is slower than have been reported for $RuCl(C_5Ph_5)(CO)_2$ itself. Kinetic resolution experiments of *rac*-**56** by its reaction with isopropenyl acetate catalyzed by either CALB or **CALB-P-55** beads, respectively, were performed under similar conditions (Na_2CO_3, toluene, Ar atm., RT). With CALB, full conversion after 3 h was observed, i.e., 50% acetylation to *R*-1-phenylethyl acetate occurred while 50% enantiopure *R*-**56** remained. Most interestingly, the reaction catalyzed by the bifunctional **CALB-P-55** beads produced (under the same reaction conditions) 28% of enantiopure *R*-1-phenylethyl acetate in 3 h (Scheme 6.13).

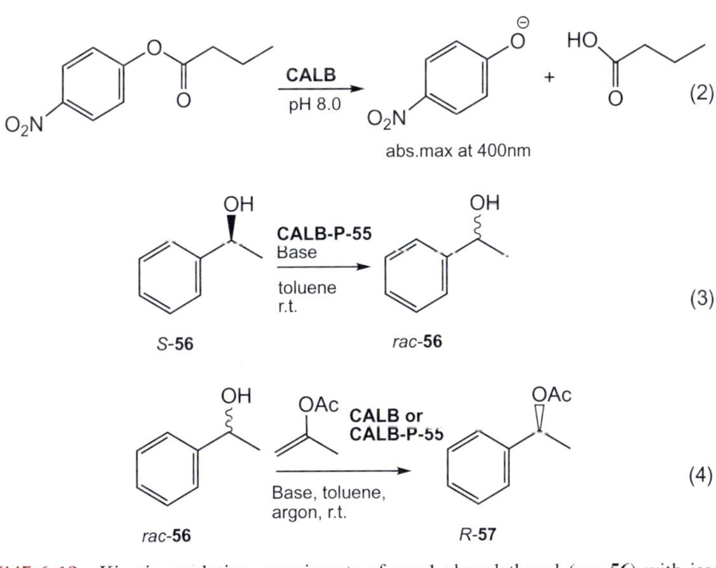

SCHEME 6.13 Kinetic resolution experiments of *rac*-1-phenylethanol (*rac*-**56**) with isopropyl acetate to *R*-1-phenylethyl acetate (*R*-**57**) using **CALB-P-55** catalytic material.[78]

Obviously, the reported results are preliminary and further optimization of the molar ratio of the ruthenium inhibitor complex **PNP-P-55** and CALB beads is needed to tune the racemization and acetylation catalytic functions of this novel bifunctional material. The fact that both catalytic functions are immobilized on a common solid support should facilitate easy separation of products at the end of the reaction.[78]

6.5 CONCLUSIONS

In spite of the plain, seemingly unspectacular nature of the structural features of NCN-pincer d^8-metal halido molecule, it displays, right from the time that we started to research its properties, unexpected features that are not encountered in typical d^8-metal complexes with a similar albeit monodentate or bidentate donor-atom sets. The amazing stability of the NCN-pincer platinum moiety is highlighted by its complete biocompatibility which enables the various applications summarized in the present report. The pincer-metal platform is remarkably versatile and is characterized by an incipient or permanent covalent metal-X_{ipso} (X is C, N, P) bond that is complemented with two donor-to-metal interactions from flanking donor-atom containing substituents.

This account has also shown that the organometallic NCN pincer–platinum platform, one of the subclasses of the "privileged" Pincer ligands, is useful for diagnostic targeting of proteins and, in combination with a serine hydrolase reactive phosphonate inhibitor opens up new activity-based protein profiling possibilities.[79] Moreover, the proof-of-principle methodology demonstrated here is by no means limited to serine hydrolases; the chemical versatility of pincer complexes also allows attachment to other biochemical probes like, for example, carbohydrates or oligopeptides. In addition, we have shown that immobilized bifunctional catalysts are accessible using, for example, CALB beads. This material, in addition to its lipase type enzymatic catalysis function, carries a racemization catalyst attached to its surface. This occurs via site-directed inhibition of certain number of these sites by a phosphonate grouping linked to a metal catalyst. This approach opens up further possibilities to make multi-catalyst systems comprising beads that have their surfaces covered with a layer of biomolecules that can be addressed separately with tethered catalysts such as that described for use in the racemization/acetylation catalysis with CALB. Designs that would also be easily re-useable and recyclable, with biocompatible catalysts loaded onto the beads, could contribute to more sustainable catalytic systems and chemistry.

Although the research presented in this chapter has been just a small part of our research efforts (see Ref. 12), it has served as an important and inspiring subject both personally and for my students. This is due to the strong connection between the organometallic NCN-pincer metal chemistry and quite relevant biochemical areas of interest. Presently, it is a subject that is rapidly expanding in many directions. As tribute to all the excellent research, that regrettably could not be cited in the framework of this chapter, three examples are now mentioned which connect the above work to cutting-edge science. These are (see also Ref. 80): (i) the synthesis of bis(NCN-pincer) iron(III) complexes (in this case the NCN-anion is 1,3-di(pyridin-2-yl)phenyl-anion) which shows a promising biological activity against various forms of cancer that are presently treated with somewhat toxic *cis*-platin[81]; (ii) the supramolecular linking of NCN-pincer palladium complexes into a ß-barrel protein host as a novel hybrid catalyst[82] and finally, (iii) the discovery that *Lactate Racemase* actually possesses an SCS-pincer nickel active site that is used by the enzyme to interconvert *D*- into *L*-lactic acid.[83] This provides the first confirmation that the pincer coordination platform plays an important role in Nature. This information has thereafter fueled "bioinspired" research on SCS-pincer nickel catalysts for the dehydrogenation of alcohols.[84]

ACKNOWLEDGMENT

Thanks are due to the many colleagues, students, and friends that were involved in the research discussed above. There excellent contributions allowed me to write this account on our collective journey to find new application for our "pincers". It has been a privilege.

Prof. Dr. R. A. Gossage (Ryerson, Toronto) is thanked for his critical reading of this manuscript.

REFERENCES

1. van Koten G. *Pure Appl Chem* 1989;**61**:1681.
2. Morales-Morales D, Jensen CM. *The chemistry of pincer compounds*. Amsterdam: Elsevier; 2007.
3. Singleton JT. *Tetrahedron* 2003;**59**:1837.
4. van der Boom ME, Milstein D. *Chem Rev* 2003;**103**:1759.
5. Albrecht M, van Koten G. *Angew Chem Int Ed* 2001;**40**:3750
6. Szabó KJ, Wendt OF. *Pincer and pincer-type complexes*. Wiley-VCH; 2013.
7. van Koten G, Milstein D, editors. *Organometallic pincer chemistry, Top. Organomet. Chem.*, 40. Heidelberg: Springer-Verlag; 2013.
8. van Koten G, Gossage RA, editors. *The privileged pincer-metal platform: Coordination chemistry & applications,* Top. Organomet. Chem., 54. Heidelberg: Springer-Verlag; 2016.
9. Rodriquez G, Albrecht M, Schoenmaker J, Ford A, Lutz M, Spek AL, et al. *J Am Chem Soc* 2002;**124**:5127.
10. McDonald AR, Dijkstra HP. *Top Organomet Chem.* 2016;**54**:335.
11. Wieczorek B, Dijkstra HP, Egmond MR, Klein Gebbink RJM, van Koten G. *J Organomet Chem* 2009;**694**:812.
12. van Koten G. *J Organomet Chem* 2017;**845**:4.
13. van Koten G. *Organometallics* 2012;**31**:7634.
14. van Koten G, Jastrzebski JTBH, Noltes JG, Spek AL, Schoone JC. *J Organometal Chem* 1978;**148**:233.

15. Jastrzebski JTBH, van Klaveren M, van Koten G. *Organometallics* 2015;**34**:2600.
16. van Koten G. *Top Organomet Chem* 2013;**40**:1.
17. Roddick DM. *Top Organomet Chem* 2013;**40**:49.
18. Moulton CJ, Shaw BL. *J Chem Soc, Dalton Trans* 1976;1020.
19. Gossage RA, Jastrzebski JTBH, van Koten G. *Angew Chem Int Ed* 2005;**44**:1448.
20. Jastrzebski JTBH, Arink AM, Kleijn H, Braam TW, Lutz M, Spek AL, et al. *J Am Chem Soc* 2013;**135**:13371.
21. Dani P, Albrecht M, van Klink GPM, van Koten G. *Organometallics* 2000;**19**:4468.
22. Valk J-M, van Belzen R, Boersma J, Spek AL, van Koten G. *J Chem Soc, Dalton Trans* 1994;2293.
23. Valk J-M, Boersma J, van Koten G. *J Organomet Chem* 1994;**483**:213.
24. Contel M, Stol M, Casado MA, van Klink GPM, Ellis DD, Spek AL, et al. *Organometallics* 2002;**21**:4556.
25. Valdes H, Gonzalez-Sebastian L, Morales-Morales D. *J Organometal Chem* 2017;**845**:229.
26. Gagliardo M, Snelders DJM, Chase PA, Klein Gebbink RJM, van Klink GPM, van Koten G. *Angew Chem Int Ed* 2007;**46**:8558.
27. Takenaka K, Minakawa M, Uoizumi Y. *J Am Chem Soc* 2005;**127**:12273.
28. Terheijden Jos, van Koten G, de Booys JL, Ubbels HJC, Stam CH. *Organometallics* 1983;**2**:1882.
29 van Beek JAM, van Koten G, Smeets WJJ, Spek AL. *J Am Chem Soc* 1986;**108**:5010.
30. Gossage RA, Ryabov AD, Spek AL, Stufkens DJ, van Beek JAM, van Eldik R, et al. *J Am Chem Soc* 1999;**121**:2488.
31. Yu. Rogachev A, Hoffmann R. *J Am Chem Soc* 2013;**135**:3262.
32. Albrecht M, Gossage RA, Frey U, Ehlers AW, Baerends EJ, Merbach AE, et al. *Inorg Chem* 2001;**40**:850.
33. Albrecht M, Gossage RA, Lutz M, Spek AL, van Koten G. *Nature* 2000;**406**:970.
34. Albrecht M, Gossage RA, Lutz M, Spek AL, van Koten G. *Chem Eur J* 2000;**6**:1431.
35. Severin K, Bergs R, Beck W. *Angew Chem Int Ed* 1998;**37**:1635.
36. Metzler-Nolte N. *Chimia* 2007;**61**:736.
37. Medicinal organometallic chemistry. In: Jaoen G, Metzler-Nolte N, editors. *Top. organomet. chem, vol. 32*, 32. Heidelberg: Springer-Verlag; 2010.
38. Gasser G, Metzler-Nolte N. *Curr Op Chem Biol* 2012;**16**:84.
39. L. Barisic, M. Dropucic, V. Rapic, H. Pritzkow, S. Kirin, N. Metzler-Nolte, *Chem Commun* (2004) 2004.
40. Barisic L, Mojca M, Khaled A, Liu Y, Kraatz H-B, Pritzkow H, et al. *Chem Eur J* 2006;**12**:4965.
41. Hillard EA, Vessières A, Jaoen G. *Top Organomet Chem* 2010;**32**:81.
42. G.D. Batema, Conjugated ECE-Pincer Metal Complexes, New Optical Materials and Bio-Conjugates, PhD Thesis, Utrecht University, 12/6/2007.
43. Davies PJ, Veldman N, Grove D m, Spek AL, Lutz BTG, van Koten G. *Angew Chem Int Ed* 1996;**35**:1959.
44. Gerhardt WW, Weck M. *J Org Chem* 2006;**71**:6333.
45. van Manen H-J, Fokkens RH, van Veggel FCJM, Reinhoudt DN. *Eur J Org Chem* 2002;3189.
46. Slagt MQ, RodrÌguez G, Grutters MMP, Klein Gebbink RJM, Klopper W, Jenneskens LW, et al. *Chem Eur J* 2004;**10**:1331.
47. Guillena G, Rodriguez G, Albrecht M, van Koten G. *Chem Eur J* 2002;**8**:5368.

48. Guillena G, Kruithof CA, Casado MA, Egmond MR, van Koten G. *J Organomet Chem* 2003;**668**:3.
49. Albrecht M, Lutz M, Schreurs AMM, Lutz ETH, Spek AL, van Koten G. *Dalton Trans* 2000;**21**:3797.
50. Okpodu CM, Alscher RG, Grabau EA, Cramer CL. *J Plant Physiol* 1996;**148**:309.
51. Guillena G, Halkes KM, Rodriguez G, Batema GD, van Koten G, Kamerling JP. *Org Lett* 2003;**5**:3064.
52. Green RJ, Frazier RA, Shakesheff KM, Davies MC, Roberts CJ, Tendler SJB. *Biomaterials* 2000;**21**:1823.
53. Rich RL, Myszka DG. *Curr Opin Biotechnol* 2000;**11**:54.
54. Beccati D, Halkes KM, Batema GD, Guillena G, Carvalho de Souza A, van Koten G, et al. *ChemBioChem* 2005;**6**:1196.
55. Guillena G, Rodriguez G, van Koten G. *Tetrahedron Lett* 2002;**43**:3895.
56. Greenfield SJ, Agarkov A, Gilbertson SR. *Org Lett* 2003;**5**:3069.
57. Agarkov A, Greenfield SJ, Ohishi T, Collibee SE, Gilbertson SR. *J Org Chem* 2004;**69**:8077.
58. Christensen CA, Meldal MA. *Chem Eur J* 2005;**11**:4121.
59. Dijkstra HP, van Klink GPP, van Koten G. *Acc Chem Res* 2002;**35**:798.
60. Kleij AW, Gossage RA, Klein Gebbink RJM, Brinkmann N, Reijerse E-J. *J Am Chem Soc* 2000;**122**:12112.
61. H.P. Dijkstra, C.A. Kruithof, N. Ronde, R. van de Coevering, D.J. Ramón, D. Vogt, et al., *J Org Chem* 68 (2203) 675-685.
62. Wilson ME, Whitesides GM. *J Am Chem Soc* 1978;**100**:306.
63. Letondor C, Humbert N, Ward TR. *Proc Nat Acad Sci USA* 2005;**102**:4683.
64. Pierron J, Malan C, Creus M, Gradinaru J, Hafner I, Ivanova A, et al. *Angew Chem Int Ed* 2008;**47**:701.
65. Steinreiber J, Ward TR. *Coord Chem Rev* 2008;**252**:751.
66. Pordea A, Creus M, Panek J, Duboc C, Mathis D, Novic M, et al. *J Am Chem Soc* 2008;**130**:8085.
67. Satake Y, Abe S, Okazaki S, Ban N, Hikage T, Ueno T, et al. *Organometallics* 2007;**26**:4904.
68. Coquière D, Feringa BL, Roelfes G. *Angew Chem Int Ed Engl* 2007;**46**:9308.
69. Guillena G, Kruithof CA, Casado MA, Egmond MR, van Koten G. *J Organomet Chem* 2003;**668**:3.
70. Kruithof CA, Casado MA, Guillena G, Egmond MR, van der Kerk-vanHoof A, Heck AJR, et al. *Chem Eur J* 2005;**11**:6869.
71. Kruithof CA, Casado MA, Guillena G, Egmond MR, van der Kerk-vanHoof A, Heck AJR, et al. *Eur J Inorg Chem* 2008;4425.
72. Dijkstra HP, Sprong H, Aerts BNH, Kruithof CA, Egmond MR, Klein Gebbink RJM. *Org Biomol Chem* 2008;**6**:523.
73. Rutten L, Wieczorek B, Mannie J-PBA, Kruithof CA, Dijkstra HP, Egmond MR, et al. *Chem Eur J* 2009;**15**:4270.
74. Kruithof CA, Dijkstra H,P, Lutz M, Spek AL, Egmond MR, Klein Gebbink RJM, et al. *Eur J Inorg Chem* 2008;4425.
75. Terheijden J, van Koten G, Grove DM, Vrieze K, Spek AL. *J Chem Soc, Dalton Trans* 1987;1359.
76. Wieczorek B, Snelders DJM, Dijkstra HP, Versluis K, Lutz M, Spek AL, et al. *Organometallics* 2012;**31**:2810.

77. Wieczorek B, Lemcke B, Dijkstra HP, Egmond MR, Klein Gebbink RJM, van Koten G. *Eur J Inorg Chem* 2010;1929.
78. Wieczorek B, Träff A, Krumlinde P, Dijkstra HP, Egmond MR, van Koten G, et al. *Tetrahedron Lett* 2011;**52**:1601.
79. Dijkstra HP, Sprong H, Aerts BNH, Kruithof CA, Egmond MR, Klein Gebbink RJM. *Org Biomol Chem* 2008;**6**:523.
80. Morales-Morales D. *Title is in print*. Amsterdam: Elsevier; 2018.
81. Estrada-Montano AS, Ryabov AD, Gries A, Gaiddon C, Lagadec RL. *Eur J Inorg Chem* 2017;1673.
82. Pocquet L, Vologdin N, Felice Mangiatordi G, Ciofini I, Nicolotti O, Thorimbert S, et al. *Eur J Inorg Chem* 2017;3622.
83. Desguin B, Zhang T, Soumillion P, Hols P, Hu J, Hausinger RP. *Science* 2015;**349**:66.
84. Xu T, Wodrich MD, Scopelliti R, Corminboeuf C, Hu X. *Proc Nat Acad Sci USA* 2017;**114**:1242.

Chapter 7

Carbon Monoxide-Releasing Molecules: Design Principles Inspired by Mechanism, Enabling Activity to be Controlled and Tuned

Ian J.S. Fairlamb and Jason M. Lynam
Department of Chemistry, University of York, York, United Kingdom

7.1 INTRODUCTION

Despite its well-deserved reputation as the "silent killer", carbon monoxide has been demonstrated to be a potential new avenue for therapeutic intervention. CO is generated in all mammals by the degradation of heme, a process catalyzed by constitutive and inducible forms of heme oxygenase enzymes. Since this discovery, it has been discovered that CO plays a key physiological role and is an important signaling molecule. CO has been shown to have a number of beneficial therapeutic effects including (but not limited to) protecting against ischemic damage and organ draft rejection, carbon monoxide is antibacterial, antimalarial and may act as a vasodilator.[1]

CO gas itself does not offer the potential to fully capitalize on these effects. Although there is an inherent risk of poisoning when administering the gas to a patent, this could, in principle, be controlled. However, CO gas treatment will always suffer from a lack of specificity, i.e., random diffusion will occur with no specific localization at the desired site of treatment. Therefore, carbon monoxide-releasing molecules (CO-RMs) have been proposed as a specific method to circumvent this problem.[2] CO-RMs are essentially prodrugs which are designed to liberate CO under specific conditions. There are a number of potential molecular species that could act as CO-RMs. For example, the photochemical reaction of organic carbonyl compounds may be employed to release CO,[3] as may the hydrolysis of boronate salts.[4] However, transition metal carbonyl complexes represent, by far, the most common structural unit

Advances in Bioorganometallic Chemistry. DOI: https://doi.org/10.1016/B978-0-12-814197-7.00007-8

employed for potential therapeutic CO-RMs. CO binds as a ligand to all transition metals and the metal-carbon interaction may be tuned through the σ-donor/π-acceptor interactions between the metal and ligand.

One of the key benefits of using transition metal carbonyl complexes as therapeutic CO-RMs is the potential ability to attenuate CO-releasing properties, solubility, stability, and potentially in vivo localization by appropriate choice of co-ligand. Furthermore, when one considers a general mechanistic picture of CO release, both the loss of carbon monoxide and any co-ligands may occur from the metal, therefore the biological effects of (1) CO, (2) the co-ligands, and (3) the liberated metal should all be considered. There are also a number of different potential triggers which may be used to liberate CO from the metal, these include thermal, photochemical, enzyme-triggered etc. In the case of thermally triggered CO-RMs there is a clear dichotomy between having CO-RMs which may be prepared in high yield and are not prone to rapid thermal decomposition during storage and purification, but will release CO at a rate appropriate for the desired biological application.

In this chapter, the contributions from our research laboratories in York to the development of an understanding of the synthesis of new CO-RMs, the mechanistic processes underpinning the CO-release process and their potential biological application are discussed.

7.2 INITIAL STUDIES – METAL PYRONE COMPLEXES

The initial studies focussed on the incorporation of 2-pyrone groups into the coordination sphere of metal complexes. Compounds containing the 2-pyrone motif exhibit a range of bioactive properties,[5] for example, they show submicromole inhibition activity against human ovarian carcinoma (A2780) and myelogenous leukemia (K562) cells, but do not show significant activity against healthly cells.[6] Therefore we envisaged that ligation of the 2-pyrone group to metals such as iron and molybdenum would result in CO-RMs which would only liberate biologically compatible species after the CO-release event (Fig. 7.1).

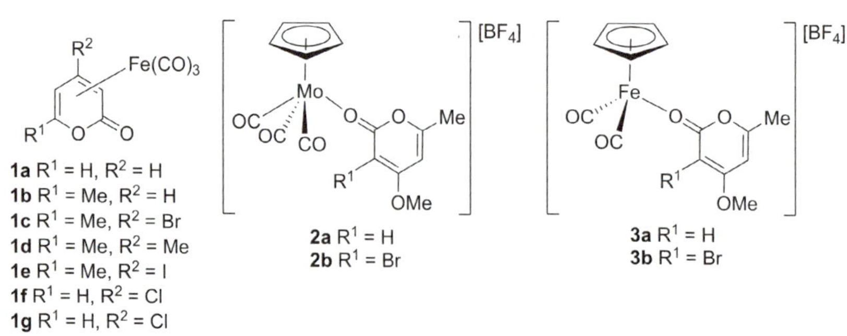

FIGURE 7.1 Preparation and CO-release behavior of pyrone complexes.

Complexes containing the 2-pyrone group coordinated in an η^4-diene fashion (**1**) to iron tricarbonyl groups may be prepared from the reaction of the pyrone with $[Fe_2(CO)_9]$ or $[Fe(coe)_2(CO)_3]$ (coe = *cis*-cyclooctene).[7] An alternative series of complexes in which the pyrone group was bound through the carbonyl group of the ester function in an η^1-O=C binding mode (**2**, **3**) may be prepared by the oxidative cleavage of dimeric complexes $[M(\eta^5\text{-}C_5H_5)(CO)_n]_2$ (M = Mo, $n = 3$; M = Fe, $n = 2$).[8]

The CO-releasing properties of the complexes was initially evaluated using as assay based on the conversion of deoxy-myoglobin (deoxyMb) to carbonmonoxy-myoglobin (MbCO). In brief, a solution of deoxyMb is treated with the CO-RM of interest and any CO-released from the metal is captured by the protein to form MbCO. The rate and extent of the conversion of deoxyMb to MbCO can be monitored by the changes to the Q-bands of the UV-vis spectrum.[9] The generation of MbCO is then assumed to reflect the kinetics of CO liberation from the CO-RM. In order to assess the rate of CO-release, the half-life of CO-release (defined as the time taken for a x μM of CO-RM to generate an Mb-CO concentration of $x/2$ μM) was introduced. The ensures that a consistent measure of CO-release which is independent of reaction mechanism can be employed to compare the behavior of a range of structurally diverse CO-RMs. Alternatively, the amount of Mb-CO formed over a set period of time may be employed as a measure of the CO-release. The deoxyMb-MbCO assay has been widely employed although care must be taken with the interpretation and implementation of the data as, for example, the sodium dithionate reducing agent employed has been shown to play a role in the CO-release process.[10]

Using this approach, it was possible to demonstrate the CO release from the metal complexes could be effectively modulated by the nature of the metal center (e.g. oxidation state, valence electron count, coordination number, and geometry) and co-ligands in the complex. CO-release data for the complexes are shown in Table 7.1. For example, in the η^4-bound pyrone series it was demonstrated that the nature of the substituents on the heterocyclic ring influenced the amount of CO-released, compounds **1a** and **1b** apparently do not release CO, whereas those with bromo- or chloro-groups did effectively release. In contrast, for the η^1-O-bound compounds, the nature of the metal had a profound effect on their behavior: the iron complexes **3** did not appear to release CO, whereas the molybdenum analogs, **2**, were effective. Notably, complex **2b**, with the 3-bromo-4-methoxy-6-methyl substituted 2-pyrone exhibited the greatest amount of CO-release.[11] The cytotoxicity of the complexes against RAW264.7 macrophages were also assessed using assays based on the release of lactate dehydrogenase (LDH) as a measure of the damage to the cell membrane and a complementary measure of cell viability was measured with Alamar blue assays. Although the behavior of the compounds depended on the nature of the substituents on the complex, on the whole the complexes were not considered to appreciably affect cell viability at concentrations below 100 μM.

TABLE 7.1 CO-Released by 2-Pyrone Metal Carbonyl Complexes as Assessed by the Conversion of Deoxy-Mb to Mb-CO. All Data are for CO-RM Solutions with Concentration 60 μM, Except 2a and 2b Which are 40 μM

Complex	CO-Released After 5 min/μM	CO-Released After 60 min/μM
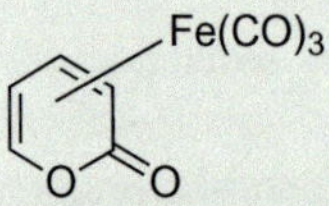 **1a**	0	0
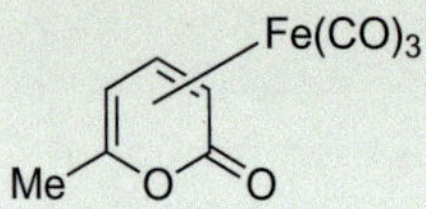 **1b**	0	0
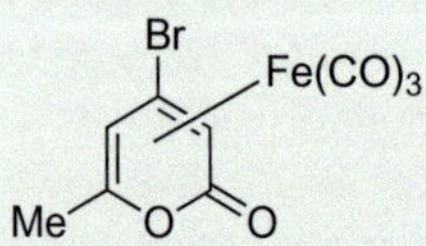 **1c**	2.3 ± 0.1	10.8 ± 0.4
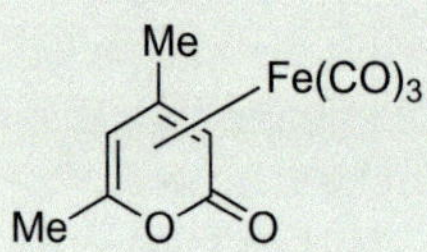 **1d**	0.8 ± 0.1	3.1 ± 0.4
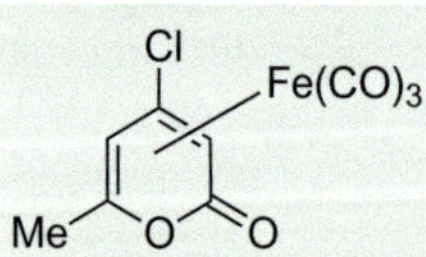 **1e**	0	0
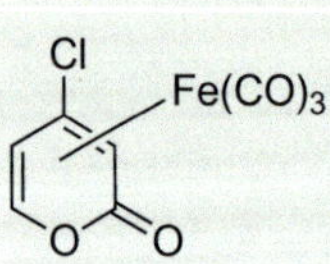 **1g**	0.8 ± 0.1	1.7 ± 0.7
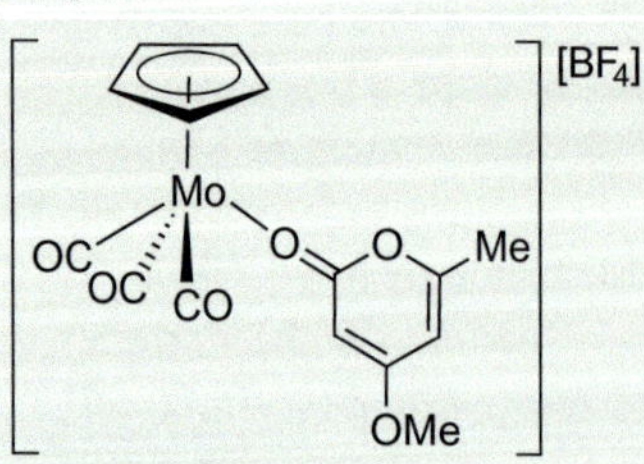 **2a**	0.6 ± 1.0	49.0 ± 0.1

(Continued)

TABLE 7.1 (Continued)

Complex	CO-Released After 5 min/μM	CO-Released After 60 min/μM
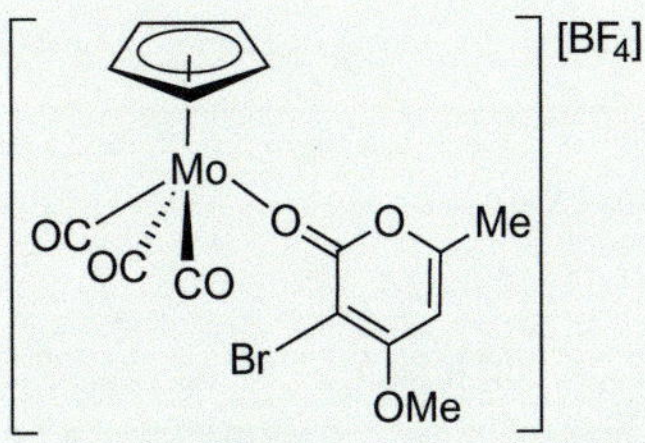 **2b**	17.0 ± 0.1	10.8 ± 0.4
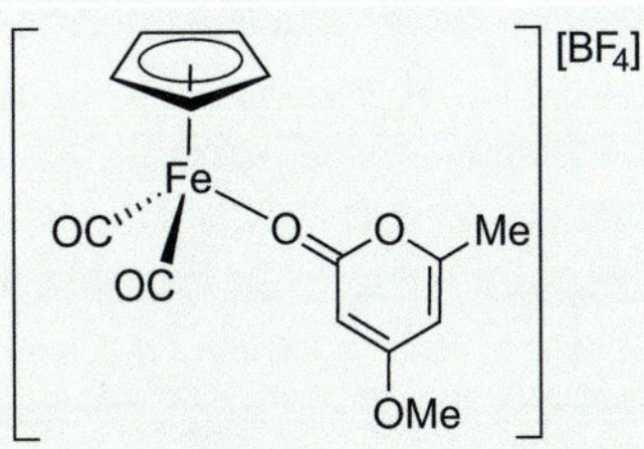 **3a**	0	0
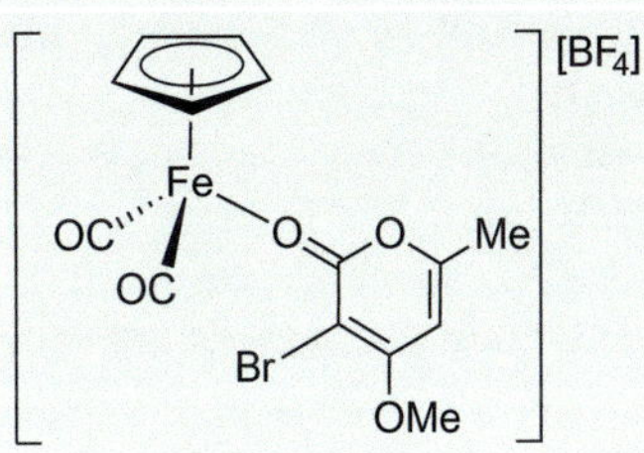 **3b**	0	0

The vasorelaxation properties of the lead compounds (**3c** and **2b**) were also evaluated by addition to a pre-contracted isolated rat aortic ring. Both were effective at relaxing the ring (by 30% and 60% respectively) demonstrating their potential as vasorelaxing agents at a concentration where no significant cytotoxicity was observed at concentrations of up to 100 μM.

7.3 MECHANISTIC STUDIES – UNDERSTANDING THE CO-RELEASE PROCESS

In order to design future generations of therapeutic CO-releasing molecules, an understanding of the fundamental processes controlling CO-dissociation

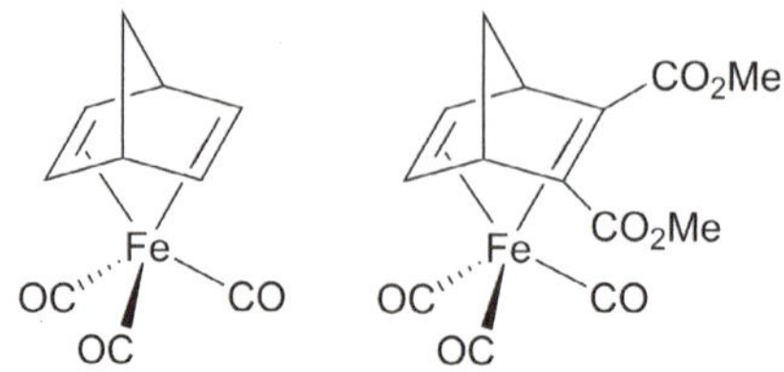

4 5

FIGURE 7.2 Fe(0)-diene complexes – the diene controls CO-loss.

from the metal is required. It is tempting to base such mechanistic proposals on the ideas of classical metal-carbonyl bonding in which σ-donation occurs from a coordinated CO ligand to a vacant metal d-orbital, with concomitant back-bonding from two filled d-orbitals to the vacant π*-orbitals of the CO ligands. In principle, such an approach is sound if the cleavage of the M-CO bond is rate controlling. Indeed, the mechanisms of substitution of CO-ligands are generally considered to be well understood. Many of these classical studies were performed using inert organic solvents in the absence of air and/or water: as such the species involved are well-defined. This is not the case when considering the implementation of a metal carbonyl complex in a therapeutic setting. Understanding behavior in aqueous media, and in vitro biological assays that are multi-component is challenging, even before considering how a CO-RM may act in vivo. Therefore far more complex mechanistic pathways, which depend on a multitude of factors beyond the M-C bond strength, may be expected in such cases.

The case of $Fe(CO)_3$ diene complexes readily illustrates the point that subtle changes in M-L bonding interactions may have pronounced effects on the mechanism of CO-release.[12] Two related iron tricarbonyl compounds **4** and **5** were prepared (Fig. 7.2), the latter bearing two ester substituents. In the case of **4** two IR bands were observed in the metal carbonyl stretching region at 2028 and 1954 cm^{-1}, whereas in the case of **5** three bands were observed at 2058, 2003 and 1976 cm^{-1}. On the basis of the arguments above on metal-carbonyl bonding, the lower frequency bands in **4** would imply that the M-CO bonds are stronger when compared to **5**. This may be readily explained as the ester-substituted alkene in **5** is a better π-acceptor, entailing less π-backbonding to the carbonyl groups. However, the trend in predicted M-C bond strength is not reflected in the CO-release data as measured by the deoxy-Mb to Mb-CO assay. Here, **4** exhibited thermal CO-release but under identical conditions no formation of Mb-CO was observed from **5**, although it could be stimulated photochemically.

An examination of the spectroscopic properties for the Fe(0) coordinated diene ligands provided a ready explanation for these observations. The coordination shift (i.e., the difference in chemical shift between the free and coordinated ligands) for the 1,4-diene ligand in **5** as far greater than in **4**, implying that the former ligand was more tightly bound to the metal and hence initial diene, rather than CO, loss is likely controlling the CO-releasing properties. Adding additional free norbornadiene to the Mb assay resulted in a retardation of the rate of CO-release, providing support for the argument that diene loss is rate-controlling in this case.

There are a number of systems in which mechanistic studies have revealed that the release of CO from metal complexes is a complex process and that the chemical medium used for the experiment plays a crucial role. Effective therapeutic CO-RMs need to function in aqueous media and, although the mechanisms of CO-substitution have been comprehensively studies under inert (O_2/H_2O-free) conditions in organic solvents, less is known about their behavior in aqueous solution. This is especially important as, it appears, that in a number of CO-RMs the interaction of the metal complex with water plays a key role in the CO-release process.[13] The metal carbonyl anions $[NEt_4][MX(CO)_5]$, **6**, (M = Cr, Mo, W; X = Cl, Br, I) are a prime example of this phenomenon. The kinetics of CO-substitution in these complexes had previously been measured by evaluating the first-order rate of formation of complexes $[NEt_4][MX(CO)(PR_3)_4]$ from the reaction of $[NEt_4]$ $[MX(CO)_5]$ with phosphines PR_3 (R = aryl). In this case, the rate of substitution was found to be Mo > Cr > W.[14] Whereas, an evaluation of the CO-release from the complexes using the Mb-assay demonstrated that the rate was Cr > Mo >> W, implying a different mechanistic process was occurring. Insight into this process can be obtained by monitoring the interaction of the complexes with water using infra-red spectroscopy. At room temperature CH_2Cl_2 solutions of the complexes $[NEt_4][MX(CO)_5]$ were indefinitely stable, however, addition of small amounts of water resulted in an instantaneous reaction and the formation of halide-bridge dimers $[NEt_4][\{M(CO)_5\}_2(\mu\text{-}X)]$. Performing this reaction in a sealed system resulted in the formation of $[M(CO)_6]$ indicating that CO was being released from the metal. It was, therefore, proposed that the key step in the CO-release from this series of complexes was hydrolysis of the M-X bond to form the halide-bridged dimers that were then responsible for the observed behavior. This demonstrated that there was a clear mechanistic shift on changing from organic to aqueous media.

Related behavior is also observed in the CO-release properties of other metal complexes based on the $[M(CO)_5L]$ framework which contain amino-esters. Reaction of $[NEt_4][MX(CO)_5]$, **6**, with amino-esters results in the loss of the halide anion and formation of neutral complexes **7**, Fig. 7.3.[15] Mechanistic studies (IR and UV-Vis) demonstrated that a common intermediate was formed during the CO-release process which corresponded to formal loss of the amino-ester ligand from the metal. In this series, it is therefore argued that M-N bond cleavage is the initial step in the CO-release process, controlled by the nature of the metal and the amino ester (Fig. 7.3).

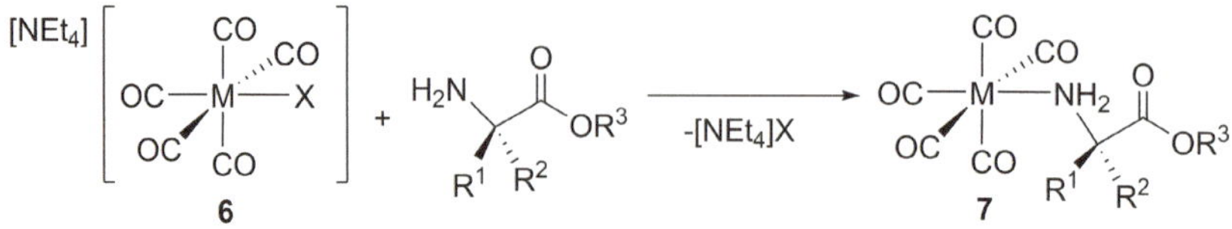

FIGURE 7.3 Reaction of $[NEt_4][MX(CO)_5]$ complexes with amino esters.

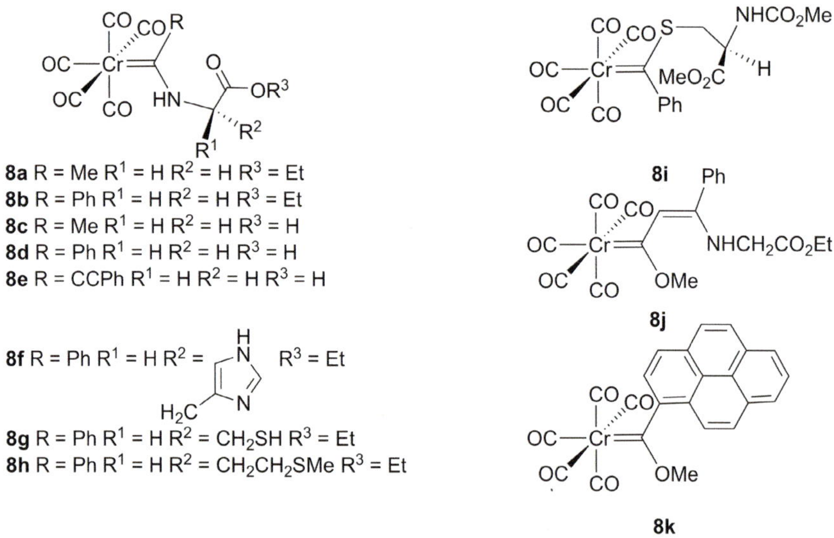

FIGURE 7.4 Structure of chromium carbene complexes.

Fischer carbene complexes $[M(CO)_5(=CR^1\{OR^2\})]$ offer further opportunities to incorporate amino esters into the coordination sphere of the metal.[16] Reaction of parent carbenes $[Cr(CO)_5(=CR\{OMe\})]$ (R = Me, Ph, -C≡CPh) with a series of amino esters results in loss of methanol and substitution at the carbene carbon to give compounds **8**, Fig. 7.4. Interesting is was possible to prepare both amino-carbenes and, by employing *N*-acetylcysteine, sulfur-substituted carbene **8i**, whereas reaction of $[M(CO)_5(=C\text{-}C{\equiv}CPh\{OMe\})]$ with $NH_2CH_2CO_2Et$ gave the Michael addition product **8j**.

The CO-release from the amino ester chromium carbene complexes was ca. 10–20 times slower than from the related complexes $[NEt_4][CrX(CO)_5]$, **6**, and **7**. The notable exception being $[Cr(CO)_5(=CR\{OMe\})]$ and **8i** examples, which show CO-release at a rate commensurate with the other derivatives. Empirically, there is a broad relationship between the ^{13}C chemical shift of the carbene carbon atom: compounds with lower field resonances exhibit more rapid CO-release. This is broadly correlated with the electrophilicity of the carbene atom and proposed that the nucleophilic attack at this site by water is the first step in the CO-release mechanism, a proposal supported by detailed mechanistic work from Bernasconi and co-workers.[17]

An important additional factor to consider when evaluating the CO-release mechanisms from metal carbonyl compounds is that the vast majority of biological assays are not performed in pure water, but rather in media such as phosphate-buffered saline (PBS), DMEM (Dulbecco's Modified Eagle's Medium) and Lysogeny broth (LB), to name but three. Also, due to the relatively poor aqueous solubility of some metal carbonyl complexes, the complexes are often administered as DMSO solutions. Therefore, evaluating the release mechanism in media[10] and evaluating the effect of the delivery vehicle on the CO-release process is vital.

One key advantage of the mechanistic studies described above is that it enables for metal systems to be targeted which may have additional functionality. For example, determining where CO-RMs are localized in a cellular environment is vital if their behavior is to be rationalized. A number of different methods, such as spectromicroscopy[18] and confocal fluorescence microscopy[19] have been employed to address this problem. The work on Fischer carbene complexes,[10] prompted the design of pyrene-containing complex **8k** (Fig. 7.4). It was envisaged the that pyrene groups could be used as a reporting group for the fate of metal carbonyl complex using confocal microscopy, coupled with fluorescence lifetime imaging (FLIM).[20] Complex **8k** could be prepared by reaction of the lithiated pyrene with $[Cr(CO)_6]$ followed by treatment with $[Me_3O][BF_4]$: a modification of Fischer's carbene synthesis. The pyrene group still exhibit strong emission, even when in the coordination sphere of the metal which allowed for its fate in HeLa cells to be monitored. Confocal microscopy and FLIM demonstrated that **8k** localized with in lipid droplets and the mitochondria of the cells. Furthermore, the imaging experiments allowed for the resulting death of the cells to be directly monitored after **8k** had been administered.

Mechanistic studies on a series of dicobalt octacarbonyl complex demonstrate the effect that the "Co-delivery vehicle" may have. Cobalt carbonyl complexes containing bridging alkyne ligands have been extensively explored as potential new anticancer agents.[21,22] As shown by the myoglobin assay, the bridged-alkyne complexes **9** (Fig. 7.5) are effective CO-releasing

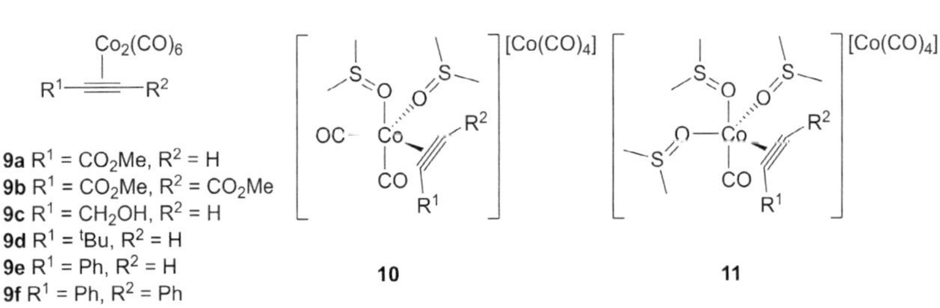

FIGURE 7.5 Structure of μ_2-alkyne cobalt carbonyl complexes.

molecules and the rate of release depends on the nature of the alkyne substituents. The interaction of these complexes with the DMSO delivery vehicle for the assays was then evaluated by the use of in situ ReactIR spectroscopy. ReactIR allows for direct reaction monitoring of the IR spectra of a solution by having a probe with an ATR crystal directly with the reaction mixture. Spectra can then be recorded over time and the changes to the speciation of the cobalt (or indeed any metal carbonyl) deduced from the changes to the vibrational modes in the region 2100–1800 cm^{-1}. Using this method it was demonstrated that **9b** converted to the cobalt anion $[Co(CO)_4]^-$, as shown by comparison with an authentic sample. In addition, monitoring the region between 1800 and 1600 cm^{-1} demonstrated that no free ester-substituted alkyne was generated, implying the formation of an alkyne-substituted counter cation. It was proposed that, on interaction of **9** with DMSO complexes such as **10** and **11** were formed, resulting in formal disproportionation of the Co(0) precursor. Such mechanistic insight could only be gained with the use of the ReactIR in situ IR instrument.

7.4 CONTROLLED CO-RELEASE MECHANISMS

The mechanistic studies described in Section 7.3 demonstrate that it is difficult to predict, a priori, the precise factors which will trigger CO-release within thermally labile metal carbonyl complexes. Although in some instances it has been demonstrate that metal carbonyls may localize within a specific organ,[23] spatial and temporal control of CO-release from a CO-RM is going to be vital to the future development of potential therapeutic agents. There has, therefore, been considerable effort in exploring different triggers for the CO-release process. For example, enzyme-triggered (ET)-CO-RMs have been designed which incorporate an ester linking which, when cleaved by an enzyme, induces CO-release.[24] Other mechanisms to trigger CO-release include electrochemistry[25] and employing magnetic heating of metal-carbonyl-functionalized nanoparticles.[26]

Light, however, remains one of the most important and flexible triggers for CO-release from metal complexes. The photochemically induced dissociation of CO ligands from metal carbonyls is a mature field,[27] and offers the potential to give the desired control needed for CO-release. In many cases light can enhance the cleavage of M-CO bonds. For example, in the case of water-soluble complex **12**,[28] CO release is sluggish in the dark (a 60 μM solution of the complex only release 2.2 μM in two hours), but on exposure to light resulted in a marked acceleration of the process ($t_{½}$ for CO release from a 60 μM solution 962 s) (Fig. 7.6).

The 2-phenylpyridyltetracarbonyl manganese complexes **13** (Fig. 7.7) were identified as a potentially highly versatile class of new photo-CORMs. These species may be readily prepared on a large scale from the reaction of $[Mn(CH_2Ph)(CO)_5]$ with a wide variety of 2-phenylpyridyl ligands.[29] They

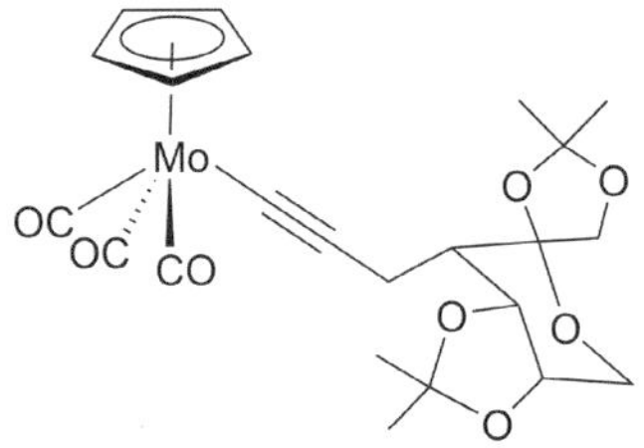

12

FIGURE 7.6 Structure of a photochemically activated molybdenum alkynyl complex.

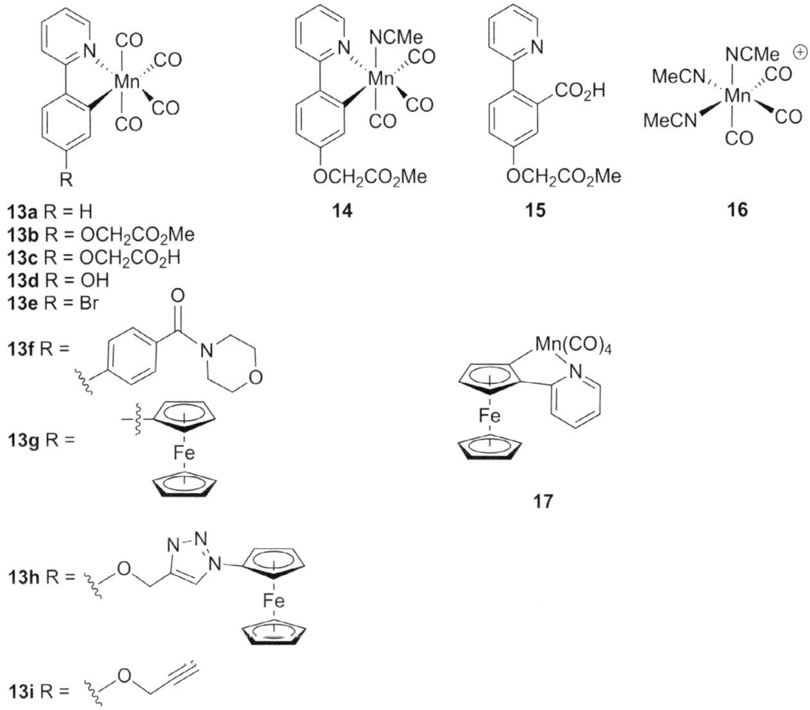

FIGURE 7.7 Structures of manganese carbonyl compounds and products from CO-release.

are attractive photo-CORMs as (1) it was anticipated that modification of the 2-phenylpyridine ligands could be used to modulate the rate of CO-release and solubility of the complex (2) the compounds are thermally robust and only undergo CO-loss on heating at high temperature (3) the presence of four carbonyl ligands offers the opportunity to release multiple molecules of CO.

Reaction of $[Mn(CH_2Ph)(CO)_5]$ with the ester- and carboxylic acid-modified 2-phenylpyridine ligands results in the formation of the water-soluble complexes **13b** and **13c** respectively.[30] In the dark, complexes **13b** and **13c** did not exhibit any CO-release, as shown by the myoglobin assay.

However on exposure to light smooth formation of Mb-CO was observed with three molecules of CO being released per Mn. An important part of this study, and vital to ensuring reproducibility between measurements, was the use of an LED system (400 nm) to control the irradiation time and the power employed. This LED system allows for the CO-release process to be monitored in NCMe solution by absorption spectroscopy and for the timing of the irradiation (on/off) to be controlled. These experiments demonstrate that substantial CO-release was only occurring when the LED was switched on, reinforcing the notation that this a photochemically initiated process.

The mechanism of CO-release was probed by mass spectrometry. A number of ions were identified in this study which gave key information about the mechanistic processes underpinning CO-release. Of particular importance was the observation of **14**, arising from the simple loss of CO and substitution by the solvent, NCMe; **15** which has apparently arisen from the migratory insertion of the 2-phenylpyridine ligand into a Mn-CO bond and **16**, which is believed to be responsible for subsequent CO release processes (Fig. 7.7).

The toxicity of the complexes was also assessed by the LDH assay both in the dark and when irradiated. The ester-substituted complex, **13b**, shows excellent cell viability even under irradiation, whereas **13c** did show a small (80%) loss in viability under irradiation at concentrations of 50 and 100 μM. Interestingly, the phenol-substituted complex, **13d**, showed marked detrimental effects on cell viability, indicating the nature of the 2-phenylpyridine is an important factor in controlling toxicity. This result highlights the importance of considering the toxicity of the ligand for metal-based CO-RMs, in addition to the metal.

One significant advantage of the manganese system is that is it possible to employ functionalized 2-phenylpyridine ligands in the initial cyclomanganation reaction and thus assess the effects of substituents on the CO-release process.[31] Furthermore, the robust nature of the manganese tetracarbonyl complex allows of post-synthetic modification on the 2-phenylpyridine ligand. For example, it is possible to perform selective Suzuki-Miyaura reactions on the bromine-substituted complex **13e** to form the morpholine derivative **13f**.

The robust nature of this manganese fragment may be further illustrated by the synthesis of the three ferrocene-modified complexes **13g**, **13h** and **17** (Fig. 7.7).[32] In addition to being able to perform the cyclometalation reactions on pre-formed ferrocene derivatives, it was also demonstrate that it was possible to copper-catalyzed azide-alkyne cycloaddition chemistry (CuAAC) to couple the manganese-modified alkyne **13i** with azidoferrocene to give **13h**.

The three ferrocenyl derivatives are photoCORMs, however **17** also exhibited slow thermal release of CO which could be traced back to protonation of the M-C bond in the organic ligand. The three complexes all underwent reversible one-electron oxidation, that may be attributed to the ferrocenyl group. The resulting oxidized products could be observed by IR spectroeletrochemistry, the bands for the carbonyl ligands in the complexes with the shorter tethers were blue shifted, consistent with an oxidation occurring. The IR spectrum of complex **13h** in the carbonyl region did not show any changes on oxidation, as presumably, the iron center is too remote to affect the manganese. Importantly, no evidence for CO-release was obtained on oxidation, demonstrating that the design of CO-RMs with different triggers is a complex issue.

7.5 PHOTOCHEMICALLY ACTIVATED CORMS WITH ANTIBACTERIAL PROPERTIES

An important part of the synthetic design strategy of CO-RMs is to consider both potential metal- and organic-based product from the CO-release process. The work on the 2-phenylpyridine Mn complexes had demonstrated that the protonated and carbonylated ligand may be lost during the photochemically induced loss of CO.[19] Therefore, manganese complexes were targeted which could release benign amino acids on CO-release. Tryptophan was employed as the amino acid of choice as, in addition to the above benefits, its intrinsic fluorescence could potentially be employed for imaging. Reaction of tryptophan with $[Mn(CH_2Ph)(CO)_5]$ results in the formation of complex **18** (Fig. 7.8), which is referred to as TrypoCORM.[33] The use of either (D) or (L) tryptophan resulted in the formation of non-racemic complexes (as shown by circular dichroism) and the tryptophan fluorescence demonstrated that it was located in a highly polar environment. As shown by the MbCO assay, TryptoCORM exhibits CO-release when exposed to light ($\lambda = 400$ nm) and the liberation of tryptophan was identified by mass spectrometry.

TryptoCORM displayed excellent cell viability properties against RAW 264.7 macrophages as measured by an Almar Blue assay, with only a small decrease in viability being observed at a concentration of 100 μM both in the dark and when exposed to ligand. At this concentration TryptoCORM was able to completely inhibit the growth of *Escherichia Coli* when exposed to light ($\lambda = 400$ nm). This is an extremely important result as it demonstrated that visible light could be employed to activate antibacterial activity in a

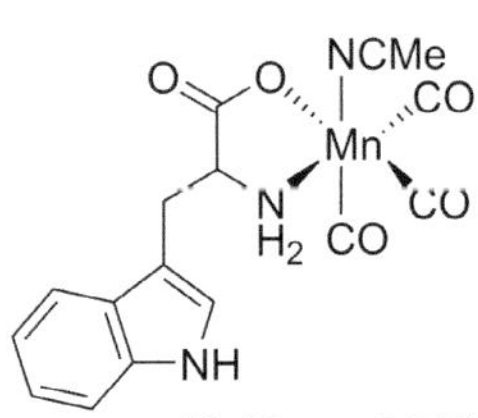

18 (TryptoCORM)

FIGURE 7.8 Structure of TryptoCORM.

photoCORM at a concentration when there was essentially no toxicity against mammalian cells.

Control experiments were performed in order to ascertain if it was the light-induced CO released from TryptoCORM that was responsible for this effect.[31] In the dark, TryptoCORM showed no activity against the *E. Coli* and exposure of the bacteria to the light source employed only resulted in a small inhibition of growth, incommensurate with the activity of the complex. However, these experiments did not unambiguously determine that it was the CO released from TryptoCORM that was responsible for this effect, it could (in principle) be a Mn-based product from the reaction or, indeed, a co-ligand. In order to rule out these possibilities a solution of TryptoCORM was irradiated for 68 mins in order to dissociate all the CO from the metal and create an inactivated CORM (or i-CORM). After purging with N_2 to remove any residual CO, the solution was added to *E. Coli*. No change in the growth of the bacteria was observed when compared to a control sample, indicating that it was not the Mn-based products or co-ligands that were responsible for the observed behavior. In a final control experiment the iCORM solution was then added to the *E. Coli* and exposed to light, again no change in the growth of the bacteria was observed. This demonstrated that photolysis of the products of CO-release was not responsible for the behavior. These data all then indicate that it is the CO-released from TryptoCORM that is responsible for the inhibition of the growth of *E. Coli*.

Even more striking data were obtained when the toxicity of TryptoCORM against *Neisseria gonorroheae* was evaluated.[34] In this case TryptoCORM was shown to be toxic to *N. gonorroheae* in the dark ($IC_{50} = 21 \pm 2$) μM (i.e., without exposure to the LED light). At first glance these data are at odds with the observation that (at least according to the MbCO assay) TryptoCORM only released CO in the dark. In order to rationalize these observations, the toxicity of TryptoCORM against *N. gonorroheae* was evaluated in the presence of leg-haemoglobin (Leg-Hb). Leg-Hb was added as a CO scavenger due to its extremely high affinity for the free molecule ($K_d = 0.9$ nm). In these cases, the activity of TryptoCORM against *N. gonorroheae* was significantly reduced implying that, at least in part, the toxicity was due to CO. It is proposed that *N. gonorroheae* are highly sensitive to the CO released slowly from TryptoCORM in the dark.

A related series of experiments with *S. aureus* demonstrated that TryptoCORM had similar effect in the dark (and the toxicity was rescued the addition of Leg-Hb). However, the culture age of both *S. aureus* and *E. Coli* affected the efficacy. When older (>5 days) cultures were employed then TryptoCORM showed increased toxicity, which is of considerable importance as persistent bacteria which are slow growing or dormant can be a problematic in the treatment of infections.

One additional challenge in the development and implementation of antibiotics in a clinical environment is bacteria populations may show significant levels of genetic variation.[35] With this in mind, the activity of $[NEt_4]$ $[MnBr_2(CO)_4]$ (Ebor-CORM-1) was evaluated against a series of clinical *Pseudomonas aeruginosa* strains.[36] Targeting *P. aeruginosa* infection is a particularly importance in the treatment of cystic fibrosis (CF). CF may lead to the development of mucus in the lungs which is difficult to clear and prone to bacterial infection. It was found that Ebor-CORM-1 was effective at clearing planktonic and biofilm cells of certain clinical strains of *P. aeruginosa* but not others. Mechanistically, Ebor-CORM-1 appears to be activated by a hydrolysis mechanism, similar to that observed for the related metal carbonyl anions $[NEt_4][MX(CO)_5]$.[8]

7.6 CHALLENGES AND FUTURE DIRECTIONS

There have been considerable advances in the design, synthesis and evaluation of metal carbonyl complexes as new therapeutic agents. However, when compared to the development of pharmaceuticals based on organic structures (Lipinski's rule of five, the use of molecular modeling of enzyme active sites to identify new structures, etc.) the understanding of metallodrugs is still somewhat lacking. It is important that methods are developed to assess the properties of metal-based CO-RMS and metal-based drugs in general as potential drug-molecules. An example of this is the oft-cited concern that complexes based on certain metals are, by definition, toxic. This has been demonstrated not to be the case, with toxicity being affected by a host of factors (ligands, oxidation state, etc.) rather than being an innate property of the metal.[37] With this in mind, identifying robust and versatile methods to incorporate functionality into organometallic complexes is an important goal as this will enable the introduction of novel properties which may then be used to effect the toxicity and key properties such as solubility, rate and mechanism of CO-release, etc. Underpinning all of these studies is the role of physical inorganic chemistry to identify the factors controlling CO-release process and then how this may be translated to the biological function of the CO-RM.

If CO-RMs are to become viable drug molecules then, in addition to the academic and scientific factors underpinning their development, it will be important to ensure that all of the steps in the drug development pipeline, which are well-established for organic pharmaceuticals, are considered. For example, ensuring that there is appropriate protection of intellectual property is vital to ensure the pathway to the clinic.

The use of CO-RMs as potential new antibiotics is an important avenue of research, especially when set against the growing problem global problems of drug-resistant bacteria. Coupled with these comments a crucial factor is to develop new species that are effective agents against bacteria but have no toxicity at the concentration employed against mammalian cells. In order to achieve this a deeper understanding of how CO is able to kill bacteria is needed. Although metalloproteins in the respiratory system of the bacteria are an obvious target for CO, it is worth considering that the mode of activity may be species-dependant, as highlighted by the difference in behavior of TryptoCORM against *E. coli* and *N. gonorroheae*. Such insight will almost certainly be needed if new antibiotic CO-RMS are to be designed. As the problem of drug-resistant bacteria continues to increase, the possibility of exploiting resistance mechanisms using metal-based CO-RMS offers an attractive proposition. For example, by incorporating CO-RMs into natural product and/or modifying existing antibiotics may offer new therapeutic avenues.

ACKNOWLEDGMENTS

We are grateful to all of our co-workers whose names are listed in the references below for their dedication and contributions to this chapter, particularly the efforts of our postgraduate students Anthony Atkin, Benjamin Aucott, Rachel Steen, and Jonathan Ward as well as PDRAs Ciara O'Brien and Wei-Qiang Zhang. We are also grateful to Professor James Moir for inspiring the work on the anti-bacterial behavior of Mn-based CO-RMs. We are grateful to the EPSRC, BBSRC, the Wellcome Trust, and the Leverhulme Trust for funding.

REFERENCES

1. Motterlini R, Otterbein LE. The therapeutic potential of carbon monoxide. *Nat Rev Drug Discov* 2010;**9**(9):728–43.
2. Johnson TR, Mann BE, Clark JE, Foresti R, Green CJ, Motterlini R. Metal carbonyls: A new class of pharmaceuticals? *Angew Chem Int Ed* 2003;**42**(32):3722–9.
3. Abeyrathna N, Washington K, Bashur C, Liao Y. Nonmetallic carbon monoxide releasing molecules (CORMs). *Org Biomol Chem* 2017;**15**(41):8692–9.
4. Alberto R, Motterlini R. Chemistry and biological activities of CO-releasing molecules (CORMs) and transition metal complexes. *Dalton Trans* 2007;**7**(17):1651–60.
5. (a) McGlacken GP, Fairlamb IJS. *Nat Prod Rep* 2005;**22**:369. (b) Dickinson JM. *Nat Prod Rep* 1993;**10**:71.
6. (a) Marrison LR, Dickinson JM, Fairlamb IJS. Bioactive 4-substituted-6-methyl-2-pyrones with promising cytotoxicity against A2780 and K562 cell lines. *Bioorg Med Chem Lett* 2002;**12**(24):3509–13. (b) Marrison LR, Dickinson JM, Ahmed R, Fairlamb IJS. *Tetrahedron Lett* 2002;**43**:8853. (c) Marrison LR, Dickinson JM, Ahmed R, Fairlamb IJS. Suzuki cross-coupling approaches to the synthesis of bioactive 3-substituted and 5-substituted-4-methoxy-6-methyl-2-pyrones. *Bioorg Med Chem Lett* 2003;**13**(16):2667–71. (d) Fairlamb IJS, Lu F-J, Schmidt J-P. *Synthesis* 2003;2564.
7. Fairlamb IJS, Duhme-Klair A-K, Lynam JM, Moulton BE, O'Brien CT, Sawle P, et al. Eta4-pyrone iron(0)carbonyl complexes as effective CO-releasing molecules (CO-RMs). *Bioorg Med Chem Lett* 2006;**16**(4):995–8.
8. Fairlamb IJS, Lynam JM, Taylor IE, Whitwood AC. *Organometallics* 2004;**21**:4964.
9. Atkin AJ, Lynam JM, Moulton BE, Sawle P, Motterlini R, Boyle NM, et al. Modification of the deoxy-myoglobin/carbonmonoxy-myoglobin UV-vis assay for reliable determination of CO-release rates from organometallic carbonyl complexes. *Dalton Trans* 2011;**40**(21):5755–61.

10. Mclean S, Mann BE, Poole R. Sulfite species enhance carbon monoxide release from CO-releasing molecules: Implications for the deoxymyoglobin assay of activity. *Anal Biochem* 2012;**427**(1):36–40.
11. Fairlamb IJS, Lynam JM, Moulton BE, Taylor IE, Duhme-Klair A-K, Sawle P, et al. Eta (1)-2-pyrone metal carbonyl complexes as CO-releasing molecules (CO-RMs): A delicate balance between stability and CO liberation. *Dalton Trans* 2007;(33):3603–5.
12. Atkin AJ, Fairlamb IJS, Ward JS, Lynam JM. *Organometallics* 2012;**31**:5894.
13. Zhang W-Q, Atkin AJ, Thatcher RJ, Whitwood AC, Fairlamb IJS, Lynam JM. Diversity and design of metal-based carbon monoxide-releasing molecules (CO-RMs) in aqueous systems: Revealing the essential trends. *Dalton Trans* 2009;(22):4351–8.
14. Allen AD, Barrett PF. *Can J Chem* 1968;**46**:1655.
15. Zhang W-Q, Whitwood AC, Fairlamb IJS, Lynam JM. Group 6 carbon monoxide-releasing metal complexes with biologically-compatible leaving groups. *Inorg Chem* 2010;**49**(19):8941–52.
16. Salmain M, Licandro E, Baldoli C, Maiorana S, Tran-Huy H, Jaouen G. *J Organomet Chem* 2001;**617**:376.
17. Bernasconi CF. *Chem. Soc. Rev* 1997;**26**:299.
18. (a) Meister K, Niesel J, Schatzschneider U, Metzler-Nolte N, Schmidt DA, Havenith M. Label-free imaging of metal-carbonyl complexes in live cells by Raman microspectroscopy. *Angew Chem Int Ed* 2010;**49**(19):3310–12. (b) Zobi F, Quaroni L, Santoro G, Zlateva T, Blacque O, Sarafimov B, et al. Live-fibroblast IR imaging of a cytoprotective PhotoCORM Activated with Visible Light. *J Med Chem* 2013;**56**(17):6719–31.
19. Pierri AE, Pallaoro A, Wu G, Ford PC. A luminescent and biocompatible photoCORM. *J Am Chem Soc* 2012;**134**(44):18197–200.
20. Arrowsmith RL, Atkin AJ, Botchway SW, Fairlamb IJS, Lynam JM, Moir JWB, et al. Confocal and fluorescence lifetime imaging sheds light on the fate of a pyrene-tagged carbon monoxide-releasing Fischer carbene chromium complex. *Dalton Trans* 2013;**44**(11):4957–62.
21. Ott I, Kircher B, Bagowshi CP, Vlecken DHW, Ott EB, Will J, et al. Modulation of the biological properties of aspirin by formation of a bioorganometallic derivative. *Angew. Chem Int Ed* 2009;**48**(6):1160–3.
22. Sergeant CD, Ott I, Sniady A, Meneni S, Gust R, Rheingold AL, et al. Metallo-nucleosides: Synthesis and biological evaluation of hexacarbonyl dicobalt 5-alkynyl-2′-deoxyuridines. *Org. Biomol. Chem* 2008;**6**(1).73–80.
23. Marques AR, Kromer L, Gallo DJ, Penacho N, Rodrigues SS, Seixas JD, et al. *Organometallics* 2012;**31**:5810.
24. Romanski S, Kraus B, Schatzschneider U, Neudörfl J M, Amslinger S, Schmalz H-G. Acyloxybutadiene iron tricarbonyl complexes as enzyme-triggered CO-releasing molecules (ET-CORMs). *Angew. Chem. Int Ed* 2011;**50**(10):2392–6.
25. McMahon S, Amirjalayer S, Buma WJ, Halpin Y, Long C, Rooney AD, et al. An investigation into the photochemistry of, and the electrochemically induced CO-loss from, [(CO)5MC(OMe)Me](M = Cr or W) using low-temperature matrix isolation, picosecond infrared spectroscopy, cyclic voltammetry, and time-dependent density functional theory. *Dalton Trans* 2015;**44**(35):15424–34.
26. Kunz PC, Meyer H, Barthel J, Sollazzo S, Schmidt AM, Janiak C. Metal carbonyls supported on iron oxide nanoparticles to trigger the CO-gasotransmitter release by magnetic heating. *Chem Commun* 2013;**49**(43):4896–8.

27. Perutz RN, Torres O, Vlcek A. *Photochemistry of Metal Carbonyls. ComprehensiveInorganic Chemistry II*, 8. Elsevier Press; 2013. p. 229 Available from: http://dx.doi.org/10.1016/B978-0-08-097774-4.00801-9.
28. Zhang W-Q, Atkin AJ, Fairlamb IJS, Whitwood AC, Lynam JM. *Organometallics* 2011;**30**:4643.
29. Bruce MI, Goodall BL, Matsuda I. *Aus J Chem* 1975;**28**:1259.
30. Ward JS, Lynam JM, Moir JWB, Sanin DE, Mountford AP, Fairlamb IJS. A therapeutically viable photo-activated manganese-based CO-releasing molecule (photo-CO-RM). *Dalton Trans* 2012;**41**(35):10514–17.
31. Ward JS, Bray JTW, Aucott BJ, Wagner C, Pridmore NE, Whitwood AC, et al. *Eur J Inorg. Chem* 2016;5044.
32. Aucott BJ, Ward JS, Andrew SG, Milani J, Whitwood AC, Lynam JM, et al. Redox-Tagged Carbon Monoxide-Releasing Molecules (CORMs): Ferrocene-Containing [Mn (C^N)(CO)4] Complexes as a Promising New CORM Class. *Inorg Chem* 2017;**56** (9):5431–40.
33. Ward JS, Lynam JM, Moir J, Fairlamb IJS. Visible-light-induced CO release from a therapeutically viable tryptophan-derived manganese(I) carbonyl (TryptoCORM) exhibiting potent inhibition against E. coli. *Chem Eur J* 2014;**20**(46):15061–8.
34. Ward JS, Morgan R, Lynam JM, Fairlamb IJS, Moir JWB. *MedChemComm* 2017;**8**:346.
35. O'Brien S, Williams D, Fothergill JL, Paterson S, Winstanley C, Brockhurst MA. High virulence sub-populations in Pseudomonas aeruginosa long-term cystic fibrosis airway infections. *BMC Microbiol* 2017;**17**(1):30 Available from: https://doi.org/10.1186/s12866-017-0941-6.
36. Flanagan L, Steen RR, Saxby K, Klatter M, Aucott BJ, Winstanley C, et al. *Front Microbio* 2018;**9**:195.
37. Egorova KS, Ananikov V. *Organometallics* 2017;**36**:4071.

Part II

Medicinal Chemistry and Imaging Probe

Chapter 8

Medicinal Chemistry: Bioorganometallic Anticancer Agents and Their Biomolecular Target Interactions

Christian G. Hartinger
School of Chemical Sciences, University of Auckland, Private Bag 92019, Auckland, New Zealand

8.1 INTRODUCTION

Since the discovery of cisplatin as the prototype anticancer metallodrug, research on metal-based anticancer agents has branched out in a wide variety of directions.[1–4] Currently investigated structures range from small coordination complexes to organometallic compounds, macromolecules decorated with metal centers and metal clusters, to name a few.[5–7] The modes of action that have been suggested for the newly designed compounds are as diverse as the structures of the drugs and drug candidates, many of which have been classified as prodrugs. The latter fact accounts for the formation of at least one covalent bond between the metal center and a donor atom of a biomolecular target which is often preceded by a ligand exchange reaction in the aqueous blood environment. Such reactions may be influenced by redox processes at the metal center that may labilize metal–ligand bonds.[1,8] In the living organism, there are many potential binding partners available and interaction with them will influence the biological properties of the metal-based drug. This requires careful analysis, often in complex matrices, and the development of new methods is paramount to trace minute concentrations of specific adducts.[9]

This chapter summarizes work that has been done in my group and with our collaborators over the last decade. It comprises the design of new compounds with novel modes of action (Fig. 8.1) and highlights how the compound design and synthesis is paralleled by mode of action studies and

Advances in Bioorganometallic Chemistry. DOI: https://doi.org/10.1016/B978-0-12-814197-7.00008-X

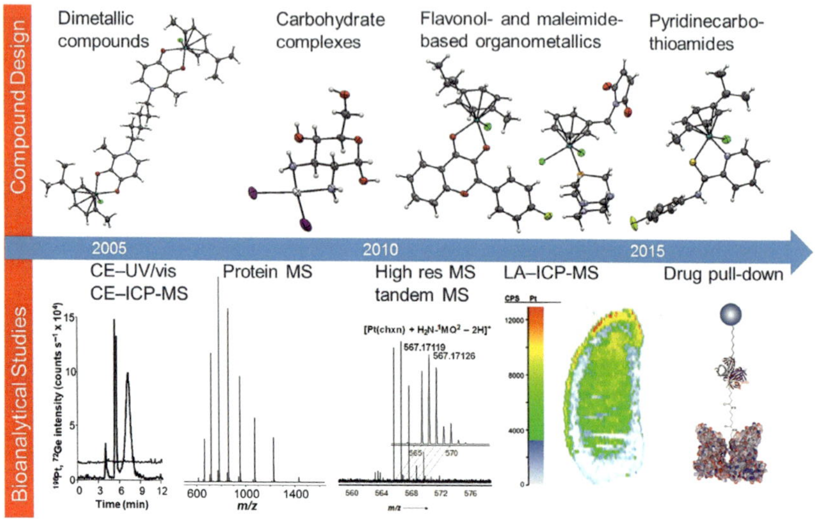

FIGURE 8.1 Overview of the development of selected compound classes paralleled by bioanalytical studies. *Partly reproduced from Refs. 10–13 with permission from Wiley, the American Chemical Society, and The Royal Chemical Society, correspondingly.*

therewith the development of analytical methods. These are often based on offline mass spectrometry (MS) or its use in conjunction with separation methods in an online setting but also considering other analytical techniques.

8.1.1 From Compound Design to Bioanalytical Investigations

We have interest in a wide variety of compound types with potential for cancer treatment and what they all have in common is that they were designed to have unusual modes of action which are different to the clinical-used platinum compounds and that they are organometallics, i.e., they contain at least one metal–carbon bond. The structural features explored range from classic ferrocenes to arene, pentamethylcyclopentadienyl, and carbene complexes, most commonly with Ru, Os, Rh, and Ir metal centers.

A few compound classes have been the standouts that we have invested a lot of time and effort in and they will be highlighted in this chapter. The first organometallic compounds we were involved in were dinuclear and trinuclear metal(arene) complexes (Fig. 8.1) that were linked through functionalized 3-hydroxy-4-pyridone moieties.[14–18] The design of the compounds was inspired by Farrell's dinuclear and trinuclear Pt compounds, with BBR3464 having been investigated clinically.[19] The compound type was shown to form unconventional DNA interactions as compared to cisplatin.[20] We were able to show that linking two organometallic units can increase the cytotoxic activity of compounds, as the mononuclear analog was moderately active.[17]

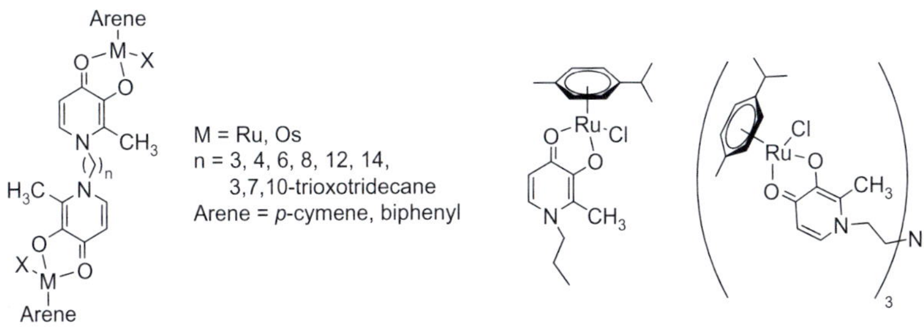

FIGURE 8.2 Chemical formulas of BBR3464-inspired dinuclear (left) and trinuclear (right) ruthenium anticancer agents, as well as a mononuclear analog (center).

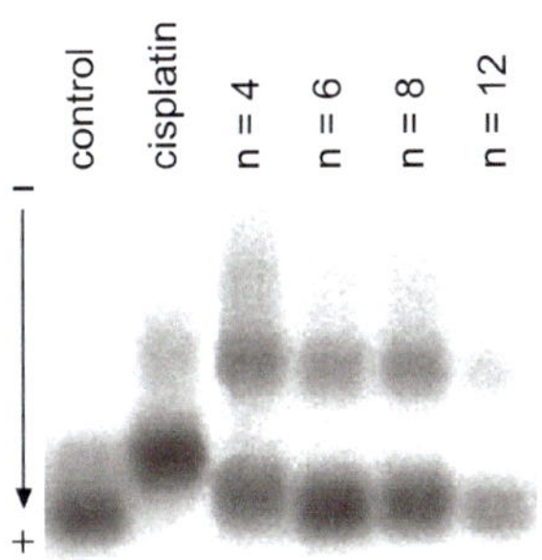

FIGURE 8.3 DNA–protein crosslinking ability of the dinuclear Ru(η^6-*p*-cymene)Cl complexes with n = 4, 6, 8 and 12 assessed by agarose gel electrophoresis. *Reproduced from Ref. 18 with permission from Elsevier.*

The cytotoxicity in cancer cells was found to be spacer length-dependent, which in turn determined their lipophilic properties.[14–17] However, when introducing a tetradecane spacer ($n = 14$; Fig. 8.2), aqueous solubility became a limiting factor for the dinuclear Ru complexes.[15] The Ru compounds were more cytotoxic than their Os analogs while mononuclear and trinuclear compounds showed similar antiproliferative potency.[17] A similar picture was found for the *p*-cymene and biphenyl derivatives which was surprising given the impact of such modification on the lipophilicity, but also points towards a more complicated mode of action rather than being only lipophilicity-dependent and therewith cell uptake-dependent.[17]

The interesting modes of action of the dimetallic compounds were first indicated by showing no cross-resistance to oxoplatin, which is a cisplatin prodrug, and even reversal of sensitivity in some oxoplatin-resistant cells. They were found to undergo rapid chlorido/aqua exchange in aqueous solution and react with proteins and with DNA.[16] Intriguingly, besides forming intrastrand and interstrand DNA crosslinks with double stranded DNA, the dinuclear compounds were found to be able to crosslink DNA duplexes as well as DNA and proteins (Fig. 8.3) to an extend which had not been observed for other Ru anticancer agents before.[18]

With the rise of non-DNA targeted anticancer agents, we became interested in the development of compounds that would interact selectively with

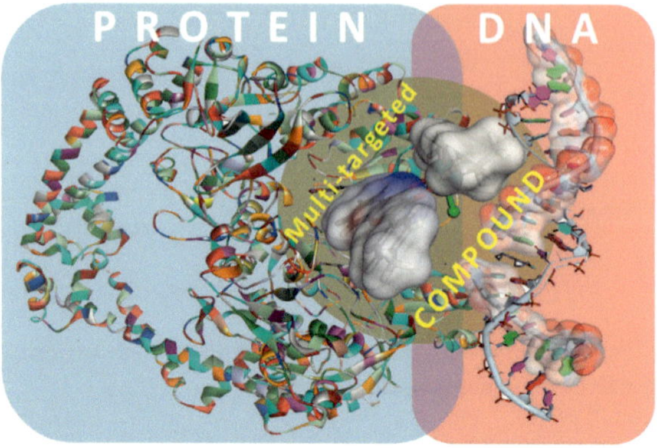

FIGURE 8.4 The concept of multi-targeted, metal-based anticancer agents. *Reproduced from Ref. 22 with permission from The Royal Society of Chemistry.*

proteins and ideally by doing so through a mechanism that involves not only the metal center, but also a bioactive ligand. This would eventually result in multitargeted, multifunctional metal-based anticancer agents (Fig. 8.4), i.e., more than one component of a molecule impacts separate targets.[21] This would result in altered pharmacological properties, metabolism and resistance development, tuneable antitumour activity, "intramolecular" combination therapy, and also selective targeted properties, to name a few.[21]

In maintaining the coordination environment introduced by the 3-hydroxy-4-pyr(id)one derived ligands in the multinuclear compounds discussed above and with experience of modifying that scaffold,[23–29] 3-hydroxyflavones were assessed as ligands to the half-sandwich organometallic metal(arene) scaffold. Flavonoids are known for their antiradical and antioxidant, antiinflammatory, estrogenic, antimicrobial, and anticancer activity. They exhibit these properties through enzyme inhibition, such as of topoisomerase and kinases, and the synthetic flavone flavopiridol has been studied in phase II clinical trials.[30] Moreover, other flavonoid–metal complexes had shown promising biological properties.[31] Inspired by these properties, we designed and synthesized a series of complexes featuring the 3-hydroxyflavone scaffold and organometallic moieties with different π-bound ligands and anionic leaving groups (Fig. 8.5) to target the DNA–topoisomerase complex.[22,32–36] Indeed, we were able to demonstrate that the complexes inhibited topoisomerase IIα while maintaining their ability to covalently interact with the nucleotide 5′-guanosine monophosphate after rapid halido/aqua ligand exchange.[32] Their inherent fluorescence was exploited for intracellular localization of the compounds in the endoplasmic reticulum, which we suggested as a reservoir, as is common for lipophilic compounds.[37] In addition, the compounds were found to inhibit cyclin-dependent kinase 2, however, without affecting the cell cycle and therefore this was excluded as a potential target.[32] Strong evidence for topoisomerase interaction as the mode of action was the correlation of the compounds'

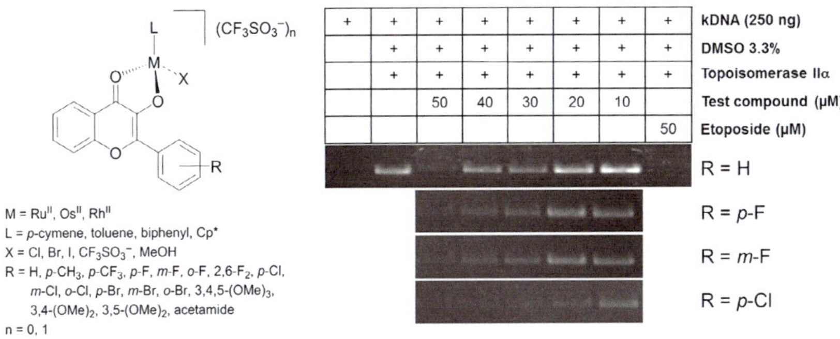

FIGURE 8.5 (left) General chemical structure of 3-hydroxyflavone-derived metal complexes and (right) concentration-dependent effect of organometallic flavone compounds (M = Ru; L = *p*-cymene; X = Cl; *n* = 0 and R as given to the right of the gel) on the catalytic activity of topoisomerase IIα, as determined by the decatenation assay. *Partly reproduced from Ref. 32 with permission from the American Chemical Society.*

topoisomerase IIα inhibitory activities with their cytotoxic potential. Furthermore, the complexes were more potent topoisomerase IIα inhibitors than the corresponding 3-hydroxyflavones, while the cytotoxicity was very similar with the *p*-chloro and *p*-fluoro derivatives being the most potent in *in vitro* anticancer activity assays.[22,35]

Structural variation allowed the determination of structure-activity relationships. As mentioned before the *p*-Cl derivative was among the most potent, and variation of the labile leaving ligand from chlorido to bromido or iodido and of the arene ligand from *p*-cymene to toluene did not affect the cytotoxicity significantly, while replacement with biphenyl decreased the antiproliferative potency.[22,32,33] Analogous (Cp*)rhodium(III) complexes (Cp* = pentamethylcyclopentadiene) featuring triflate counterions were too insoluble to be assayed for their anticancer activity,[34] however, the analogous Rh(Cp*)Cl were determined to show similar cytotoxic activity, as did the Os(*p*-cymene)Cl derivatives.[36] The metal center had an impact on the stability of the complexes with Rh(Cp*) organometallics being more stable under physiological conditions, than the Os(*p*-cymene) and Ru(*p*-cymene) derivatives, which behaved similarly.[36] Preference for interaction with L-histidine and ubiquitin was observed while the reactivity to cytochrome c was low and a preference for protein over DNA binding was found. However, out of the DNA building blocks, the organorhodium complexes bound preferably to 5′-dATP, and the Ru(cym) and Os(cym) compounds to 5′-dGTP, which reflects the different hardness of the metal centers.

Inspired by the redox activity of the quinone structure and maintaining again the chemical environment around the organometallic center, we introduced lapachol-derived ligands for organometallic anticancer agents.[38,39] We found an interesting impact of the metal center, i.e., Ru, Os, or Rh, on the biological activity which was traced back to the induction of oxidative stress

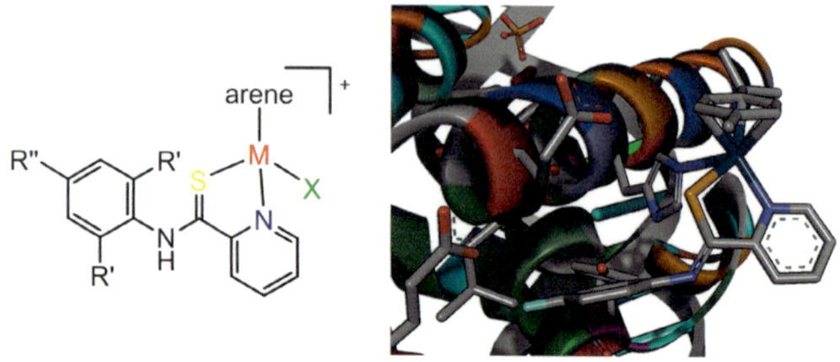

FIGURE 8.6 (left) General chemical structure of 2-pyridinecarbothioamide metal complexes and (right) coordination of an Os(*p*-cymene) derivative to the NCP through binding to a His residue.

in cancer cells.[38] Differences in their properties were also linked to their biomolecule interactions, which was studied by top–down ESI-MS on ubiquitin adducts. These studies revealed Met1 as the binding site on ubiquitin, however, after cleavage of the bioactive ligand.[38] Research to improve the stability through use of an *N,O*-bidentate ligand led to the attempts of introducing oxime derivatives[39] and the use of quinoline-5,8-dione ligands.[40] In both cases, surprising reactions were observed. While in the former case nitrosonaphthalene complexes were obtained, in the latter the quinoline-5,8-dione scaffold was converted into an *ortho*-quinone structure rather than the desired product.

With the aim of introducing a bioactive ligand with stronger coordination to the organometallic Ru(arene) scaffold, 2-pyridinecarbothioamide (PCA) complexes were designed as orally active anticancer agents (Fig. 8.6).[41–44] PCAs are bioactive substances that act as gastric mucosal protectants and exhibit low acute toxicity *in vivo*.[45] Coordination of PCA ligands to metal centers led to cytotoxic metallodrugs and the most lipophilic and smallest congeners were the most potent derivatives. Their interesting chemical properties include exceptional stability in hydrochloric acid, which suppresses chlorido/aqua ligand exchange reactions and low reactivity towards biological nucleophiles. In aqueous solution, they form transient thione-bridged dimers upon hydrolysis, which may explain their low reactivity towards biological nucleophiles, while the biological effect is probably induced by the monomer. It was also the monomer which was found in crystallographic studies with the nucleosome core particle (NCP) to form adducts with the histone proteins (Fig. 8.6).

In terms of predictability of oral bioavailability, the original molecular mass constraint[46] postulated in "Lipinski's rule of 5" seemed inappropriate for metal-based compounds with high atomic weight elements. A more holistic treatment is required which was termed the quantitative estimate of drug-likeness (QED).[47] We introduced this concept in metallodrug research[41,44,48] which will, however, need more data for orally active metal-based drugs in clinical studies to establish criteria analogous to organic drugs. This will

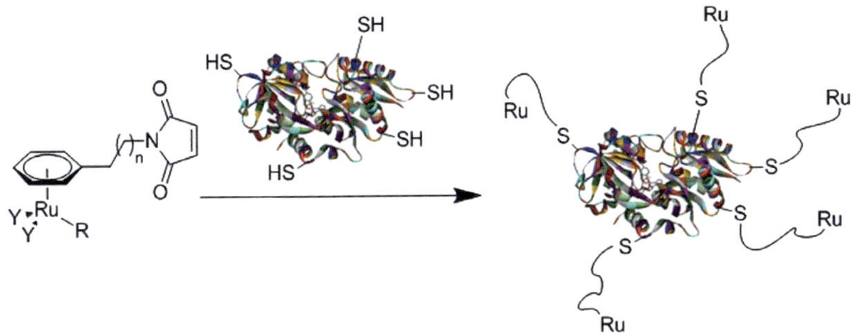

FIGURE 8.7 Covalent modification strategy of protein thiol residues with organometallic maleimide moieties. *Reproduced from Ref. 50 with permission from The Royal Society of Chemistry.*

eventually provide access to unused chemical space with novel modes of action, which cannot be obtained by organic molecules.[49] Using this concept for the PCA complexes demonstrated that the QED was similar as for the established anticancer agents erlotinib, tamoxifen, imatinib, and sorafenib.[41] The promising biological activity observed in *in vitro* assays also translated to *in vivo* models, where oral administration resulted in response of a murine colon carcinoma tumor model and an invasive melanoma model.[43] The mode of action of the compounds was determined to involve interaction with plectin, a scaffold protein and cytolinker, and therefore the compound class was termed plecstatins (Fig. 8.7).

In order to equip the compound class with targeting properties to achieve higher tumor accumulation, we modified the PCA scaffold with a maleimide functional group.[42] We had shown earlier that such modification of organometallic compounds allows for their selective reaction with thiol-containing biomolecules.[50] To enhance drug delivery, we are especially interested in loading our compounds on the biological vector human serum albumin (HSA), which features in Cys34 a free thiol.[42,50,51] Biological macromolecules such as HSA can extravasate into the tumor tissue and accumulate. This effect is called the enhanced permeability and retention (EPR) effect and is based on the formation of leaky vessels during angiogenesis and a dysfunctional lymphatic drainage in tumors.[52] While the functionalization of the PCA ligand led to compounds that were very reactive to L-cysteine, their *in vitro* cytotoxicity was low.[42] In contrast, when we modified the arene moiety of Ru(arene) anticancer agents, the antiproliferative activity of the pharmacophore was largely maintained.[50]

8.1.2 Bioanalytical Method Development and Application to Mode of Action Studies

The development of drugs is closely associated with the understanding of their modes of action. Most commonly, metal-based drugs are considered as prodrugs that undergo ligand exchange reactions and interact with the biomolecular targets by the formation of dative bonds. We are interested in identifying the type of adducts formed, the kinetics of the bond formation, the targets in biological environment and the distribution in living organisms. Usually we use mass spectrometric methods to study the behavior of the compounds prepared. We exploit the presence of metal centers in our molecules by employing element-specific methods of analysis that complement separation methods and molecular mass spectrometry approaches. In particular, electrospray ionization (ESI) and inductively-coupled plasma (ICP) mass spectrometry (MS) have been widely used, in some studies together with matrix-assisted laser desorption/ionization (MALDI) MS.[11,53–58] The former has been employed to study the types of adducts formed with amino acids and model proteins. ESI-MS would not only provide us with the nature and stoichiometry of the products formed in the reaction, but also enable the identification of binding sites on biological macromolecules, such as proteins and DNA.[11,54–56] For example, we identified in a top–down MS study the amino acid binding sites of cisplatin and oxaliplatin on the model protein ubiquitin which gave feedback on the preference of the metal compounds for donor atoms and binding sites and supports the design of the next generation of compounds. In case of oxaliplatin, the amino acid residues Met1 and His68 of ubiquitin were unambiguously identified as the binding sites using electron transfer dissociation, which compared with other tandem mass spectrometry fragmentation techniques.[56]

The same concept was also used to characterize a click reaction product between a Ru(*p*-cymene) complex of a 3-hydroxy-4-pyridone derivative and a modified peptide for enhanced drug delivery (Fig. 8.8).[59] We use this now in combination with other methods, for example, protein crystallography which gave additional insights into the local environment experienced by the metal conjugates and on the structure and stability of the metalated protein.[60]

Often offline MS is not sufficient to obtain the information sought and separation of different species in solution is required. We have focused on the use of electrophoretic methods in order to gain additional information. In particular, capillary electrophoresis (CE) is an interesting method as it allows for example in capillary zone electrophoresis (CZE) mode to use exclusively aqueous conditions and mimicking of the biological environment. Moreover, unlike in high performance liquid chromatography (HPLC), there is no stationary phase the reactive metallodrugs could interact with. This provides us with a high level of confidence that the speciation of the sample does not change during analysis. We have used CZE to study the aquation of metallodrugs in solution, i.e., the exchange of labile leaving ligands with water molecules, to elucidate the kinetics of DNA binding by employing nucleotides as DNA models, and to characterize the binding of metallodrugs to serum proteins.[61,62] In particular being able to separate human serum albumin from transferrin was found very useful as they have been suggested to be the main reaction partners in human serum for metal-based anticancer agents.[55,58,63–65]

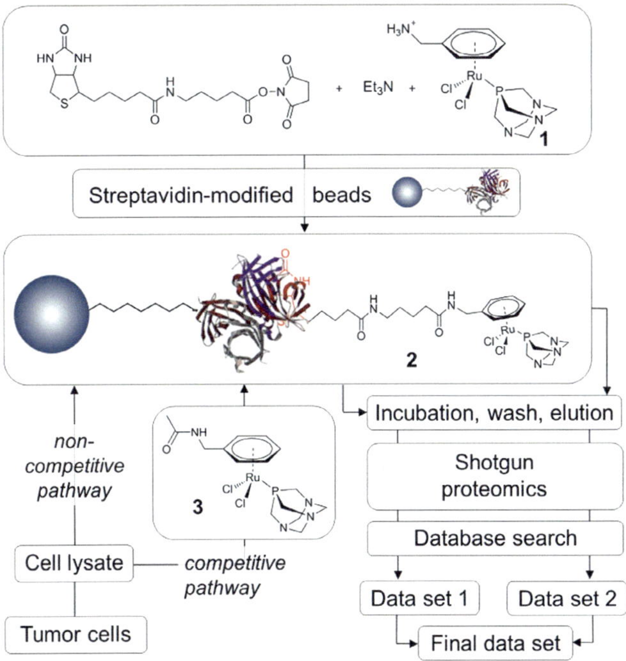

FIGURE 8.10 Work-flow used in the metallodrug pull-down experiments. *Reproduced from Ref. 13 with permission from The Royal Society of Chemistry.*

The concept was subsequently developed further and its combination with target validation for the plecstatin compound class revealed the protein plectin as a selective target.[43] This was the first time that direct molecular target identification for a metallodrug was achieved for a target other than DNA. This method now provides a tool to answer one of the most limiting questions raised by industry for the translation of metallodrugs into clinical trials.

8.2 CONCLUSIONS

Anticancer drug development is a highly interdisciplinary field of research and requires skill sets hardly found in a single research group. The approaches followed together with our collaborators not only delivered compounds with interesting biological properties but also helped us to identify surprising chemical transformations. This allowed us to contribute to improving the understanding of the behavior of metal-based drugs in biological systems and to translate this understanding into the design of new pharmacophores with novel modes of action.

ACKNOWLEDGMENT

I would like to thank all the students and postdoctoral researchers who have worked on these and other projects as well as our collaborators that contributed ideas and complemented our expertise. We are grateful to the many organizations, foundations and funding agencies who have supported our work financially.

REFERENCES

1. Jakupec MA, Galanski M, Arion VB, Hartinger CG, Keppler BK. Antitumour metal compounds: More than theme and variations. *Dalton Trans* 2008;183–94.
2. Komeda S, Casini A. Next-generation anticancer metallodrugs. *Curr Top Med Chem* 2012;**12**:219–35.
3. Barry NPE, Sadler PJ. 100 years of metal coordination chemistry: From Alfred Werner to anticancer metallodrugs. *Pure Appl Chem* 2014;**86**:1897–910.
4. Mjos KD, Orvig C. Metallodrugs in medicinal inorganic chemistry. *Chem Rev* 2014;**114**:4540–63.
5. Jaouen G, Dyson PJ. Med Organomet Chem. Elsevier Ltd; 2007. p. 445–64.
6. Govender P, Therrien B, Smith GS. Bio-metallodendrimers - emerging strategies in metal-based drug design. *Eur J Inorg Chem* 2012;**2012**:2853–62.
7. Allardyce CS, Dyson PJ. Metal-based drugs that break the rules. *Dalton Trans* 2016;**45**:3201–9.
8. Renfrew AK, O'Neill ES, Hambley TW, New EJ. Harnessing the properties of cobalt coordination complexes for biological application. *Coord Chem Rev* 2018. Ahead of Print.
9. Hartinger CG, Groessl M, Meier SM, Casini A, Dyson PJ. Application of mass spectrometric techniques to delineate the modes-of-action of anticancer metallodrugs. *Chem Soc Rev* 2013;**42**:6186–99.
10. Timerbaev AR, Aleksenko SS, Polec-Pawlak K, Ruzik R, Semenova O, Hartinger CG, et al. Platinum metallodrug-protein binding studies by capillary electrophoresis-inductively coupled plasma-mass spectrometry: Characterization of interactions between Pt(II) complexes and human serum albumin. *Electrophoresis* 2004;**25**:1988–95.
11. Hartinger CG, Tsybin YO, Fuchser J, Dyson PJ. Characterization of platinum anticancer drug protein-binding sites using a top-down mass spectrometric approach. *Inorg Chem* 2008;**47**:17–19.
12. Egger AE, Theiner S, Kornauth C, Heffeter P, Berger W, Keppler BK, et al. Quantitative bioimaging by LA-ICP-MS: A methodological study on the distribution of Pt and Ru in viscera originating from cisplatin- and KP1339-treated mice. *Metallomics* 2014;**6**:1616–25.
13. Babak MV, Meier SM, Huber KVM, Reynisson J, Legin AA, Jakupec MA, et al. Target profiling of an antimetastatic RAPTA agent by chemical proteomics: Relevance to the mode of action. *Chem Sci* 2015;**6**:2449–56.
14. Mendoza-Ferri MG, Hartinger CG, Eichinger RE, Stolyarova N, Severin K, Jakupec MA, et al. Influence of the spacer length on the *in vitro* anticancer activity of dinuclear ruthenium-arene compounds. *Organometallics* 2008;**27**:2405–7.
15. Mendoza-Ferri MG, Hartinger CG, Nazarov AA, Kandioller W, Severin K, Keppler BK. Modifying the structure of dinuclear ruthenium complexes with antitumor activity. *Appl Organomet Chem* 2008;**22**:326–32.

16. Mendoza-Ferri MG, Hartinger CG, Mendoza MA, Groessl M, Egger AE, Eichinger RE, et al. Transferring the Concept of multinuclearity to ruthenium complexes for improvement of anticancer activity. *J Med Chem* 2009;**52**:916–25.
17. Mendoza-Ferri MG, Hartinger CG, Nazarov AA, Eichinger RE, Jakupec MA, Severin K, et al. Influence of the arene ligand, the number and type of metal centers, and the leaving group on the *in vitro* antitumor activity of polynuclear organometallic compounds. *Organometallics* 2009;**28**:6260–5.
18. Nováková O, Nazarov AA, Hartinger CG, Keppler BK, Brabec V. DNA interactions of dinuclear RuII arene antitumor complexes in cell-free media. *Biochem Pharmacol* 2009;**77**:364–74.
19. Hartinger CG, Phillips AD, Nazarov AA. Polynuclear ruthenium, osmium and gold complexes. The quest for innovative anticancer chemotherapeutics. *Curr Top Med Chem* 2011;**11**:2688–702.
20. Komeda S, Moulaei T, Woods KK, Chikuma M, Farrell NP, Williams LD. A third mode of DNA binding: Phosphate clamps by a polynuclear platinum complex. *J Am Chem Soc* 2006;**128**:16092–103.
21. Zimmermann GR, Lehár J, Keith CT. Multi-target therapeutics: When the whole is greater than the sum of the parts. *Drug Discovery Today* 2007;**12**:34–42.
22. Kurzwernhart A, Kandioller W, Bartel C, Bachler S, Trondl R, Muhlgassner G, et al. Targeting the DNA-topoisomerase complex in a double-strike approach with a topoisomerase inhibiting moiety and covalent DNA binder. *Chem Commun* 2012;**48**:4839–41.
23. Kandioller W, Hartinger CG, Nazarov AA, Kasser J, John RO, Jakupec MA, et al. Tuning the anticancer activity of maltol-derived ruthenium complexes by derivatization of the 3-hydroxy-4-pyrone moiety. *J Organomet Chem* 2009;**694**:922–9.
24. Kandioller W, Hartinger CG, Nazarov AA, Bartel C, Skocic M, Jakupec MA, et al. Maltol-derived ruthenium-cymene complexes with tumor inhibiting properties: The impact of ligand-metal bond stability on anticancer activity *In vitro*. *Chem Eur J* 2009;**15**:12283–91.
25. Kandioller W, Hartinger CG, Nazarov AA, Kuznetsov ML, John RO, Bartel C, et al. From pyrone to thiopyrone ligands-rendering maltol-derived ruthenium(II)-arene complexes that are anticancer active *In vitro*. *Organometallics* 2009;**28**:4249–51.
26. Hanif M, Henke H, Meier SM, Martic S, Labib M, Kandioller W, et al. Is the reactivity of M(II)-arene complexes of 3-hydroxy-2(1H)-pyridones to biomolecules the anticancer activity determining parameter? *Inorg Chem* 2010;**49**:7953–63.
27. Kasser JH, Kandioller W, Hartinger CG, Nazarov AA, Arion VB, Dyson PJ, et al. Mannich products of kojic acid and N-heterocycles and their Ru(II)-arene complexes: Synthesis, characterization and stability. *J Organomet Chem* 2010;**695**:875–81.
28. Kandioller W, Kurzwernhart A, Hanif M, Meier SM, Henke H, Keppler BK, et al. Pyrone derivatives and metals: From natural products to metal-based drugs. *J Organomet Chem* 2011;**696**:999–1010.
29. Henke H, Kandioller W, Hanif M, Keppler BK, Hartinger CG. Organometallic ruthenium and osmium compounds of pyridin-2-and-4-ones as potential anticancer agents. *Chem Biodiversity* 2012;**9**:1718–27.
30. Middleton Jr E, Kandaswami C, Theoharides TC. The effects of plant flavonoids on mammalian cells: implications for inflammation, heart disease, and cancer. *Pharmacol Rev* 2000;**52**:673–751.
31. Grazul M, Budzisz E. Biological activity of metal ions complexes of chromones, coumarins and flavones. *Coord Chem Rev* 2009;**253**:2588–98.

32. Kurzwernhart A, Kandioller W, Bächler S, Bartel C, Martic S, Buczkowska M, et al. Structure–activity relationships of targeted $Ru^{II}(\eta^6$-*p*-Cymene) anticancer complexes with flavonol-derived ligands. *J Med Chem* 2012;**55**:10512–22.
33. Kurzwernhart A, Kandioller W, Enyedy EA, Novak M, Jakupec MA, Keppler BK, et al. 3-Hydroxyflavones vs. 3-hydroxyquinolinones: Structure-activity relationships and stability studies on Ru^{II}(arene) anticancer complexes with biologically active ligands. *Dalton Trans* 2013;**42**:6193–202.
34. Schwarz MB, Kurzwernhart A, Roller A, Kandioller W, Keppler BK, Hartinger CG. Rhodium(Cp*) compounds with flavone-derived ligand systems: Synthesis and characterization. *Z Anorg Allg Chem* 2013;**639**:1648–54.
35. Kubanik M, Tu JK-Y, Söhnel T, Hejl M, Jakupec MA, Kandioller W, et al. Expanding on the structural diversity of flavone-derived ruthenium$^{II}(\eta^6$-arene) anticancer agents. *Metallodrugs* 2015;**1**:24–35.
36. Kurzwernhart A, Mokesch S, Klapproth E, Adib-Ravazi MS, Jakupec MA, Hartinger CG, et al. Flavonoid-based organometallics with different metal centers - investigations of the effects on reactivity and cytotoxicity. *Eur J Inorg Chem* 2016;240–6.
37. Marko D, Merz K-H, Kunz C, Müller A, Tarasova N, Eisenbrand G. Intracellular localization of 7-benzylamino-6-chloro-2-piperazino-4-pyrrolidino-pteridine in membrane structures impeding the inhibition of cytosolic cyclic AMP-specific phosphodiesterase. *Biochem Pharmacol* 2002;**63**:669–76.
38. Kandioller W, Balsano E, Meier SM, Jungwirth U, Goschl S, Roller A, et al. Organometallic anticancer complexes of lapachol: Metal centre-dependent formation of reactive oxygen species and correlation with cytotoxicity. *Chem Commun* 2013;**49**:3348–50.
39. Kubanik M, Kandioller W, Kim K, Anderson RF, Klapproth E, Jakupec MA, et al. Towards targeting anticancer drugs: Ruthenium(II)-arene complexes with biologically active naphthoquinone-derived ligand systems. *Dalton Trans* 2016;**45**:13091–103.
40. Kubanik M, Lam NYS, Holtkamp HU, Sohnel T, Anderson RF, Jamieson SMF, et al. Quinoline-para-quinones and metals: Coordination-assisted formation of quinoline-ortho-quinones. *Chem Commun* 2018;**54**:992–5.
41. Meier SM, Hanif M, Adhireksan Z, Pichler V, Novak M, Jirkovsky E, et al. Novel metal (II) arene 2-pyridinecarbothioamides: A rationale to orally active organometallic anticancer agents. *Chem Sci* 2013;**4**:1837–46.
42. Hanif M, Moon S, Sullivan MP, Movassaghi S, Kubanik M, Goldstone DC, et al. Anticancer activity of Ru- and Os(arene) compounds of a maleimide-functionalized bioactive pyridinecarbothioamide ligand. *J Inorg Biochem* 2016;**165**:100–7.
43. Meier SM, Kreutz D, Winter L, Klose MHM, Cseh K, Weiss T, et al. An organoruthenium anticancer agent shows unexpected target selectivity for plectin. *Angew Chem Int Ed Engl* 2017;**56**:8267–71.
44. Arshad J, Hanif M, Movassaghi S, Kubanik M, Waseem A, Söhnel T, et al. Anticancer Ru (η^6-p-cymene) complexes of 2-pyridinecarbothioamides: a structure–activity relationship study. *J Inorg Biochem* 2017;**177**:395–401.
45. Kinney WA, Lee NE, Blank RM, Demerson CA, Sarnella CS, Scherer NT, et al. N-phenyl-2-pyridinecarbothioamides as gastric-mucosal protectants. *J Med Chem* 1990;**33**:327–36.
46. Lipinski CA, Lombardo F, Dominy BW, Feeney PJ. Experimental and computational approaches to estimate solubility and permeability in drug discovery and development settings. *Adv Drug Delivery Rev* 1997;**23**:3–25.

47. Bickerton GR, Paolini GV, Besnard J, Muresan S, Hopkins AL. Quantifying the chemical beauty of drugs. *Nat Chem* 2012;**4**:90–8.
48. Kubanik M, Holtkamp H, Söhnel T, Jamieson SMF, Hartinger CG. Impact of the halogen substitution pattern on the biological activity of organoruthenium 8-hydroxyquinoline anticancer agents. *Organometallics* 2015;**34**:5658–68.
49. Bregman H, Carroll PJ, Meggers E. Rapid access to unexplored chemical space by ligand scanning around a ruthenium center: discovery of potent and selective protein kinase inhibitors. *J Am Chem Soc* 2006;**128**:877–84.
50. Hanif M, Nazarov AA, Legin A, Groessl M, Arion VB, Jakupec MA, et al. Maleimide-functionalised organoruthenium anticancer agents and their binding to thiol-containing biomolecules. *Chem Commun* 2012;**48**:1475–7.
51. Moon S, Hanif M, Kubanik M, Holtkamp H, Söhnel T, Jamieson SMF, et al. Organoruthenium and osmium anticancer complexes bearing a maleimide functional group: Reactivity to cysteine, stability, and cytotoxicity. *ChemPlusChem* 2015;**80**:231–6.
52. Allardyce CS, Dyson PJ, Ellis DJ, Heath SL. [Ru(.eta.6-p-cymene)Cl2(pta)] (pta = 1,3,5-triaza-7-phosphatricyclo[3.3.1.1]decane): A water soluble compound that exhibits pH dependent DNA binding providing selectivity for diseased cells. *Chem Commun* 2001;1396–7.
53. Hartinger CG, Ang WH, Casini A, Messori L, Keppler BK, Dyson PJ. Mass spectrometric analysis of ubiquitin-platinum interactions of leading anticancer drugs: MALDI versus ESI. *J Anal At Spectrom* 2007;**22**:960–7.
54. Egger AE, Hartinger CG, Hamidane HB, Tsybin YO, Keppler BK, Dyson PJ. High resolution mass spectrometry for studying the interactions of cisplatin with oligonucleotides. *Inorg Chem* 2008;**47**:10626–33.
55. Groessl M, Tsybin YO, Hartinger CG, Keppler BK, Dyson PJ. Ruthenium versus platinum: Interactions of anticancer metallodrugs with duplex oligonucleotides characterised by electrospray ionisation mass spectrometry. *J Biol Inorg Chem* 2010;**15**:677–88.
56. Meier SM, Tsybin YO, Dyson PJ, Keppler BK, Hartinger CG. Fragmentation methods on the balance: Unambiguous top-down mass spectrometric characterization of oxaliplatin-ubiquitin binding sites. *Anal Bioanal Chem* 2012;**402**:2655–62.
57. Meier SM, Babak MV, Keppler BK, Hartinger CG. Efficiently detecting metallodrug-protein adducts: Ion trap versus time-of-flight mass analyzers. *ChemMedChem* 2014;**9**:1351–5.
58. Bytzek AK, Koellensperger G, Keppler BK, G Hartinger C. Biodistribution of the novel anticancer drug sodium trans [tetrachloridobis(1H-indazole)ruthenate(III)] KP-1339/IT139 in nude BALB/c mice and implications on its mode of action. *J Inorg Biochem* 2016;**160**:250–5.
59. Meier SM, Novak M, Kandioller W, Jakupec MA, Arion VB, Metzler-Nolte N, et al. Identification of the Structural determinants for anticancer activity of a ruthenium arene peptide conjugate. *Chem Eur J* 2013;**19**:9297–307.
60. Sullivan MP, Groessl M, Meier SM, Kingston RL, Goldstone DC, Hartinger CG. The metalation of hen egg white lysozyme impacts protein stability as shown by ion mobility mass spectrometry, differential scanning calorimetry, and X-ray crystallography. *Chem Commun* 2017;**53**:4246–9.
61. Bytzek AK, Hartinger CG. Capillary electrophoretic methods in the development of metal-based therapeutics and diagnostics: New methodology and applications. *Electrophoresis* 2012;**33**:622–34.

62. Holtkamp H, Grabmann G, Hartinger CG. Electrophoretic separation techniques and their hyphenation to mass spectrometry in biological inorganic chemistry. *Electrophoresis* 2016;**37**:959–72.
63. Bytzek AK, Boeck K, Hermann G, Hann S, Keppler BK, Hartinger CG, et al. LC- and CZE-ICP-MS approaches for the *in vivo* analysis of the anticancer drug candidate sodium trans-[tetrachloridobis(1H-indazole)ruthenate(III)] (KP1339) in mouse plasma. *Metallomics* 2011;**3**:1049–55.
64. Groessl M, Bytzek A, Hartinger CG. The serum protein binding of pharmacologically active gallium(III) compounds assessed by hyphenated CE-MS techniques. *Electrophoresis* 2009;**30**:2720–7.
65. Groessl M, Hartinger CG, Polec-Pawlak K, Jarosz M, Keppler BK. Capillary electrophoresis hyphenated to inductively coupled plasma-mass spectrometry: A novel approach for the analysis of anticancer metallodrugs in human serum and plasma. *Electrophoresis* 2008;**29**:2224–32.
66. Bytzek AK, Reithofer MR, Galanski M, Groessl M, Keppler BK, Hartinger CG. The first example of MEEKC-ICP-MS coupling and its application for the analysis of anticancer platinum complexes. *Electrophoresis* 2010;**31**:1144–50.
67. Giringer K, Holtkamp HU, Movassaghi S, Tremlett W, Lam NYS, Kubanik M, et al. Analysis of ruthenium anticancer agents by MEEKC-UV and MEEKC–ICP-MS: Impact of structural motifs on lipophilicity and biological activity. *Electrophoresis* 2018. in press.
68. Murray BS, Babak MV, Hartinger CG, Dyson PJ. The development of RAPTA compounds for the treatment of tumors. *Coord Chem Rev* 2016;**306**:86–114.

Chapter 9

Antimicrobial Activity of Organometal Compounds: Past, Present, and Future Prospects

Ulrich Schatzschneider
Institut für Anorganische Chemie, Julius-Maximilians-Universität Würzburg, Am Hubland, Würzburg, Germany

9.1 INTRODUCTION

In the chemotherapy of cancer, metal-based agents such as cisplatin, carboplatin, and oxaliplatin (Fig. 9.1A), have been an essential part of many therapeutic approaches since the approval of cisplatin for clinical applications in the mid-1970s, often in the form of a combination therapy together with other drugs.[1,2] Annual sales of platinum drugs amount to several hundred million US$ per year and have enabled fully curative treatment of certain forms of malignancies, in particular testicular cancer, although many others remain much less responsive.[3] Since then, a wide range of other metal compounds across the periodic table has been screen, admittedly with largely variable success, for similar indications.[4,5]

In contrast, significantly less attention has been paid to the antimicrobial potential of metal-based drug candidates. However, with the steadily increasing emergence of drug-resistant pathogenic microorganisms and the risk to enter a "post-antibiotic era", in which infectious diseases become lethal again due to a lack of potent drugs, this part of the chemical space cannot be ignored.

One can only speculate why just a rather small part of the medicinal inorganic chemistry community, which is otherwise well-connected, and quite visible, focuses on the development of novel antimicrobial metal complexes. One reason is possibly in the difficult handling of pathogenic microorganisms, which is more tightly regulated than experiments with cancer cells. Furthermore, parasites are often more difficult to culture than simple adherent cell lines. Thus, screening opportunities are a notable – bottleneck, although there are now non-profit platforms for antimicrobial drug discovery which do not set limits to the compounds accepted for

Advances in Bioorganometallic Chemistry. DOI: https://doi.org/10.1016/B978-0-12-814197-7.00009-1

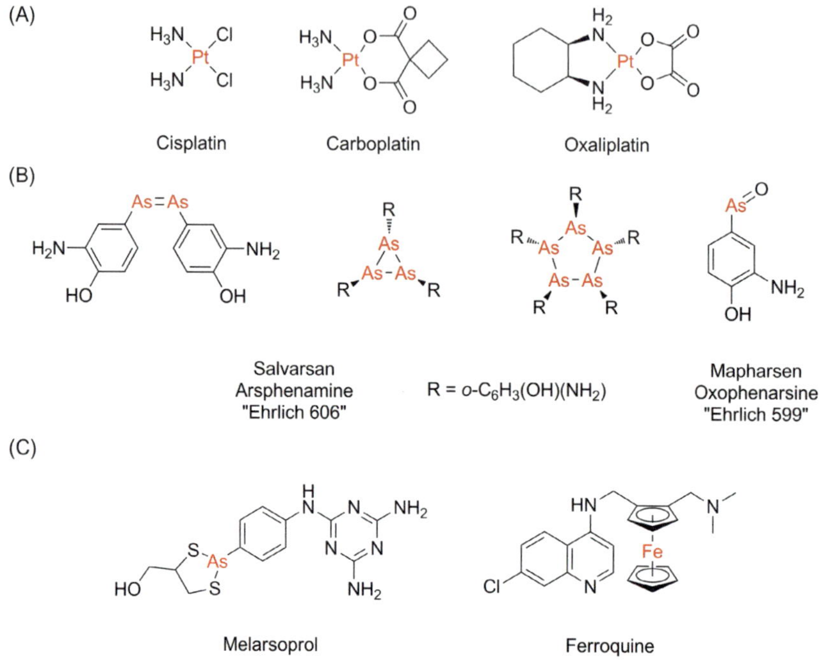

FIGURE 9.1 (A) Structures of the most important clinically approved metal-based anticancer drugs. (B) Different structures proposed for Salvarsan ("Ehrlich 606"), the first chemotherapeutic agent, developed by Ehrlich for the treatment of syphilis. While initially, a structure with an As = As double bond was assumed, later investigations showed a main composition of three- and five-membered ring structures. The active species, however, is most likely Mapharsen, which was separately prepared as "Ehrlich 599". (C) Examples of metal-based antiparasitic drugs and drug candidates.

screening, beyond solubility in a proper solvent such as dimethyl sulfoxide (DMSO).[1] In addition, in the pharmaceutical industry and among some pharmacists, there seems to be a rather persistent perception that "metals" are generally toxic. This is, of course, a blatantly oversimplified view as it does not distinguish the elemental form of a certain metal from its different molecular compounds. Thus, much more educational efforts are urgently need to make it generally known that the toxicity of a metal is not a constant property, but strongly dependent on its oxidation state, coordination geometry, type of ligands, and ligand exchange kinetics when challenged with different biological donor groups from amino acid side chains, nucleotides, and small molecules, as well as solubility in aqueous medium. Furthermore, uptake, and speciation in complex biosystems will strongly depend on environmental

1. For example, CO-ADD, Community for Open Antimicrobial Drug Discovery, http://www.co-add.org

conditions, membrane composition, and active versus passive influx and efflux systems.[6] For example, even for a metal generally assumed to be highly toxic such as chromium, LD_{50} values vary by several orders of magnitude depending on the chemical form, ranging from about 50 mg/kg for chromium (VI) oxide (CrO_3) to more than 15 g/kg for chromium(III) oxide (Cr_2O_3), with the only difference in the metal oxidation state (oral LD_{50} values on rats).[6] Thus, very clearly, generalized statements on "metal toxicity" are not warranted. Instead, it is strongly believed that coordination and organometal compounds with a well-defined coordination sphere and known ligand exchange properties can be subject to standard drug development procedures as any other organic drug candidate alike.

Interestingly, the founder of modern chemotherapy, Nobel Prize winner Paul Ehrlich, did not restrict his search for compounds capable of killing microorganisms invading the human body based on such considerations. Possibly inspired by the centuries-old knowledge on the toxicity of arsenic compounds, going back to antiquity, his synthesis and screening program also involved organoarsenics. Finally, after preparing and screening several hundred compounds, "Ehrlich 606" emerged with potent activity against *Treponema pallidum*, the causative agent of syphilis.[7] Initially thought to be a bimetallic compound with a central As-As double bond, it was later shown to be mostly a mixture of tricyclic and pentacyclic polyarsenic compounds,[8] while the active ingredient is actually oxophenarsine (Mapharsen, Fig. 9.1B). Interestingly, the latter compound was also separately prepared by Ehrlich and co-workers and listed as "Ehrlich 599", but discarded due to high toxicity.[7] Although such arsenicals have fallen out of favour due to toxicity concerns for a long time, compounds such as melarsoprol (Fig. 9.1C), are still indispensable in the treatment of African sleeping sickness and thus listed on the World Health Organisation's (WHO), List of Essential Medicines.

Despite the ground-breaking developments by Ehrlich and his contemporaries, many medicinal inorganic chemists will likely name the discovery of cisplatin by Rosenberg in the early 1960s as the initial emergence of the field.[9] However, already about a decade earlier, in his seminal publication on the "Biological activity of complex ions", Dwyer and co-workers, in 1952, reported on the bacteriostatic and bactericidal properties of some prototypical coordination compounds such as $[Ru(phen)_3](ClO_4)_2$, with phen = 1,10-phenanthroline.[10]

Before moving from this historical perspective to more recent developments, some terminological clarification is required as there is often a confusing use of terms in the literature. Thus, any compound with a direct action on microorganisms, prokaryotic or eukaryotic, regardless of its chemical origin, and used for the treatment or prevention of infections, is called "antimicrobial" (Fig. 9.2). Depending on the type of organism targeted, these are then further subdivided into antibacterials, antifungals, antiprotozoals, and antiparasitics. Antiviral agents are often also considered as a part of this class, although viruses have no life due to the lack of a metabolism

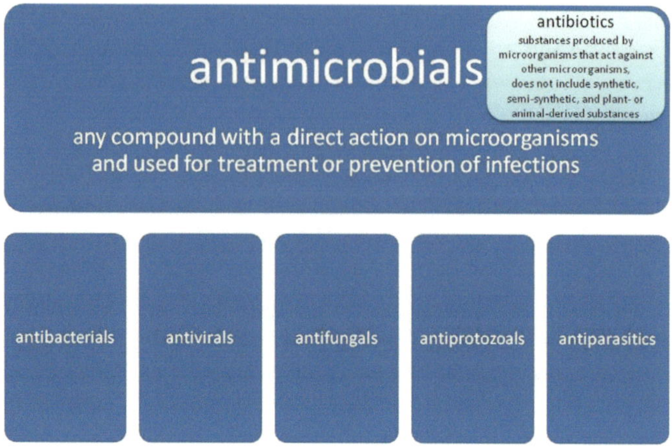

FIGURE 9.2 Different classes of antimicrobials, which are synthetic, semi-synthetic, or natural compounds with a direct action on microorganisms and used for treatment or prevention of infectious diseases. In contrast, antibiotics in a strict sense are only such substances that are produced by microorganisms as chemodefense agents against other microbes.

independent from the host. In contrast, the term "antibiotics" indicates substances produced by microorganisms that act against other microorganisms, and thus does not include synthetic, semi-synthetic, and plant-derived or animal-derived substances.[11] Furthermore, antibacterials can also be classified according to their cellular target structures, including cell wall synthesis inhibition, cell membrane disruption, nucleic acid synthesis inhibition, and protein synthesis inhibition.

9.2 ANTIMALARIAL ACTIVITY OF ORGANOMETAL COMPOUNDS

Malaria is possibly the most widespread infectious disease, with approximately one half of the world population, mostly in tropical and subtropical countries, at risk. The causative agents are unicellular eukaryotic parasites from the genus *Plasmodium*, which are transferred to humans by *Anopheles* mosquitos as the vector. Since the invention of alkaloid drugs such as quinine (Fig. 9.3A), based on native American herbal medicines, a vast development program has focused on novel antiplasmodial drug candidates, to combat drug-resistant strains of the parasite, which already started to emerged in the early 1970s. Two of the most important new antimalarials are artesunate and chloroquine (CQ, Fig. 9.3A), with the discovery of the former honoured with the Nobel Prize in Medicine to Youyou Tu in 2015.[12] Unfortunately, it has not been recorded what initially inspired researchers to prepare and study metal complexes for their antiplasmodial activity, but certainly the quinoline ring as a central structural element in quinine, and

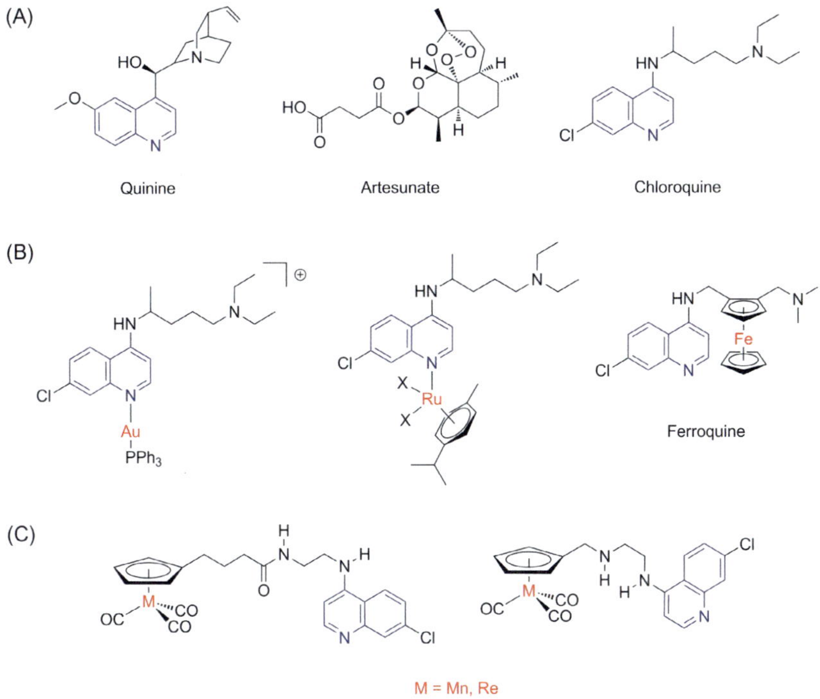

FIGURE 9.3 (A) Organic antimalarial drugs. (B) Organometallic chloroquine complexes and ferroquine, a metal-based antimalarial drug candidate in advanced clinical testing. (C) Metal half-sandwich conjugates of chloroquine evaluated for antimalarial and antitrypanosomal activity.

chloroquine would have fired the imagination of coordination chemists. Therefore, a significant number of metal and organometal fragments have been coordinated to the quinoline ring nitrogen atom as well as the pendant diethylamino group (Fig. 9.3B).[13–15] One of the earliest such works are the palladium(II) and platinum(II) complexes with mefloquine as the ligand reported by Beck and co-workers, although no bioactivity data was reported on these compounds.[16] Albeit lacking structural data, there is also a number of even earlier reports on a wide range of other metals using amodiaquine and primaquine as the ligand.[17] However, the term "metalloantimalarials" first appeared in 2003, when Sharma et al. reported on the antiplasmodial activity of iron(III) and gallium(III) complexes with salen-derived hexadentate ligands. Interestingly, mid-nanomolar IC_{50} values were reported for these early systems on CQ-responsive parasite strains, but even the CQ-resistant Dd2 line gave IC_{50} values in the very low micromolar range.[18] Unfortunately, besides 15 years of research since then, with one notable exception further discussed below, the state of metalloantimalarials has not advanced much beyond empirically guided "synthesize and screen" approaches.

The one exception is organometallic chloroquine-derived ferroquine, in which the 4-amino-7-chloroquinoline core is retained but the side chain replaced by a bulky and rigid 1,2-disubstituted ferrocene moiety (Fig. 9.3B). Pioneered by Biot, a large number of related organometal complexes has been screened for antiplasmodial activity on both CQ-responsive and resistant cells lines, but still ferroquine (FQ, SSR97193) has remained the most promising lead structure.[14] Major in vitro metabolization involves the subsequent loss of the two methyl groups on the distal amino group while the quinoline-ferrocene core of the compound remains intact. It has been hypothesized that the weaker base properties and higher lipophilicity of FQ relative to CQ, as well as a special hydrogen bond between the NH and $N(CH_3)_2$ groups are key factors responsible for the high antiplasmodial potential of ferroquine and allow it to traverse to the digestive vacuole of the parasite, where it interferes with hematin biocrystallization, which results in the production of toxic hydroxyl radicals.[19] Several phase 1/2 clinical studies up to a dose-escalation safety and activity investigation were carried out so far with support from industry,[20] and the bioorganometallics community eagerly awaits approval of ferroquine for clinical use.

Other research groups, including our own, have also ventured to explore alternative organometal moieties attached to the periphery of chloroquine.[13,14,21,22] Conjugation of bulky neutral cymantrene or cyrhetrene groups to the terminal aminoethyl moiety of chloroquine by amide bond formation (Fig. 9.3C), for example, lead to sub-micromolar antimalarial activity on the CQ-resistant Dd2 strain while a cymantrenylmethyl group instead of the dimethylamino moiety in CQ resulted in complete loss of activity at 2 μM.[23] With a resistance index (RI), of approximately 1 and selectivity index (SI) of 20−30, the most potent compounds are equally active against CQ-responsive and resistant plasmodial strains and significantly less toxic to humans compared to the parasite cells, but still do not meet WHO criteria for further lead development.[24,25]

9.3 ANTITRYPANOSOMAL AND ANTILEISHMANIAL ACTIVITY OF ORGANOMETAL COMPOUNDS

Tropical countries, in particular in Africa, are also plagued by other parasitic infections, with human African trypanosomiasis (HAT), commonly also referred to as "sleeping sickness" due to the symptoms which appear when parasites have spread to the brain,[26] possibly the second most prevalent after malaria. This protozoal disease is caused by infections with two different subspecies of *Trypanosoma brucei*, which are transmitted by the bite of the tsetse fly (*Glossina*). About 70 million people are at risk in different parts of Africa, about 10% of the total population of that continent.[27] Only a very limited number of drugs are available for the treatment of HAT, and in particular the secondary brain stage disease often requires application of melarsoprol (Fig. 9.1C). Approximately 5%–10% of the treated patients die due to toxic side effects such as encephalopathies, but for this otherwise fatal disease, there are few other options to date. Related parasites of the species *Trypanosoma cruzi* are also prevalent in South America, where they are transmitted to humans by *Triatoma* bugs ("kissing bugs"), causing Chagas disease.[28]

For decades, only some nitrofuran and nitroimidazole drugs, such as nifurtimox and benznidazole (Fig. 9.4A), were available for the treatment of trypanosomal infections. However, about two decades ago, Sánchez-Delgado and co-workers noticed that azole antifungal agents[29,30] are also able to block the proliferation of *T. cruzi* and different species of *Leishmania*, a closely related family of kinetoplastid parasites. The organic compounds themselves act as sterol biosynthesis inhibitors (SBIs) as they block the

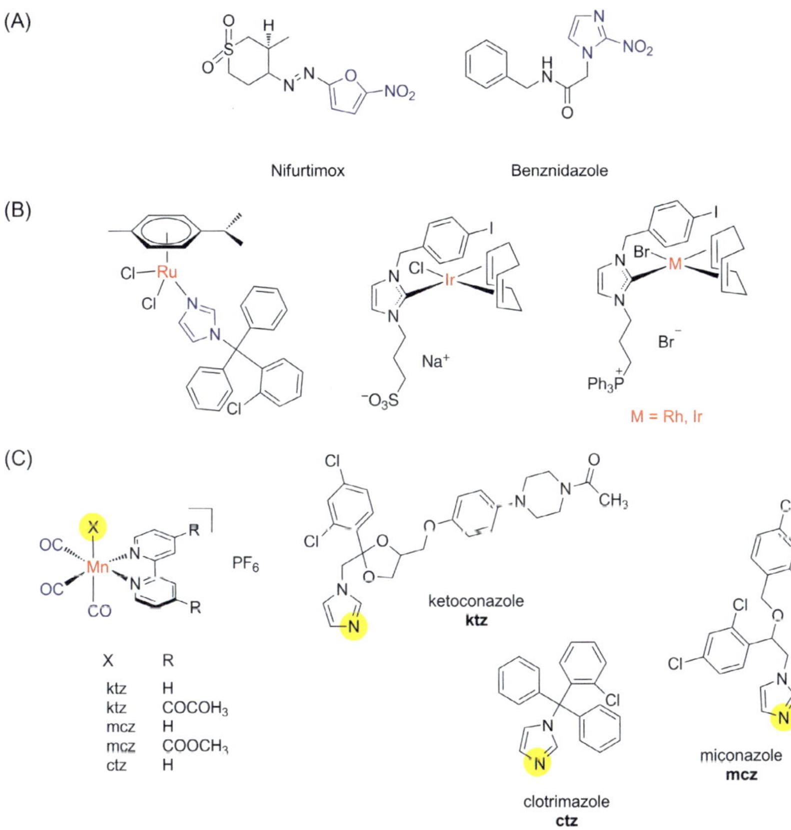

FIGURE 9.4 (A) Organic antitrypanosomal drugs. (B) Organometallic Ru(II) arene as well as rhodium(I) and iridium(I) NHC complexes explored against kinetoplastide parasites. (C) Manganese(I) tricarbonyl complexes with axial azole ligands tested for antitrypanosomal and antibacterial activity.

cytochrome P450-depenent demethylation of lanosterol to ergosterol. When viewed through the eyes of an inorganic chemist, the imidazole group of azole agents such as clotrimazole (CTZ, Fig. 9.4C), immediately captures one's imagination and suggests modification by metal coordination. Thus, a number of ruthenium(II), rhodium(I), platinum(II), gold(III), and copper(II), CTZ complexes were reported in an initial publication and screened for anti-trypanosomal activity on *T. cruzi*.[31] Although no IC_{50} values were given, some of the metal complexes still showed higher growth inhibition on epimastigotes of this parasite than CTZ alone, in particular the ruthenium(II) complex $[RuCl_2(ctz)_2]$. Interestingly, an X-ray structure analysis showed the compound to feature a rare four-coordinated Ru(II) centre, at least in the solid state, while [RhCl(cod)(ctz)] is possibly the first organometallic drug candidate evaluated.[31] Going further back in the literature, even more simple metal salts, such as dimeric rhodium(II), acetate, $[RuCl_2(dmso)_4]$, and cisplatin, were also evaluated against *T. brucei rhodiense*, for example.[32–35] However, significant research efforts have, in the past couple of years, shifted towards the evaluation of organometallic antitrypanosomal drug candidates, due to their robust coordination sphere and facile variation of key structural features in a modular fashion. For example, the ruthenium(II), arene moiety has been utilized, by Maldonado and Sánchez-Delgado, as a scaffold for the attachment of CTZ, as in $[Ru(p\text{-cym})(ctz)\ (X,\ Y)]^n$, with either X = Y = Cl, $n = 0$, X, Y = bpy or en, $n = 2$, or X, Y = acac, $n = 1$, with bpy = 2,2′-bipyridine, en = 1, 2-ethylenediamine, or acac = acetylacetonate (Fig. 9.4B).[36,37] Some of the abovementioned compounds showed sub-micromolar LD_{50} values against *L. major* and *T. cruzi* while the biological activity on human osteoblasts was one to two orders of magnitude lower, highlighting the promise of organometal antiparasitic drug candidates. We also screened some of the cymantrene and cyrhetrene chloroquine conjugates shown in Fig. 9.3C for their anti-leishmanial and anti-trypanosomal activity. Although IC_{50} values on both species were in the low micromolar range, insufficient differentiation between parasite and human cells was observed in this case and, therefore, our focus switched to other lead structures.[23] As cell membranes constitute a formidable barrier for drugs aimed at intracellular target structures, we also set out to explore the role of charged substituents on the antiparasitic activity. To expand the range of metal/ligand combinations, sulfonate and phosphonium groups were attached to the periphery of NHC ligands coordinated to rhodium (I) and iridium(I) centres (Fig. 9.4B).[38] A comparison of [MX(cod)(nhc)] with M = Rh, Ir, X = Cl, Br, cod = 1,4-cyclooctadiene, and an *N*-heterocyclic carbene ligand (nhc) completing the coordination sphere, revealed that the anionic species were totally inactive at up to 100 μM on both parasite and human cells lines, likely due to an incapability to cross the membrane. In contrast, the cationic complexes showed low micromolar to sub-micromolar biological activity on *L. major* and *T. brucei*. Interestingly, the latter parasite species was about

10 to 100 times more responsive to the compounds, with the most potent iridium compound showing an IC_{50} value on *T. brucei* of 150 nM, which is, however, still by a factor of 30 less potent than the pentamidine standard drug evaluated as a control under similar conditions. Still, with a selectivity index of about 20 in the base case, the complexes hold some promise for further optimization. In particular, the NHC ligand will offer a facile handle for further modification and bioconjugation, to improve selective targeting of parasites over host cells.[38]

In an alternative line of investigation, we have expanded the range of azole ligands from clotrimazole (CTZ) to ketoconazole (ktz) and miconazole (MCZ) and introduced these to the axial coordination site in $[Mn(azole)(bpy^{R,R})(CO)_3]^+$, in which the substituent of the 2,2-bipyridine (bpy) coligand was varied from H to $COOCH_3$ for further modulation of the bioavailability (Fig. 9.4C).[39] While the free azoles only showed mid-molar IC_{50} values on *L. major* as well as *T. brucei*, metal coordination to the manganese (I), tricarbonyl moiety resulted in a significant increase of the biological potency. All five organometal compounds evaluated exhibited IC_{50} values below 15 μM on both parasite species, with *T. brucei* in general somewhat more responsive than *L. major*. The complexes featuring the non-substituted bpy ligand were somewhat more potent than the methyl ester ones while the influence of the azole (CTZ versus KTZ versus MCZ) was negligible.[39] The most active complex $[Mn(bpy)(CO)_3(mcz)]PF_6$ showed an IC_{50} value of 0.4 μM on *T. brucei* with a selectivity index of about five versus two different mammalian cells lines and was about 40 times more potent than miconazole alone. Further studies are now aimed to elucidate the role of the metal by comparison of the Mn(I), and Re(I) complexes, as well as inclusion of additional azole as the axial ligand.

9.4 OTHER ANTIPARASITIC ACTIVITIES ORGANOMETAL COMPOUNDS

In addition to unicellular parasites which cause malaria (*P. falciparum*), and African sleeping sickness (*T. brucei*), multicellular organisms are also of considerable importance to public health. This is especially the case with helminths, a class of parasitic worms that switch host between humans and other species. Second only to malaria, schistosomiasis (also called bilharzia), occurs in more than 70 countries, in particular in tropical and subtropical Africa, and affects more than half a billion people. Transmission takes place in fresh water from snails to humans, where the worms reside in the blood vessels and feed on red blood cells.[40] Initially quite effective therapeutic approaches were based on the application of antimony tartrate $K_2[Sb_2(OOC(CHO))_2]$, but later discontinued due to sometimes fatal systemic reactions, which did not improve with other Sb(III)-based drugs. Today, the chemotherapy of schistosomiasis is based on praziquantel (PZQ, Fig. 9.5A) introduced

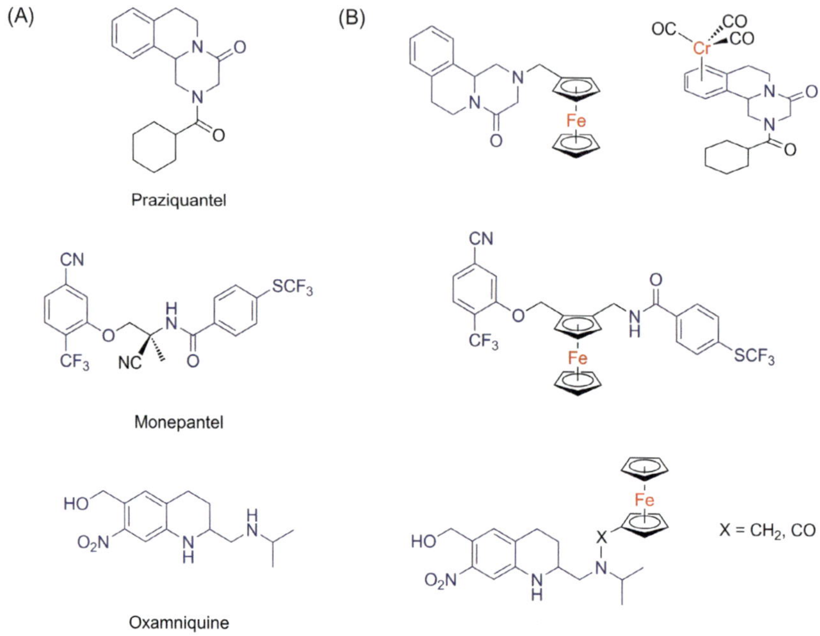

FIGURE 9.5 (A) Organic anthelminthic drugs. (B) Organometallic derivatives of the compounds shown on the left, with ferrocene or chromium(0) tricarbonyl modifications or replacements.

in the 1980s. A particular highlight of this oral drug is the effectiveness against all clinically relevant species of schistosomes with a single one-day treatment at very low cost (US$ 0.20). However, resistances to praziquantel can be induced in laboratory settings for *S. mansoni* and *S. japnocum.*[40]

Thus, inspired by earlier attempts to utilize organometallic moieties such as ferrocene as a surrogate for phenyl and cyclohexyl groups, in 2012, the group of Gasser and co-workers reported on the first organometal praziquantel derivatives (Fig. 9.5B).[41] However, the initial generation of compounds was inactive against *S. mansoni* with IC_{50} values in the mid-micromolar range, while PZQ had an IC_{50} of 0.1 μM under similar conditions. Thus, in a second attempt, the aromatic part of the PZQ core was targeted for modification instead of the cyclohexyl group. Indeed, introduction of a chromium(0), tricarbonyl group to this ring gave rise to new organometal derivatives (Fig. 9.5B), which were however isolated as a mixture of stereoisomers due to addition of the $Cr(CO)_3$ moiety on the same or opposite side of the 11b proton.[42] This modification now lead to sub-micromolar anthelmintic activity, with the most potent organometal compound exhibiting an IC_{50} of 0.25 μM against *S. mansoni*. The selectivity index for these complexes was also very high, with IC_{50} on human cells above 70 μM. However, the potency did not exceed that of praziquantel tested as a reference under similar conditions (IC_{50} of 0.10 μM). Still, these results clearly demonstrate that organometallic drug candidates can be rationally modified and their biological potency improved, as is the case with "normal" organic compounds. Furthermore, the inherent spectroscopic signature of the $Cr(CO)_3$ moiety allowed its detection in worm slices by IR mapping of the carbonyl vibrational band in a window at around 1980–1945 cm^{-1} where cells are essentially transparent while the presence of the chromium(0) was revealed by X-ray fluorescence mapping.[43]

Some ruthenium(II) arene complexes with different azole ligands such as clotrimazole (CTZ), tioconazole (tcz), and miconazole (mcz, Fig. 9.4C), were also evaluated for their antischistosomal activity by Turel and co-workers,[44] but significantly less active than the organometal PZQ compounds described above.

In a hybrid approach akin to the design of ferroquine, Gasser and co-workers also explored a second class of anthelmintic drugs, in which monepantel (Fig. 9.5A), served as the template for the synthesis of organometal derivatives, with the central chiral $CH_2C(CN)(CH_3)$, linker was replaced by a 1,2-disubstituted ferrocene moiety.[45] While the biological activity on mammalian cell lines was generally rather low, some of the monepantel organometallic drug candidates had EC_{50} values of less than 5 μg/mL on parasitic worms, in particular *T. colubriformis* and *D. immitis*, but were unfortunately not given in molar concentrations. Still, compared to this first generation of metal-based drug candidates, organic reference compounds were at least two orders of magnitude more potent.[45] A wide range of other organometal derivatives, replacing ferrocene by cymantrene ($CpMn(CO)_3$) and ruthenocene, were also screened but activity did not improve much.[46]

However, it is mostly the imagination of the bioorganometallic chemist which limits the range of novel lead structures for antimicrobial drug discovery. For example, oxamniquine (Fig. 9.5A), is another anthelmintic drug which was widely used in Brazil against infections with *S. mansoni*, but issues with resistance appeared rather quickly, the first in helminths. Based on knowledge of the specific target of oxamniquine, the parasite-specific sulfotransferase SmSULT, different ferrocene derivatives of the organic parent compound were prepared and evaluated, mostly by conjugation to the easily accessible secondary amine groups.[47] Some of the new organometal compounds lead to significantly higher reduction in parasite viability compared to the reference drug (29% versus 67%) and had low activity on mammalian cells. In particular, the worm burden could be reduced in mice with an efficiency comparable to the organic compound. Although further mechanistic studies will be required to elucidate the mode of action of these complexes, bioorganometallic chemistry certainly holds great promise for further antiparasitic lead discovery.

9.5 ANTIBACTERIAL ORGANOMETAL COMPOUNDS

As mentioned in the introduction, more than 60 years ago, Dwyer and co-workers reported on the bacteriostatic and bactericidal properties of some prototypical coordination compounds such as $[Ru(phen)_3](ClO_4)_2$ with phen = 1,10-phenanthroline.[10] However, it took over four decades until first reports on the antibacterial activity of organometal compounds appeared in the literature. For example, in 1997, Cetinkaya reported on the biological activity of rhodium(I) and ruthenium(II) carbene complexes on a wide range of strains of *Staphylococcus aureus* (Fig. 9.6, top left).[48] Interestingly, low minimum inhibitory concentrations (MICs) of around 5 μg/mL were determined for the title compounds, but this data unfortunately received very little attention as it appeared in a rather obscure journal. Even earlier, Edwards, Epton, and Marr described ferrocenyl conjugates with established penicillin and cephalosporin antibacterials, but only crude qualitative statements ("high antibiotic activity") were given in their publication without MIC values or bacterial strains tested reported.[49] Only just over a decade ago, the principle of ferrocene conjugation for modulation of the antibacterial activity was picked up again and transferred to the field of antibacterial peptides by Metzler-Nolte and co-workers.[50] Elaborate design principles were established

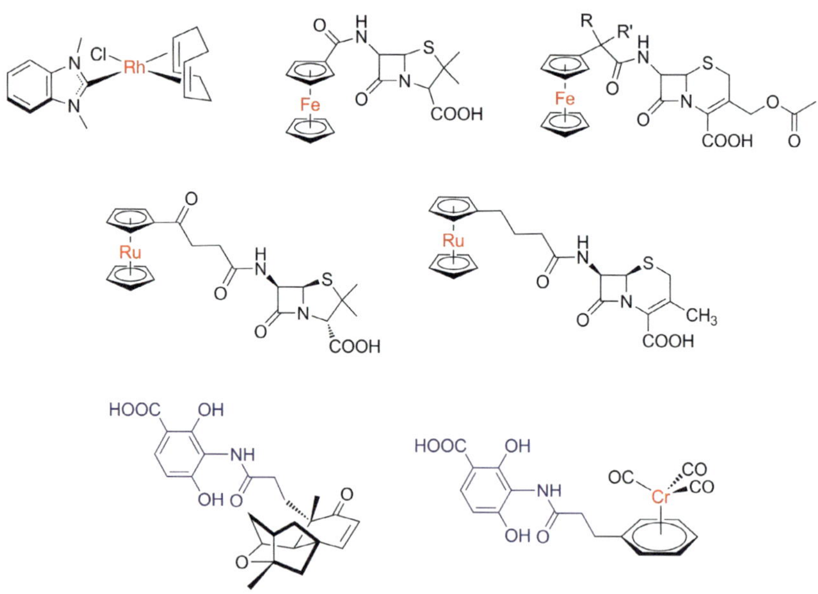

FIGURE 9.6 Organometallic antibacterials based on rhodium(I) NHC complexes and ferrocene as well as ruthenocene penicillin and cephalosporin conjugates (top and centre), together with new organic drug candidate platensimycin and its chromium(0) tricarbonyl derivative (bottom).

since then and the role of the metal centre (Fe versus Ru versus Os) investigated, with low micromolar MIC values (<5 μM), achieved for third-generation antibacterial organometal-peptide conjugates.[51] In very detailed mechanistic studies, finally the activity was traced back to disturbances in the distribution of peripheral membrane proteins.[52] For the design of small-molecule organometal compounds for use as antibacterials, the concept of replacement of bulky organic groups by organometal moieties was also applied to platensimycin, a natural product isolated from bacteria and shown to exhibit high activity against Gram-positive bacteria, including vancomycin-resistant *Enterococcus faecalis*, due to inhibition of a key enzyme (FabF), in essential fatty acid biosynthesis. Thus, the intricate tetracyclic right hemisphere of platensimycin was replaced by a chromium(0) arene tricarbonyl moiety (Fig. 9.6 bottom right), to check whether it is mostly steric bulk that is required for high efficiency.[53] Initial screening of the first-generation compounds was rather disappointing, with all compounds active on *Bacillus subtilis* only at mid-micromolar concentrations and above. So far, even assistance from computer docking experiments did not allow to improve the antibacterial potency of these interesting compounds, indicating that possibly other obstacles, such as low general membrane permeability, are not properly addressed so far.[54]

The great value of structural data on protein binding by (organo)metal antibacterials was demonstrated by the group of Kowalski, who studied ruthenocene conjugates with penicillin derivatives (Fig. 9.6 centre), and were able to obtain a co-crystal structure with CTX-M β-lactamase, which is important in resistance mechanisms, as it inactivates β-lactam drugs by hydrolysis of the four-membered ring.[55] MIC values as low as 2.0 μg/mL against *S. aureus* and *S. epidermidis* were determined for the most potent compounds and values were in a similar range as those for ampicillin used as a control. Interestingly, the X-ray structural data revealed the presence of the hydrolysis product of the Ru conjugate, in which the bulky ruthenocenyl group contributes to stabilization via stacking with the peptide bonds as well as some other non-polar interactions.[55] Careful optimization of the linker and the organic β-lactam moiety allowed these researchers to identify the key structural factors for strong inhibition of DD-carboxypeptidase 64-575 as well as β-lactamase CTX-M-14. In particular, replacement of the five-membered ring in penicillin by a six-membered one in cephalosporin and introduction of a methylene instead of a keto group in the linker adjacent to the Cp_2Ru moiety lead to highest inhibitory power, with IC_{50} values of the best compound at 27 nM. Unfortunately, the in vitro potency did not follow this trend and only mid-micromolar IC_{50} (or rather MIC?), values were determined.[56] Still, additional co-crystal structures revealed interesting details of the organometal-protein interaction, with the ruthenium centre in close contact to a protein sidechain residue of Tyr241. In addition to organometal compounds, various gold complexes have also been evaluated for their antibacterial activity, based on initial result by Robert Koch on the activity of $K[Au(CN)_2]$, on *Mycobacterium tuberculosis*.[57]

Our own contributions to this field were based on the antibacterial effects of carbon monoxide[58] and, therefore, we set out to explore the application of some of our *CO-releasing molecules* (CORMs),[59–61] originally prepared in a different context, against pathogenic bacteria. In a very fruitful collaboration with the groups of Robert Poole (Sheffield) and Roberto La Ragione (Guildford), it was shown that $[Mn(CO)_3(tpa\text{-}\kappa^3N)]Br$ with tpa = tris(2-pyridylmethyl)amine, a dark-stable CO prodrug which releases carbon monoxide only when triggered by light (Fig. 9.7), induced typical changes in the UV/Vis difference spectra of *E. coli* cell cultures that are indicative of CO transfer to heme centres of the bacterial terminal oxidases.[62] Inspired by this observation of in vitro carbon monoxide transfer from a CORM to a bacterial heme protein, cell growth was studied with a UV/Vis assay with succinate as the only carbon source. While bacterial growth was undisturbed by up to 500 μM of metal complex when kept in the dark for up to 1 h, 5 min illumination at 365 nm lead to a significant growth retardation already at 50 μM, while a control sample without metal complex added was not affected by the light exposure in spite the rather high energy excitation wavelength.[62] The work was then extended to clinically more relevant *E. coli* strains which cause urinary tract infections. Such applications are readily compatible with photoactivatable prodrugs due to the availability of waveguide techniques which can be introduced to the bladder, for example, via catheter or even solid organs such as pancreas and prostate through brachytherapy needles.[63] Interestingly, in our in vitro setting, we found that pre-illuminated $[Mn(CO)_3(tpa\text{-}\kappa^3N)]Br$, which has already lost part of the CO ligands, was a more effective antimicrobial agents than CO gas or the non-activated CORM prodrug.[64] Therefore, it appears as if the carbonyl ligands more functions as a photolabile protective group to the Mn(tpa) moiety, which is then mainly responsible for the biological activity observed, rather than the metal coligand fragment acting as an "innocent" carrier for carbon monoxide

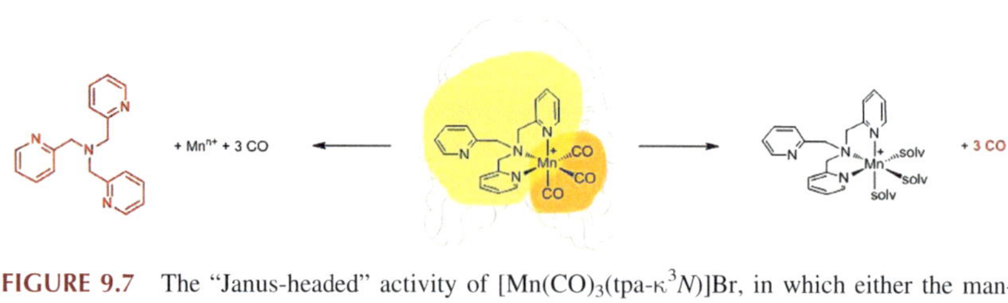

FIGURE 9.7 The "Janus-headed" activity of $[Mn(CO)_3(tpa\text{-}\kappa^3N)]Br$, in which either the manganese(I) tricarbonyl moiety acts as a photolabile protective group for the organic ligand, which later on binds other biometals, or the Mn(tpa) ligand fragment serves as a carrier for carbon monoxide, an important small molecule bioregulator (SMB).

(Fig. 9.7). In follow-up work, this "Janus-headed" activity of $[Mn(CO)_3(tpa\text{-}\kappa^3N)]Br$ was also observed in co-administration experiments, in which only the metal complex but not CO gas was able to potentiate the antibacterial potential of colistin (CST) and doxorubicin (DOX).[65] Finally, the in vivo toxicity and antibacterial potential of this manganese(I) tricarbonyl complex was evaluated in an insect model using larvae of *Galleria mellonella*, a species of wax moth. Relevant toxicity was only observed at or above 200 mg/kg applied metal complex while 40 mg/kg showed no lethality at up to 48 h of incubation.[66] Furthermore, the metal complex in combination with CST largely saved *G. mellonella* from infections with an avian pathogenic strain of *E. coli* (APEC), and completely cleared haemolymph from bacteria at 20 mg/kg Mn complex plus 0.625 mg/kg colistin (CST).

The cationic rhodium and iridium NHC complexes shown in Fig. 9.4C also exhibited antibacterial activity in the low micromolar (5–20 μM), range on Gram-positive pathogens such as *S. aureus*, *S. epidermidis*, *E. faecalis*, and *E. faecium*, while Gram-negative bacteria (*E. coli*, *P. aeruginosa*, *Y. pseudotuberculosa*, and *Y. pestis*), were mostly non-responsive even at the highest tested concentration of 40 μM and the analogous compounds with anionic sulfonate groups were totally inactive on all eight bacterial strains.[38] This again highlights the importance of membrane permeability of metal-based drug candidates, which needs to be tightly controlled and optimized in any rational drug design.

Due to the successful redirection of azole drugs from antifungal to anti-parasitic applications as outlined above, we also studied the antibacterial activity of a series of manganese(I) tricarbonyl complexes of the general formula $[Mn(azole)(bpy^{R,R})(CO)_3]PF_6$, in which $bpy^{R,R}$ is a derivative of 2,2′-bipyridine with R = H, $COOCH_3$, and azoles were clotrimazole (ctz), ketoconazole (ktz), and miconazole (mcz, Fig. 9.4C).[39] The panel of evaluated bacteria was as mentioned above for the Rh and Ir NHC complexes. Again, all Gram-negative organisms were essentially non-responsive at up to 40 μM. In contrast, growth of the Gram-positive bacteria was significantly inhibited by the organometal compounds, with MIC values below 5 μM against all four bacterial strains tested for the most active complexes. The most potent compound with R = H and clotrimazole as the axial monodentate ligand exhibited a sub-micromolar MIC of 0.625 μM on *S. aureus* and *S. epidermidis*. In contrast, the azoles themselves were significantly less potent, with MIC values significantly higher than those observed when coordinated to the metal centre. As a general trend, the non-substituted bpy ligand and lower molecular weight lead to higher antibacterial activity, which increased in the order of ktz < mcz < ctz.[39] In future work, the role of the metal will be explored by evaluation of the rhenium congeners and mechanistic studies undertaken to elucidate the cellular target structures of these compounds.

9.6 CONCLUSION AND OUTLOOK

Although the first chemotherapeutic approaches developed by Paul Ehrlich more than a hundred years ago were based on arsenic main group organometal compounds, the potential of metal complexes for the treatment of infectious diseases has been neglected for a long time. With a dramatic increase of pathogen resistance against antimicrobial drugs and insufficient treatment options for common tropical parasitic infections, there is an urgent need to also explore those parts of the chemical space that involve coordination and organometal compounds. In particular, metal-inherent properties might help to circumnavigate common resistance mechanisms, as they are not present in nature.

In contrast to common misbelief, metals and metal complexes are not generally toxic, as this is a function of oxidation state, coordination geometry, and nature of the ligands, and thus toxicity has to be evaluated for each new compound separately. Rather, metal complexes with well-defined ligand exchange rates can be subject to medicinal chemistry approaches to tune their biological potential as any organic compound alike. Privileged structures for such approaches might be based on elaborate chelating ligands or, given the significant stability of metal–carbon bonds, organometallic compounds.

Even the rather limited research efforts of the past 10–20 years have revealed a wide range of potent organometal complexes with remarkable activity against parasitic and bacterial infections. With ferroquine, incorporating a ferrocene moiety as a prototypical organometallic structure, now in advanced clinical studies, there is hope that metal-based antimicrobial agents might see a similar rise as metal anticancer agents have experienced since their humble start with the work of Rosenberg on cisplatin in the 1960s.

What will be required to really tap into this treasure chest of potential lead structures is a concerted effort by the medicinal inorganic and bioorganometallic chemistry community to promote this avenue of research. Key bottlenecks for future progress are the limited access to screening facilities, in particular with difficult to culture organisms, and a lack of common standards to report bioactivity data. Antibacterial activity, for example, is often reported as the "zone of inhibition" in millimetre based on the agar diffusion Kirby-Bauer test, which might be fine for a quick screening, but lacks comparability, as is the reporting of MIC values in μg/mL or μg/kg, as it will be dependent on the molecular weight of the tested compound. Only reporting the minimum inhibitory concentration (MIC), in molar concentrations (μM), ensures that data can be compared across different publications. Furthermore, although very demanding, for promising compounds, key toxicological data needs to be collected and then supplemented with mechanistic studies, while mere "synthesize-and-screen" approaches might reveal additional interesting lead structures, but will not further advance the field if not supplemented with information on the target structures. Rather, based on initial hits, insights in cellular uptake and intracellular accumulation as well as protein target structures will be required for a rational optimization of promising complexes. As mentioned above, this is possible for skilled organometallic chemists at a level routinely done in organic pharmaceutical chemistry. Given the urgent medical needs to come up with next-generation antimicrobials, there should be a bright future for new organometallic lead structures.

REFERENCES

1. Lippert B. *Cisplatin - Chemistry and Biochemistry of a Leading Anticancer Drug.* Weinheim: VCH; 1999.
2. Kelland L. The resurgence of platinum-based cancer chemotherapy. *Nature Rev Cancer* 2007;**7**:573–84.
3. Wong E, Giandomenico CM. Current status of platinum-based antitumor drugs. *Chem Rev* 1999;**99**:2451–66.
4. Hartinger CG, Dyson PJ. Bioorganometallic chemistry - from teaching paradigms to medicinal applications. *Chem Soc Rev* 2009;**38**:391–401.
5. Barry NE, Sadler PJ. Exploration of the medical periodic table: Towards new targets. *Chem Commun* 2013;**49**:5106–31.
6. Egorova KS, Ananikov VP. Toxicity of metal compounds: Knowledge and myths. *Organomet* 2017;**36**:4071–90.
7. Riethmiller S. From atoxyl to salvarsan: Searching for the magic bullet. *Chemotherapy* 2005;**51**:234–42.
8. Lloyd NC, Morgan HW, Nicholson BK, Ronimus RS. The composition of ehrlich's salvarsan: Resolution of a century-old debate. *Angew Chem Int Ed* 2005;**44**:941–4.
9. Rosenberg B, VanCamp L, Krigas T. Inhibition of cell division in escherichia coli by electrolysis products from a platinum electrode. *Nature* 1965;**205**:698–9.
10. Dwyer FP, Gyarfas EC, Rogers WP, Koch JP. Biological activity of complex ions. *Nature* 1952;**170**:190–1.
11. Li F, Collins JG, Keene FR. Ruthenium complexes as antimicrobial agents. *Chem Soc Rev* 2015;**44**:2529–42.
12. Tu Y. Artemisinin - a gift from traditional chinese medicine to the world (Nobel Lecture). *Angew Chem Int Ed* 2016;**55**:10210–26.
13. Biot C, Castro W, Botte CY, Navarro M. The therapeutic potential of metal-based antimalarial agents: Implications for the mechanism of action. *Dalton Trans* 2012;**41**:6335–49.
14. Navarro M, Castro W, Biot C. Bioorganometallic compounds with antimalarial targets: Inhibiting hemozoin formation. *Organometallics* 2012;**31**:5715–27.
15. Salas PF, Herrmann C, Orvig C. Metalloantimalarials. *Chem Rev* 2013;**113**:3450–92.
16. Hubel R, Polborn K, Knizek J, Nöth H, Beck W. Palladium(II)- und Platin(II)-Komplexe mit dem Antimalariamittel Mefloquin als Liganden. *Z Anorg Allg Chem* 2000;**626**:1701–8.
17. Wasi N, Singh HB, Gajanana A, Raichowdhary AN. Synthesis of metal complexes of antimalarial drugs and *in vitro* evaluation of their activity against *Plasmodium falciparum. Inorg Chim Acta* 1987;**135**:133–7.
18. Harpstrite SE, Beatty A, Collins SD, Oksman A, Goldberg DE, Sharma V. Metalloantimalarials: Targeting of *P. falciparum* strains with novel iron(III) and gallium (III) complexes of an amine phenol ligand. *Inorg Chem* 2003;**42**:2294–300.

19. Biot C, Nosten F, Fraisse L, Ter-Minassian D, Khalife J, Dive D. The antimalarial ferroquine: From bench to clinic. *Parasite* 2011;**18**:207–14.
20. McCarthy JS, Ruckle T, Djeriou E, Cantalloube C, Ter-Minassian D, Baker M, et al. Pilot trial to evaluate safety and efficacy of ferroquine against early *Plasmodium falciparum* in an induced blood-stage malaria infection study. *Malarial J* 2016;**15**:469.
21. Ekengard E, Kumar K, Fogeron T, de Kock C, Smith PJ, Haukka M, et al. Pentamethylcyclopentadienyl-rhodium and iridium complexes containing (N^N and N^O) bound chloroquine analogue ligands: Synthesis, characterization and antimalarial properties. *Dalton Trans.* 2016;**45**:3905–17.
22. Adams M, Li Y, Khot H, De Kock C, Smith PJ, Land K, et al. The synthesis and antiparasitic activity of aryl- and ferrocenyl-derived thiosemicarbazone ruthenium(II)–arene complexes. *Dalton Trans* 2013;**42**:4677–85.
23. Glans L, Hu W, de Kock C, Smith PJ, Haukka M, Bruhn H, et al. Synthesis and biological activity of cymantrene and cyrhetrene 4-aminoquinoline conjugates against malaria, leishmaniasis, and trypanosomiasis. *Dalton Trans* 2012;**41**:6443–50.
24. Nwaka S, Hudson A. Innovative lead discovery strategies for tropical diseases. *Nature Rev Drug Discov* 2006;**5**:941–55.
25. Nwaka S, Ramirez B, Brun R, Maes L, Douglas F, Ridley R. Advancing drug innovation for neglected diseases - criteria for lead progression. *PLOS Negl Trop Dis* 2009;**3**:e440.
26. Barrett MP, Burchmore RJS, Stich A, Lazzari JO, Frasch AC, Cazzulo JJ, et al. The trypanosomiases. *Lancet* 2003;**362**:1469–80.
27. Njoroge M, Njuguna NM, Mutai P, Ongarora DSB, Smith PW, Chibale K. Recent approaches to chemical discovery and development against malaria and the neglected tropical diseases human African trypanosomiasis and schistosomiasis. *Chem Rev* 2014;**114**:11138–63.
28. Stuart K, Brun R, Croft SL, Fairlamb A, Gürtler RE, McKerrow J, et al. Kinetoplastids: Related protozoan pathogens, different diseases. *J Clin Invest* 2008;**118**:1301–10.
29. Kathiravan MK, Salake AB, Chothe AS, Dudhe PB, Watode RP, Mukta MS, et al. The biology and chemistry of antifungal agents: A review. *Bioorg Med Chem* 2012;**20**:5678–98.
30. Heeres J, Meerpoel L, Lewi P. Conazoles. *Molecules* 2010;**15**:4129–88.
31. Sanchez-Delgado RA, Navarro M, Lazardi K, Atencio R, Capparelli M, Vargas F, et al. Toward a novel metal based chemotherapy against tropical diseases 4. Synthesis and characterization of new metal-clotrimazole complexes and evaluation of their activity against Trypanosoma cruzi. *Inorg Chim Acta* 1998;**275–276**:528–40.
32. Farrell NP, Williamson J, McLaren DJM. Trypanocidal and antitumour activity of platinum-metal and platinum-metal-drug dual-function complexes. *Biochem Pharmacol* 1984;**33**:961–71.
33. Osuna A, Ruiz-Perez LM, Lopez MC, Castanys S, Gamarro F, Craciunescu DG, et al. Antitrypanosomal action of cis-diamminedichloroplatinum(II) analogs. *J Parasitol* 1987;**73**:272–7.
34. Farrell NP, Hacker MP. Properties of a chloroquine adduct of rhodium acetate. *Inorg Chim Acta* 1989;**166**:35–7.
35. Sanchez-Delgado RA, Lazardi K, Rincon L, Urbina JA, Hubert AJ, Noels AN. Toward a novel metal-based chemotherapy against tropical diseases. 1. Enhancement of the efficacy of clotrimazole against trypanosoma cruzi by complexation to ruthenium in $RuCl_2$(clotrimazole)$_2$. *J Med Chem* 1993;**36**:2041–3.

36. Martinez A, Carreon T, Iniguez E, Anzellotti A, Sanchez A, Tyan M, et al. Searching for new chemotherapies for tropical diseases: Ruthenium – clotrimazole complexes display high in vitro activity against *Leishmania major* and *Trypanosoma cruzi* and Low toxicity toward normal mammalian cells. *J Med Chem* 2012;**55**:3867–77.
37. Iniguez E, Sanchez A, Vasquez MA, Martinez A, Olivas J, Sattler A, et al. Metal–drug synergy: new ruthenium(II) complexes of ketoconazole are highly active against Leishmania major and Trypanosoma cruzi and nontoxic to human or murine normal cells. *J Biol Inorg Chem* 2013;**18**:779–90.
38. Simpson PV, Schmidt C, Ott I, Bruhn H, Schatzschneider U. Synthesis, cellular uptake and biological activity against pathogenic microorganisms and cancer cells of rhodium and iridium N-heterocyclic carbene metal complexes bearing. *Eur J Inorg Chem* 2013;5547–54.
39. Simpson PV, Nagel C, Bruhn H, Schatzschneider U. Antibacterial and antiparasitic activity of manganese(I) tricarbonyl complexes with ketoconazole, miconazole, and clotrimazole ligands. *Organometallics* 2015;**34**:3809–15.
40. Thetiot-Laurent SAL, Boissier J, Robert A, Meunier B. Schistosomiasis chemotherapy. *Angew Chem Int Ed* 2013;**52**:7936–56.
41. Patra M, Ingram K, Pierroz V, Ferrari S, Spingler B, Keiser J, et al. Ferrocenyl derivatives of the anthelmintic praziquantel: Design, synthesis, and biological evaluation. *J Med Chem* 2012;**55**:8790–8.
42. Patra M, Ingram K, Pierroz V, Ferrari S, Spingler B, Gasser RB, et al. [(η^6-Praziquantel)Cr $(CO)_3$] derivatives with remarkable *in vitro* anti-schistosomal activity. *Chem Eur J* 2013;**19**:2232–5.
43. Clede S, Cowan N, Lambert F, Bertrand HC, Rubbiani R, Patra M, et al. Bimodal X-ray and infrared imaging of an organometallic derivative of praziquantel in *Schistosoma mansoni*. *Chem Bio Chem* 2016;**17**:1004–7.
44. Kljun J, Scott AJ, Rizner TL, Keiser J, Turel I. Synthesis and biological evaluation of organoruthenium complexes with azole antifungal agents. First crystal structure of a tioconazole metal complex. *Organometallics* 2014;**33**:1594–601.
45. Hess J, Patra M, Pierroz V, Spingler B, Jabbar A, Ferrari S, et al. Synthesis, characterization, and biological activity of ferrocenyl analogues of the anthelmintic drug monepantel. *Organometallics* 2016;**35**:3369–77.
46. Hess J, Patra M, Rangasamy L, Konatschnig S, Blacque O, Jabbar A, et al. Organometallic derivatization of the nematocidal drug monepantel leads to promising antiparasitic drug candidates. *Chem Eur J* 2016;**22**:16602–12.
47. Hess J, Panic G, Patra M, Mastrobuoni L, Spingler B, Roy S, et al. Ferrocenyl, ruthenocenyl, and benzyl oxamniquine derivatives with cross-species activity against *Schistosoma mansoni* and *Schistosoma haematobium*. *ACS Infect Dis* 2017;**3**:645–52.
48. Durmaz R, Kücükbay H, Cetinkaya E, Cetinkaya B. Antimicrobial activity of rhodium(I) and ruthenium(II) carbene complexes derived from benzimidazole against *Staphylococcus aureus* isolates. *Turk J Med Sci* 1997;**27**:59–61.
49. Edwards EI, Epton R, Marr G. New class of semisynthetic antibiotics - Ferrocenylpenicillins and ferrocengylcephalosporins. *J Organomet Chem* 1976;**107**:351–7.
50. Chantson JT, Verga Falzacappa MV, Crovella S, Metzler-Nolte N. Antibacterial activities of ferrocenoyl- and cobaltocenium-peptide bioconjugates. *J Organomet Chem* 2005;**690**:4564–72.
51. Albada B, Metzler-Nolte N. Highly potent antibacterial organometallic peptide conjugates. *Acc Chem Res* 2017;**50**:2510–18.

52. Wenzel M, Chiriac AI, Otto A, Zweytick D, May C, Schumacher C, et al. Small cationic antimicrobial peptides delocalize peripheral membrane proteins. *Proc Natl Acad Sci* 2014;**111**:1409–18.
53. Patra M, Gasser G, Pinto A, Merz K, Ott I, Bandow JE, et al. Synthesis and biological evaluation of chromium bioorganometallics based on the antibiotic platensimycin lead structure. *Chem Med Chem* 2009;**4**:1930–8.
54. Patra M, Gasser G, Metzler-Nolte N. Small organometallic compounds as antibacterial agents. *Dalton Trans* 2012;**41**:6350–8.
55. Lewandowski EM, Skiba J, Torelli NJ, Rajnisz A, Solecka J, Kowalski K, et al. Antibacterial properties and atomic resolution X-ray complex crystal structure of a ruthenocene conjugasted β-lactam antibiotic. *Chem Commun* 2015;**51**:6186–9.
56. Lewandowski EM, Szczupak L, Wong S, Skiba J, Guspiel A, Solecka J, et al. Antibacterial properties of metallocenyl-7-ADCA derivatives and structure in complex with CTX-M β-lactamase. *Organometallics* 2017;**36**:1673–6.
57. Glisic BD, Djuran MI. Gold complexes as antimicrobial agents: An overview of different biological activites in relation to the oxidation state of the gold ion and the ligand structure. *Chem Soc Rev* 2014;**43**:5950–69.
58. Wilson JL, Jesse HE, Poole RK, Davidge KS. Antibacterial effects of carbon monoxide. *Curr Pharm Des* 2012;**13**:760–8.
59. Niesel J, Pinto A, Peindy N'Dongo HW, Merz K, Ott I, Gust R, et al. CO release, cellular uptake, and cytotoxicity of a tris(pyrazolyl)methane manganese tricarbonyl complex. *Chem Commun* 2008;1798–800.
60. Schatzschneider U. Novel lead structures and activation mechanisms for CO-releasing molecules (CORMs). *Br J Pharmacol* 2015;**172**:1638–50.
61. Schatzschneider U. PhotoCORMs: Light-triggered release of carbon monoxide from the coordination sphere of transition metal complexes for biological applications. *Inorg Chim Acta* 2011;**374**:19–23.
62. Nagel C, McLean S, Poole RK, Braunschweig H, Kramer T, Schatzschneider U. Introducing $[Mn(CO)_3(tpa\text{-}\kappa^3N)]^+$ as a novel photoactivatable CO releasing molecule with well-defined iCORM intermediates - synthesis, spectroscopy, and antibacterial activity. *Dalton Trans* 2014;**43**:9986–97.
63. Moore CM, Mosse A, Allen C, Payne H, Emberton M, Bown SG. Light penetration in the human prostate: A whole prostate clinical study at 763 nm. *J Biomed Opt* 2011;**16** 015003–015001.
64. Tinajero-Trejo M, Rana N, Nagel C, Jesse HE, Smith TW, Wareham LK, et al. Antimicrobial activity of the manganese photoactivated carbon monoxide-releasing molecule $[Mn(CO)_3(tpa\text{-}\kappa^3N)]^+$ against a pathogenic *Escherichia coli* that causes urinary infections. *Antioxid Redox Signal* 2016;**24**:765–80.
65. Rana N, Jesse HE, Tinajero-Trejo M, Butler JA, Tarlit JD, von und zur Mühlen ML, Nagel C, Schatzschneider U, Poole RK. , A manganese photosensitive tricarbonyl molecule $[Mn(CO)_3(tpa\text{-}\kappa^3N)]Br$ enhances antibiotic efficacy in a multi-drug-resistant *Escherichia coli*. *Microbiology* 2017;**163**:1466–76.
66. Betts Nagel J, Nagel C, Schatzschneider U, Poole RK, La Ragione RM. Antimicrobial activity of carbon monoxide-releasing molecule $[Mn(CO)_3(tpa\text{-}\kappa^3N)]Br$ versus multidrug-resistant isolates of Avian Pathogenic *Escherichia coli* and its synergy with colistin. *PLOS ONE* 2017;**12**:e0186359.

Chapter 10

Targeting Sensitive-Strain and Resistant-Strain Malaria Parasites Through a Metal-Based Approach*

Nadia Baartzes, Tameryn Stringer and Gregory S. Smith
Department of Chemistry, University of Cape Town, Cape Town, South Africa

10.1 INTRODUCTION

The prevalence of parasitic diseases in developing countries poses a huge health risk with a narrow spectrum of effective drugs available for treatment. Of greater concern is the rising widespread resistance that renders current therapies useless. Malaria is a highly infectious parasitic disease that affects the lives of over 40% of the world's population, mainly in tropical and subtropical regions of Africa and Asia.[1] The causative agent is a parasite of the genus *Plasmodium*, the most virulent strain being *Plasmodium falciparum*. The parasite is transmitted by an infected female *Anopheles* mosquito and is extremely pathogenic when it invades the red blood cells of its host. A malarial infection often results in bouts of fever and, in severe cases, death. Conventional chemotherapeutics involve the use of quinoline-based drugs,[2–7] as exemplified by chloroquine, but increased drug resistance has reduced the efficacy of the drugs and these therapies have been replaced largely by fixed-dose combination therapies that contain a derivative of the Chinese natural product, artemisinin, that is more commonly known as artemisinin combination therapy, ACT.[8,9]

With the emergence of resistance to ACT, the search for novel drug regimens pursued, exploring the use of transition metals to derivatize existing

*__Author Contributions__: The manuscript was written through contributions of all authors. All authors have given approval to the final version of the manuscript. ‡These authors contributed equally.

Advances in Bioorganometallic Chemistry. DOI: https://doi.org/10.1016/B978-0-12-814197-7.00010-8

drugs of known biological significance, in an attempt to combat rising resistance. This is a strategy we have used in recent years and the compound, ferroquine, is a prime example in the field of Bioorganometallic Chemistry, of an organometallic compound incorporated into a compound of known therapeutic value.[10–14]

This combination has seen ferroquine overcome the resistance experienced by the parent drug chloroquine through its higher lipophilicity, allowing for greater trans-membrane permeation and higher accumulation in the parasite.[15] It is particularly pleasing to note that ferroquine has re-entered clinical studies and is currently in Phase IIb combination clinical trials with artefenomel, a fast-killing trioxalane.[16]

The incorporation of metals into various organic scaffolds and the synthesis of various metal-based antimalarials has been the subject of several excellent reviews and books.[15,17–26] In keeping with the scope of this book, this chapter is aimed at providing a comprehensive coverage of our research largely on platinum group metal (PGM)-based antimalarials, comprising our development of new compounds against *Plasmodium falciparum* and efforts to target resistant strains of the malaria parasite. The chapter is divided into sections based on different organometallic platforms, namely small molecule and multinuclear transition metal-containing platforms.

10.2 TRANSITION METAL-BASED SMALL MOLECULE PLATFORMS

10.2.1 Thiosemicarbazones

The application of thiosemicarbazones (TSCs) in drug discovery and design has grown since the discovery of thiacetazone (amithiozone), a thiosemicarbazone-based oral antibiotic, previously used in the treatment of tuberculosis. In addition to their known antimycobacterial activity,[27–30] other pharmacological properties including antitumour,[31,32] antibacterial[33,34] and antiparasitic[35–39] activities have also been established in past and current research.

The coordination chemistry of TSCs is well-studied and has been extensively reviewed.[40,41] Transition metal (TM) complexes of thiosemicarbazones incorporating cobalt,[42] copper,[43] nickel,[44] silver,[45] and vanadium[46] have been reported in the literature. Within the context of the application of thiosemicarbazone complexes in antiparasitic studies, our work has focused on transition metal-based complexes containing palladium, platinum, gold, ruthenium, rhodium, and iron (in the form of ferrocene).

10.2.1.1 Palladium and Platinum

Palladium and platinum ions preferentially coordinate to TSCs in bidentate or tridentate chelating modes. Condensation of a salicylaldehyde entity with

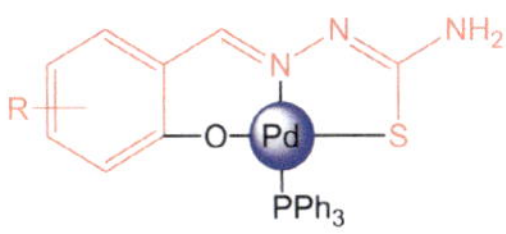

10.1 (R = H)
10.2 (R = 3-*t*-Bu)
10.3 (R = 3-OMe)
10.4 (R = 5-Cl)

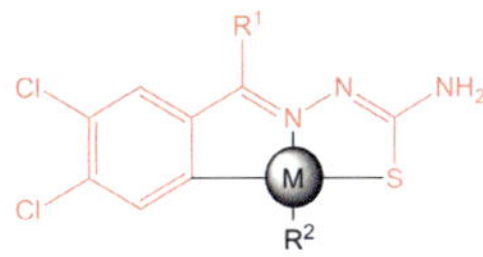

10.5 (M = Pd, R^1 = Me, R^2 = PPh_3)
10.6 (M = Pd, R^1 = Et, R^2 = PPh_3)
10.7 (M = Pd, R^1 = Me, R^2 = PTA)
10.8 (M = Pd, R^1 = Et, R^2 = PTA)
10.9 (M = Pd, R^1 = Me, R^2 = *N*-(2-(diphenylphosphaneyl)benzyl)propan-1-amine)
10.10 (M = Pt, R^1 = Me, R^2 = PPh_3)
10.11 (M = Pt, R^1 = Me, R^2 = PTA)

FIGURE 10.1 Examples of tridentate thiosemicarbazone complexes. *Adapted from Ref. 47 with permission from John Wiley and Sons. Adapted from Refs. 48,49 with permission* from *Elsevier. Adapted with permission from Chellan P, Land KM, Shokar A, Au A, An SH, Clavel CM, Dyson PJ, de Kock C, Smith PJ, Chibale K, Smith GS, Organometallics. 2012;**31**(16): 5791–9. Copyright 2012 American Chemical Society.*

a TSC results in the availability of three donor atoms [O,N,S] and thus the possibility of a tridentate ligand. An example of this was reported by Chellan et al.,[47] in which a series of palladium functionalized salicylaldimine thiosemicarbazones was prepared (Fig. 10.1, **10.1–10.4**). These complexes and corresponding ligands were evaluated for their in vitro antiplasmodial activity against the chloroquine-sensitive (CQS) D10 and chloroquine-resistant (CQR) W2 strains of *P. falciparum*, with the complexes displaying enhanced activity compared to the ligands.[47] In a subsequent study, a tridentate chelate formed by complexation of a Pd(II) ion to 3,4-dichloroacetophenone thiosemicarbazone ligands generated cyclopalladated [C,N,S] complexes via C-H activation.[48] The mononuclear cyclopalladated complexes of this kind (Fig. 10.1, **10.5–10.6**) were screened against the CQS 3D7 and CQR K1 *P. falciparum* strains,[48] and were found to be at least two-fold more active than their corresponding ligands, acting as moderate inhibitors of both parasitic strains.[48] Complexes **10.5** and **10.6** showed comparable inhibition in the sensitive strain, while **10.6** was more potent in the resistant strain.[48]

Adams et al.[49] expanded on this cyclopalladate architecture by varying the phosphorous-containing ligands, as shown in Fig. 10.1 (**10.7–10.9**). In the sensitive NF54 strain, the mononuclear complexes significantly inhibited the parasite ($IC_{50} \sim 2\ \mu M$), with activity maintained in the resistant Dd2 strain.[49] In an analogous way to these Pd(II) complexes, a series of cycloplatinated Pt(II) complexes were synthesised (Fig. 10.1, **10.10–10.11**) and their in vitro antiplasmodial activity tested against the CQS D10 and CQR Dd2 strains.[50] In the sensitive strain, these complexes showed enhanced activity compared to the thiosemicarbazone ligand, with values in the micromolar range.[50] This activity was maintained (**10.11**) and improved (**10.10**) in the resistant parasitic strain.[50]

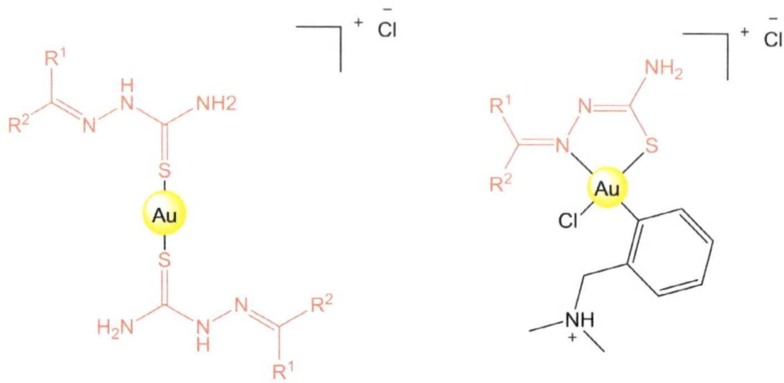

10.12 (R^1 = Me, R^2 = 3-bromobenzene)
10.13 (R^1 = Et, R^2 = 3-bromobenzene)
10.14 (R^1 = Et, R^2 = 3,4-dichlorobenzene)
10.15 (R^1 = H, R^2 = 4-hydroxy-3-methoxybenzene)
10.16 (R^1 = H, R^2 = ferrocene)
10.17 (R^1 = Me, R^2 = 3,4-dichlorobenzene)
10.18 (R^1 = Me, R^2 = 3-bromobenzene)
10.19 (R^1 = Et, R^2 = 3-bromobenzene)
10.20 (R^1 = Et, R^2 = 3,4-dichlorobenzene)
10.21 (R^1 = Me, R^2 = 4-bromobenzene)

FIGURE 10.2 Examples of monodentate and bidentate thiosemicarbazone complexes. *Adapted from Refs. 51,52 with permission from Elsevier.*

10.2.1.2 Gold

The mode of coordination of ligands to a gold ion depends on the oxidation state of the metal. Structures are largely limited to Au(I)-TSC complexes, in which the Au(I) ion is linearly coordinated to two TSC ligands. Further examples include bidentate [N,S] and tridentate [N,N,S] Au(III) complexes of TSCs. Khanye et al.[51] reported the in vitro antiplasmodial activity of Au(I)-TSC complexes (Fig. 10.2, **10.12–10.16**), tested against the D10 and W2 strains. The complexes are good inhibitors of parasite growth and, overall, had enhanced activity in the resistant W2 strain (IC_{50}~3 μM).[51] With the introduction of a ferrocene core, the heterobimetallic (Fe–Au) analogue **10.16**, which was inactive in the CQS strain, showed the most potent inhibition in the CQR strain for the entire series.[51] This supports the notion that the incorporation the organometallic ferrocenyl moiety may be favorable in combatting resistant parasitic strains, as exemplified by ferroquine. Bidentate Au(III) complexes (Fig. 10.2, **10.17–10.21**) of structurally similar TSC ligands were evaluated against the W2 parasitic strain.[52] These complexes have moderate antiplasmodial activity (IC_{50}~7 μM) and have enhanced activity in comparison with the corresponding ligands. Complex **10.18** has superior activity in the series (>2-fold), with an IC_{50} value similar to that of the Au(I) complexes in the same strain (W2).[51,52]

10.2.1.3 Ruthenium and Rhodium

To add diversity, Adams et al.[53] reported the synthesis and antiplasmodial activity (NF54 and Dd2) of cationic half-sandwich ruthenium(II)-arene

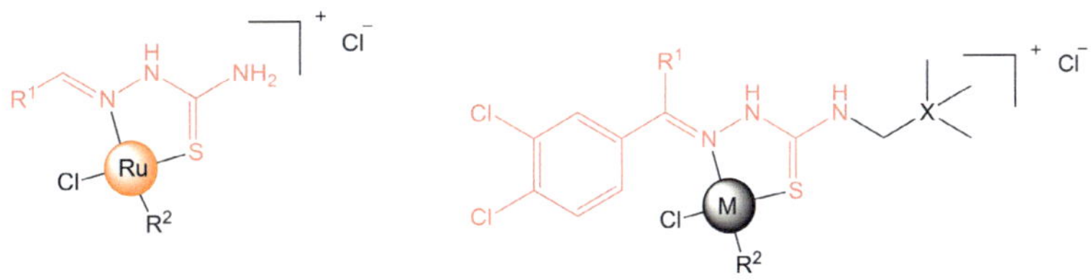

10.22 (R^1 = benzene, R^2 = benzene)
10.23 (R^1 = benzene, R^2 = *p*-cymene)
10.24 (R^1 = 4-chlorobenzene, R^2 = benzene)
10.25 (R^1 = 4-chlorobenzene, R^2 = *p*-cymene)
10.26 (R^1 = naphthalene, R^2 = benzene)
10.27 (R^1 = naphthalene, R^2 = *p*-cymene)

10.28 (M = Ru, R^1 = H, R^2 = p-cymene, X = Si)
10.29 (M = Ru, R^1 = Me, R^2 = p-cymene, X = Si)
10.30 (M = Ru, R^1 = Me, R^2 = p-cymene, X = C)
10.31 (M = Rh, R^1 = H, R^2 = Cp*, X = Si)
10.32 (M = Rh, R^1 = Me, R^2 = Cp*, X = Si)
10.33 (M = Rh, R^1 = Me, R^2 = Cp*, X = C)

FIGURE 10.3 Various monodentate, bidentate, and tridentate thiosemicarbazone complexes. *Adapted from Refs. 53,54 with permission from The Royal Society of Chemistry.*

complexes of aryl-functionalized thiosemicarbazones (Fig. 10.3, **10.22–10.27**). The complexes displayed significantly enhanced activity compared to corresponding ligands. Of the complexes studied, the arylthiosemicarbazone (**10.24–10.27**) complexes showed moderate activity in the low micromolar range ($IC_{50} \leq 6.5\ \mu M$).[53]

Modification of a series of bidentate [N,S] arylthiosemicarbazone complexes containing half-sandwich Ru(II)- and Rh(III) entities (Fig. 10.3, **10.28–10.33**) was achieved by varying the imine R-substituents and incorporating a terminal organosilane moiety.[54] The rationale was that organosilane entities are known to increase the lipophilicity of drug molecules and in turn provide several physiological benefits, which may result in an increase in permeability and bioavailability. Most of the Ru(II) and Rh(III) complexes showed good antiplasmodial activity, with IC_{50} values as low as 1 μM in selected cases, and the antiplasmodial activity either being maintained or improved in the resistant strain.[54] Furthermore, complex **10.32** was screened for its ability to inhibit β-haematin formation (synthetic haemozoin, a target of the malaria parasite), displaying potent inhibition (38 μM) superior to that of chloroquine (74 μM).[54]

10.2.2 Aminoquinolines

Quinoline-based antimalarials represent one of the oldest and highly utilized classes of antimalarials to date. An important example of this class of drugs is chloroquine, a 4-aminoquinoline and one of the most potent antimalarials in use[55] until the emergence of resistant strains of the parasite, which has now rendered it ineffective. Chloroquine (CQ) and other drugs containing the quinoline scaffold, act to inhibit haemozoin formation, the parasite's haem detoxification mechanism, by accumulating in the digestive vacuole.[55–57]

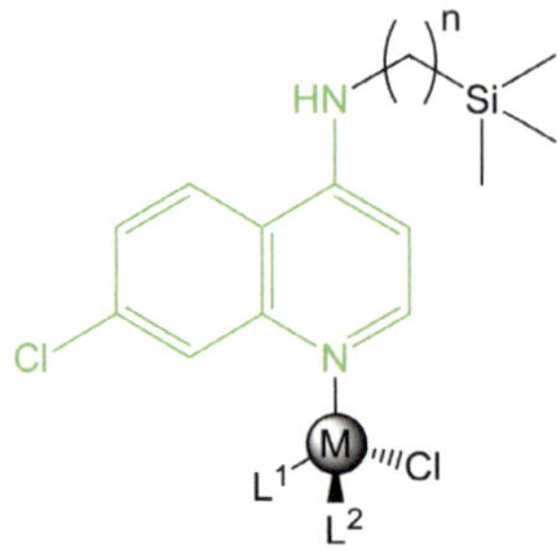

10.34 (n = 1, M = Ru, L^1 = benzene, L^2 = Cl)
10.35 (n = 1, M = Ru, L^1 = *p*-cymene, L^2 = Cl)
10.36 (n = 1, M = Ru, L^1 = $C_6H_5C_2H_4OH$, L^2 = Cl)
10.37 (n = 1, M = Rh, L^1 = L^2 = COD)
10.38 (n = 1, M = Rh, L^1 = Cp*, L^2 = Cl)
10.39 (n = 3, M = Ru, L^1 = benzene, L^2 = Cl)
10.40 (n = 3, M = Ru, L^1 = *p*-cymene, L^2 = Cl)
10.41 (n = 3, M = Ru, L^1 = $C_6H_5C_2H_4OH$, L^2 = Cl)
10.42 (n = 3, M = Rh, L^1 = L^2 = COD)
10.43 (n = 3, M = Rh, L^1 = Cp*, L^2 = Cl)

FIGURE 10.4 Transition metal-based complexes of organosilane-derived quinolines. *Adapted with permission from Li Y, de Kock C, Smith PJ, Guzgay H, Hendricks DT, Naran K, Mizrahi V, Warner DF, Chibale K, Smith GS, Organometallics. 2013;**32**(1):141–50. Copyright 2013 American Chemical Society.*

The formation of coordination complexes of the quinoline moiety has been well explored in the interest of exploiting the broad pharmacological properties of this scaffold. On the heterocycle itself, the quinoline nitrogen is a potential site of metal coordination. A series of Ru(II)-, Rh(III)-, and Rh(I)-aminoquinoline complexes incorporating an organosilane moiety in the lateral side chain (Fig. 10.4, **10.34–10.43**) was prepared by Li et al.[58] The varied ligands were evaluated against the CQS NF54 and CQR Dd2 strains of *P. falciparum*. All of the ligands and their corresponding organometallic complexes were active in vitro antiplasmodial agents, with IC_{50} values in the nanomolar range. In selected cases, the introduction of the transition metal resulted in enhanced antiplasmodial activity (**10.34**, **10.39**, **10.40**) with the Ru(II) complexes generally showing superior activity over the Rh(III) and Rh(I) analogues. In the NF54 strain, selected ruthenium complexes (**10.34**, **10.39**, **10.40**) revealed IC_{50} values less than 100 nM.[58] In general, the activity of the compounds was not maintained in the resistant Dd2 strain, suggesting cross-resistance in a similar way to that of chloroquine. Additionally, for the β-haematin inhibition studies, the ligands and complexes showed comparable or enhanced inhibition activity to chloroquine, suggesting a similar mechanism of antiplasmodial action.[58]

Derivatization of the aminoquinoline with substituted salicylaldehydes generated ligands with multiple coordination sites in the lateral side chain, as reported by Ekengard et al.[59] Bidentate [N,O] Ru(II) and Os(II) complexes of the salicylaldimine ligands (Fig. 10.5, **10.44–10.56**) were extremely active in vitro against the NF54 and Dd2 strains.[59] In the CQ-sensitive (NF54) strain, all ligands and complexes showed good antiplasmodial activity in the low micromolar range ($IC_{50} < 0.73\ \mu M$), with the activity of the ligands being comparable to that of chloroquine.[59] In all cases, activity was not maintained in the resistant Dd2 strain, however, the ligands maintained comparable activity to chloroquine with one exception.[59] This trend suggests

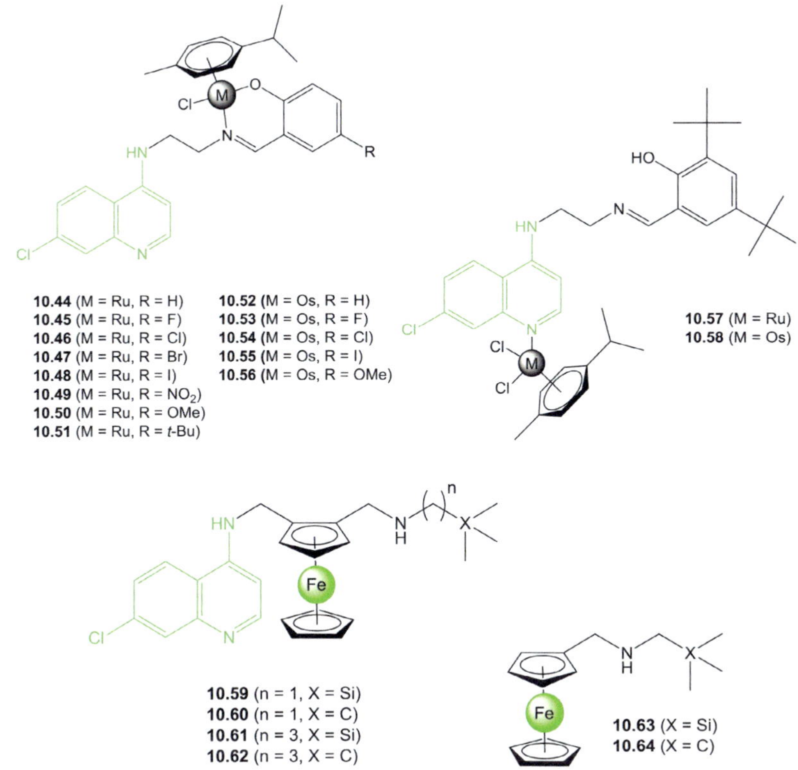

FIGURE 10.5 Transition metal-based complexes of salicylaldimine-derived quinolines and ferrocenyl bioisosteres. *Adapted from Ref. 59 with permission from The Royal Society of Chemistry.*

that these compounds may experience cross-resistance similar to that of chloroquine. Overall, metal complexation resulted in decreased parasite growth inhibition and two ruthenium complexes were completely inactive in the resistant strain.[59] Comparing the activity of the ruthenium complexes against the osmium complexes, the Ru(II) complexes generally proved to be more potent growth inhibitors of *P. falciparum*.[59] Tert-butyl substituted salicylaldimine-aminoquinoline conjugates with metal coordination at the quinoline nitrogen (**10.57–10.58**) were also reported and evaluated against the CQS D10 and CQR Dd2 strains.[59] These complexes had potent inhibition activity in the sensitive strain (<0.06 μM) but displayed decreased activity in the resistant strain, as seen for other Ru(II) and Os(II) quinoline complexes.[59]

In the search for ferroquine-like compounds that evade resistance mechanisms, Adams et al. reported the antiparasitic activity of bioisosteric ferrocenyl-containing quinolines (**10.59–10.62**) and ferrocenylamines (**10.63–10.64**), incorporating organosilanes in their lateral side chains.[60]

Against the CQS NF54 *P. falciparum* strain, the ferrocenyl quinolines are potent inhibitors with values in the nanomolar range, with a high selectivity towards parasites over mammalian cells. The solid-state structure of **10.61** reveals a folded conformation, stabilised by an intramolecular hydrogen bond (analogous to ferroquine) which presumably contributes to its antimalarial activity. The antimalarial activity was maintained (for the silicon analogues – **10.59** and **10.61**) or slightly decreased (for the carbon analogues –**10.60** and **10.62**) in the CQR Dd2 strain.[60] The ferrocenylamines were not as potent, displaying low (**10.63** – 3 μM) and moderate (**10.64** – 24 μM) micromolar IC_{50} values in the sensitive strain.[60] This trend demonstrates the effect of the quinoline entity on the antiplasmodial activity, likely as a result of haemozoin inhibition activity in the parasite and the importance of including a ferrocene core.[60] Despite the good antimalarial activity, these compounds were rapidly metabolised when incubated with both human and mouse liver microsomes. Although similar metabolites of ferroquine were identified, further studies are needed to evaluate the major metabolites.

10.3 TRANSITION METAL-BASED MULTINUCLEAR PLATFORMS

10.3.1 Multinuclearity and Biology

Various approaches have been used to enhance the biological properties of metal-based compounds, particularly in anticancer research. One such strategy is the use of "multinuclearity" to access different properties not exhibited by mononuclear complexes. This involves incorporating two or more metal moieties within one compound generating a multinuclear platform. The complexes could be either homometallic or heterometallic. A plethora of multinuclear complexes have been screened as potential anticancer agents.[61–69] These complexes have shown alternative redox behavior compared to mononuclear systems and have also shown altered reactivity towards biological targets, often with the multinuclear complexes showing enhanced activity.[64,70,71] Despite this promising approach, research of multinuclear complexes as antiparasitic agents has been sparsely explored.

10.3.2 Multinuclear Heterometallic Complexes

10.3.2.1 Thiosemicarbazones

As discussed previously in this chapter, monomeric/mononuclear TSCs were explored, generating interesting biological activities.[33–37,47,72–79] Further to this, multinuclear platinum-based TSCs (Fig. 10.6), specifically tetranuclear (**10.65**–homometallic), dinuclear (**10.66**–homometallic), and trinuclear (**10.67**-hetero-metallic) derivatives were investigated for antiparasitic activity by Chellan et al.[50] The platinum complexes (**10.65**–**10.67**) were tested

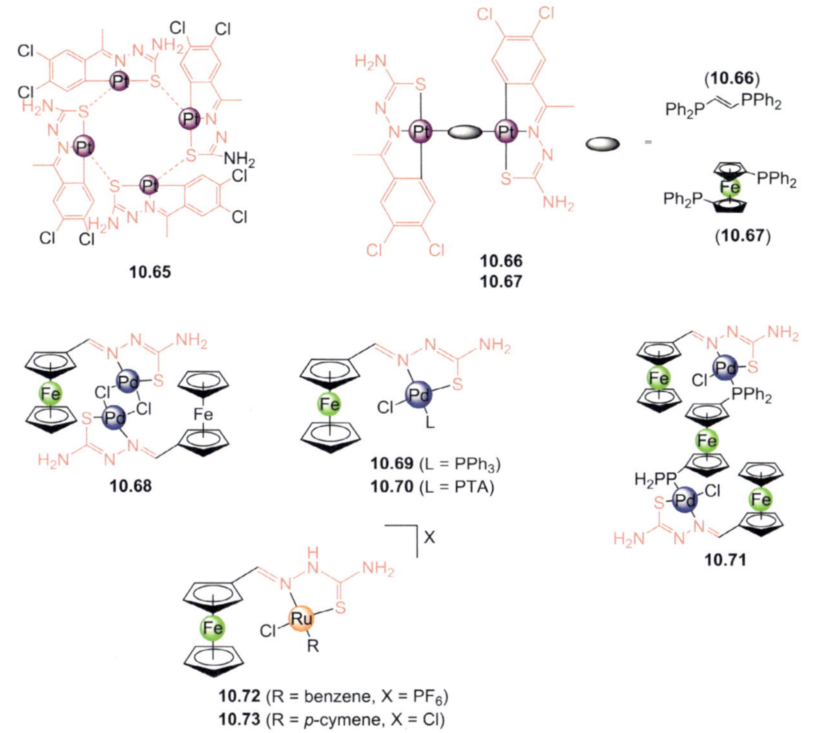

FIGURE 10.6 Homometallic and heterometallic thiosemicarbazones evaluated as antiparasitics. *Adapted with permission from Chellan P, Land KM, Shokar A, Au A, An SH, Clavel CM, Dyson PJ, de Kock C, Smith PJ, Chibale K, Smith GS, Organometallics. 2012;**31**(16):5791–9. Copyright 2012 American Chemical Society. Adapted from Ref. 80 with permission from Elsevier. Adapted from Ref. 53 with permission from The Royal Society of Chemistry.*

against the D10 strain of *P. falciparum*.[50] Comparatively, the tetranuclear complex (**10.65**) is more active than its dinuclear (**10.66**) and trinuclear (**10.67**) analogues, with an IC_{50} value of 32 μM. Complexes **10.66** and **10.67** were not active at the tested concentration, even up to 100 μg/ml. Although **10.65** showed promising activity, it was not as active as chloroquine.[50] Heterometallic ferrocenyl ruthenium and palladium complexes (Fig. 10.6) evaluated by Adams et al.,[53,80] including tetranuclear (**10.68**), dinuclear (**10.69**, **10.70**), and pentanuclear (**10.71**) complexes,[80] were screened for antiplasmodial activity against the chloroquine-sensitive (CQS) (NF54) and chloroquine-resistant (CQR) (Dd2) *P. falciparum* strains.[80] The dinuclear triphenylphosphine complex (**10.69**) showed the most promising activity with IC_{50} values of 3.89 and 4.94 μM for the NF54 and Dd2 strains, respectively. The pentanuclear complex (**10.71**) showed decreased activity in the Dd2 strain compared to the dinuclear PPh_3 complex **10.69**.[80] In a parallel study, cationic ruthenium arene complexes (**10.72** and **10.73**) containing a benzene

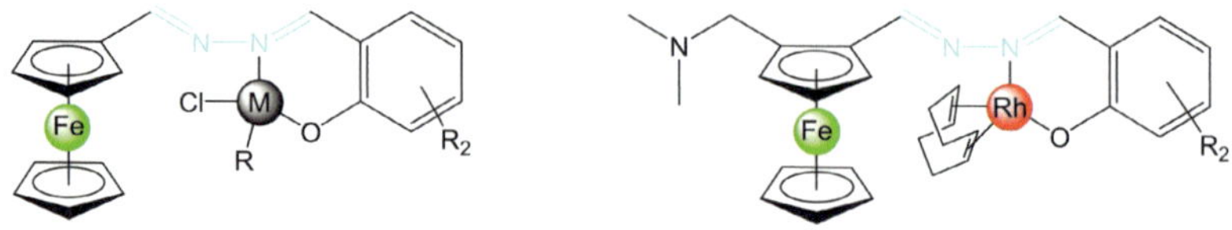

10.74 (R_2 = H; M = Ru; R = *p*-cymene)
10.75 (R_2 = H; M = Rh, R = Cp*)
10.76 (R_2 = H; M = Ir; R = Cp*)
10.80 (R_2 = 5-SO_3^-; M = Ru, R = *p*-cymene)
10.81 (R_2 = 5-SO_3^-; M = Rh, R = Cp*)
10.82 (R_2 = 5-SO_3^-; M = Ir, R = Cp*)

10.77 (R_2 = 5-Cl)
10.78 (R_2 = 3-OMe)
10.79 (R_2 = H)

FIGURE 10.7 Ferrocenyl azine bimetallic complexes screened for antiparasitic activity. *Adapted from Refs. 81,82 with permission* from *Elsevier.*

or *p*-cymene co-ligand were also investigated.[53] Complexes **10.72** and **10.73** displayed IC_{50}s values in the range 6.6–16.5 μM, with the ruthenium-*p*-cymene complex (**10.73**) showing better activity than the ruthenium-benzene analogue (**10.72**).

10.3.2.2 Azines

A series of hetero-bimetallic ferrocenyl azine complexes (**10.74–10.79**) containing either a Ru-*p*-cymene, M-Cp* (where M = Ir, Rh)[81] or Rh-COD (where COD = cyclooctadiene)[82] (Fig. 10.7) moiety were prepared together with water-soluble analogues containing a sulfonated salicylaldimine moiety (**10.80–10.82**), to compare the effect on biological activity.[83] Complexes (**10.74–10.76**) were screened against the NF54 strain of *P. falciparum.*[81] The ruthenium complex (**10.74**) showed slightly enhanced activity in comparison with the rhodium (**10.75**) and iridium (**10.76**) analogues. When compared with the water-soluble analogues, the neutral complexes were found to be more active.[83] Both series were investigated for their ability to inhibit β-haematin formation. The neutral complexes showed moderate inhibition activity, while the sulfonated complexes did not exhibit activity at the tested concentration. The rhodium (COD) complexes **10.77–10.79**, were evaluated against both the NF54 and CQ-resistant K1 strains of *P. falciparum.*[82] All of the complexes showed moderate activity, with the 5-Cl complex (**10.77**) the most active (IC_{50} = 3.9 μM) in the resistant strain.

10.3.2.3 Quinolines and Isoniazid

Incorporating quinoline moieties is a common strategy in bioorganometallic chemistry, however, the study of bimetallic quinoline complexes has been explored to a lesser degree. Li et al. reported on the activity of Ru(II), Rh (III), and Rh(I) carbosilane complexes derived from ferroquine (**10.83–10.87**, Fig. 10.8) against the NF54 and Dd2 strains of *P. falciparum.*[84] These complexes were potent β-haematin inhibitors, a consequence of the presence of

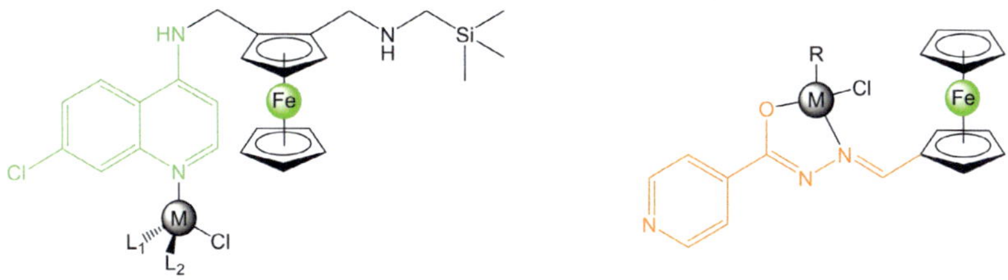

10.83 (M = Ru; L_1 = benzene; L_2 = Cl)
10.84 (M = Ru; L_1 = *p*-cymene; L_2 = Cl)
10.85 (M = Ru; L_1 = $C_6H_5OC_2H_4OH$; L_2 = Cl)
10.86 (M = Rh; L_1 = L_2 = COD)
10.87 (M = Rh; L_1 = Cp*; L_2 = Cl)

10.88 (M = Ru; R = *p*-cymene)
10.89 (M = Rh; R = Cp*)
10.90 (M = Ir; R = Cp*)

FIGURE 10.8 Hetero-bimetallic quinoline and isoniazid complexes. *Adapted with permission from Li Y, De Kock C, Smith PJ, Chibale K, Smith GS, Organometallics. 2014;**33**:4345–8. Copyright 2014 American Chemical Society. Adapted from Ref. 85 with permission from The Royal Society of Chemistry.*

the quinoline moiety. In the NF54 strain, the ruthenium (L_1 = *p*-cymene and $C_6H_5OC_2H_4OH$) and rhodium (L_1 = Cp*) complexes were most active, with IC_{50} values lower than 10 nM[84], as well as greater activity than ferroquine (IC_{50} = 43 nM) in the sensitive strain. This activity was not maintained in the resistant strain, suggesting cross-resistance. All the compounds were more active than chloroquine (IC_{50} = 108 nM), with lower resistance index (RI) values. The rhodium-COD complex (**10.86**) had the highest β-haematin inhibition activity (IC_{50} = 4.84 μM).[84] With a view to repurposing an existing tuberculosis drug, bimetallic complexes containing a ferrocenyl isoniazid ligand (Fig. 10.8) have been studied as antiparasitics.[85] The complexes (**10.88–10.90**) showed moderate activity against NF54 *P. falciparum*, with the rhodium complex (**10.89**) more active (IC_{50} = 2.99 μM) than **10.88** and **10.90**.[85] The rhodium complex (**10.89**) also showed β-haematin inhibition activity comparable to CQ.

10.3.3 Multinuclear Homometallic Polyamines, Polyesters, and Polyethers

10.3.3.1 Polyamines

A promising strategy to target sensitive and resistant strains of *P. falciparum* is to use polyamine-based multimeric/multinuclear scaffolds. Polyamines are low molecular weight molecules, usually aliphatic in nature, that are present in all living cells. Synthetic polyamines have subsequently attracted attention as potential pharmaceuticals[86,87] as studies have shown that polyamines are able to transport drugs into tumorous cells.[88,89] *Plasmodium*-infected erythrocytes show enhanced polyamine uptake compared to uninfected erythrocytes.[90] This attribute prompted us to investigate the synthesis of metal-containing polyamines as potential antiparasitic agents. The multimeric ferrocenyl thiosemicarbazones (**10.91–10.96**, Fig. 10.9),[91–93] ferrocenyl imino/amino (**10.97–10.102**, Fig. 10.9),[94] and quinoline-derived polyamines (**10.103–10.111**, Fig. 10.10)[95,96] were evaluated as potential antiparasitics. Multimeric ferrocenyl TSCs (**10.91–10.96**, Fig. 10.9) were prepared from polyamines such as tris(2-aminoethyl)amine, DAB-Am-4 (polypropylenimine tetraamine dendrimer generation 1), or DAB-AM-8 (polypropylenimine octaamine dendrimer generation 2).[91–93] The complexes (**10.91–10.96**) show modest antiplasmodial activities against NF54 *P. falciparum*, with enhanced activities against resistant strains including Dd2 and W2.[91–93] Compounds **10.91** and **10.92** are also capable of inhibiting synthetic haemozoin formation.[92]

In addition to TSC moieties, ferrocenyl complexes solely possessing the polyamine backbone were also designed (Fig. 10.9).[94] Tris- (**10.97**), tetra- (**10.98**), and octa-ferrocenyl (**10.99**) imino and amino complexes (**10.100–10.102**) were prepared and screened.[94] The ferrocenylimine complexes (**10.97–10.99**) had poor activity in the NF54 strain of *P. falciparum*, however, their ferrocenylamine counterparts (**10.100–10.102**) have enhanced activity in the same strain.[94] With the same idea in mind, quinoline moieties were incorporated into polyamine scaffolds. Dinuclear rhodium-COD

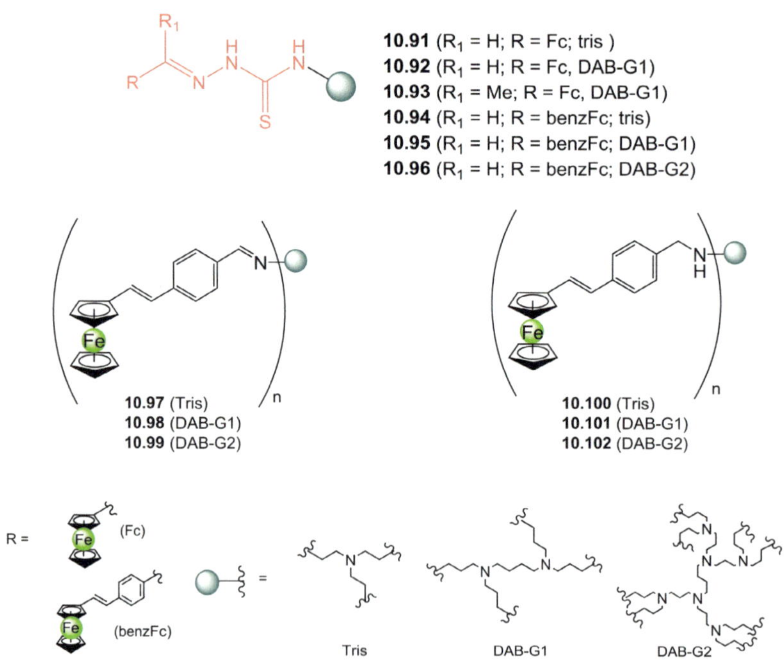

FIGURE 10.9 Multimeric ferrocenyl polyamine complexes. *Adapted from Refs. 91–94 with permission from Elsevier.*

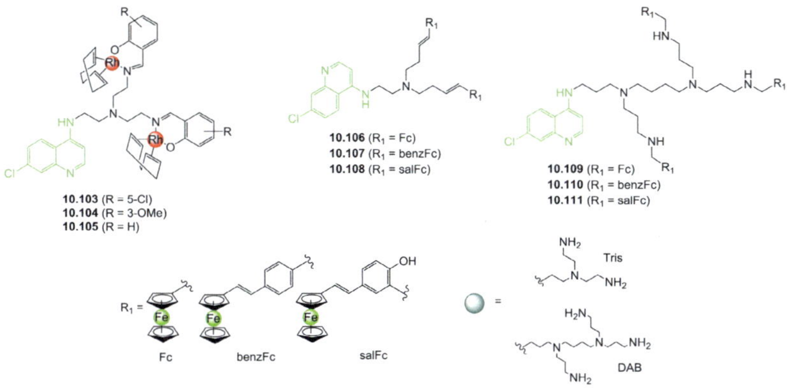

FIGURE 10.10 Multinuclear quinoline complexes with antiparasitic activity. *Adapted from Refs. 95,96 with permission from The Royal Society of Chemistry.*

complexes (**10.103–10.105**, Fig. 10.10) were afforded from bis-salicylaldimines.[95] Since the quinoline nitrogen is of utmost importance for activity, the metal fragment was incorporated into the lateral side chain of the ligands rather than at the quinoline nitrogen. Another strategy focussed on incorporating ferrocene, affording multifunctional ferrocenyl aminoquinoline complexes (**10.106–10.111**, Fig. 10.10).[96] Incorporating the quinoline moiety greatly increased activities in the NF54 strain, while the ferrocenyl polyamines (**10.109–10.111**) showed promising activities in both the CQS (NF54) and CQR (K1) strains.[95,96] The biological activity was maintained in the CQR strain in some cases, further supporting the benefits of incorporating polyamines to combat resistance mechanisms. The complexes were inhibitors of β-haematin formation, as anticipated, due to the presence of the 7-chloroquinoline moiety.

10.3.3.2 Polyesters and Polyethers

The aromatic ester moiety has been vastly investigated in medicinal chemistry due to its lipophilic nature, which is an attractive property for diffusion across cell membranes.[97,98] A variety of biological applications of these compounds exist, ranging from antihypertensive, anti-inflammatory, antitumor, and antimicrobial to antileishmanial.[99–105] Poly(aryl) ethers represent another class of compounds that has been studied as polymeric supports in biology and catalysis.[106–109] This led us to investigate polyester and polyether scaffolds as potential antiparasitic agents. Polynuclear PGM pyridyl ester (**10.112–10.117**) and pyridyl ether (**10.118–10.123**) complexes (Fig. 10.11) containing di- and tri-pyridyl ester/ether ligands acting as monodentate donors were screened for antiparasitic activity.[110,111] The polyester complexes **10.112–10.117** were screened against the NF54 CQS strain of

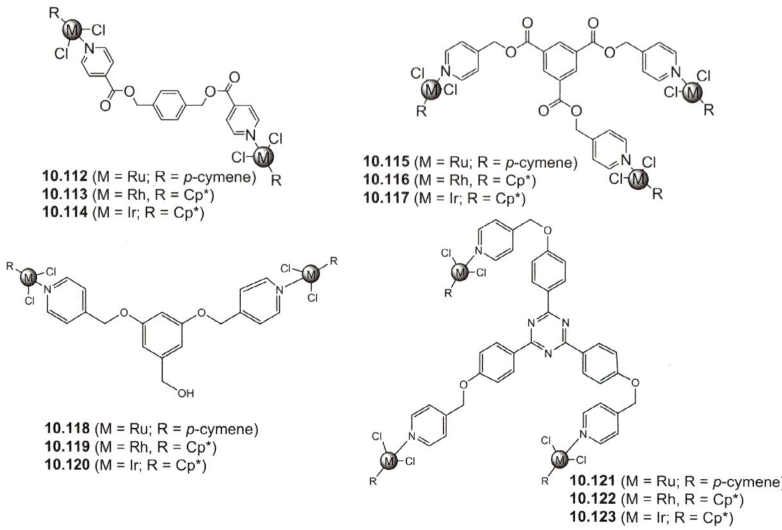

FIGURE 10.11 Polynuclear pyridyl ester and ether PGM complexes. *Adapted from Ref. 110 with permission from The Royal Society of Chemistry. Adapted with permission from Chellan P, Land KM, Shokar A, Au A, An SH, Taylor D, Smith PJ, Chibale K, Smith GS, Organometallics. 2013;***32***(17):4793–804. Copyright 2013 American Chemical Society.*

P. falciparum.[110] The trinuclear polyester complexes (**10.115–10.117**) were more active than their dinuclear counterparts (**10.112–10.114**) and showed superior activity in comparison with their respective metal precursors, suggesting a cooperative effect between the polyester ligand and the metal. The polyether complexes (**10.118–10.123**) were evaluated against the NF54 strain of *P. falciparum*, and the most active complexes evaluated against the Dd2 CQR strain as well.[111] The dinuclear complexes (**10.118–10.120**) had moderate activity in the NF54 strain, while the trinuclear polyethers (**10.121–10.123**) exhibited potent activity in both the NF54 and Dd2 strains, with the trinuclear ruthenium (**10.121**) and rhodium (**10.122**) complexes displaying the most promising activities. Both series were found to be inhibitors of β-haematin formation.

10.4 CONCLUSIONS AND FUTURE OUTLOOK

The increase in resistance to conventional therapies has led to a drive to stem the spread of drug-resistant infections. This has been based largely on scaffolds of known biological significance and subsequently there has been a prolific interest in bioorganometallic-based platforms like small molecule metal complexes and metallomacromolecules, as alternatives to combat resistant microbes. This chapter describes our investigations of organometallic antiplasmodial complexes, some of which show remarkable activity particularly against resistant strains of the *Plasmodium falciparum* parasite. Through optimization of several series over the years, we have prepared hybrid molecules, combining known pharmacophores with organosilanes and polyamines (to name a few) to target drug-resistant strains of the malaria parasite. Invariably, we have demonstrated that the activities can be enhanced in the presence of sandwich (ferrocene) and half-sandwich organometallic complexes containing platinum group metals. The biological space that requires further extensive investigation and elucidation is the mode of action of these complexes. Understanding this would lead to more promising drug leads for application in the realm of bioorganometallic chemistry.

ACKNOWLEDGMENTS

The support from the University of Cape Town and the National Research Foundation of South Africa (Grant No: 111707) are acknowledged.

REFERENCES

1. World Health Organization, *World Malaria Report 2016*, Geneva, Switzerland; 2016).
2. Chavain N, Vezin H, Dive D, Touati N, Paul JF, Buisine E, et al. Investigation of the redox behavior of ferroquine, a new antimalarial. *Mol Pharm* 2008;**5**(5):710–16.
3. Bellot F, Cosledan F, Vendier L, Brocard J, Meunier B, Robert A. Trioxaferroquines as new hybrid antimalarial drugs. *J Med Chem* 2010;**53**(10):4103–9.
4. Cann A. Malaria. Microbiologybytes. <https://microbiologybytes.wordpress.com/2009/04/03/nature-collections-malaria/> 2009.
5. Cheng JJ, Zeidan R, Mishra S, Liu A, Pun SH, Kulkarni RP, et al. Structure - function correlation of chloroquine and analogues as transgene expression enhancers in nonviral gene delivery. *J Med Chem* 2006;**49**(22):6522–31.
6. Olliaro PL, Taylor WRJ. Antimalarial compounds: From bench to bedside. *J Exp Biol* 2003;**206**(21):3753–9.
7. Foley M, Tilley L. Quinoline antimalarials: Mechanisms of action and resistance and prospects for new agents. *Pharmacol Ther* 1998;**79**(1):55–87.
8. Dondorp AM, Nosten F, Yi P, Das D, Phyo AP, Tarning J, et al. Artemisinin resistance in *Plasmodium falciparum* malaria. *New Engl J Med* 2009;**361**(5):455–67.
9. Noedl H, Se Y, Schaecher K, Smith BL, Socheat D, Fukuda MM, et al. Evidence of Artemisinin-Resistant Malaria in Western Cambodia. *New Engl J Med* 2008;**359**(24):2619–20.
10. Daher W, Biot C, Fandeur T, Jouin H, Pelinski L, Viscogliosi E, et al. Assessment of *Plasmodium falciparum* resistance to ferroquine (SSR97193) in field isolates and in W2 strain under pressure. *Malaria J* 2006;**5**:11
11. Daher W, Pelinski L, Klieber S, Sadoun F, Meunier V, Bourrie M, et al. *In vitro* metabolism of ferroquine (SSR97193) in animal and human hepatic models and antimalarial activity of major metabolites on *Plasmodium falciparum*. *Drug Metab Dispos* 2006;**34**(4):667–82.
12. Biot C, Delhaes L, Abessolo H, Domarle O, Maciejewski LA, Mortuaire M, et al. Novel metallocenic compounds as antimalarial agents. Study of the position of ferrocene in chloroquine. *J Organomet Chem* 1999;**589**(1):59–65.

13. Biot C, Delhaes L, N'Diaye CM, Maciejewski LA, Camus D, Dive D, et al. Synthesis and antimalarial activity *in vitro* of potential metabolites of ferrochloroquine and related compounds. *Bioorg Med Chem* 1999;**7**(12):2843–7.
14. Biot C, Taramelli D, Forfar-Bares I, Maciejewski LA, Boyce M, Nowogrocki G, et al. Insights into the mechanism of action of ferroquine. Relationship between physicochemical properties and antiplasmodial activity. *Mol Pharm* 2005;**2**(3):185–93.
15. Dubar F, Egan TJ, Pradines B, Kuter D, Ncokazi KK, Forge D, et al. The antimalarial ferroquine: Role of the metal and intramolecular hydrogen bond in activity and resistance. *Acs Chem Biol* 2011;**6**(3):275–87.
16. Mombo-Ngoma G, Supan C, Dal-Bianco MP, Missinou MA, Matsiegui PB, Salazar CLO, et al. Phase I randomized dose-ascending placebo-controlled trials of ferroquine - a candidate anti-malarial drug - in adults with asymptomatic *Plasmodium falciparum* infection. *Malaria J.* 2011;**10**.
17. Navarro M, Castro W, Madamet M, Amalvict R, Benoit N, Pradines B. Metal-chloroquine derivatives as possible anti-malarial drugs: evaluation of anti-malarial activity and mode of action. *Malaria J* 2014;**13**:471.
18. Salas PF, Herrmann C, Orvig C. Metalloantimalarials. *Chem Rev* 2013;**113**(5):3450–92.
19. Navarro M, Castro W, Biot C. Bioorganometallic compounds with antimalarial targets: Inhibiting haemozoin formation. *Organometallics* 2012;**31**(16):5715–27.
20. Gambino D, Otero L. Perspectives on what ruthenium-based compounds could offer in the development of potential antiparasitic drugs. *Inorg Chim Acta* 2012;**393**:103–14.
21. Biot C, Castro W, Botte CY, Navarro M. The therapeutic potential of metal-based antimalarial agents: Implications for the mechanism of action. *Dalton Trans* 2012;**41**(21):6335–49.
22. Quirante J, Dubar F, Gonzalez A, Lopez C, Cascante M, Cortes R, et al. Ferrocene-indole hybrids for cancer and malaria therapy. *J Organomet Chem* 2011;**696**(5):1011–17.
23. Navarro M, Gabbiani C, Messori L, Gambino D. Metal-based drugs for malaria, trypanosomiasis and leishmaniasis: Recent achievements and perspectives. *Drug Discov Today* 2010;**15**(23–24):1070–8.
24. Biot C, Dive D. Bioorganometallic chemistry and malaria. *Top Organomet Chem* 2010;**32**:155–93.
25. Navarro M. Gold complexes as potential anti-parasitic agents. *Coord Chem Rev* 2009;**253**(11–12):1619–26.
26. Kouznetsov VV, Gomez-Barrio A. Recent developments in the design and synthesis of hybrid molecules based on aminoquinoline ring and their antiplasmodial evaluation. *Eur J Med Chem* 2009;**44**(8):3091–113.
27. Collins FM, Klayman DL, Morrison NE. Correlations between structure and antimycobacterial activity in a series of 2-acetylpyridine thiosemicarbazones. *Microbiology* 1982;**128**(6):1349–56.
28. Coxon GD, Craig D, Corrales RM, Vialla E, Gannoun-Zaki L, Kremer L. Synthesis, antitubercular activity and mechanism of resistance of highly effective thiacetazone analogues. *PLoS One* 2013;**8**(1):e53162.
29. Sriram D, Yogeeswari P, Thirumurugan R, Pavana RK. Discovery of new antitubercular oxazolyl thiosemicarbazones. *J Med Chem* 2006;**49**(12):3448–50.
30. Mashayekhi V, Tehrani KHME, Azerang P, Sardari S, Kobarfard F. Synthesis, antimycobacterial and anticancer activity of novel indole-based thiosemicarbazones. *Arch Pharm Res* 2013;1–13.

31. Quiroga AG, Ranninger CN. Contribution to the SAR field of metallated and coordination complexes: Studies of the palladium and platinum derivatives with selected thiosemicarbazones as antitumoral drugs. *Coord Chem Rev* 2004;**248**(1):119–33.
32. Pandeya S, Dimmock J. Recent evaluations of thiosemicarbazones and semicarbazones and related compounds for antineoplastic and anticonvulsant activities. *Pharmazie* 1993;**48**(9):659–66.
33. Khan SA, Asiri AM, Yusuf M. Synthesis and biological evaluation of some thiazolidinone derivatives of steroid as antibacterial agents. *Eur J Med Chem* 2009;**44**(6):2597–600.
34. Costello C, Karpanen T, Lambert PA, Mistry P, Parker KJ, Rathbone DL, et al. Thiosemicarbazones active against *Clostridium difficile*. *Bioorg Med Chem Lett* 2008;**18**(5):1708–11.
35. de Oliveira RB, de Souza-Fagundes EM, Soares RP, Andrade AA, Krettli AU, Zani CL. Synthesis and antimalarial activity of semicarbazone and thiosemicarbazone derivatives. *Eur J Med Chem* 2008;**43**(9):1983–8.
36. Abid M, Agarwal SM, Azam A. Synthesis and antiamoebic activity of metronidazole thiosemicarbazone analogues. *Eur J Med Chem* 2008;**43**(9):2035–9.
37. Walcourt A, Loyevsky M, Lovejoy DB, Gordeuk VR, Richardson DR. Novel aroylhydrazone and thiosemicarbazone iron chelators with anti-malarial activity against chloroquine-resistant and -sensitive parasites. *Int J Biochem Cell Biol* 2004;**36**(3):401–7.
38. Fujii N, Mallari JP, Hansell EJ, Mackey Z, Doyle P, Zhou Y, et al. Discovery of potent thiosemicarbazone inhibitors of rhodesain and cruzain. *Bioorg Med Chem Lett* 2005;**15**(1):121–3.
39. Beraldo H, Gambino D. The wide pharmacological versatility of semicarbazones, thiosemicarbazones and their metal complexes. *Mini Rev Med Chem* 2004;**4**(1):31–9.
40. Lobana TS, Sharma R, Bawa G, Khanna S. Bonding and structure trends of thiosemicarbazone derivatives of metals—an overview. *Coord Chem Rev* 2009;**253**(7):977–1055.
41. Ali MA, Livingstone S. Metal complexes of sulphur-nitrogen chelating agents. *Coord Chem Rev* 1974;**13**(2–3):101–32.
42. El-Shazly RM, Al-Hazmi G, Ghazy S, El-Shahawi M, El-Asmy A. Synthesis and spectroscopic characterization of cobalt (II) thiosemicarbazone complexes. *J Coord Chem* 2006;**59**(8):845–59.
43. Ferrari MB, Capacchi S, Pelosi G, Reffo G, Tarasconi P, Albertini R, et al. Synthesis, structural characterization and biological activity of helicin thiosemicarbazone monohydrate and a copper (II) complex of salicylaldehyde thiosemicarbazone. *Inorg Chim Acta* 1999;**286**(2):134–41.
44. Leovac VM, Jovanović LS, Jevtović VS, Pelosi G, Bisceglie F. Transition metal complexes with thiosemicarbazide-based ligand—Part LV: Synthesis and X-ray structural study of novel Ni (II) complexes with pyridoxal semicarbazone and pyridoxal thiosemicarbazone. *Polyhedron* 2007;**26**(13):2971–8.
45. Laly S, Parameswaran G. Synthesis and Characterisation of Some thiosemicarbazone complexes of Ag (I), Pt (II) and Pd (II). *Asian J Chem* 1993;**5**(3):712.
46. Mangalam NA, Kurup MP. Synthesis and spectral investigations of vanadium (IV/V) complexes derived from an ONS donor thiosemicarbazone ligand. *Spectrochim Acta A Mol Biomol Spectrosc* 2009;**71**(5):2040–4.
47. Chellan P, Shunmoogam-Gounden N, Hendricks DT, Gut J, Rosenthal PJ, Lategan C, et al. Synthesis, structure and *in vitro* biological screening of palladium(II) complexes of functionalised salicylaldimine thiosemicarbazones as antimalarial and anticancer agents. *Eur J Inorg Chem* 2010;**22**:3520–8.

48. Chellan P, Nasser S, Vivas L, Chibale K, Smith GS. Cyclopalladated complexes containing tridentate thiosemicarbazone ligands of biological significance: Synthesis, structure and antimalarial activity. *J Organomet Chem* 2010;**695**(19):2225–32.
49. Adams M, de Kock C, Smith PJ, Chibale K, Smith GS. Synthesis, characterization and antiplasmodial evaluation of cyclopalladated thiosemicarbazone complexes. *J Organomet Chem* 2013;**736**:19–26.
50. Chellan P, Land KM, Shokar A, Au A, An SH, Clavel CM, et al. Exploring the versatility of cycloplatinated thiosemicarbazones as antitumor and antiparasitic agents. *Organometallics* 2012;**31**(16):5791–9.
51. Khanye SD, Smith GS, Lategan C, Smith PJ, Gut J, Rosenthal PJ, et al. Synthesis and *in vitro* evaluation of gold (I) thiosemicarbazone complexes for antimalarial activity. *J Inorg Biochem* 2010;**104**(10):1079–83.
52. Khanye SD, Wan BJ, Franzblau SG, Gut J, Rosenthal PJ, Smith GS, et al. Synthesis and in vitro antimalarial and antitubercular activity of gold(III) complexes containing thiosemicarbazone ligands. *J Organomet Chem* 2011;**696**(21):3392–6.
53. Adams M, Li YQ, Khot H, De Kock C, Smith PJ, Land K, et al. The synthesis and antiparasitic activity of aryl- and ferrocenyl-derived thiosemicarbazone ruthenium(II)-arene complexes. *Dalton Trans* 2013;**42**(13):4677–85.
54. Adams M, de Kock C, Smith PJ, Land KM, Liu N, Hopper M, et al. Improved antiparasitic activity by incorporation of organosilane entities into half-sandwich ruthenium (II) and rhodium (III) thiosemicarbazone complexes. *Dalton Trans* 2015;**44**(5):2456–68.
55. Egan TJ. Quinoline antimalarials. *Expert Opin Ther Pat* 2001;**11**(2):185–209.
56. Olliaro P. Mode of action and mechanisms of resistance for antimalarial drugs. *Pharmacol Ther* 2001;**89**(2):207–19.
57. Tilley L, Loria P, Foley M. Chloroquine and other quinoline antimalarials. In: Rosenthal P, editor. *Antimalarial chemotherapy*. Humana Press; 2001. p. 87–121.
58. Li YQ, de Kock C, Smith PJ, Guzgay H, Hendricks DT, Naran K, et al. Synthesis, characterization,and pharmacological evaluation of silicon-containing aminoquinoline organometallic complexes as antiplasmodial, antitumor, and antimycobacterial agents. *Organometallics* 2013;**32**(1):141–50.
59. Ekengard E, Glans L, Cassells I, Fogeron T, Govender P, Stringer T, et al. Antimalarial activity of ruthenium (II) and osmium (II) arene complexes with mono-and bidentate chloroquine analogue ligands. *Dalton Trans* 2015;**44**(44):19314–29.
60. Adams M, Stringer T, De Kock C, Smith PJ, Land KM, Liu N, et al. Bioisosteric ferrocenyl-containing quinolines with antiplasmodial and antitrichomonal properties. *Dalton Trans* 2016;**45**(47):19086–95.
61. Farrell N. Polynuclear platinum drugs. *Met Ions Biol Syst* 2004;**42**:251–96.
62. Mendoza-Ferri MG, Hartinger CG, Eichinger RE, Stolyarova N, Severin K, Jakupec MA, et al. Influence of the spacer length on the *in vitro* anticancer activity of dinuclear ruthenium-arene compounds. *Organometallics* 2008;**27**(11):2405–7.
63. Mendoza-Ferri MG, Hartinger CG, Mendoza MA, Groessl M, Egger AE, Eichinger RE, et al. Transferring the concept of multinuclearity to ruthenium complexes for improvement of anticancer activity. *J Med Chem* 2009;**52**(4):916–25.
64. Govender P, Antonels NC, Mattsson J, Renfrew AK, Dyson PJ, Moss JR, et al. Anticancer activity of multinuclear arene ruthenium complexes coordinated to dendritic polypyridyl scaffolds. *J Organomet Chem* 2009;**694**(21):3470–6.
65. Auzias M, Gueniat J, Therrien B, Suss-Fink G, Renfrew AK, Dyson PJ. Arene-ruthenium complexes with ferrocene-derived ligands: Synthesis and characterization of complexes of

the type [Ru(η^6-arene)($NC_5H_4CH_2NHOC$-$C_5H_4FeC_5H_5$)Cl_2] and [Ru(η^6-arene)(NC_3H_3N $(CH_2)_2O_2C$-$C_5H_4FeC_5H_5$)Cl_2]. *J Organomet Chem* 2009;**694**(6):855–61.

66. Casini A, Cinellu MA, Minghetti G, Gabbiani C, Coronnello M, Mini E, et al. Structural and solution chemistry, antiproliferative effects, and DNA and protein binding properties of a series of dinuclear gold(III) compounds with bipyridyl ligands. *J Med Chem* 2006;**49**(18):5524–31.
67. Gabbiani C, Casini A, Messori L, Guerri A, Cinellu MA, Mlnghetti G, et al. Structural characterization, solution studies, and DFT calculations on a series of binuclear gold(III) oxo complexes: Relationships to biological properties. *Inorg Chem* 2008;**47**(7):2368–79.
68. Cinellu MA, Maiore L, Manassero M, Casini A, Arca M, Fiebig HH, et al. [$Au_2(phen^{2Me})_2(\mu$-$O)_2](PF_6)_2$, a novel dinuclear gold(III) complex showing excellent antiproliferative properties. *Acs Med Chem Lett* 2010;**1**(7):336–9.
69. Vergara E, Cerrada E, Casini A, Zava O, Laguna M, Dyson PJ. Antiproliferative activity of gold(I) alkyne complexes containing water-soluble phosphane ligands. *Organometallics* 2010;**29**(11):2596–603.
70. Pelletier F, Comte V, Massard A, Wenzel M, Toulot S, Richard P, et al. Development of bimetallic titanocene-ruthenium-arene complexes as anticancer agents: Relationships between structural and biological properties. *J Med Chem* 2010;**53**(19):6923–33.
71. Govender P, Riedel T, Dyson PJ, Smith GS. Regulating the anticancer properties of organometallic dendrimers using pyridylferrocene entities: Synthesis, cytotoxicity and DNA binding studies. *Dalton Trans* 2016;**45**(23):9529–39.
72. Kovala-Demertzi D, Demertzis MA, Filiou E, Pantazaki AA, Yadav PN, Miller JR, et al. Platinum(II) and palladium(II) complexes with 2-acetyl pyridine 4N-ethyl thiosemicarbazone able to overcome the cisplatin resistance. Structure, antibacterial activity and DNA strand breakage. *Biometals* 2003;**16**(3):411–18.
73. Kolesar JM, Schelman WR, Geiger PG, Holen KD, Traynor AM, Alberti DB, et al. Electron paramagnetic resonance study of peripheral blood mononuclear cells from patients with refractory solid tumors treated with Triapine. *J Inorg Biochem* 2008;**102**(4):693–8.
74. Liu MC, Lin TS, Sartorelli AC. Synthesis and antitumor activity of amino derivatives of pyridine-2-carboxaldehyde thiosemicarbazone. *J Med Chem* 1992;**35**(20):3672–7.
75. Dilovic I, Rubcic M, Vrdoljak V, Kraljevic Pavelic S, Kralj M, Piantanida I, et al. Novel thiosemicarbazone derivatives as potential antitumor agents: Synthesis, physicochemical and structural properties, DNA interactions and antiproliferative activity. *Bioorg Med Chem* 2008;**16**(9):5189–98.
76. Khan SA, Kumar P, Joshi R, Iqbal PF, Saleem K. Synthesis and *in vitro* antibacterial activity of new steroidal thiosemicarbazone derivatives. *Eur J Med Chem* 2008;**43**(9):2029–34.
77. Duffy KJ, Shaw AN, Delorme E, Dillon SB, Erickson-Miller C, Giampa L, et al. Identification of a pharmacophore for thrombopoietic activity of small, non-peptidyl molecules. 1. Discovery and optimization of salicylaldehyde thiosemicarbazone thrombopoietin mimics. *J Med Chem* 2002;**45**(17):3573–5.
78. Du X, Guo C, Hansell E, Doyle PS, Caffrey CR, Holler TP, et al. Synthesis and structure-activity relationship study of potent trypanocidal thio semicarbazone inhibitors of the trypanosomal cysteine protease cruzain. *J Med Chem* 2002;**45**(13):2695–707.
79. Chellan P, Stringer T, Shokar A, Dornbush PJ, Vazquez-Anaya G, Land KM, et al. Synthesis and *in vitro* evaluation of palladium(II) salicylaldiminato thiosemicarbazone complexes against *Trichomonas vaginalis*. *J Inorg Biochem* 2011;**105**(12):1562–8.

80. Adams M, de Kock C, Smith PJ, Malatji P, Hutton AT, Chibale K, et al. Heterobimetallic ferrocenylthiosemicarbazone palladium(II) complexes: Synthesis, electrochemistry and antiplasmodial evaluation. *J Organomet Chem* 2013;**739**:15–20.
81. Nkoana W, Nyoni D, Chellan P, Stringer T, Taylor D, Smith PJ, et al. Heterometallic half-sandwich complexes containing a ferrocenyl motif: Synthesis, molecular structure, electrochemistry and antiplasmodial evaluation. *J Organomet Chem* 2014;**752**:67–75.
82. Stringer T, Guzgay H, Combrinck JM, Hopper M, Hendricks DT, Smith PJ, et al. Synthesis, characterization and pharmacological evaluation of ferrocenyl azines and their rhodium(I) complexes. *J Organomet Chem* 2015;**788**:1–8.
83. Baartzes N, Stringer T, Chellan P, Combrinck JM, Smith PJ, Hutton AT, et al. Synthesis, characterization, antiplasmodial evaluation and electrochemical studies of water-soluble heterobimetallic ferrocenyl complexes. *Inorg Chim Acta* 2016;**446**:111–15.
84. Li Y, De Kock C, Smith PJ, Chibale K, Smith GS. Synthesis and evaluation of a carbosilane congener of ferroquine and its corresponding half-sandwich ruthenium and rhodium complexes for antiplasmodial and β-haematin inhibition activity. *Organometallics* 2014;**33**:4345–8.
85. Stringer T, Seldon R, Liu N, Warner DF, Tam C, Cheng LW, et al. Antimicrobial activity of organometallic isonicotinyl and pyrazinyl ferrocenyl-derived complexes. *Dalton Trans* 2017;**46**(30):9875–85.
86. Karigiannis G, Papaioannou D. Structure, biological activity and synthesis of polyamine analogues and conjugates. *Eur J Org Chem* 2000;**10**:1841–63.
87. Wang JH, Xie SQ, Li YJ, Guo YJ, Ma YF, Zhao J, et al. Synthesis and evaluation of unsymmetrical polyamine derivatives as antitumor agents. *Bioorg Med Chem* 2008;**16**(14):7005–12.
88. Wang CJ, Delcros JG, Cannon L, Konate F, Carias H, Biggerstaff J, et al. Defining the molecular requirements for the selective delivery of polyamine conjugates into cells containing active polyamine transporters. *J Med Chem* 2003;**46**(24):5129–38.
89. Tian ZY, Xie SQ, Mei ZH, Zhao J, Gao WY, Wang CJ. Conjugation of substituted naphthalimides to polyamines as cytotoxic agents targeting the Akt/mTOR signal pathway. *Org Biomol Chem* 2009;**7**(22):4651–60.
90. Muller IB, Das Gupta R, Luersen K, Wrenger C, Walter RD. Assessing the polyamine metabolism of *Plasmodium falciparum* as chemotherapeutic target. *Mol Biochem Parasitol* 2008;**160**(1):1–7.
91. Khanye SD, Gut J, Rosenthal PJ, Chibale K, Smith GS. Ferrocenylthiosemicarbazones conjugated to a poly(propyleneimine) dendrimer scaffold: Synthesis and *in vitro* antimalarial activity. *J Organomet Chem* 2011;**696**(21):3296–300.
92. Stringer T, Taylor D, de Kock C, Guzgay H, Au A, An SH, et al. Synthesis, characterization, antiparasitic and cytotoxic evaluation of thioureas conjugated to polyamine scaffolds. *Eur J Med Chem* 2013;**69**:90–8.
93. Baartzes N, Stringer T, Okombo J, Seldon R, Warner DF, de Kock C, et al. Mono- and polynuclear ferrocenylthiosemicarbazones: Synthesis, characterisation and antimicrobial evaluation. *J Organomet Chem* 2016;**819**:166–72.
94. Baartzes N, Stringer T, Seldon R, Warner DF, de Kock C, Smith PJ, et al. Synthesis, characterization and antimicrobial evaluation of mono- and polynuclear ferrocenyl-derived amino and imino complexes. *J Organomet Chem* 2016;**809**:79–85.
95. Stringer T, Taylor D, Guzgay H, Shokar A, Au A, Smith PJ, et al. Polyamine quinoline rhodium complexes: synthesis and pharmacological evaluation as antiparasitic agents against *Plasmodium falciparum* and *Trichomonas vaginalis*. *Dalton Trans* 2015;**44**(33):14906–17.

96. Stringer T, De Kock C, Guzgay H, Okombo J, Liu J, Kanetake S, et al. Mono- and multimeric ferrocene congeners of quinoline-based polyamines as potential antiparasitics. *Dalton Trans* 2016;**45**(34):13415–26.
97. Ritchie TJ, Macdonald SJF. The impact of aromatic ring count on compound developability - are too many aromatic rings a liability in drug design? *Drug Discov Today* 2009;**14**(21–22):1011–20.
98. Ritchie TJ, Macdonald SJF, Peace S, Pickett SD, Luscombe CN. The developability of heteroaromatic and heteroaliphatic rings - do some have a better pedigree as potential drug molecules than others? *Medchemcomm* 2012;**3**(9):1062–9.
99. Fogt SW, Scozzie JA, Heilman RD, Powers LJ. Lipophilic and hydrophilic esters of 4-acetyl-2-(2-hydroxyethyl)-5,6-bis(4-chlorophenyl)-2H-pyridazin-3-one as antihypertensive agents. *J Med Chem* 1980;**23**(12):1445–8.
100. Kumar S, Arya P, Mukherjee C, Singh BK, Singh N, Parmar VS, et al. Novel aromatic ester from Piper longum and its analogues inhibit expression of cell adhesion molecules on endothelial cells. *Biochemistry-Us* 2005;**44**(48):15944–52.
101. Bandgar BP, Sarangdhar RJ, Fruthous K, Mookkan J, Chaudhary S, Chavan HV, et al. Synthesis and biological evaluation of ester prodrugs of benzafibrate as orally active hypolipidemic agents. *Eur J Med Chem* 2012;**57**:217–24.
102. Aissa I, Sghair RM, Bouaziz M, Laouini D, Sayadi S, Gargouri Y. Synthesis of lipophilic tyrosyl esters derivatives and assessment of their antimicrobial and antileishmania activities. *Lipids Health Dis* 2012;**11**.
103. Hess-Stumpp H, Bracker TU, Henderson D, Politz OMS. -2.7.5., a potent orally available inhibitor of histone deacetylases - The development of an anticancer agent. *Int J Biochem Cell B* 2007;**39**(7–8):1388–405.
104. De Jesus OLP, Ihre HR, Gagne L, Frechet JMJ, Szoka FC. Polyester dendritic systems for drug delivery applications: *In vitro* and *in vivo* evaluation. *Bioconjugate Chem* 2002;**13**(3):453–61.
105. Xiao YL, Hong H, Javadi A, Engle JW, Xu WJ, Yang YA, et al. Multifunctional unimolecular micelles for cancer-targeted drug delivery and positron emission tomography imaging. *Biomaterials* 2012;**33**(11):3071–82.
106. Bonaplata E, Ding H, Hanson BE, Mcgrath JE. Hydroformylation of octene-1 with a poly(arylene ether triaryl phosphine) rhodium complex. *Polymer* 1995;**36**(15):3035–9.
107. Natarajan B, Jayaraman N. Synthesis and studies of Rh(I) catalysts within and across poly(alkyl aryl ether) dendrimers. *J Organomet Chem* 2011;**696**(3):722–30.
108. Bedos-Belval F, Rouch A, Vanucci-Bacque C, Baltas M. Diaryl ether derivatives as anticancer agents - a review. *Medchemcomm* 2012;**3**(11):1356–72.
109. Wang XF, Tian XT, Ohkoshi E, Qin B, Liu YN, Wu PC, et al. Design and synthesis of diarylamines and diarylethers as cytotoxic antitumor agents. *Bioorg Med Chem Lett* 2012;**22**(19):6224–8.
110. Chellan P, Land KM, Shokar A, Au A, An SH, Taylor D, et al. Synthesis and evaluation of new polynuclear organometallic Ru(II), Rh(III) and Ir(III) pyridyl ester complexes as *in vitro* antiparasitic and antitumor agents. *Dalton Trans* 2014;**43**(2):513–26.
111. Chellan P, Land KM, Shokar A, Au A, An SH, Taylor D, et al. Di- and trinuclear ruthenium-, rhodium-, and iridium-functionalized pyridyl aromatic ethers: A new class of antiparasiticagents. *Organometallics* 2013;**32**(17):4793–804.

Chapter 11

Technetium and Rhenium Complexes with Aromatic Hydrocarbons as Ligands: An Entry into Biomimetic Imaging

Roger Alberto, Giuseppe Meola and Daniel Hernández Valdés
Department of Chemistry, University of Zurich, Zurich Switzerland

11.1 INTRODUCTION

Since the commercialization of the ^{99}Mo/^{99m}Tc generator in the early 1960s, the relevance of the daughter radionuclide for radiopharmacy and nuclear medicine has grown.[1,2] Over a long period, basic chemistry with ^{99}Tc, the ground state of metastable ^{99m}Tc, was motivated by potential applications in these fields. Since ^{99m}Tc is present in minute quantities, chromatographic analytical methods are the only way of "characterizing" once-formed compounds. Rarely and only in the recent past with more and more sensitive mass spectrometers, MS became an additional tool. Suspected complexes are prepared with long-lived, macroscopic amounts of ^{99}Tc and their HPLC retention times compared to the products from analogous preparations with ^{99m}Tc. To proceed from ^{99}Tc to ^{99m}Tc was one direction but the reverse procedure was even more frequent; for interesting new complexes with ^{99}Tc, procedures were sought for the preparation of the ^{99m}Tc analogues.[3–5] The transfer for a metal complex preparation between isotopes, or not even isotopes in the case of ^{99m}Tc and ^{99}Tc, seems weird since both in principle possess the same nuclei and electronic structure. However, since ^{99m}Tc solutions are million-fold more diluted, kinetics comes into play which might direct a reaction on a different path than with macroscopic amounts. The preparation of ^{99m}Tc-DTPA (Technescan, Pentetate), of which the structure is still unknown, is a prototype example.[6,7] Furthermore, the preparation of a promising complex developed with ^{99}Tc must be brought down to a level at which the synthesis is done in one-step, in a short time and in close to quantitative

Advances in Bioorganometallic Chemistry. DOI: https://doi.org/10.1016/B978-0-12-814197-7.00011-X

yields. This reduces the accessible numbers and types of complexes substantially, but does not limit complexes to pure coordination compounds as will be shown in this article. Cardiolite, probably the most prominent radiopharmaceutical and an organometallic low valent ^{99m}Tc complex (of known structure) developed by Davison and coworkers in the early 1990s is a very good example.[8] This interplay between macroscopic–microscopic remains a great challenge in technetium chemistry. It shall be emphasized that a potential outcome of new radiopharmaceuticals is not the only thrust but was paralleled for a long time by pure fundamental research. Pure research declined over the past decade for different reasons and only a handful of groups remain active in this field.

For in vivo application of technetium compounds, stability is a core issue. For stabilizing a ^{99m}Tc core, multidentate ligands were thought essential to provide thermodynamic stability.[9] Consequently, the largest number of complexes were reported for $^{99}Tc^{V}$ complexes, stabilized by tetradentate N,S,O ligands as reviewed many times in the past.[1,10–13] Technescan MAG3 developed by Fritzberg in the Eighties[14,15] or ^{99m}Tc-PnAO by Volkert[16,17] are the standard examples for In vivo stable $^{99m}Tc^{V}$ radiopharmaceuticals comprising the $[^{99m}Tc=O]^{3+}$ or $[^{99m}TcO_2]^{+}$ core respectively.[14,18] In the advent of ^{99m}Tc radiopharmaceutical chemistry, the focus was on so-called "perfusion agents" or "technetium-essential" compounds. These terms refer to de-novo complexes with no immediately obvious biological functions. For one reason or the other, they accumulate in specific organs or allow for visualization of functions, but comprise neither targeting nor otherwise bioactive functionalities. As the field advanced, these known and potent chelators were conjugated to targeting, bioactive moieties such as peptides, proteins, or nucleotides to accumulate selectively ^{99m}Tc at a desired target.[2,19] The chelator with its integrated coupling function is referred to "bifunctional chelator" whereas the entire conjugate as second generation radiopharmaceutical (Scheme 11.1, third generation radiopharmaceuticals are discussed below).

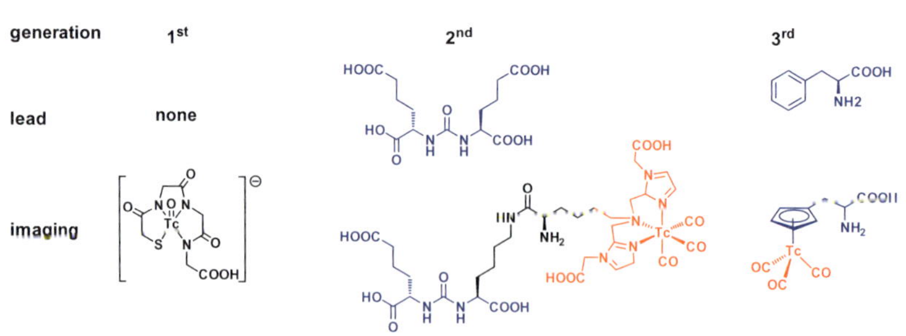

SCHEME 11.1 1st to 3rd generation radiopharmaceuticals; 1st generation de novo radiopharmaceutical Technescan® with no relation to biological function[14,15], 2nd generation target specific radiopharmaceutical, targeting part in blue, pendent group for imaging in red[24–26] and 3rd generation radiopharmaceutical with ligand in blue mimicking a biological structure.[27,28]

The very same strategy is applied to other radionuclides as well and dominates all strategies nowadays, e.g., with the "3 + family" of radionuclides such as $^{6x}Ga^{3+}$ or $^{17x}Lu^{3+}$. Peptides for numerous targets are the favored biovectors of second generation radiopharmaceuticals, but the market success or the benefit for health care is comparably limited.[2,20–23]

Since molecular imaging is shifting more and more towards PET radionuclides, exclusively imaging with SPECT radionuclides such as ^{99m}Tc is probably not sufficient anymore to maintain its leading role, despite the application advantages it has over, e.g., ^{68}Ga. An additional function is desired which leads into multi-modality imaging agents. Multi-modality means two complementary imaging methods, such as MRI and SPECT, with the same compound or compounds, which combine diagnosis and therapy, so-called theranostic agents.[29,30] Medicinal inorganic chemistry is the field searching for metal-containing complexes with favorable, typically therapeutic properties.[13,31] Cisplatin is the lead compound, which inspired generations of medicinal inorganic chemists or, more recently in extension to coordination compounds, organometallic chemists in bioorganometallic chemistry.[32] For a theranostic "inorganic" concept, compounds containing a d-element with isotopes for imaging (radioactive) and for therapy (inactive) are essential. This is rarely the case for the core therapeutic elements such as Fe, Ru, Pt, and Os in medicinal inorganic chemistry since these elements do not have appropriate radioisotopes for the (radio)imaging part. The concept is, however, common in drug discovery and development of pharmaceutical chemistry. Novel pharmaceuticals often comprise a fluorine, which can be replaced by ^{18}F. This allows for non-invasively seeing and treating the same time. Since ^{99m}Tc cannot be complemented with large amounts of ^{99}Tc for obvious reasons, the concept is not applicable along the classical theranostic concept to technetium. ^{99m}Tc, generally considered as a pure diagnostic radionuclide due to its γ-emission, emits 4–5 Auger electrons. If sufficient amounts of ^{99m}Tc can thus be carried in the intimate vicinity of DNA, these Auger electrons can break double-strand DNA and thus exert a therapeutic effect as shown in different publications by our and other research groups.[33–39] The difficulty with such molecular architectures is manifold; the compound should target only the desired receptors, be internalized through the cell and then the nuclear membrane, and stably intercalate into DNA. This requires essentially tri-functional or even tetra-functional compounds with the related difficulties of synthesis, labeling, and design (Scheme 11.2).

Whereas radiopharmaceutical chemistry relies on the bifunctional chelator approach, one can learn from medicinal inorganic chemistry with stable elements that mimicking an existing pharmaceutical is a very promising and advanced strategy. In this pharmaco-mimetic approach, a substructure of an organic molecule is replaced by a metal complex, in which ligands take over the topology of the replaced part. The new, metal-containing molecule might exert different mode of actions on the original target and helps,

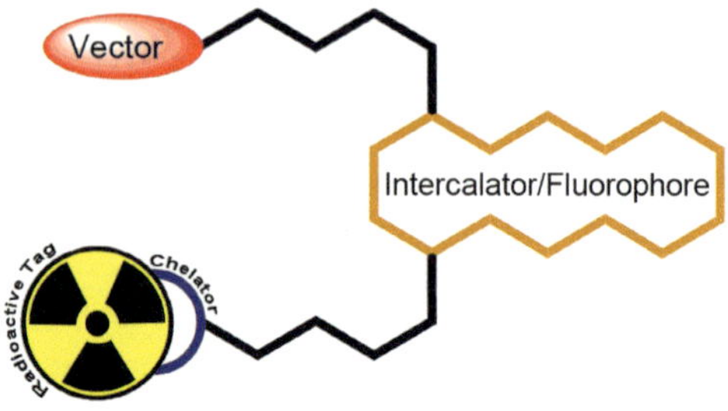

SCHEME 11.2 Concept for a multi-functional ^{99m}Tc radiopharmaceutical for imaging and Auger electron induced therapy; receptor specific function (red), DNA intercalating part eventually with fluorescent properties (yellow) and chelator for stably coordination to a ^{99m}Tc core.[33–39]

e.g., to overcome drug resistance, often encountered in cancer therapy. Ferrocifen and ferroquine are probably the most prominent examples for this kind of bioorganometallic pharmaco-mimetics.[40,41] Subsequently, Meggers and coworkers showed in a conceptually similar approach how a part of a protein kinase inhibiting lead structure is replaced by metal complexes.[42,43] These complexes do not have much in common with the lead, apart from the inhibition on these enzymes, but allow to direct the new compounds, being composed of a maintained and a new substructure, with high selectivity to distinct subtypes of protein kinases with remarkable inhibition constants (Scheme 11.3).[44]

In all of these examples, the core complexes are organometallic in nature, an incentive to develop similar compounds with technetium for providing its chemistry with a new thrust and direction different from the classical bifunctional chelator approach. For an understanding of the following more specific part, an important difference between technetium and, e.g., ruthenium and other stable element's chemistry has to be taken into account. Since any useful molecular imaging agent has to be synthesized from water in short time and in quantitative yields, the structural flexibilities of the participating compounds and the product are not as large as with cold, stable elements. For the latter, essentially any structure can be prepared under user-defined reaction conditions, subsequent purification steps are not prohibitive limitations due to, e.g., time constraints. Therefore, a building block concept is preferred, for which one part of the ^{99m}Tc complex is kept constant whereas another part is variable. Macroscopic amounts of long-lived ^{99}Tc as therapeutic agent are obviously not an option; the isostructural rhenium complexes must, thus, be available as surrogates for the theranostic concept.

To keep the structure of a metal complex as close as possible to features of pharmaceutically active lead structures, aromatic ligands with higher hapticity are probably the only option as evident in the case of ferrocifen or ferroquine and related bioorganometallic compounds with ruthenium-arene substructures. Aromatic ligands are thereby mainly the cyclopentadienyl ligand or arenes, both six-electron donors occupying three coordination sites

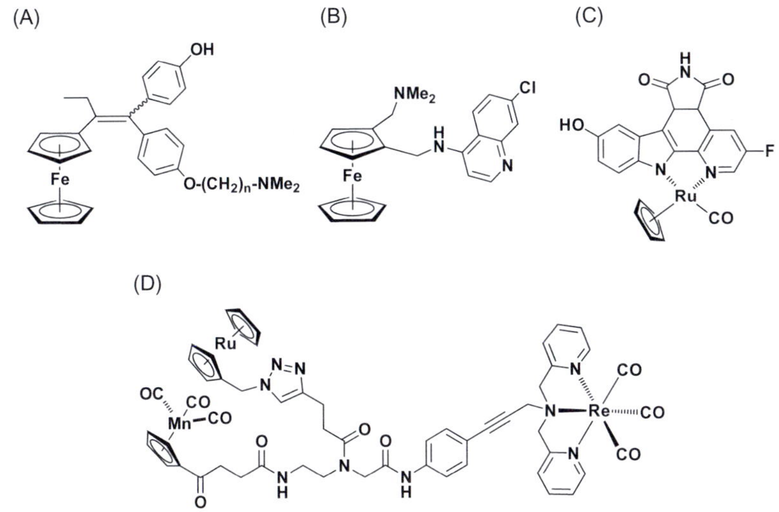

SCHEME 11.3 Pharmacomimetic compounds from bioorganometallic chemistry; Ferrocifen (A),[40] Ferroquine (B),[45] a protein kinase inhibitor (C)[46] and a multinuclear biomimetic compound (D).[47]

in a facial mode. The following more specific sections 11.2 and 11.3 will review the possibilities of preparing ^{99m}Tc complexes with these classical organometallic ligands under conditions as outlined above. Complexes with cyclopentadienyl, or arene ligands, offer many perspectives for the preparation of multi-modality imaging agents for pure diagnostic, but also for theranostic, approaches according to a biomimetic approach.

It is important to note that only a few research groups are still active in this section of bioorganometallic chemistry with ^{99}Tc and ^{99m}Tc. Thus, the numbers of quotes are limited and the advances in ^{99m}Tc-based bioorganometallic chemistry is more intended towards complementing rhenium chemistry than being considered as a standalone field.

11.2 THE CYCLOPENTADIENYL LIGAND AS PHENYL MIMICKING SUBSTRUCTURE

The cyclopentadienyl (Cp^-, $C_5H_5^-$) is a strong, mono-anionic ligand the complexes of which are generally water stable and therefore biocompatible. Cp^- is a key ligand, which found entry into all fields of fundamental and applied organometallic chemistry. For group 7 elements, piano-stool complexes are prototypical and have been pivotal since the advent of organometallic chemistry as mentioned in the introduction. Due to the high stability of $[(\eta^5\text{-}C_5H_5)^{99m}Tc(CO)_3]$ complexes (cytectrene) towards water and air, derivatives of the cyclopentadienyl ligands have been considered as versatile bifunctional chelators for coordination and conjugation to targeting functions. Keeping the simplicity of quantitative complex preparation or labeling in water in mind was, however, always an obstacle preventing systematic exploration of this ligand system for molecular imaging. In addition to the demanding synthetic procedures for complex formation, derivatizations of cyclopentadiene with biological functionalities is also not routine. Since second generation radiopharmaceuticals must comprise targeting functions, chemistry becomes even more complex. As early as 1992, Wenzel and coworkers presented a non-aqueous synthesis for the preparation of Cp-derivatized estradiol with the corresponding ferrocene complex as the source for the latter [(estradiol-Cp)^{99m}Tc(CO)$_3$] imaging agent. [MnBr(CO)$_5$] served as the CO source for the completion of the coordination sphere.[48,49] Although this procedure is not compatible with clinical requests for routine application, a proof of principle for the preparation of cytectrene, derivatized with a biologically active moiety, was given. The compounds presented follow the pendent approach for second generation radiopharmaceuticals and the Cp-ligand was not supposed to mimic any substructure of the lead. The synthetic methodology has later been improved by Katzenellenbogen and coworkers in the Double-Ligand-Transfer (DLT) protocol, still in organic solvents but adaptable to a wider variety of biomolecules, according to the pendent approach.[50–55] The two strategies are shown in Scheme 11.4.

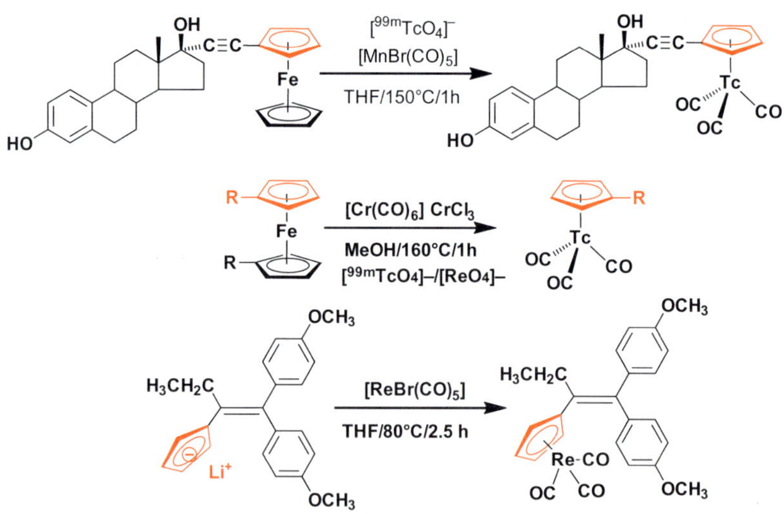

SCHEME 11.4 Preparation of piano-stool complexes from [^{99m}TcO$_4$]$^-$ according to Wenzel et al.,[48,49] The main disadvantage of this approach was the concerted formation of the manganese homologue of the product (top equation). The improved double ligand transfer (DLT) approach by Katzenellenbogen et al., allows conjugation of a large variety of functionalities "R" (middle).[52,56] Jaouen's route to [(Tam-Cp)Re(CO)$_3$] bearing tamoxifen derivative starting from [ReBr(CO)$_5$].[57,58]

These early preparations mirror very well the difficulties encountered when preparing piano-stool complexes of rhenium and ^{99m}Tc. For ^{99m}Tc, it is mandatory to start from $[^{99m}TcO_4]^-$ as shown in Scheme 11.4. This leads to rather unexpected synthetic conditions (such as $Cr(CO)_6$ as CO source) and starting materials. Although the approach of Katzenellenbogen et al., is also exemplified for the synthesis starting with $[ReO_4]^-$, this is ultimately not necessary since, e.g., $[ReBr(CO)_5]$ represents a convenient starting material comprising the CO ligands already. The synthesis arrives then at the same products along classical organometallic preparations. Jaouen et al. showed the "labeling" of biomolecules with $[ReBr(CO)_5]$ very clearly and in many cases. Some early examples of complexes with tamoxifen (Tam) and steroid analogues or other receptor targeting biomolecules bearing a $[(Cp)Re(CO)_3]$ structure were presented.[57–61] More recently, tamoxifen was labeled via a Cp ring bound through a carbonylamino linker to a phenyl ring in 1-position. The rhenium homologues of the ^{99m}Tc complexes, obtained after exchange from the ferrocenyl precursor and in the presence of $[MnBr(CO)_5]$ at high temperature, showed substantial antiproliferative effects on some breast cancer tumor cell lines.[62]

Although ultimately "only" a reaction between LiCp and $[ReBr(CO)_5]$, Jaouen's approach reveals another challenge in the chemistry of piano-stool complex derivatized biomolecules; simply the synthesis of the organic Cp-precursor without a metal. The preparation of the cyclopentadiene analogue of tamoxifen (Scheme 11.4) for instance was achieved by an elegant but not straightforward approach, namely the initial synthesis of the manganese homologue $[(Tam\text{-}Cp)Mn(CO)_3]$ followed by photolytic cleavage of the $[Mn(CO)_3]$ fragment from the Cp ring. The resulting metal-free Cp-Tam was deprotonated to its lithium salt, which only then could be reacted with the rhenium precursor. It should be noted that cyclopentadiene is generally reactive and tends to undergo Diels-Alder reactions and/or to polymerize. The derivatization with structurally complex biomolecules complicates the situation even more. Thus, for ^{99m}Tc, not only the labeling is a challenge, but the preparation of the organic Cp-derivatized biomolecule sometimes even more so. In some examples, the problem with instabilities of cyclopentadiene bound to bioactive functionalities was bypassed by using ferrocene, i.e., a protected form of cyclopentadiene. For targeting multi-drug resistance for instance, Chen et al., derivatized a phosphonium salt with ferrocene, conjugated over a pentyl linker. The reaction of this derivative directly with $[^{99m}Tc(OH_2)_3(CO)_3]^+$ at 145°C for 45 min gave $[(\eta^5\text{-}C_5H_4CO(CH_2)_5\text{-}PPh_3)^{99m}Tc(CO)_3]^+$ in 40%–50% yield (Scheme 11.5).[63] This preparation started from the convenient $[^{99m}Tc(OH_2)_3(CO)_3]^+$ precursor and did not apply chromium salts or carbonyl complexes as in the DLT approach. As will be shown below, the Cp precursor of this phosphonium salt comprises a carbonyl function adjacent to the Cp-ring. This carbonyl function acidifies the Cp sufficiently to deprotonate it in water around physiological pH values.

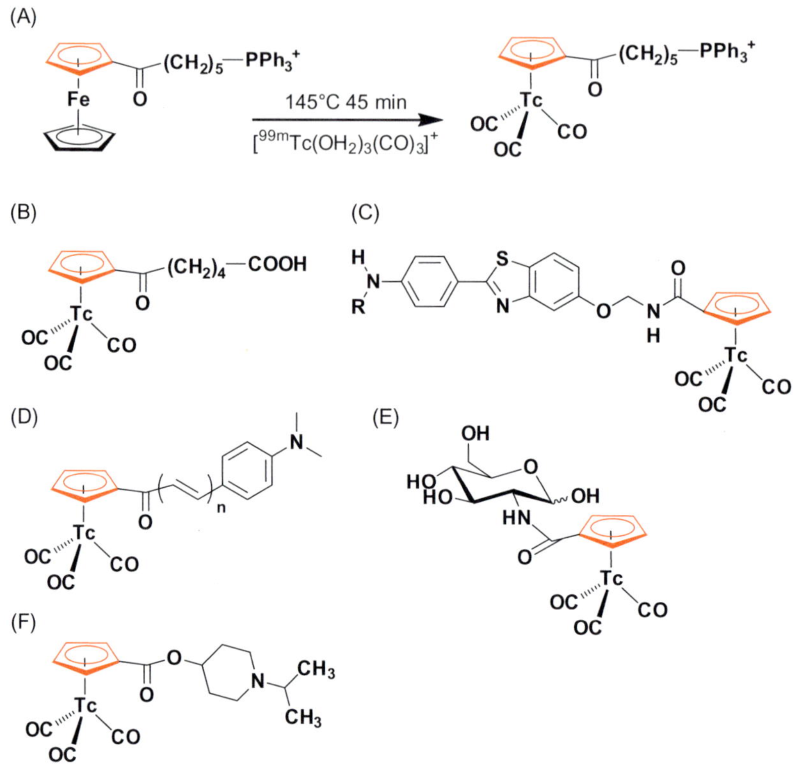

SCHEME 11.5 Synthesis of a piano-stool compound of ^{99m}Tc, starting from the iron-protected Cp precursor (top-line) and a series of complexes prepared along the DLT reaction approach: For targeting multi-drug resistance (A),[63] for myocardial imaging (B),[64,65] for targeting amyloid plaques (C and D),[66,67] as ^{18}F-DG surrogate (E),[68] and for serotonergic receptor 5-HT1A targeting (F).[69]

Cyclopentadiene derivatized in this way is stable and can be directly used without ferrocene protection and in one-step from $[^{99m}TcO_4]^-$.

Despite the precariousness of chromium salts and the availability of more convenient routes (vide infra), the DLT approach is till applied sometimes. Zeng et al., for instance prepared fatty acid conjugates along the DLT approach aiming at new myocardial imaging agents (Scheme 11.5). To arrive at the product, generator eluate was first dried under a stream of nitrogen at 70°C followed by reaction at 170°C in methanol and in the presence of Cr $(CO)_6$ and $CrCl_3$ and several purification steps.[64,65] It is obvious that such a procedure is far from any applicability and more convenient routes have to be found should piano-stool complexes ever be considered as viable tools. An interesting study with fatty acids was presented by Lee et al., In analogy to the previously described procedure, they synthesized a cyclopentadienyl-16-oxo-hexadecanoic acid and the corresponding rhenium and ^{99m}Tc complex. In vivo investigations with these complexes showed clearly that the fatty acid side chain was metabolized via β-oxidation in the myocardium to the corresponding cyclopentadienyl-oxo-butyric acid complex. The Cp complex was left intact, underlining not only the chemical, but also the biological stability of the $[Cp^{99m}Tc(CO)_3]$ arrangement.[70] To replace the working horse in PET radiopharmacy, 2-^{18}F-DG by a ^{99m}Tc labeled glucose is a long-standing issue, which has challenged research for a long time. In the course of the research, cyclopentadienyl complexes of ^{99m}Tc have also been considered as a label for glucose although the combination of a highly hydro- with a highly lipophilic organometallic complex is not immediately intuitive. Ferreira et al., chose glucose-2-amine and coupled the ferrocene derivative $[Fe(\eta^5\text{-}C_5H_4COOH)_2]$ via an amide to the glucose. Labeling was performed via a modified DLT reaction at 160°C for 1 h or by direct reaction with the $[^{99m}Tc(OH_2)_3(CO)_3]^+$ precursor at 80°C. The latter procedure was called Single Ligand Transfer reaction (SLT). The rhenium and ^{99m}Tc complexes (Scheme 11.5E) were subjected to the standard hexokinase assay. The K_i value for the enzyme was substantial which would make the compound an excellent inhibitor. However, the difficulty with ^{18}F-DG analogues is the transport by GLUT2 rather than the inhibition of hexokinase.[68] In a follow-up study, some structural alterations were introduced, i.e., the link between the glucose unit and the $[Cp^{99m}Tc(CO)_3]$ moiety was extended and comprised two amide groups. In contrast to the previous study, none of the complexes showed significant phosphorylation by hexokinase, which contrasts similar complexes from the same study but with non-organometallic tridentate ligands. These derivatives of glucose again depicted substantial hexokinase inhibition. It might well be that the negative results with the cyclopentadienyl label were due to solubility issues. Remarkably, for these studies only the SLT procedure was applied with satisfying yields and at moderate temperatures.[71]

The DLT approach was applied for the labeling of a number of other biomolecules as well and labeling with the "exotic" radionuclide ^{94m}Tc belongs to this series.[72] Since the cyclopentadienyl ligand is small and its complexes lipophilic, $[Cp^{99m}Tc(CO)_3]$ was conjugated to phenylbenzothiazole, a structural entity which is known to bind to amyloid plaques. The reaction of $[^{99m}TcO_4]^-$ with the ferrocene precursor and $[MnBr(CO)_5]$ at 150°C in DMF for 1 h gave a series of compounds in about 50% yields (Scheme 11.5C).[66] Apart from benzothiazoles and other highly conjugated structures, chalcones are prominent structural features for targeting amyloid plaques.[67] For labeling chalcones with a different number of ethylene units between the Cp^- ring and the aniline group (Scheme 11.5D), the authors used a slightly different approach. In a first step, they prepared the acetyl-cyclopentadienyl (ACp) complex $[(\eta^5\text{-}C_5H_4COCH_3)^{99m}Tc(CO)_3]$ from the corresponding ferrocene precursor at high temperature and then subjected this fragment to a Claisen condensation with the reminder of the chalcone structure. Purification lead to the desired molecules. Although K_i values for Aβ1 − 42 where only in the low nM range, these compounds showed promise for visualizing amyloid plaques. The compounds with chalcones are good examples for the integrated approach in which the piano-stool complex mimics a part of the structure of the lead compound.[73] Whereas the lipophilicity of the [Cp^{99m}Tc(CO)$_3$] complex represents a disadvantage in, e.g., the labeling of glucose, it is an advantage when aiming at the targeting of brain receptors for passing the blood brain barrier. Pietzsch et al., prepared cytectrene I and II, which comprised a function for targeting the serotonin 5-HT1A receptor in the brain (Scheme 11.5F). Following the DLT or the SLT approach, they started from the ferrocene precursor N-methylpiperidino-4[(bispentahaptocyclopentadienyl) iron] carboxylate to obtain the corresponding ^{99m}Tc complex. In autoradiograms, both complexes showed substantial binding to regions rich in the 5-HT$_{1A}$ receptor and one of the complexes showed high uptake in the rat brain. Clearly, high lipophilicity supported this beneficial in vivo behavior.[69] There are a few examples in which proteins and peptides were labeled with [Cp^{99m}Tc(CO)$_3$]. After the introduction of the DLT approach by Katzenellenbogen and coworkers, acetyl-cyclopentadienyl was transferred from acetyl-ferrocene to the [^{99m}Tc(CO)$_3$]$^+$ fragment.[52] After ester hydrolysis and carboxylate activation, the [(η^5-C_5H_4-COOR)^{99m}Tc(CO)$_3$] was coupled to amino-groups of peptides and small molecules, namely leucine encephalin, oxytocin, and substance P.[74] Although the procedure is time consuming, the option of introducing the classical piano-stool complex even to larger biomolecules was nicely exemplified. Finally, we note that the piano-stool rhenium homologues where conjugated to many more biomolecules along classical organometallic procedures and without a direct implication of preparing the respective ^{99m}Tc counterparts. The motivation behind this was often a comparison of the bioactivities with the ferrocene derivatives rather than to combine an eventual biological action of cold, non-radioactive rhenium with ^{99m}Tc for imaging purposes.

These examples describe the preparations of bioorganometallic compounds mainly in organic solvents, mostly with subsequent purification and transfer into an aqueous environment. For attempted routine application, a synthesis in water and directly from [^{99m}TcO$_4$]$^-$ and in the absence of other metallic additives is mandatory. The stabilities and reactivities of cyclopentadiene and its derivatives are major obstacles toward this objective. For solving the challenge of a fully aqueous synthesis, our group concluded that the cyclopentadiene precursors should be water-soluble and be completely, or at least partially, deprotonated around the physiological pH value. Acetylation of Na[C_5H_5] leads to acetyl-cyclopentadiene ACp. Its pK_a value is about 8.5 and it exhibits increased water solubility due to the carbonyl group adjacent to the Cp-ring and concomitant deprotonation. With this model cyclopentadiene compound, a fully aqueous preparation of the piano-stool type complex [(ACp)^{99m}Tc(CO)$_3$] in one step and directly from [^{99m}TcO$_4$]$^-$ generator eluate became feasible for the first time. Relatively high concentrations of the

ACp precursor were necessary to achieve quantitative ^{99m}Tc complex formation in less than 60 min.[75,76] We noted that excess borano-carbonate as reductant and CO source lead to partial reduction of the carbonyl to a secondary hydroxyl group if conditions were not carefully controlled. Since the keto-carbon is enantiotopic, two enantiomers were generated in this step after reduction. Formation of racemates is not favorable keeping the stereoselectivity of many receptors in mind. Still, no organic solvents and no carbonyl complexes as CO sources were required. The reaction thus presented a true advantage towards fully aqueous preparation of $[(\eta^5\text{-}C_5H_4\text{-}R)^{99m}Tc(CO)_3]$ type bioorganometallics. The approach as outlined in Scheme 11.6 is highly flexible with respect to the substituents R or biological moieties, conjugated to the cyclopentadiene system. The obtained Cp-derivatives show some water solubility and the procedure can be transferred to any biomimetic system.[28,77–79] This procedure was at that time the synthetic method, which matched routine clinical demands best, fully aqueous, one-step, quantitative, and short reaction time. The main disadvantage was the relatively high ligand/bioconjugate concentration, which ensued eventual purification from unlabeled biovectors after labeling. Some examples for the labeling of biomolecules along this synthetic pathway are given in Scheme 11.6.

Improving all the odds of cyclopentadiene chemistry for the preparation of cytectrene and functionalized analogues like radiopharmaceuticals was finally possible via an inverse Diels-Alder reaction. Cyclopentadiene derivatives undergo often Diels-Alder reactions to the corresponding dimeric compounds such as di-cyclopentadiene $C_{10}H_{12}$ as the standard example. Such products can be cleaved thermally into their respective monomers at high temperatures, e.g., for di-cyclopentadiene at temperatures higher than 150°C.

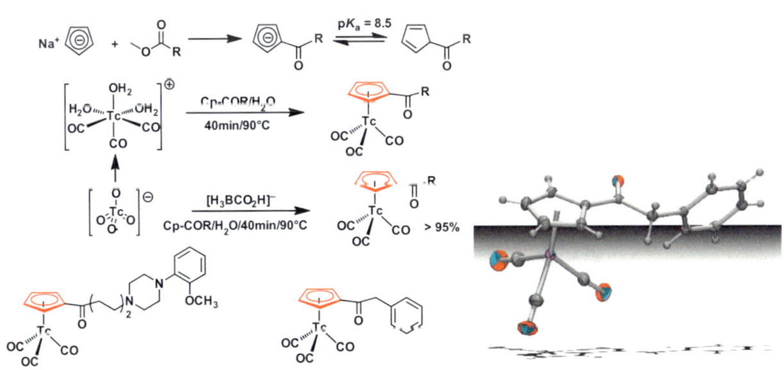

SCHEME 11.6 Cyclopentadiene with an adjacent carbonyl group has a pK_a in the physiological region (top line), reaction with the $[^{99m}Tc(OH_2)_3(CO)_3]^+$ complex gives the corresponding piano-stool complexes in quantitative yields. The reaction can be performed in a one-pot procedure directly from $[^{99m}TcO_4]^-$. Two examples of complexes prepared along these pathways and ORTEP of a ^{99}Tc model complex.[75,76]

It seems, thus, that such dimers are useless for radiopharmaceutical chemistry since preparations should be performed below 100°C. At these temperatures, no cracking of the dimers into the cyclopentadiene monomers takes place to a measurable degree. Attempting to introduce a more flexible linking group in the Cp framework, we prepared cyclopentadiene-carboxylic acid $C_5H_5(COOH)$ which rapidly dimerizes to so-called Thiele's acid $(C_5H_5(COOH))_2$. Thiele's acid can be cleaved again into its monomer at temperatures well above 150°C.[80] Although it did not seem likely to prepare from this precursor a cytectrene-like complex, we performed the reaction in water at 100°C and obtained $[(\eta^5\text{-}C_5H_4COOH)^{99m}Tc(CO)_3]$ in close to quantitative yields.[81] As Thiele's acid does not cleave under these conditions, we concluded that the mechanism leading to $[(\eta^5\text{-}C_5H_4COOH)^{99m}Tc(CO)_3]$ follows a metal-mediated retro-Diels-Alder reaction. The carboxylate-groups may act as weak anchors for the technetium-core, followed by thermal breakage into the monomers and paralleled by coordination to the $[^{99m}Tc(CO)_3]^+$ fragment. Although the mechanism is not elucidated, the reaction is one pot and directly from $[^{99m}TcO_4]^-$, in water and in the presence of the dimerized Cp-ligand and boranocarbonate as the reducing agent/CO source. Since the Diels-Alder products are water-stable and air-stable precursors, their handling is convenient and they are prepared in large amounts with essentially any biovector bound to it. We emphasize that the concentrations of Cp-derivatized biomolecules are in the sub-mM range and, thus, low enough not to require purification for many targeting molecules. The fundamental reactions are shown in Scheme 11.7.

The two carboxylic acids in Thiele's acid can be conjugated via, e.g., amide formation to almost any kind of targeting molecules or bioactive compounds. These structures may be part of an overall structure following the third generation radiopharmaceutical or biomimetic concept or just be pendent to a targeting vector (second generation). The reactions shown in Scheme 11.7 can thus be performed with almost any kind of tag along a general reaction scheme. We have introduced a variety of sulphonamide structures for targeting and inhibiting carbonic anhydrases, an important target in cancer therapy and diagnosis.[83–85] Labeling follows the outlined procedure and inhibition constants are in the low nanomolar range.[82] The receptor subtype selectivities of some of these bioorganometallic drugs are remarkable and much better when compared to the purely organic compounds. They thus confirm the 3D space occupation concept formulated by Meggers et al.[42]

Apart from the sulphonamides, a number of other molecules were attached to the Cp framework according to the concept of mimicking phenyl subunits in lead structures. Due to its structural simplicity, HDAC inhibitors, e.g., of the hydroxamato type are ideal lead structures.[86,87] The phenyl ring can be replaced by a cytectrene unit or its rhenium homologue. Vorinostat® is an example of a successfully applied pharmaceutical in clinical routine for quite some time.[88] Scheme 11.8 shows a series of biomimetic rhenium and

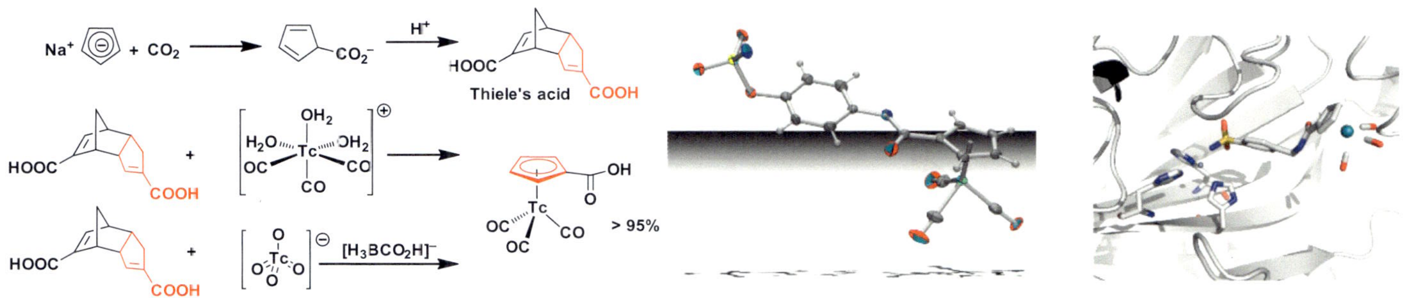

SCHEME 11.7 Preparation of piano-stool complexes in one-step from Thiele's acid precursor in water, at 90–100°C and in less than 30 min[81] (left). An ORTEP of a carbonic anhydrase model complex with rhenium and bound to CA II (right).[82]

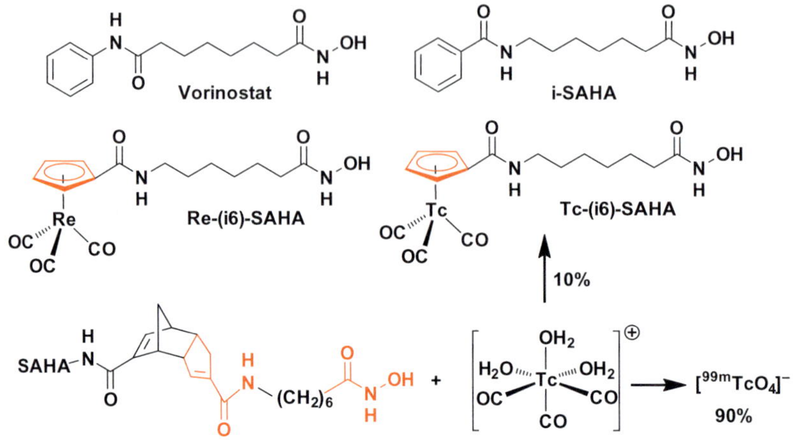

	MCF-7	A431	HeLa	A375	B16F1
Re-(i7)SAHA	9.5	13.7	14.4	14.7	9.8
Re-(i6)SAHA	15.2	17.1	13.3	23.3	12.6
Re-SAHA	11.4	17.3	8.3	12.5	15.2
(i7)SAHA	1.7	2.5	1.65	2.6	3.1
(i6)SAHA	7.2	5.2	5.8	4.8	8.9
SAHA	3.7	4.4	4.4	4.6	3.7

SCHEME 11.8 The clinically applied pharmaceutical Vorinostat and its form inverted at the amide group i-SAHA, ^{99m}Tc and Re mimics for the inverted forms. **TABLE 1** IC_{50} values (μM) of the compounds towards different cell lines (unpublished results).

technetium analogues of Vorinostat together with their IC_{50} values determined for these biomimetic complexes. The values are close to those of the lead compounds and the ^{99m}Tc and Re complexes therefore act as surrogates. However, labeling the dimerized precursors is hampered by the hydroxamato groups. The reactions with $[^{99m}Tc(OH_2)_3(CO)_3]^+$ achieve yields of only around 10% under optimized conditions besides $[^{99m}TcO_4]^-$ as the "main product". Our research showed that the hydroxamato group as a weakly oxidizing functionality decomposes the $^{99m}Tc^I$ precursor entailing low labeling yields and $[^{99m}TcO_4]^-$. To pursue this principally interesting biomimetic compound, protection of the hydroxamato group and concomitant deprotection and labeling needs to be further explored. The model compounds and the labeling procedures are depicted in Scheme 11.8.

Inspecting the structures of Thiele's acid derivatives for the labeling of biomolecules entails the question about the mechanism. Which one of the two cyclopentadiene systems coordinates to the $[^{99m}Tc(CO)_3]^+$ fragment, the diene or the dienophile, or is there a distinction at all? To obtain a deeper insight, we prepared substituted Thiele's acids with two different functionalities bound to the two carboxylate groups and performed labeling reactions according to the standard procedure. If one of the sites was preferred in $[(C_5H_5CONHR)(C_5H_5CONHR')]$, exclusive formation of $[(\eta^5\text{-}C_5H_4CONHR)^{99m}Tc(CO)_3]$ or $[(\eta^5\text{-}C_5H_4CONHR')^{99m}Tc(CO)_3]$ would be expected (Scheme 11.9). Indeed, the reaction gave both products in ratios close to 1:1 if the link between the function R/R' and Thiele's acid scaffold was similar, e.g., an alkyl chain. If the linkers were different, the ratio was in favor of the dienophile site but not to a significant extent, still indicating some preference. This non-selectivity can be regarded in a positive sense; one reaction delivers too different radiopharmaceuticals in one-step.

In conclusion, aromatic hydrocarbons of the cyclopentadienyl type have over the years proven to be versatile ligand systems for rhenium and for ^{99m}Tc. They provide very high stabilities to the metal centers in the + I oxidation state and are decorated with many promising targeting and/or cytotoxic groups. Cyclopentadienyl was used for preparing second generation radiopharmaceuticals but also for the biomimetic approach (third generation) in which the cytectrene moiety replaces, e.g., a phenyl group in a pharmaceutically active lead structure. Examples for both are described in the previous section. As outlined in detail, rhenium complexes were mostly prepared along classical organometallic reaction pathways whereas for ^{99m}Tc new strategies had to be developed. Originally, the Double Ligand Transfer led to the desired cytectrene compounds but required high temperatures and organic solvents as well as a metal-containing CO source. This DLT approach was

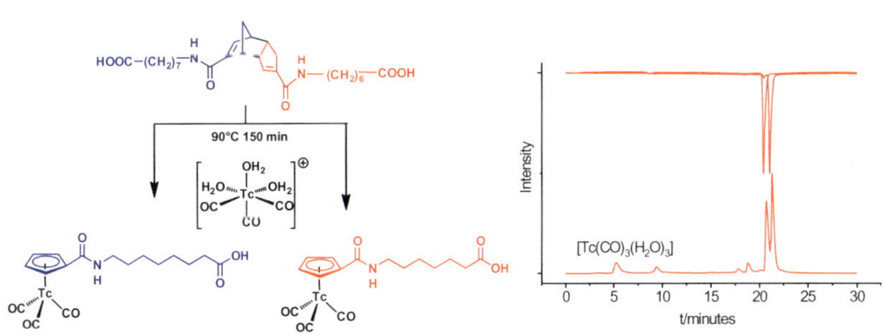

SCHEME 11.9 Diene or dienophile? The reaction with differently substituted Thiele's acid gives both products in a ratio of about 1/1, HPLC traces of the ^{99m}Tc complexes (γ-Detection bottom−up) and comparison with rhenium compounds (UV/vis detection top−down).

improved by starting with acetyl-cyclopentadiene-like compounds which deprotonate in water, providing an entry into aqueous cyclopentadienyl chemistry. Finally, dimerization of cyclopentadiene precursors led to a kind of protected cyclopentadienyl which split into their monomers in the presence of a metal precursor like the $[^{99m}Tc(CO)_3]^+$ fragment, finally allowing preparations directly from $[^{99m}TcO_4]^-$ in water. Since a large variety of functionalities may be attached to these "$(Cp)_2$" precursors, cytectrene became a versatile pendent group to targeting molecules or can easily be integrated in lead structures. Both schemes are promising for discovering or developing novel radiopharmaceuticals and molecular imaging agents. In combination with biologically active but non-radioactive rhenium homologues, a track towards a molecule-based theranostic strategy can be envisaged.

11.3 AROMATIC HYDROCARBON LIGANDS OF THE ARENE-TYPE

Cyclopentadienyl is the classical aromatic hydrocarbon and a six-electron donor. Its "multidenticity" and the anionic charge makes it a very tightly binding ligand as outlined above. Electronically distinctly different and uncharged, are arene systems, with benzene being the parent of this class of ligands. In contrast to cyclopentadienyl, benzene is a weaker donor and a stronger acceptor. Since benzene-type ligands are generally neutral in charge, they bind more weakly to metal centers. In contrast to cyclopentadienyl systems, arenes are substantially less studied with rhenium and technetium in particular. Without going into details, the only field where arenes play a decisive role is in the development of ruthenium-based organometallic drugs. The core structural feature is the $[Ru(\eta^6\text{-arene})]^{2+}$ moiety, mainly introduced and developed by Sadler and Dyson et al., into numerous highly active and in-depth studied complexes.[89–93] With respect to the concepts outlined in the introduction for molecular imaging agents, arene ligands are favorable over cyclopentadienyl since they are "true" phenyls with six carbon atoms rather than just five as in cyclopentadiene. The organic chemistry of phenyls is generally better defined than the one of cyclopentadiene since derivatives of the former are stable and do not undergo reactions such as Diels-Alder dimerization. Thus, phenyls occurring in essentially all pharmaceuticals (in contrast to cyclopentadiene) can conceptually be taken directly as ligands in lead structures, if appropriate syntheses are found. In organometallic chemistry with non-radioactive metals, preparations generally follow the classical Fischer-Hafner reaction, metal precursors such as binary halides or oxides, a reductant like aluminum or zinc and the arene as solvent. An excellent review comprehensively describes these synthetic strategies in detail.[94] These conditions are incompatible with the far-reaching goal of preparing imaging agents with ^{99m}Tc. However, it must be kept in mind that a synthesis along whatever path allows the initial evaluation of ^{99m}Tc compounds under in vitro/in vivo conditions. If pharmacology and biological behavior is favorable, a synthetic approach must be developed subsequently. For the group 7 elements manganese, technetium, and rhenium, organometallic chemistry with arene ligands is scarce. Whereas the classical bis-arene complexes of the type $[M(\eta^6\text{-}C_6R_6)_2]^+$ have been prepared in the 1960s by Fischer et al., follow up studies about their chemistry in general, and on the coordinated arene rings in particular, are even rarer.[95–97] Only five years ago, Kudinov et al., presented a reasonable strategy for $[Re(\eta^6\text{-}C_6H_6)_2]^+$ and some hydrocarbon derivatives of the parent benzene.[98] Follow-up chemistry with these precursors towards applications have not yet been done. Due to the conditions of preparation, they do not exist since Fischer-Hafner conditions are incompatible for arenes with additional functionalities such as carboxylates, amines, or other biologically relevant groups. Stability studies under aqueous conditions for application in the life sciences are absent, with one remarkable exception. On the search for new myocardial imaging agents, Wester et al., in the early 1990s, prepared $[^{99m}Tc(\eta^6\text{-arene})_2]^+$ along the Fischer-Hafner procedure. These cationic and lipophilic complexes were thought to accumulate in the heart as Cardiolite® did, thereby assessing the function of this organ. The preparation was time consuming, including evaporation of water from the generator eluate and re-dissolution in organic solvents prior to the reduction/coordination process. Still, a large number of complexes were prepared and studied with animals in vivo. Heart uptake could not compete with the established complexes and the development of $[^{99m}Tc(\eta^6\text{-arene})_2]^+$ type myocardial imaging agents was therefore abandoned.[99] However, the important message from this study was the stability assessment of $[^{99m}Tc(\eta^6\text{-arene})_2]^+$ under in vivo conditions. Analyses corroborated that these complexes were excreted unchanged and not metabolized. Altogether, the biological data together with the difficulties of preparations resulted in this class not being pursued. The original preparation of $[^{99m}Tc(\eta^6\text{-arene})_2]^+$ type complexes is shown in Scheme 11.10.

The stabilities of bis-arene complexes of group 7 elements are unexpected since the analogue complexes of the neighboring elements behave differently; group 6 bis-arene complexes are sensitive towards oxygen and the ones of group 8 tend to cleave at least one arene ring rather easily. In order to come closer to a true biomimetic approach, we became interested in bis-arene complexes of the manganese triad, not only due to potential applications but also because they might show a very rich chemistry, explored to a very small extent only. First, the syntheses of $[Re(arene)_2]^+$ was improved with purely aliphatic derivatives of benzene such as tetraline, hexamethylbenzene, and many other to yields higher than 60%. Trans-alkylation reactions, as often found under Fischer-Hafner conditions, were observed to a minor extent. With the exception of the original preparation by Fischer and coworkers and the single publication of Wester et al., $[^{99m}Tc(\eta^6\text{-}C_6H_6)_2]^+$ was not studied at all. Initially we prepared these sandwich complexes by

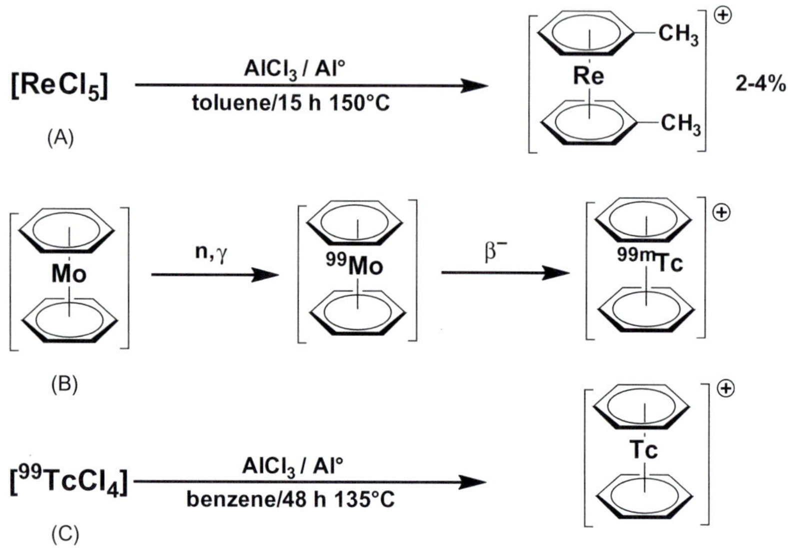

SCHEME 11.10 Standard preparations for rhenium and technetium bis-arene complexes; standard Fischer-Hafner conditions (A),[97] synthesis of microscopic amounts with ^{99m}Tc by element transmutation (B)[95] and the first preparation of weighable amounts of the ^{99}Tc complex (C).[96]

just copying the conditions of rhenium, which gave low yields and revealed great difficulties in separating the products. If the synthesis was performed in the absence of Zn° as reductant, yields as high as 80% were achieved with facile isolation. Rhenium does not react without reductant, a distinct difference in the chemistry of the two elements. Since ^{99}Tc as a 4d element has a higher oxidation potentials than the 5d element rhenium, we assume that the arene itself acts as a reductant and no Zn° is necessary.[100] The reaction conditions together with the ORTEPs of some complexes with ^{99}Tc are shown in Scheme 11.11.

Electrochemical investigations with cyclovoltammetry evidenced the extremely high redox stability of these sandwich complexes; the redox potential Re^{I}/Re^{II} and Tc^{I}/Tc^{II} are at around +1.3 V and +1.5 V respectively versus Fc/Fc^{+} in acetonitrile and fully reversible. Irreversible reductions are found below −2 V under the same conditions. Thermodynamically, O_2 is thus not able to oxidize either of these bis-arene complexes.

The facile availability is now an entry into the more basic chemistry with $[Re(\eta^6\text{-}C_6H_6)_2]^+$ and $[^{99}Tc(\eta^6\text{-}C_6H_6)_2]^+$ as starting materials. For instance, the protons on the coordinated benzenes are rather acidic and reaction with $Li[N(^{i}Pr)_2]$ leads to deprotonation/lithiation. Subsequent reaction with CO_2 gives the bis-benzoic acid complex $[Re(\eta^6\text{-}C_6H_5\text{-}COOH)_2]^+$. Lithiation with $^{n}BuLi$ did not result in deprotonation but in butylation of the benzene ring, mirroring together with the strongly high-field shifted ^{1}H NMR resonances

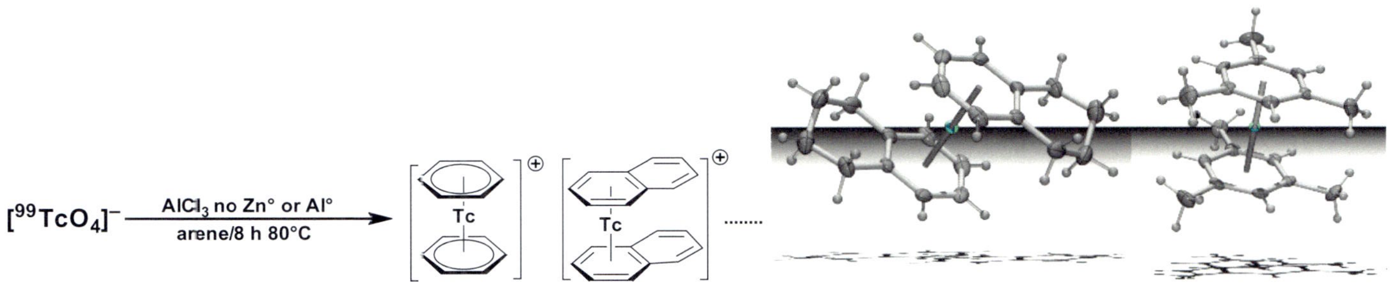

SCHEME 11.11 Synthetic conditions for the preparation of ^{99}Tc bis arene complexes directly from $[^{99}TcO_4]^-$ (left), ORTEPs of $[^{99}Tc(\eta^6\text{-tetraline})_2]^+$ and $[^{99}Tc(\eta^6\text{-}C_6H_3(CH_3)_3)_2]^+$ (right).[100]

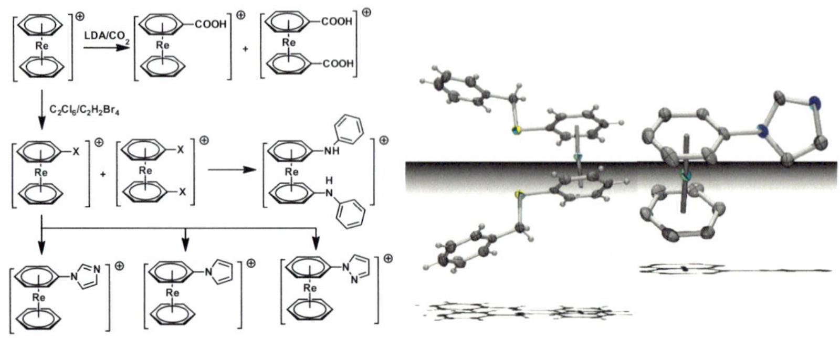

SCHEME 11.12 Derivatizations of basic $[Re(\eta^6\text{-}C_6H_6)_2]^+$ and some subsequent reactions (left) and ORTEPs of compounds with pendent thioether groups or imidazole.[101,102]

the now electrophilic nature of the coordinated arenes. Bromination with tetrabromo-ethene or chlorination with hexachloro-ethane resulted in mono-halide complexes $[Re(\eta^6\text{-}C_6H_6)(\eta^6\text{-}C_6H_5\text{-}X)]^+$. Halogenation on each of the arenes to $[Re(\eta^6\text{-}C_6H_5\text{-}X)_2]^+$ was facile, but higher halogenation up to a total of five halides were found as well. These complexes are perfect starting materials for the conjugation of bioactive molecules, either by peptide formation with amines (in case of $[Re(\eta^6\text{-}C_6H_5\text{-}COOH)_2]^+$) or by nucleophilic aromatic substitutions of the halides in coordinated halo-benzene (in case of $[Re(\eta^6\text{-}C_6H_6)(\eta^6\text{-}C_6H_5\text{-}X)]^+$).[101] We emphasize that the chloro substituents are much more reactive than the bromo groups. Numerous derivatives of $[Re(\eta^6\text{-}C_6H_6)_2]^+$ have been prepared along either route and some examples are shown in Scheme 11.12.

Sandwich complexes of the $[M(C_6H_6)_2]^+$ type are cationic in nature which might be a disadvantage for biological applications. However, a number of examples showed that replacing ferrocene by positively charged cobaltocenium even improved the biological characteristics in some cases, thus, the charge must not necessarily be a disadvantage.[103,104] At least, water solubility improves when compared to uncharged ferrocene derived biomolecules. Inspecting the $[M(C_6H_6)_2]^+$ structure, it is tempting to think about the substitution of one of the arenes by monodentate or multidentate ligands bound to bioactive functionalities of any kind. However, all attempts to do so failed due to the extremely high stability of the coordinated benzene or toluene and similar ligands. Higher conjugated aromatics such as naphthalene are weaker donors and acceptors and, therefore, weaker ligands due to stabilization over a larger conjugated π-System. When reacting $[ReO_4]^-$ in benzene in the presence of naphthalene, the complex $[Re(\eta^6\text{-}C_{10}H_8)_2]^+$ and the mixed-ring sandwich complex $[Re(\eta^6\text{-}C_6H_6)(\eta^6\text{-}C_{10}H_8)]^+$ were formed. One or both of the naphthalene ligands were susceptible for substitution by mono- or multidentate ligands. Both reactivities resemble the pattern of the ones found in the $[^{99m}Tc(OH_2)_3(CO)_3]^+$ complexes where the three H_2O ligands (here naphthalene) are replaced, whereas the three CO's (here benzene) remain untouched. Conceptually advancing the field of carbonyl chemistry, the $[Tc/Re(\eta^6\text{-}C_6H_6]^+$ moiety can be considered as an isoelectronic $[Tc/Re(CO)_3]^+$ analogue (Scheme 11.13).[102]

Attempts to obtain uncharged rhenium or technetium with mixed-ring sandwich complexes of the ferrocene type by substitution of, e.g., naphthalene with cyclopentadienyl failed so far. A number of such sandwich complexes were described in literature but only obtained along complicated procedures. Reports about their properties are rare and its behavior in water completely unknown.[105–108] We produced a mixed-ring sandwich complex along a novel and uncommon route. When reacting $[Re(\eta^6\text{-}C_6H_6)(\eta^6\text{-}C_6H_5\text{-}Br)]^+$ under alkaline conditions and at 60–80°C in water, the bromide hydrolyzed and the mixed phenol/benzene complex $[Re(\eta^6\text{-}C_6H_6)(\eta^6\text{-}C_6H_5\text{-}OH)]^+$ formed. At even higher pH and at $T > 80°C$, we found a ring contraction of one benzene to a cyclopentadienyl unit, an unprecedented reaction scheme which yields up to 80% product.[109] The neutral mixed-ring sandwich complex $[Re(\eta^5\text{-}C_5H_4CHO)(\eta^6\text{-}C_6H_6)]$ is stable in water and displays a redox potential for the Re^I/Re^{II} couple which is essentially identical to Fc/Fc^+. Thus, the donating cyclopentadienyl ligand shifts the potential by about 0.8 V relative to the $[Re(\eta^6\text{-}C_6H_6)_2]^+$ parent. With amines, the aldehyde was converted to an imine and subsequently reduced to an, e.g., amine linked biomolecule. Reduction leads to the corresponding alcohol $[Re(\eta^5\text{-}C_5H_4CH_2OH)(\eta^6\text{-}C_6H_6)]$ and following protonation with a strong acid to the first structurally characterized fulvene complex $[Re(\eta^5\text{-}C_5H_4{=}CH_2)(\eta^6\text{-}C_6H_6)]^+$. The fulvene complex is water stable over a broad range of pH values, but is susceptible for nucleophilic attack by thiols and to a lower extent by amines. It therefore has the potential to react directly and site-specifically with, e.g., cysteine in proteins or peptides (Scheme 11.14).

These examples demonstrate the plethora of fundamental chemistry directed towards application in bioorganometallic chemistry with the $[Re(\eta^6\text{-}C_6H_6)_2]^+$ building blocks and structurally related complexes. Whereas their

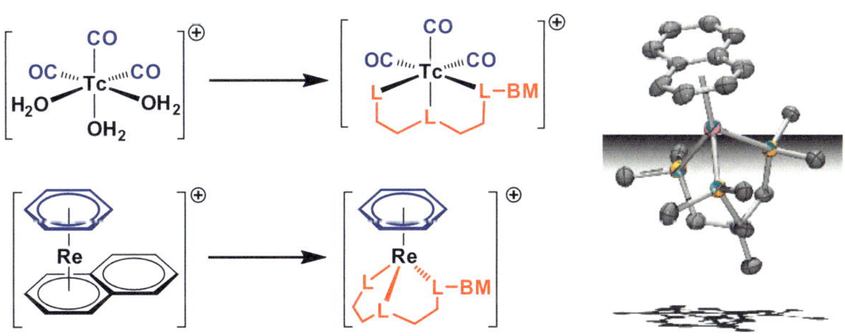

SCHEME 11.13 Conceptual parallels between labeling chemistry with $[^{99m}Tc(OH_2)_3(CO)_3]^+$ and $[Re(\eta^6\text{-}C_6H_6)(\eta^6\text{-}C_{10}H_8)]^+$. In both building blocks, the blue parts are invariable whereas the black parts can be substituted.[101,102]

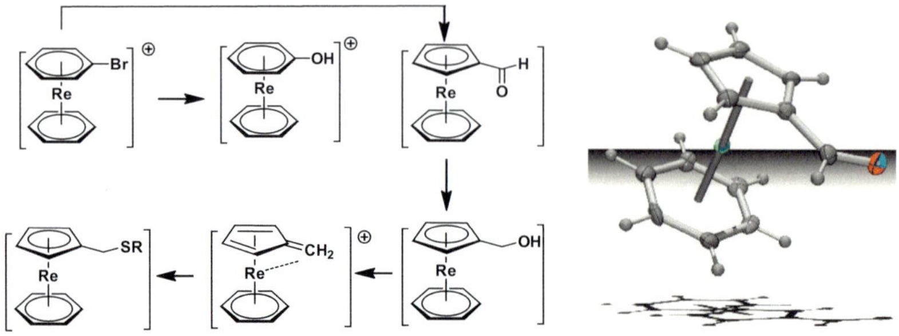

SCHEME 11.14 Behavior of $[Re(\eta^6\text{-}C_6H_6)(\eta^6\text{-}C_6H_5\text{-}Br)]^+$ under strongly alkaline conditions and formation of the mixed-ring sandwich complex $[Re(\eta^5\text{-}C_5H_4CHO)(\eta^6\text{-}C_6H_6)]$. Subsequent reduction and protonation leads to a fulvene complex which is prone to nucleophilic attack. ORTEP of $[Re(\eta^5\text{-}C_5H_4CHO)(\eta^6\text{-}C_6H_6)]$.[109]

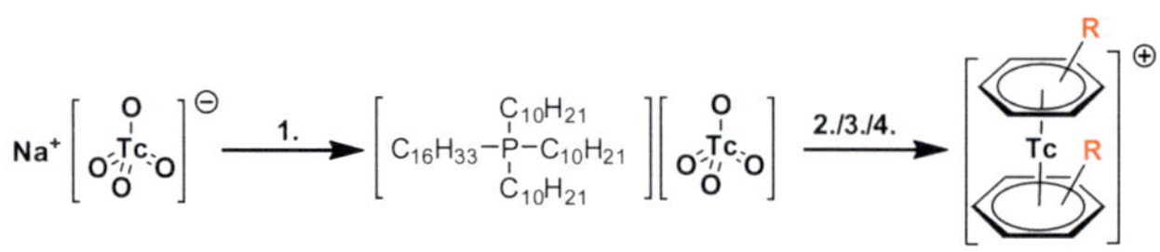

SCHEME 11.15 Synthesis of $[^{99m}Tc(\eta^6\text{-arene})_2]^+$ from generator eluted $[^{99m}TcO_4]^-$ in saline: 1. Extract $[^{99m}TcO_4]^-$ with a few μl of ionic liquid from saline, separate aqueous phase; 2. Add 100 μl of arene and 1–5 mg $AlCl_3$; 3. Microwave for 10 min at 100°C; 4. Extract into buffered solution.[100]

stabilities prevent efficient and direct interaction with the rhenium center, decoration of the arene rings with a multitude of functionalities is possible under essentially any reaction condition. This is an ideal synthetic base for the preparation of biomimetic compounds for therapeutic purposes. It mirrors the statement that the potential role of rhenium in medicinal inorganic or organometallic chemistry is strongly underestimated.[110] None of the bis-arene complexes has been investigated in vitro or in vivo so far, but the statement is certainly true for $[Re(CO)_3]^+$ based compounds.[111–114]

What remains at this point is the important question about ^{99m}Tc homologues for imaging purposes and for complementing rhenium chemistry for therapeutic purposes with a diagnostic ^{99m}Tc tool. As mentioned earlier, Wester et al. synthesized complexes of the $[^{99m}Tc(\eta^6\text{-arene})_2]^+$ type along a multi-step procedure, including evaporation of water from the generator eluate. This was a demanding and probably limiting step since evaporation of solvents in the presence of radionuclides requires strong safety measures. We chose an alternative procedure by quantitative extraction of $[^{99m}TcO_4]^-$ from the generator eluate into a tiny amount of ionic liquid. Dissolving the residue in the arene and reaction with $AlCl_3$ but without Zn° resulted in a more or less quantitative formation of, e.g., $[^{99m}Tc(\eta^6\text{-}C_6H_6)_2]^+$. Extraction into a buffer solution and separation over a SepPak column finally gave a pure complex in a physiological environment (Scheme 11.15).[100] This procedure was still off routine application directly from water, but it bypassed evaporation steps and allowed access to the physico-chemical properties of ^{99m}Tc bis-arene complexes. The very high chemical stability, as found for rhenium homologues, was confirmed with the ^{99m}Tc complexes; heating up to 160°C in the presence of air and over a broad pH range did not affect the complex at all, a surprising and promising finding when compared to the behavior of other ^{99m}Tc complexes on the tracer level. It is that a fully aqueous preparation must be developed for making $[^{99m}Tc(\eta^6\text{-}C_6H_6)_2]^+$ complexes accessible to imaging studies.

11.4 SUMMARY AND CONCLUSION

For advancing bioorganometallic chemistry, the study and preparation of receptor binding complexes based on known organometallic building blocks, and their refinements into, e.g., inhibitors with ever better affinities is a major track. Progress in tamoxifen chemistry as presented by Jaouen et al., is an excellent example here; the organometallic unit was kept essentially as ferrocene (among others) and the organic lead structure was altered.[41] Furthermore, changing complex structures while leaving parts of an organic lead untouched is another strategy. An excellent example for progress along this line is the 3D-space occupation concept by Meggers and coworkers—substantial alteration of the coordination sphere around Ru^{II} centers led to tremendous differences in receptor specificities and affinities.[43,115] At the same time, advancement also means discovering and developing new organometallic complexes, which have not yet received attention from the community, and there are many. Because phenyls of any kind are substructures in most pharmaceuticals, we started to explore arene complexes with the scope of coordinating technetium or rhenium fragments directly to these potential binding sites. This led straight to the exploration of new building blocks in bioorganometallic chemistry, the advancement of which may well complement research done with ferrocene or other cyclopentadienyl units. Although at the beginning, new fundamental and applied perspectives already became evident at this point. In the field of $^{99(m)}$Tc chemistry, advancement does not only mean the discovery of promising building blocks and bioactive compounds, but also the development of synthetic procedures for their preparation in one step and in quantitative yields. This is a major challenge for introducing new technetium complexes in imaging and, as a far-reaching goal, in theranostics. Arenes and, to a smaller extent, cyclopentadienyl complexes inherently carry the potential of mimicking pharmaceutically active compounds. Since affecting such established structures, the least is one way to be successful in bioorganometallic chemistry, biomimetic concepts with surrogating complex structures based on integrated ligands are key.

REFERENCES

1. Morais GR, Paulo A, Santos I. *Organometallics* 2012;**31**(16):5693–714.
2. Bartholoma MD, Louie AS, Valliant JF, Zubieta J. *Chem Rev* 2010;**110**(5):2903–20.
3. Alberto R. *J Organomet Chem* 2007;**692**:1179–86.
4. Alberto R. Technetium. In: Mc Cleverty JA, Meer TS, editors. *Compr. Coord. Chem. II.* Amsterdam: Elsevier Science; 2003. p. 127–271.
5. Alberto R. Radiopharmaceuticals. In: Jaouen G, editor. *Bioorganometallics.* Weinheim: Wiley-VCH; 2006. p. 97–124.
6. Eckelman WC, Richards P, Meinken G. *J Nucl Med* 1972;**13**(8):577–81.
7. Eckelman W. *Nucl Med Biol* 2011;**38**(5):613–16.
8. Abrams MJ, Davison A, Jones AG, Costello CE, Pang H. *Inorg Chem* 1983;**22**:2798–800.
9. Coogan MP, Doyle RP, Valliant JF, Babich JW, Zubieta J. *J Labelled Compd Rad* 2014;**57**(4):255–61.
10. Jurisson SS, Luxdon JD. *Chem Rev* 1999;**99**:2205–18.
11. Jurisson S, Cutler C, Smith SV, Nucl QJ. *Med Mol Im* 2008;**52**(3):222–34.
12. Liu S. *Chem Soc Rev* 2004;**33**(7):445–61.
13. Mjos KD, Orvig C. *Chem Rev* 2014;**114**(8):4540–63.
14. Fritzberg AR, Kasina S, Eshima D, Johnson DL. *J Nucl Med* 1986;**27**(1):111–16.
15. Nosco DL, Manning RG, Fritzberg A. *J Nucl Med* 1986;**27**(6). 939-939.
16. Troutner DE, Volkert WA, Hoffman TJ, Holmes RA. *J Nucl Med* 1983;**24**(5). P10-P10.
17. Troutner DE, Volkert WA, Hoffman TJ, Holmes RA. *Int J Appl Radiat Isot* 1984;**35**(6):467–70.
18. Chaplin SB, Oberle PO, Hoffman TJ, Volkert WA, Holmes RA, Nowotnik DP, et al. *J Nucl Med* 1985;**26**(5). P18-P18.
19. Morais M, Paulo A, Gano L, Santos I, Correia JDG. *J. Organomet. Chem.* 2013;**744**:125–39.
20. Bolzati C, Refosco F, Marchiani A, Ruzza P. *Curr Med Chem* 2010;**17**(24):2656–83.
21. Prasanphanich AF, Lane SR, Figueroa SD, Ma LX, Rold TL, Sieckman GL, et al. *In Vivo* 2007;**21**(1):1–16.
22. Okarvi SM. *Med Res Rev* 2004;**24**(5):685–6.
23. Chakraborty S, Liu S. *Curr Top Med Chem* 2010;**10**(11):1113–34.
24. Maresca KP, Hillier SM, Lu GL, Marquis JC, Zimmerman CN, Eckelman WC, et al. *Inorg. Chim. Acta* 2012;**389**:168–75.
25. Vallabhajosula S, Nikolopoulou A, Babich JW, Osborne JR, Tagawa ST, Lipai I, et al. *J Nucl Med* 2014;**55**(11):1791–8.
26. Afshar-Oromieh A, Babich JW, Kratochwil C, Giesel FL, Eisenhut M, Kopka K, et al. *J Nucl Med* 2016;**57**:79s–89s.
27. Liu Y, Pak J-K, Schmutz P, Bauwens M, Mertens J, Knight H, et al. *J Am Chem Soc* 2006;**128**:15996–7.
28. Sulieman S, Can D, Mertens J, N'Dongo HWP, Liu Y, Schmutz P, et al. *Organometallics* 2012;**31**(19):6880–6.
29. Svenson S. *Mol. Pharmaceut.* 2013;**10**(3):848–56.
30. Kelkar SS, Reineke TM. *Bioconjugate Chem.* 2011;**22**(10):1879–903.
31. Gasser G, Ott I, Metzler-Nolte N. *J Med Chem* 2011;**54**(1):3–25.
32. Jaouen S, Top S, Vessières A. Organometallics targeted to specific biological sites: The development of new therapies. In: Jaouen G, editor. *Bioorganometallics.* Weinheim: Wiley-VCH; 2006. p. 65–90.

33. Imstepf S, Pierroz V, Raposinho P, Bauwens M, Felber M, Fox T, et al. *Bioconjugate Chem.* 2015;**26**(12):2397–407.
34. Agorastos N, Borsig L, Renard A, Antoni P, Viola G, Spingler B, et al. *Chem Eur J* 2007;**13**(14):3842–52.
35. Haefliger P, Agorastos N, Renard A, Giambonini-Brugnoli G, Marty C, Alberto R. *Bioconjugate Chem* 2005;**16**(3):582–7.
36. Häfliger P, Agorastos N, Spingler B, Georgiev O, Viola G, Alberto R. *Chembiochem* 2005;**6**(2):414–21.
37. Wendisch M, Freudenberg R, Drechsel J, Runge R, Wunderlich G, Kotzerke J. *Nuklearmedizin* 2010;**49**(4):154–60.
38. Ferl S, Wunderlich G, Smits R, Hoepping A, Naumann A, Kotzerke J. *Medchemcomm* 2015;**6**(5):887–97.
39. Reissig F, Mamat C, Steinbach J, Pietzsch HJ, Freudenberg R, Navarro-Retamal C, et al. *PLoS ONE* 2016;**11**(9).
40. Jaouen G. *Chem Brit* 2001;**37**(6):36–8.
41. Jaouen G, Vessieres A, Top S. *Chem Soc Rev* 2015;**44**(24):8802–17.
42. Meggers E. *Chem Commun* 2009;**9**:1001–10.
43. Meggers E. *Angew Chem Int Edit* 2011;**50**(11):2442–8.
44. Qin J, Rajaratnam R, Feng L, Salami J, Barber-Rotenberg JS, Domsic J, et al. *J Med Chem* 2015;**58**(1):305–14.
45. Dubar F, Slomianny C, Khalife J, Dive D, Kalamou H, Guerardel Y, et al. *Angew Chem Int Ed* 2013;**52**(30):7690–3.
46. Debreczeni JE, Bullock AN, Atilla GE, Williams DS, Bregman H, Knapp S. E. Meggers. *Angew Chem Int Ed* 2006;**45**(10):1580–5.
47. Patra M, Gasser G, Bobukhov D, Merz K, Shtemenko AV, Metzler-Nolte N. *Dalton T.* 2010;**39**(24):5617–19.
48. Wenzel M. *J Labelled Compd Radiopharm* 1992;**31**(9):641–50.
49. Wenzel M, Klinge C. *J. Labelled Compd. Radiopharm.* 1994;**34**(10):981–7.
50. Hom RK, Katzenellenbogen JA. *Nucl Med Biol* 1997;**24**:485–98.
51. Katzenellenbogen JA, Welch MJ, Dehdashti F. *Anticancer Res* 1997;**17**(3B):1573–6.
52. Spradau TW, Katzenellenhogen JA. *Organometallics* 1998;**17**(10):2009–17.
53. Cesati IRR, Katzenellenbogen JA. *J Am Chem Soc* 2001;**17**:4093–4.
54. Cesati IRR, Tamagnan G, Baldwin RM, Zoghbi SS, Innis RB, Kula NS, et al. *Bioconjugate Chem.* 2002;**13**(1):29–39.
55. Luyt LG, Bigott HM, Welch MJ, Katzenellenbogen JA. *Bioorgan. Med. Chem.* 2003;**11**(23):4977–89.
56. Spradau TW, Edwards WB, Anderson CJ, Welch MJ, Katzenellenbogen JA. *Nucl Med Biol* 1999;**26**:1–7.
57. Top S, Kaloun EB, Toppi S, Herrbach A, McGlinchey MJ, Jaouen G. *Organometallics* 2001;**20**:4554–61.
58. Le Bideau F, Kaloum EB, Haquette P, Kernbach U, Marrot J, Stephan E, et al. *Chem. Commun.* 2000;**3**:211–12.
59. Top S, Kaloun EB, Jaouen G. *J Am Chem Soc* 2000;**122**(4):736–7.
60. Top S, Elhafa H, Vessieres A, Quivy J, Vaissermann J, Hughes DW, et al. *J Am Chem Soc* 1995;**117**(32):8372–80.
61. Top S, Vessieres A, Jaouen G. *J Chem Soc Chem Commun* 1994;**4**:453–4.
62. Dallagi T, Saidi M, Vessieres A, Huche M, Jaouen G, Top S. *J. Organomet. Chem.* 2013;**734**:69–77.

63. Chen ST, Zhang Y, Li XY, Jia HM, Lu J. *Bioorg Med Chem Lett* 2017;**27**(15):3551–4.
64. Zeng HH, Zhang HB. *Eur J Med Chem* 2014;**72**:10–17.
65. Zeng HH, Zhang HB, Wu XX, Chao FF, Yu G, Zhang LQ, et al. *J. Labelled Compd. Rad.* 2013;**56**(1):1–5.
66. Jia JH, Cui MC, Dai JP, Liu BL. *Dalton T.* 2015;**44**(14):6406–15.
67. Ono M, Hori M, Haratake M, Tomiyama T, Mori H, Nakayama M. *Bioorg Med Chem* 2007;**15**(19):6388–96.
68. Ferreira CL, Ewart CB, Bayly SR, Patrick BO, Steele J, Adam MJ, et al. *Inorg Chem* 2006;**45**(17):6979–87.
69. Saidi M, Seifert S, Kretzschmar M, Bergmann R, Pietzsch HJ. *J. Organomet. Chem.* 2004;**689**(25):4739–44.
70. Lee BC, Kim DH, Lee I, Choe YS, Chi DY, Lee KH, et al. *J Med Chem* 2008;**51**(12):3630–4.
71. Bowen ML, Chen ZF, Roos AM, Misri R, Hafeli U, Adam MJ, et al. *Dalton T.* 2009;**42**:9228–36.
72. Bigott HM, Parent E, Luyt LG, Katzenellenbogen JA, Welch MJ. *Bioconjugate Chem.* 2005;**16**(2):255–64.
73. Li ZJ, Cui MC, Dai JP, Wang XD, Yu PR, Yang YP, et al. *J Med Chem* 2013;**56**(2):471–82.
74. Spradau TW, Katzenellenbogen JA. *Bioconjugate Chem.* 1998;**9**(6):765–72.
75. Wald J, Alberto R, Ortner K, Candreia L. *Angew. Chem., Int. Ed.* 2001;**40**:3062–6.
76. Bernard J, Ortner K, Spingler B, Pietzsch HJ, Alberto R. *Inorg Chem* 2003;**42**(4):1014–22.
77. N'Dongo HWP, Liu Y, Can D, Schmutz P, Spingler B, Alberto R. *J. Organomet. Chem.* 2009;**694**(6):981–7.
78. N'Dongo HWP, Raposinho PD, Fernandes C, Santos I, Can D, Schmutz P, et al. *Nucl Med Biol* 2010;**37**(3):255–64.
79. Ursillo S, Can D, N'Dongo HWP, Schmutz P, Spingler B, Alberto R. *Organometallics* 2014;**33**(23):6945–52.
80. Top S, Lehn J-S, Morel P, Jaouen G. *J. Organomet. Chem.* 1999;**000**:63–8.
81. Liu Y, Spingler B, Schmutz P, Alberto R. *J Am Chem Soc* 2008;**130**(5):1554–6.
82. Can D, Spingler B, Schmutz P, Mendes F, Raposinho P, Fernandes C, et al. *Angew. Chem. Int. Edit.* 2012;**51**(14):3354–7.
83. Supuran CT. *Curr. Pharm. Design* 2008;**14**(7):601–2.
84. Supuran CT. *Nat Rev Drug Discov* 2008;**7**(2):168–81.
85. Supuran CT, Di Fiore A, De Simone G. *Expert Opin. Emerg. Dr.* 2008;**13**(2):383–92.
86. Kenny RG, Chuah SW, Crawford A, Marmion CJ. *Eur. J. Inorg. Chem.* 2017;**12**:1596–612.
87. Griffith D, Morgan MP, Marmion CJ. *Chem. Commun.* 2009;**44**:6735–7.
88. Xu WS, Parmigiani RB, Marks PA. *Oncogene* 2007;**26**(37):5541–52.
89. Romero-Canelon I, Sadler PJ, Natl P. *Acad. Sci. USA* 2015;**112**(14):4187–8.
90. Soldevila-Barreda JJ, Sadler PJ. *Curr Opin Chem Biol* 2015;**25**:172–83.
91. Zhang PY, Sadler PJ. *J. Organomet. Chem.* 2017;**839**:5–14.
92. Murray BS, Babak MV, Hartinger CG, Dyson PJ. *Coord Chem Rev* 2016;**306**:86–114.
93. Allardyce CS, Dyson PJ. *Dalton T.* 2016;**45**(8):3201–9.
94. Pampaloni G. *Coord Chem Rev* 2010;**254**(5–6):402–19.
95. Baumgartner F, Fischer EO, Zahn U. *Chem. Ber. Recl.* 1961;**94**(8):2198–203.
96. Palm C, Fischer EO. *Tetrahedron Lett* 1962;**6**:253–4.
97. Fischer EO, Wirzmuller A. *Chem Ber* 1957;**90**(9):1725–30.

98. Trifonova EA, Perekalin DS, Lyssenko KA, Kudinov AR. *J. Organomet. Chem.* 2013;**727**:60–3.
99. Wester DW, Coveney JR, Nosco DL, Robbins MS, Dean RT. *J Med Chem* 1991;**34**:3284–90.
100. Benz M, Braband H, Schmutz P, Halter J, Alberto R. *Chem Sci* 2015;**6**(1):165–9.
101. Meola G, Braband H, Schmutz P, Benz M, Spingler B, Alberto R. *Inorg Chem* 2016;**55**(21):11131–9.
102. Meola G, Braband H, Jordi S, Fox T, Blacque O, Spingler B, et al. *Dalton T.* 2017;**46**(42):14631–7.
103. Noor F, Wustholz A, Kinscherf R, Metzler-Nolte N. *Angew. Chem. Int. Ed.* 2005;**44**(16):2429–32.
104. Noor F, Kinscherf R, Bonaterra GA, Walczak S, Wolfl S, Metzler-Nolte N. *Chembiochem* 2009;**10**(3):493–502.
105. Fischer EO, Wehner HW. *Chem Ber* 1968;**101**(2):454–62.
106. Herberhold M, Hofmann T, Milius W, Wrackmeyer B. *J. Organomet. Chem.* 1994;**472**(1–2):175–83.
107. Green MLH, Lowe ND, Ohare D. *J. Chem. Soc. Chem. Comm.* 1986;**20**:1547–8.
108. Green MLH, Ng DKP. *Chem Rev* 1995;**95**(2):439–73.
109. Meola G, Braband H, Hernandez-Valdes D, Gotzmann C, Fox T, Spingler B, et al. *Inorg Chem* 2017;**56**(11):6297–301.
110. Leonidova A, Gasser G. *ACS Chem Biol* 2014;**9**(10):2180–93.
111. Patra M, Wenzel M, Prochnow P, Pierroz V, Gasser G, Bandow JE, et al. *Chem Sci* 2015;**6**(1):214–24.
112. Imstepf S, Pierroz V, Rubbiani R, Felber M, Fox T, Gasser G, et al. *Angew. Chem. Int. Edit.* 2016;**55**(8):2792–5.
113. Kitanovic I, Can SZ, Alborzinia H, Kitanovic A, Pierroz V, Leonidova A, et al. *Chem-Eur J* 2014;**20**(9):2496–507.
114. Knopf KM, Murphy BL, MacMillan SN, Baskin JM, Barr MP, Boros E, et al. *J Am Chem Soc* 2017;**139**(40):14302–14.
115. Meggers E. *Angew. Chem. Int. Edit.* 2017;**56**(21):5668–75.

Chapter 12

Phosphorescent Ru(II) Complexes as Probes of Model Membrane Systems

Edward Rosenberg[1], **J.B. Alexander Ross**[1] **and Ayesha Sharmin**[2]

[1]Department of Chemistry and Biochemistry, University of Montana, Missoula, MT, United States, [2]Department of Chemistry, Bangladesh University of Engineering and Technology (BUET), Dhaka, Bangladesh

12.1 INTRODUCTION

Phosphorescent metal complexes have found many applications in the study of biomolecular assemblies. These complexes have been used in cell imaging,[1] studying biomolecular dynamics using covalently bound conjugates,[2] intercalation in double-stranded DNA[3] and selective labeling of peptides and nucleotides[4] and for molecular imaging to name a few.[5] The appeal of these complexes for these applications is their long excited-state lifetimes, their large Stokes shifts and their lack of reactivity with biomolecules unless they are specifically designed for selective binding to a particular site on a biomolecule. In general, there are two types of metal complex that offer these properties: (1) transition metal diimine chelates of the second and third row that exhibit metal-to-ligand charge transfer (MLCT) absorption transitions in the near UV and emit in the red; and (2) lanthanide chelates of aromatic amides that transfer ligand excitation energy to *f* orbital electrons resulting in *f*–*f* transitions in the visible region.[6]

For the transition metal complexes significant singlet–triplet intersystem crossing resulting from spin–orbit coupling is required to obtain the long excited-state lifetimes and red-shifted emissions. The classic example of this type of complex is $Ru(bpy)_3{}^{2+}$ which exhibits an MLCT at 436 nm and emits at 600 nm with lifetimes of 200–700 ns (depending on the environment) but has a quantum yield of only 0.06.[7] In this overview of our work for the period 2009–16 we have designed complexes that have higher quantum yields than $Ru(bpy)_3{}^{2+}$ and then bioconjugated them for incorporation

Advances in Bioorganometallic Chemistry. DOI: https://doi.org/10.1016/B978-0-12-814197-7.00012-1

into liposomes and nanodiscs with the goal of probing the dynamics of the bioconjugates using the polarized emission of these complexes.[8–12]

The same series of phosphorescent complexes has also been covalently tethered to the surface of an inorganic/organic hybrid material composed of amorphous silica gel coated with a covalently bound polyamine. These silica polyamine composites serve as models for biomacromolecules and afford the opportunity for investigating how local environments, ancillary ligands on the complex, and metal ions in solution influence their photophysical properties.[13,14] The details of these studies have been reported elsewhere.[13,14]

Most,[8,9,12] but not all, of the studies discussed here have been published. This chapter will demonstrate the usefulness of time-resolved, polarized emission as a tool for studying biomolecular dynamics and to elucidate the fundamental causes of the observed variations in lifetime, quantum yield and emission anisotropy as they occur in response to the changes in the environment of a given complex in different phases.

12.2 SYNTHESIS AND PHOTOPHYSICAL PROPERTIES OF [RU (*TRANS/CIS*-$PR_3)_2(CO)(L_2)X][PF_6]$ (X = H, TRIFLUORACETATE, L_2 = BIPYRIDYL, PHENANTHROLINE)

It has been known for quite some time that substituting CO in luminescent diimine complexes with other ancillary ligands can lead to significant changes in their photophysical properties.[8] In the particular case of trialkylphosphines, enhanced electron donor ability and delocalization of electron density in the excited state can lead to longer lifetimes and higher quantum yields by increasing the contribution of the triplet state to the excited state ensemble. To address this issue in a systematic way we varied the geometry of the phosphine,[8] its electron donor ability,[12] and examined the contribution of the phosphine[9] to the excited state while keeping the degree of phosphine substitution constant at two.

In our initial synthetic studies we relied on an easy-to-prepare ruthenium tricarbonyl complex that could be made in high yield from triruthenium dodecacarbonyl that gave only bis-substituted phosphine complexes or in one step by reaction of the diimine with $Ru(Ph_3P)_3(CO)(Cl)H$ (Scheme 12.1). The one-step reaction is, of course, more convenient but the starting tris-phosphine complex is more expensive than the parent carbonyl cluster and has a shorter shelf life.

Several trends are obvious for the data reported (Table 12.1). The bpy complexes have significantly longer excited-state lifetimes than the analogous phenanthroline complexes. As might be expected, the exited-state lifetimes are much longer in the absence of oxygen and the large difference observed here means that oxygen is a very effective quencher of these complexes in solution (**1**, **2**, and **4** in Table 12.1). In addition, the ancillary ligands, H versus TFA, have an extraordinary effect on excited-state lifetime (**1** vs **2** in Table 12.1). We have previously noted[8] that the TFA ligand tends

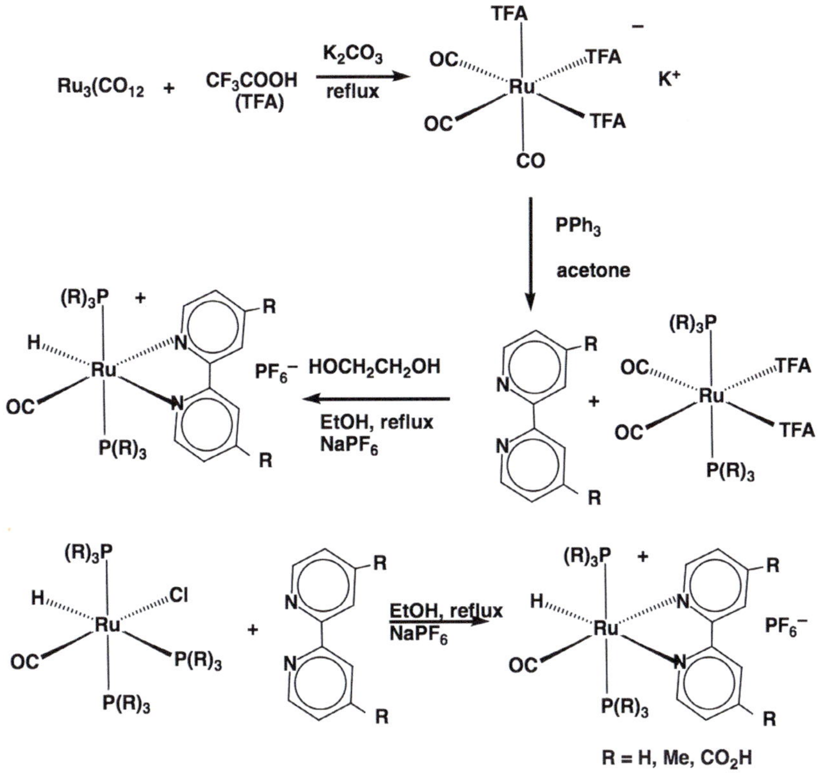

SCHEME 12.1 Synthesis of the bis-phosphine Ru(II) diimine complexes by two different routes.

to dissociate, making it useful for substituting with anions, but also opening a coordination site that would allow transient coordination of quenchers such as oxygen. The most important result of these initial studies is the significant increase in quantum yield for the bioconjugable complexes (**1**, **3**, and **4** in Table 12.1). We attribute this to the participation of the phosphine ligands in the composition of the excited state. Evidence for this comes from the excitation spectrum of **1**, which clearly shows the contribution of the phosphine to the composition of the excited state and would be expected to delocalize electron density leading to a longer lifetime (Fig. 12.1).[13] Interestingly, the tris-bpy complex $Ru(bpy)_2$(5-amino-penanthroline)$^{2+}$ (**5**) has very similar photophysical properties to the bis-phosphine analog **4**, suggesting that in the phenanthroline complexes there is less participation of the phosphine in the composition of the excited state relative to bpy.[15] It is also interesting to note that the less bulky and more electron donating $PPh(Me)_2$ in **6** (Table 12.1) has a shorter excited-state lifetime than its PPh_3 analog.[12] Thus steric bulk and electron donor ability of the phosphine has less of an effect on excited-state lifetime than the ability of the phosphine to delocalize electron density.

TABLE 12.1 Photophysical Properties of Selected $Ru(II)X(L-L)CO^+$ Diimine Complexes

#	Diimine	L–L	*X*	MLCT (nm)	λ_{excit} (nm)	λ_{em} (nm)	*T* (ns) (No Oxygen)	Φ
1	4,4′-CO_2H bpy	$2PPh_3$	H	450	450	647	720(2630)	0.30
2	4,4′-CO_2H bpy	$2PPh_3$	TFA	450	450	630	185(670)	
3	bpy	$2PPh_2$-$CH_2CH_2CO_2H$ dppa	H	460	460	608	266	0.5
4	5-Amino-phen	$Ph_2PCHCHPPh_2$ dppene	H	445	470	607	250 (570)	0.25
5	*5-Amino-phen	$(bpy)_2$	–	445	470	625	220	–
6	Bpy	$PPh(Me)_2$	H	480	470	600	330	

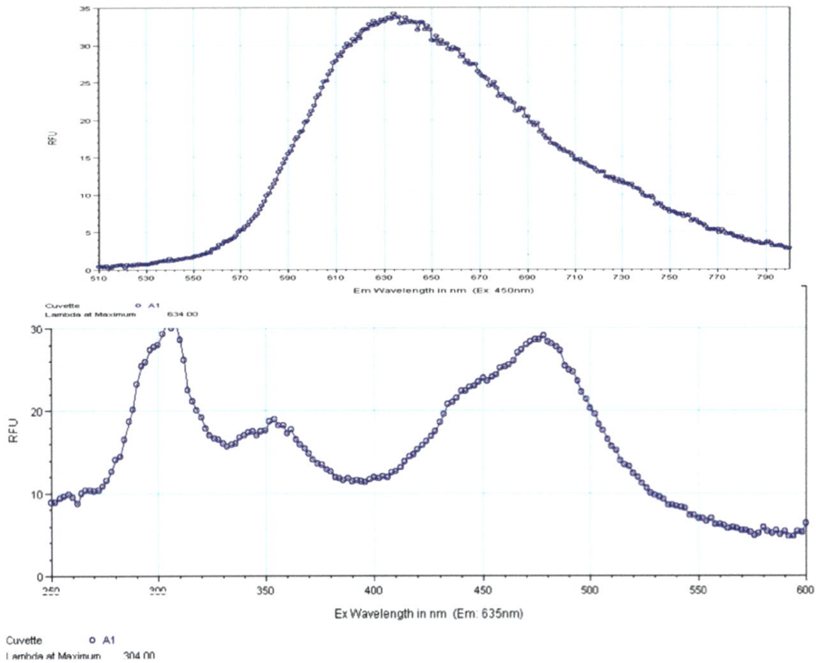

FIGURE 12.1 Excitation spectrum of **1** irradiated 450 nm (lower trace) and emission spectrum of 1 irradiated at 450 nm showing the contribution of the phosphine to the composition of the excited state.

In summary, the goal of the synthetic work was to make complexes with large Stokes shifts and long excited-state lifetimes and higher quantum yields than their tris-bpy analogs. Compounds **1**, **3**, and **4** provide these photophysical properties and we can now discuss their bioconjugation and incorporation into membrane and models. In the entire series, the emission frequency is relatively insensitive to ancillary ligand composition indicating relatively small changes in HOMO–LUMO gaps in this series of compounds.

12.3 BIOCONJUGATION OF THE RU(II) COMPLEXES

In targeting the bioconjugation of the high quantum yield Ru(II) diimine complexes **1**, **3**, and **4** we decided to focus on how these probes would respond to incorporation into liposomes. Proteins and small molecules often initiate reorganization of membrane surfaces.[16] The classic example of this phenomenon is the formation of "lipid rafts" where cholesterol in a lipid membrane aggregates around a surface-binding molecule on incorporation into the lipid surface.[16] Cholesterol-rich regions have been identified in cells and in liposomes. We, therefore, thought it would be useful to conjugate a cholesterol molecule to the diimine complexes **4** and **5** where the amino group could be readily bound to the cholesterol molecule via the commercially available chloroacetyl cholesterol (Scheme 12.2). The resulting complexes were fully characterized by spectroscopic techniques.[9]

An additional goal of the bioconjugation of the phosphorescent Ru(II) complexes was to follow lipid motion/reorganization on binding of a biomolecule to the liposome. The conjugation of a lipid to **1** and **3** was accomplished by reaction of the amino group of dipalmitoyl-phosphatidylethanolamine with the carboxylate groups of **1**and **3** via the activated ester *N*-hydroxy succinimide followed by peptide coupling with dicyclohexyl carbodiimide (Scheme 12.3). This protocol gave the bis-lipid derivatives of

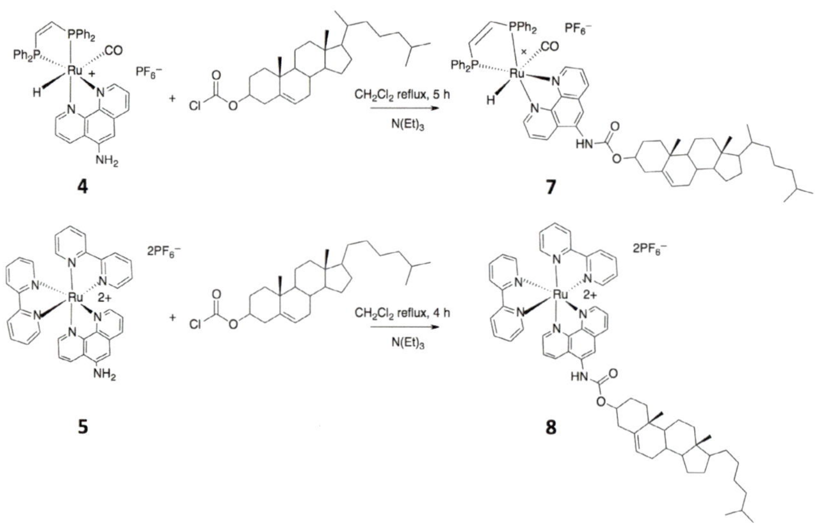

SCHEME 12.2 Covalent binding of Ru(II) diimine complexes to cholesterol.

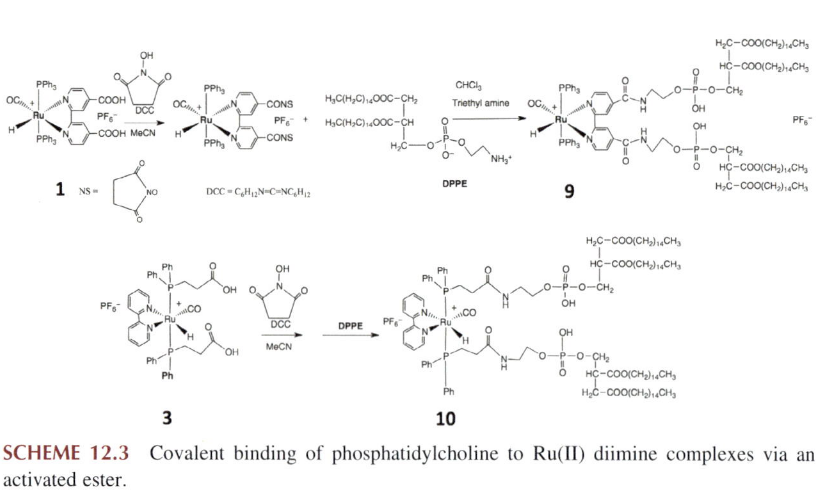

SCHEME 12.3 Covalent binding of phosphatidylcholine to Ru(II) diimine complexes via an activated ester.

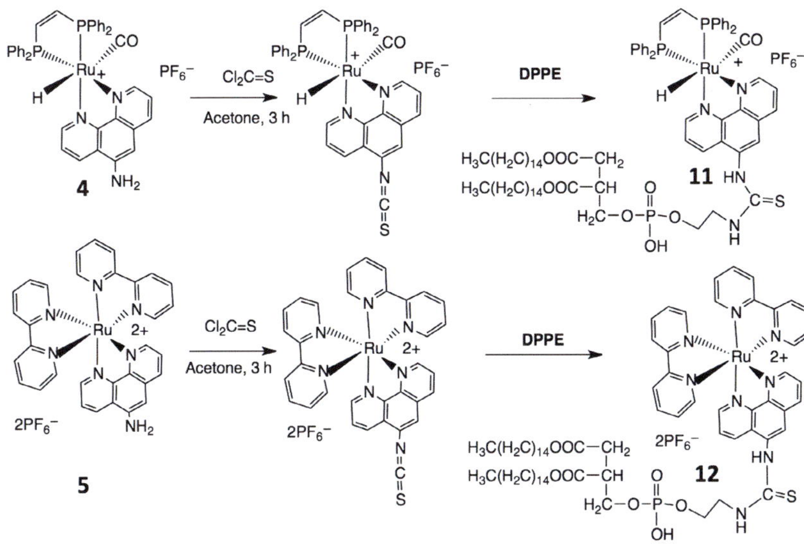

SCHEME 12.4 Covalent binding of phosphatidylcholine to Ru(II) diimine complexes via an isothiocyanate.

1 and **3**, **9**, and **10**, in low to moderate yields. Complex **3** was designed to test how the site of bioconjugation affected the photophysical properties of the lipid derivatives and their interaction with lipid models.

The amino-phenanthroline complexes **4** and **5** offer the opportunity to synthesize mono-lipid derivatives via conversion to an isothiocyanate followed by reaction with the amino group of phosphatidylethanolamine (Scheme 12.4). Complex **5**, which has been previously reported, provides the opportunity to evaluate the relative impact of two bpy ligands on the photophysical parameters of the acceptor phenanthroline compared with the *trans*-phosphine ligands in **4**.

In addition to the cholesterol and lipid bioconjugates of the complexes in Table 12.1 we have also synthesized sulfhydryl specific complexes [Os(II)(bpy)$_2$(CO)(NH$_2$CH$_2$CH$_2$NH(C(O)CH$_2$I))]Cl$_2$ and [Ru(II)(dppene)H(CO)(5-amino-phen)]PF$_6$ utilizing iodoacetamide as the sulfhydryl selective linker for binding to proteins and peptides.[10,17]

12.4 PHOTOPHYSICAL PROPERTIES AND POLARIZED EMISSION OF THE BIOCONJUGATED CHOLESTEROL COMPLEXES IN LIPOSOMES

The photophysical parameters of interest in phosphorescent metal complexes are Stokes shift, excited-state lifetime, and quantum yield. These parameters vary with the nature of the metal, the ancillary ligands, and the environment.

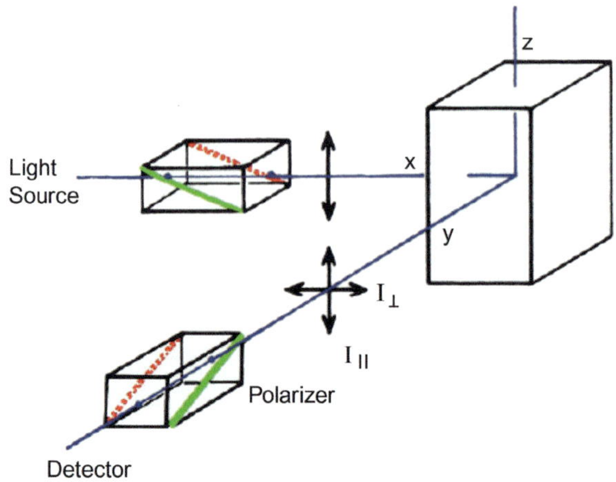

FIGURE 12.2 Schematic drawing of the apparatus used for detecting emission anisotropy.

None give direct evidence of the dynamic behavior of the luminophore. Measuring the polarization or anisotropy of the emission, however, can provide direct access to rotational correlation times, a parameter particularly useful in assessing restricted molecular motion in constrained environments.[18] This is done by passing polarized light through the sample and then detecting the emission perpendicular to the incident radiation as is typical with most emission experiments (Fig. 12.2). The emitted light is then passed through a second polarizer where the parallel and perpendicular components are measured by a photon-counting detector (Fig. 12.2).

The measured anisotropy is the given by the difference in intensity according to the formula (Eq. 12.1):

$$r(t) = \frac{I_{VV}(t) - I_{VH}(t)}{I_{VV}(t) + 2I_{VH}(t)} \tag{12.1}$$

I_{VV} and I_{VH} are the emission intensities in the vertical plane and the horizontal planes respectively with vertical excitation. The fundamental anisotropy, r_0 is related to the measured anisotropy r by Eq. (12.2) and is the anisotropy prior to the onset of molecular rotation:

$$r_0 = r\left(1 + \frac{\tau}{\theta}\right) \tag{12.2}$$

τ is the excited-state lifetime and θ is the rotational correlation time. The rotational correlation time θ depends on the viscosity of the medium, the molecular volume and temperature and can be obtained from the time-dependent decay of the fundamental anisotropy (Eq. 12.3):

$$r(t) = r_0 e^{-t/\theta} \tag{12.3}$$

Thus measurement of the time-dependent decay provides a direct link to molecular motion on the time sale of the emission. If the time-dependent decay is done at the magic angle (54.7 degrees) it is only due to the lifetime of the excited state (Fig. 12.3).

One of the fundamental questions we wanted to answer for the cholesterol bioconjugates **7** and **8** (Scheme 12.2) is whether the overall molecular symmetry is directly related to the polarization of the emission and related photophysical properties. We have measured the photophysical properties of **7** and **8** and their fundamental anisotropies, r_0; the data are given in Fig. 12.4. Bioconjugate **8** has a pseudo-C_2 axis, ignoring the positional

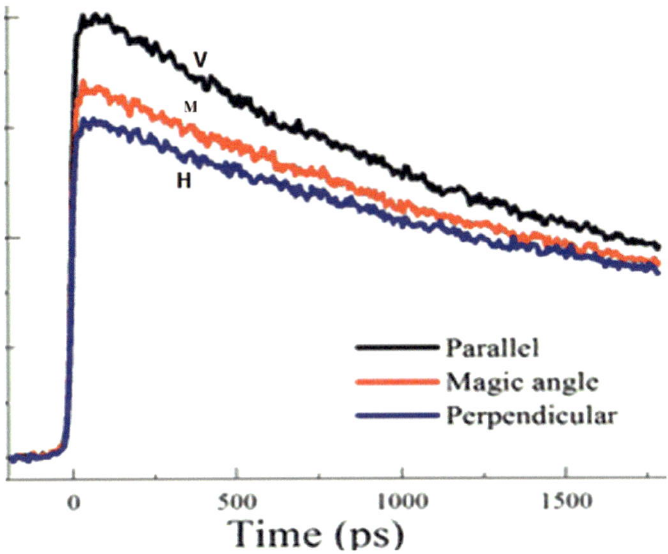

FIGURE 12.3 Time-dependent decay of the anisotropic emitted light from a sample using the configuration shown in Fig. 12.2.

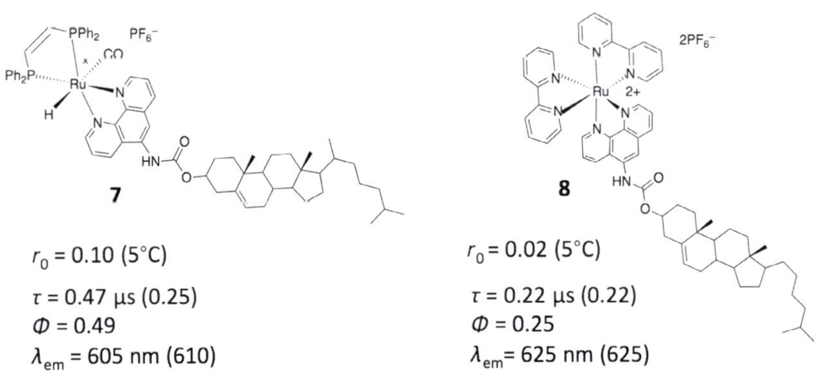

FIGURE 12.4 Photophysical properties and fundamental anisotropy of cholesterol bioconjugates **7** and **8** in ethanol (numbers in parentheses are for the precursor complexes).

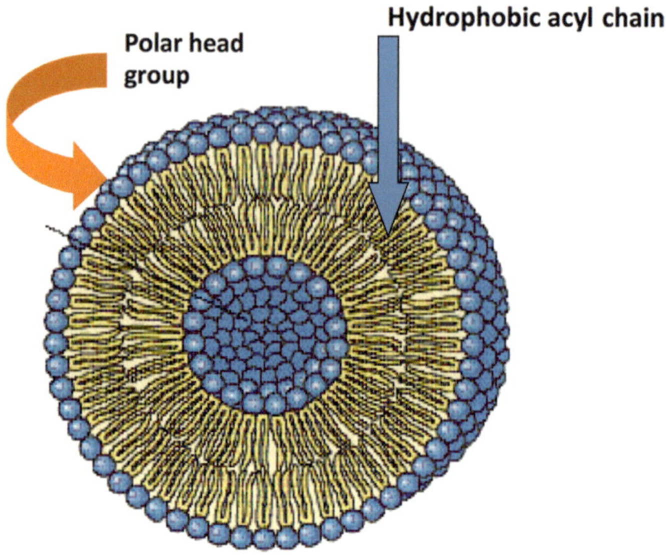

FIGURE 12.5 Schematic drawing of lipid bilayer vesicles.

asymmetry of the cholesterol moiety and **7** lacks this element and is much less symmetric. Bioconjugate **7** has a fundamental anisotropy of 0.1 while **8**, has a much smaller anisotropy of 0.02. It is dangerous to extrapolate from this one example, but the data certainly suggest a relationship between molecular symmetry and r_0, at least for chemically similar complexes. The presence of the phosphine ligands boosts quantum yield, but has little effect on emission frequency. Conjugation of cholesterol has a much bigger effect on increasing the excited-state lifetime in the case of **7** and more so than in **8**.

The time-dependent anisotropy of **7** was measured after incorporation into lipid vesicles of 100 nm diameter prepared by extruding a suspension of the Ru-complex-bioconjugate, **7**, and phosphatidylcholine (1:99) in HEPES buffer through a sizing membrane. Two types of lipid vesicles were used. Egg-PC, which has variable chain length lipids and DPPC (1,2-dipalmitoyl-*sn*-glycero-3-phosphatidylcholine), which has uniform 15 carbon lipid chains and is more ordered. An idealized drawing of the vesicles is shown in Fig. 12.5.[9]

The fundamental anisotropy r_0 of 0.1 for **7** and its relatively long excited-state lifetime suggest that this complex could be used for monitoring molecular motion on a membrane surface. This is indeed born out by the measurements made in the temperature range of 10–50°C (Table 12.2). The decreases in rotational correlation times with increasing temperature indicate that this cholesterol conjugate can be used for tracking dynamic changes in the local environment. For example, addition of more cholesterol to the liposome (which can result in information of lipid rafts) will affect θ by

TABLE 12.2 Variable Temperature Anisotropy Data for 7

Temperature (°C)	$<\tau>$ (μs)	r_∞	θ (ns)
10	0.956	0.031	112
20	0.862	0.030	112
30	0.710	0.019	40
40	0.540	0.013	19
50	0.448	0.009	14

decreasing viscosity. Binding of a protein to the liposome should also affect θ as a result of surface reorganization. In this case, θ could increase or decrease depending on surface composition. The residual anisotropy, r_∞, is a calculated parameter that is an estimate of restricted molecular motion and residual motion that could occur after emission at infinite time; r_∞ provides insight into anisotropic molecular motion at time scales that are long compared to the anisotropy decay.

12.5 PHOTOPHYSICAL PROPERTIES OF LIPID CONJUGATES IN LIPOSOMES

The bis-lipid conjugates **9** and **10** (Scheme 12.3) gave surprising results when their time-resolved polarized emissions were measured. In solution (EtOH), **9** showed no emission while **10** showed a very weak, blue-shifted, short-lived emission ($\tau = 0.004$ μs, $\phi = 0.019$, $\lambda = 505$ nm) relative to its precursor, **3** ($\tau = 0.266$ μs, $\phi = 0.5$, $\lambda = 608$). We attribute these drastic changes to the presence of the conjugated lipid, which introduces many new vibrational modes, increasing internal conversion, and quenching or reducing emission intensity. Similarly, the mono-lipid conjugate **11** (Scheme 12.4) showed no emission likely for the same reasons. However, **12** showed the usual red-shifted emission (625 nm) and long lifetime (0.22 μs) as its precursor, **5** (Table 12.1).[15] Again, this points to the special role of the phosphine ligands in the composition of the excited states while for **5** only the unsubstituted bpy rings (bpy is a better electron acceptor than phen) contribute to the photophysical properties.

The lipid conjugates **9** and **11**, which showed no emission in solution, exhibited dramatically different behavior when incorporated into 100 nm lipid vesicles. Both showed intense blue-shifted emissions at 534 and 545 nm respectively and very short excited-state lifetimes of 0.011 and 0.008 μs respectively (Fig. 12.6).

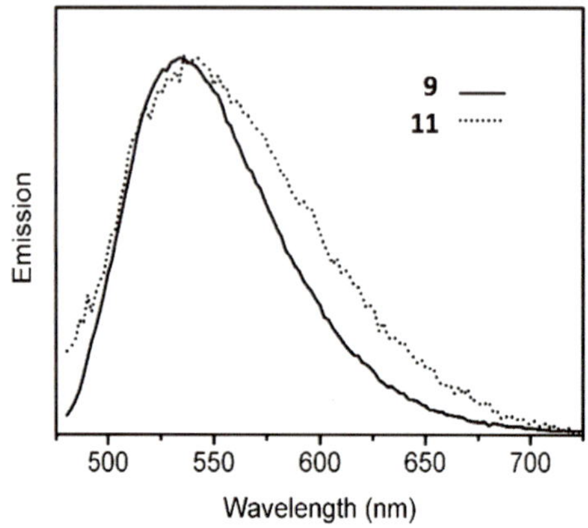

FIGURE 12.6 Blue-shifted emissions **9** and **11** after incorporation into lipid vesicles.

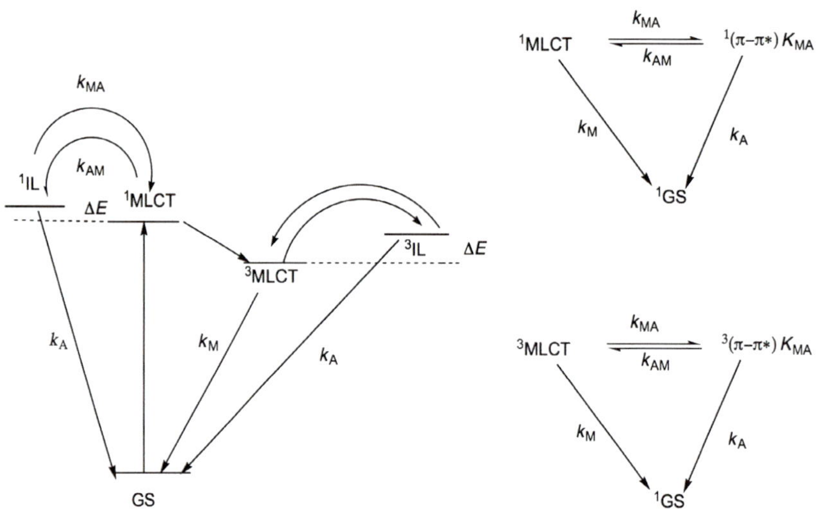

FIGURE 12.7 Energy transfer from Ru dπ−π* to π−π* states, in **9** and **11**.

The blue-shifted emissions and the short lifetimes for **9** and **11** relative to their precursor complexes **1** and **4** (Table 12.1) suggest a crossover to singlet states based on intraligand transitions. This could occur by geometric distortions in the excited states resulting in significant changes in their energies and bringing their excited-state energies close to those of intraligand transitions and resulting in energy transfer from the MLCT^1 to the L−L^1 excited-state manifold (Ru dπ−π* and π−π* states, Fig. 12.7).

There is precedent for this phenomenon with related Os diimine complexes in the work of Schmehl et al.[19]

TABLE 12.3 Variable Temperature Emission Anisotropy for 9 in Ethanol

T (°C)	r_0	r_∞	θ_1 (ns)	θ_2 (ns)
10	0.36	0.07	7.6	0.7
25	0.36	0.03	5.2	0.6
40	0.36	0.03	3.5	0.5
55	0.35	0.02	1.1	0.1

The variable temperature anisotropy measurements in the range +10°C to +55°C. Table 12.3 shows that r_0 is invariant with temperature and had an extremely high fundamental anisotropy (0.36), close to the maximum value of 0.4.[18] Along with the short ns lifetime, this suggests a singlet emission. The T independent correlation times are indicative of a very unrestricted motion, most likely, a wobble. The mono-lipid is very similar ($r_0 = 0.24$). These results would not be useful for measuring slow dynamics (τ values are very short) but could be applied to tracking lipid reorganization in a bilayer or nanodisc (vide infra). The two different rotational correlation times are reflected by a double exponential decay and suggest two distinct motions of the molecule.

Accompanying the short excited-state lifetimes for **9** and **11** (11 and 8 ns respectively), the lipid conjugates have high fundamental anisotropy and temperature insensitive rotational correlation times, which are helpful for studying faster, local motions (up to 33 ns) in the vesicles. Interestingly, complexes **9** and **11**, which have two and one lipid conjugates respectively, showed almost the same longer rotational correlation times (7–8 ns at 10°C, Table 12.3) and in egg-PC vesicles.[10] This timescale and its insensitivity to the number of lipid anchors suggests that this motion could reflect restricted diffusion, commonly referred as "wobble in a cone,"[20–23] and not due to axial rotation. The axial rotation involves rotation of the entire lipid about its long axis. "Wobble in a cone" can be described as the restricted diffusive reorientation of the lipid long axis with respect to the bilayer normal.[20,23]

Kinosita et al.[20] have developed two ways for analyzing fluorescence anisotropy decay data to model restricted diffusion of lipid probes in membranes. In one, a fluorophore wobbles uniformly within a cone (square-well potential). In the other, the probe rotational distribution is Gaussian. The wobbling-cone model provides a simple but useful way to assess the impact of interactions of the lipid probe with the neighboring lipids. In the wobbling-cone model, it is assumed that the major axis of the probe wobbles uniformly within a cone of semiangle θ_c (i.e., probe's major axis is confined within a cone of angle θ_c). The angle θ_c with respect to the symmetry axis of the cone can be estimated using the following relationship

$$\frac{r_\infty}{r_0} = \left[\frac{1}{2\cos\,\theta_c(1+\cos\,\theta_c)}\right]^2 \tag{12.4}$$

The motion of the lipid probes in egg-PC vesicles were analyzed using this model. The cone angle θ_c varied from 44 to 72 degrees for complex **9** and from 55 to 74 degrees for complex **11** over the temperature range from 10 to 50°C. The shorter correlation times are significantly different for complex **9** and **11** being 1.9 and 0.7 ns, respectively. This timescale and dependence on the numbers of anchoring lipids indicate that these shorter rotational correlation times are dominated by the axial motion of the lipid incorporated in the vesicle.[22] Thus these probes could be useful for studying the lipid head-group motions.

The most important result from the liposome studies is the sensitivity of the emission anisotropy to the nature of the conjugate (cholesterol vs lipid bioconjugation) suggesting different applications for the same luminophore with different bioconjugates.

12.6 PHOTOPHYSICAL PROPERTIES OF BIOCONJUGATES IN NANODISCS

Given the above results it would be interesting to investigate the behavior of these Ru-based luminophores incorporated into protein-belted nanodiscs. These membrane models differ significantly from liposomes in that the two lipid surfaces are exposed to the environment and do not have the internal cavity associated with the lipid bilayer (Fig. 12.5 vs Fig. 12.8).

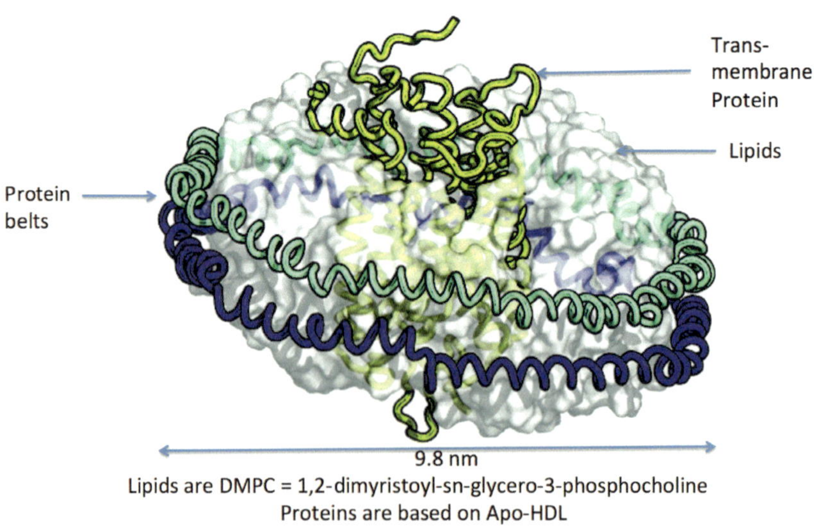

FIGURE 12.8 Schematic drawing of a protein-belted nanodisc.

The nanodiscs are constructed with Apo-lipoprotein A1 in the presence of the detergent, cholate and other lipids such as DMPC.[24] The amphipathic α-helical proteins from the Apo-A_1, termed membrane scaffold proteins (MSPs), make the nanodisc water soluble by virtue of having both lipophilic and lyophilic amino acids in their sequence. All of the components are solubilized by detergents. Then when treated with bio-beads that have selective and high affinity for the detergent, the belt protein and lipids self-assemble into nanodiscs. These membrane models have found important applications in structural biology because:[24]

1. Proteins have better stability and activity inside nanodiscs.
2. Nanodiscs allow controlled and precise control of the lipid compositions around the membrane protein.
3. Nanodiscs allow access to both extra and intracellular sides of membrane protein.

The nanodiscs afford the opportunity for incorporation of a luminescent probe into the lipid or bioconjugation to the protein belt, two very different but potentially complimentary probes of nanodisc global and segmental motion. Our initial experiments involved introducing two different concentrations of the cholesterol probe **7** (5% and 20%) and measuring the excited-state lifetime.[11] This could inform on any probe–probe interactions as the lifetime should be different in the two different concentrations of **7** if probe–probe interactions occur. This experiment would also provide information on global motions of the nanodisc. A summary of these measurements in the temperature range 5-30°C is given in Table 12.4. Note that the excited-state lifetime is relatively insensitive to the concentration of **7,** indicating that there are no probe–probe interactions, a fact that could be useful for future studies with membrane-bound biomolecules.

TABLE 12.4 Excited-State Lifetime Data for 7 Incorporated Into Nanodiscs at Two Concentrations

T (°C)	5% $<\tau>$	20% $<\tau>$
5	2.42	2.33
10	1.89	2.12
15	1.69	1.73
20	1.33	1.51
25	1.17	1.26
30	0.84	0.93

The rotational diffusion time, θ for the oblate shaped $9.8 \times 5.5\ nm^2$ diameter nanodisc can be calculated using the molecular volume, V and the solvent viscosity, η, for water at 20°C (Eq. 12.5):

$$\theta = \frac{\eta V}{kT} \tag{12.5}$$

The value of 130 ns is in excellent agreement with the value obtained from analysis of the anisotropy data for the emission at 590 nm: 129.6 ns. The limiting anisotropy, $r_0 = 0.11$, is the same as for the probe in solution (0.10). This means that the angle between the absorption and emission transition moments does not change. This result assures accurate determination of global motions when membrane-binding molecules are added to the nanodisc.

We then went on to covalently attach the thioisocyanate derivative of complex **4** (Scheme 12.4). This probe would be expected to randomly react with primary amines of lysine in the MSP. The luminescence lifetime measurements revealed that the excited-state decay of randomly labeled protein in 2 M Guanidine hydrochloride solution is 898 ns whereas the starting thioisocyanate complex had a lifetime of 250 ns, the same as **4.** The bioconjugated **4** gave an emission at 600 nm and the excitation spectrum showed involvement of both the MLCT and the phosphine ligand. Simply mixing **4** with the MSP gave only a weak emission whose intensity precluded lifetime measurement (Fig. 12.9).

Unfortunately, this conjugate did not form stable nanodiscs, perhaps due to multiple labeling at a single MSP. Instead we did a site-directed mutagenesis substituting cysteine for serine 136, which is located in a hydrophobic region of the MSP (Fig. 12.10). A sulfhydryl selective probe was synthesized from the isothiocyanate derivative of **4** using diiodo-acetic anhydride.

This sulfhydryl selective probe, **13** was then reacted with the cysteine 136 mutant of the MSP after reduction and purification. Bioconjugation followed standard procedures for sulfhydryl selective probes[17] using a probe: protein ratio of 10:1 (Fig. 12.11).

As for the isothiocyanate derivative of **4** conjugated to the unmodified MSP a severalfold increase in lifetime was realized relative to the unconjugated probe **13** (840 vs 157 ns for **13**).

The cysteine 136 MSP mutant readily formed stable nanodiscs using 1,2-dimyristoyl-*sn*-glycero-3-phosphocholine and sodium cholate as the detergent. Size exclusion chromatography was used to separate the nanodiscs from higher order aggregates. The yield of the $10 \times 5.5\ nm^2$ nanodiscs was about 70%.[25]

A steady-state luminescence study of cysteine 136 labeled protein in the detergent suspension showed the usual emission at 590 nm. After nanodisc formation and purification the emission is blue shifted to 520 nm

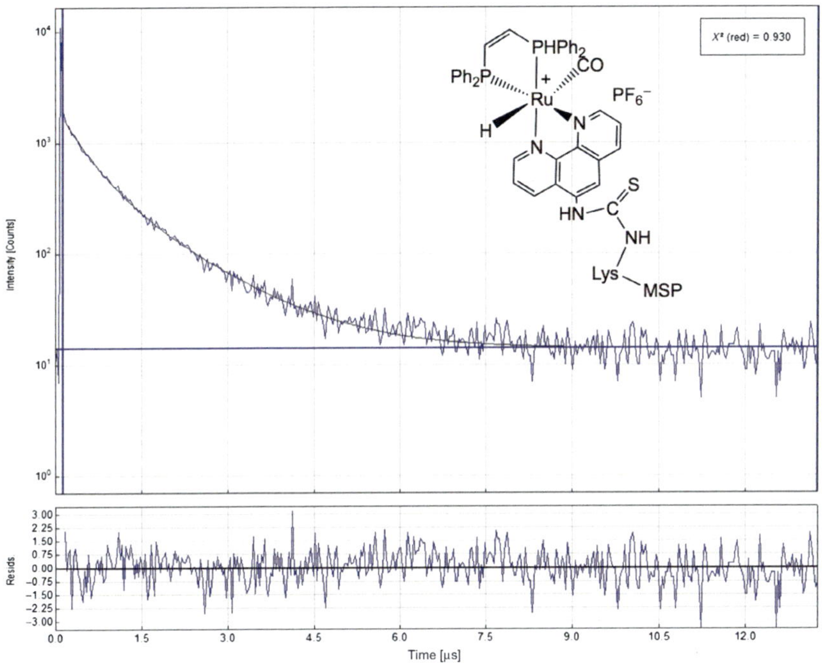

FIGURE 12.9 Time-dependent anisotropy measurement of randomly labeled amine conjugate.

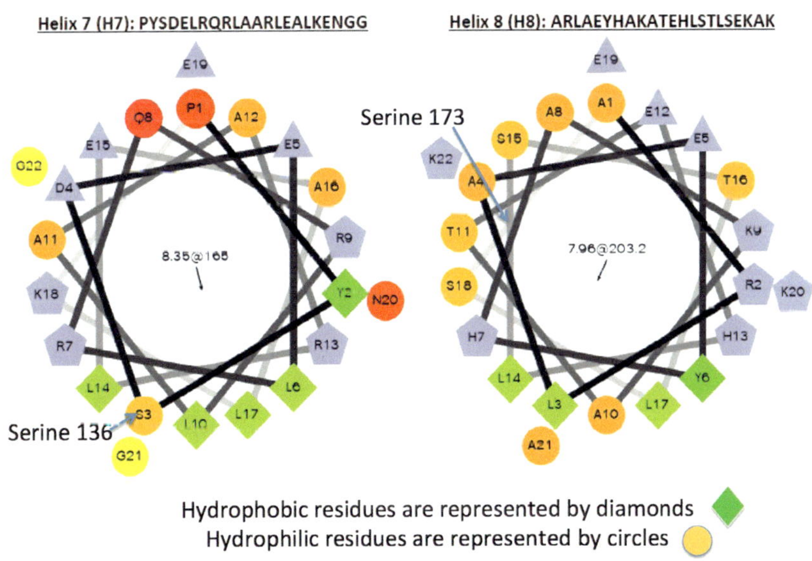

FIGURE 12.10 Helical wheel projections: serine 136 where the cysteine substitution was made is in a hydrophobic region of the protein, while serine 173 is in a hydrophilic region of the MSP.

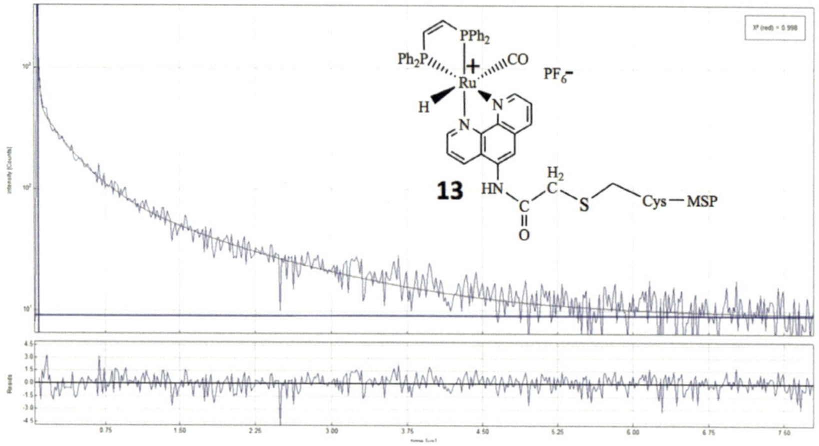

FIGURE 12.11 Excited-state lifetime of cysteine 136 iodoacetamide conjugate in 20 mM tris buffer (pH = 7.4).

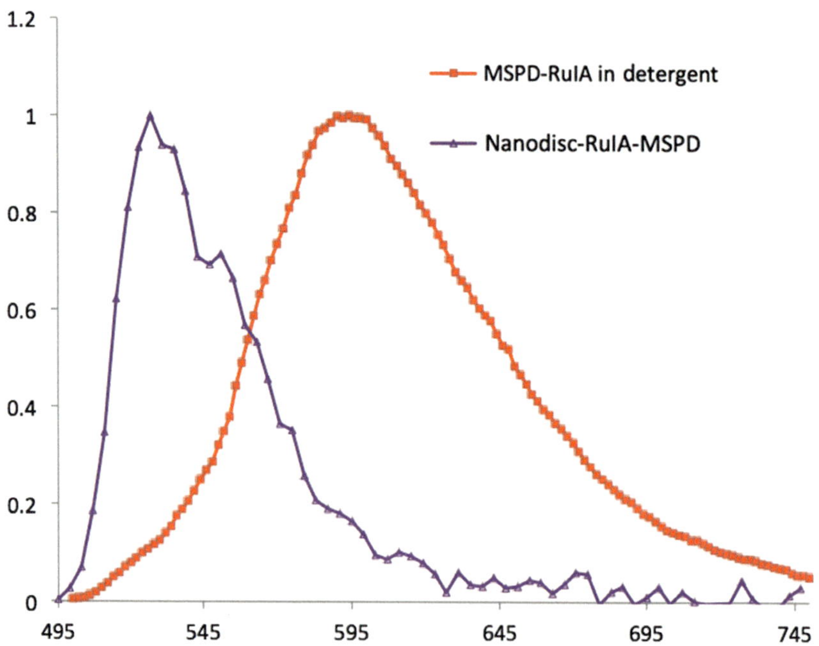

FIGURE 12.12 The sulfhydryl selective probe, **13**, shows the usual red-shifted emission at 590 nm when dispersed in detergent. In the nanodiscs the emission is blue shifted to 520 nm.

(Fig. 12.12). This result is reminiscent of what was previously observed when the lipid conjugates, **9** and **11** were incorporated into liposomes.[10]

Most interestingly, as excess cholate is added to the thiol mutant nanodiscs labeled with **13** the emission red shifts and after standing overnight in 60 mM cholate the red-shifted emission is restored (Fig. 12.13).

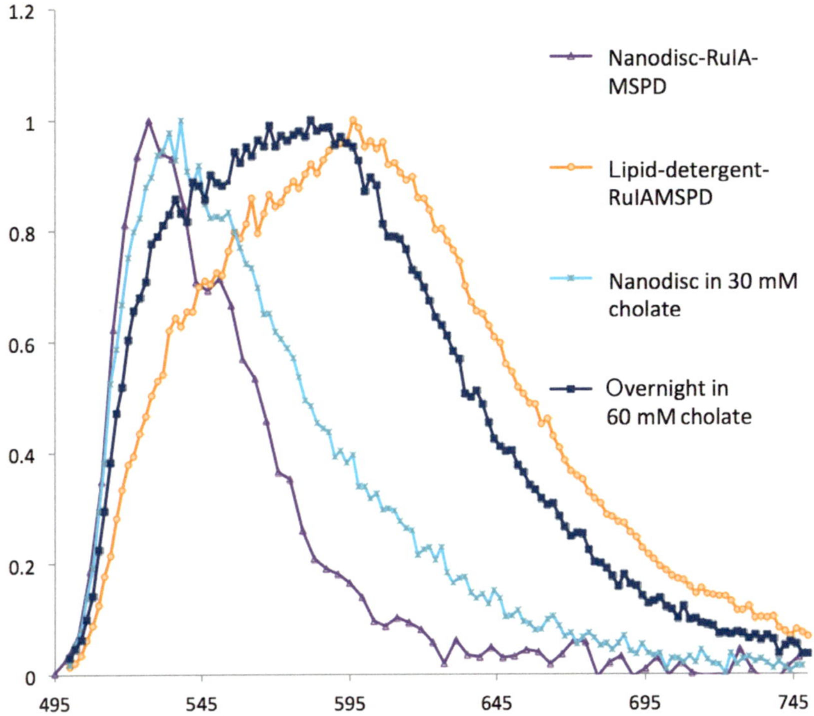

FIGURE 12.13 The effect of added detergent and standing overnight on the nanodiscs formed containing **13**.

The excited-state lifetime of the **13** labeled nanodiscs gave a singlet-type value of 4 ns and a fundamental anisotropy 0.31. These data are very similar to those obtained from the incorporation of **9** and **11** in the 100 nm liposomes (Figs. 12.5 12.7). It would appear that incorporating these Ru phosphorescent probes into either the lipophilic environment of a liposome or the lipophilic region of the thiol mutant MSP results in a change from a long-lived and red-shifted emission to a short lived and blue-shifted emission. Distortions in the excited state and restricted motion in a lipophilic environment are likely responsible for these changes.[9,19] Complexes **9**, **11**, and **13** all have relatively long nonrigid points of attachment to the luminophore (Fig. 12.14).

Complex **10** would be expected to show the same behavior, but its emission was too weak to proceed with labeling and time-resolved measurements. The cholesterol conjugate, **7**, is more rigid and does not exhibit this apparently triplet-to-singlet interconversion. Further evidence for this interpretation will come from an investigation of the photophysical properties of the conjugation of **13** with the cysteine mutant of the MSP at serine 173, located in a hydrophilic region of the protein where interactions of the probe

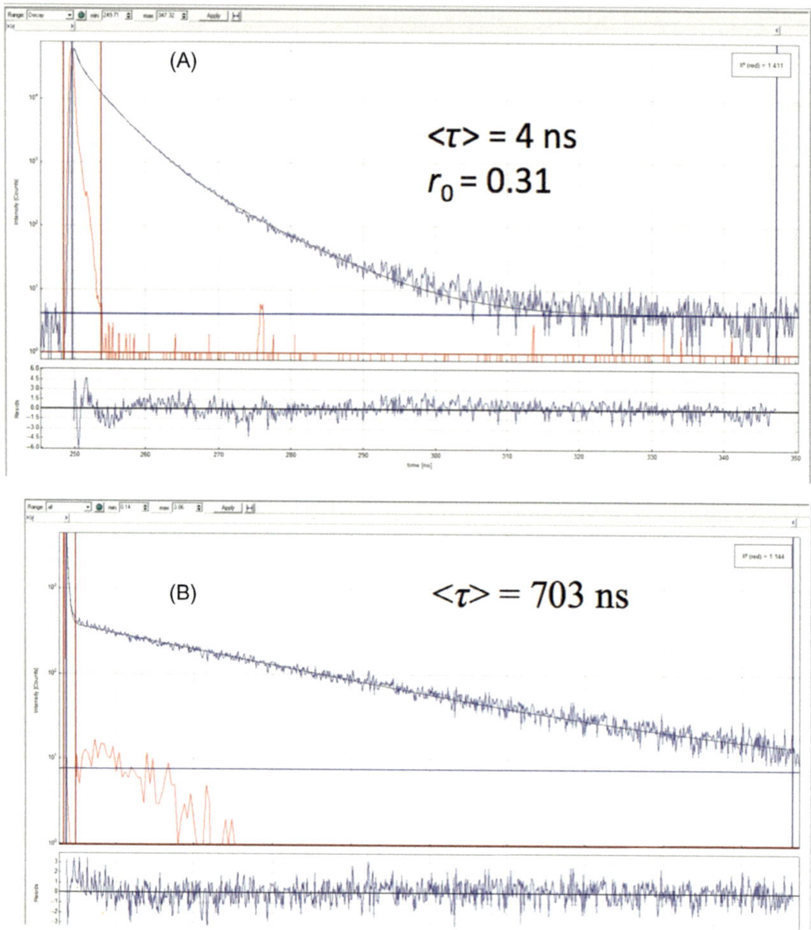

FIGURE 12.14 (A) Time-dependent anisotropy measurement of thiol-labeled nanodiscs. (B) Time-dependent anisotropy measurement of thiol-labeled nanodiscs in 60 mM cholate after standing overnight.

with the surrounding residues are expected to be weaker, leading to less intimate involvement of the conjugate chain with the environment.

12.7 CONCLUSIONS

The lipid conjugates in the liposome and the thiol-labeled protein in the nanodisc are both "buried" in a constrained lipophilic environment. This causes molecular distortions in the excited state that result in orbital energy changes leading to short-lived, blue-shifted emissions. The more rigid cholesterol probe appears to be only partially incorporated into the lipid environment in both membrane models, and thus the photophysical properties, with exception of the lifetime, are essentially unchanged from **7**. Both results will prove useful in tracking the structural dynamics of the membrane models. Only the lipid conjugates with phosphine ligands exhibit the short excited-state lifetimes. The tris-diimine probe, **12** photophysics are essentially unaffected after incorporation into the liposomes an indication that the lipophilic phosphines play in important role in supporting the observed changes in photophysical properties. Labeling of both the lipid and the protein in the nanodiscs opens the door to study Förster energy transfer and the selective measurement of global or segmental motions before and after target protein binding as well as incorporation of lipid conjugates in nanodiscs and liposomes to compare the effect of adding cholesterol to each.

ACKNOWLEDGMENTS

We gratefully acknowledge the support of the National Science Foundation (CHE CHE-1049569). Luminescence measurements were performed in the BioSpectroscopy Core Research Laboratory, which is supported by NIH Centers of Biomedical Research Excellence (CoBRE) Award P20GM103546 to the Center for Biomolecular Structure and Dynamics and by the vice president of Research and Creative Scholarship at the University of Montana.

REFERENCES

1. Hemmila I. Progress in delayed fluorescence immunoassay. In: Wolfbeis OS, editor. *Fluorescence spectroscopy, new methods and applications*. New York: Springer-Verlag; 1993. p. 259–66.
2. Guo X-Q, Castellano FN, Li L, Lakowicz JR. Use of a long-lifetime Re(I) complex in fluorescence polarization immunoassays of high molecular weight analytes. *Anal Chem* 1998;**20**:632–7.
3. Turro C, Bossmann SH, Jenkins Y, Barton JK, Turro NJ. Proton transfer quenching of the MLCT excited-state of $Ru(phen)_2dppz^{2+}$ in solution and bound to DNAJ. *Am Chem Soc* 1995;**117**:9026–32.
4. Szmacinski H, Terpetschnig E, Lakowicz JR. Synthesis and evaluation of Ru-complexes as labeling as anisotropy probes for protein hydrodynamics and immunoassays of high molecular weight antigens. *Biophys Chem* 1996;**62**:109–20.
5. Watson A, Wu X, Bruchez M. Lighting up cells with quantum dots. *Biotechnology* 2003;**34**:296–303.
6. Sabbatini N, Guardigli M, Manet I. Lanthanide complexes of encapsulating ligands as luminescent devices. In: Neckers DC, Volman DH, von Bünau G, editors. *Advances in photochemistry*, Vol. 23. New York: John Wiley & Sons; 1997. p. 213–78.
7. Juris A, Balzani V, Barigelletti F, Campagna S, Belser P, Von Zelewsky A. Ru polypyridine complexes: photophysics, photochemistry, electrochemistry and chemiluminescence. *Coord Chem Rev* 1988;**84**:85–277.

8. Sharmin A, Darlington RC, Hardcastle KI, Ravera M, Rosenberg E, Ross JBA. Tuning photophysical properties with ancillary ligands in Ru(II) mono-diimine complexes. *J Organmetal Chem* 2009;**694**:988–1000.
9. Sharmin A, Salassa L, Rosenberg E, Ross JBA, Abbott G, Black L, et al. Photophysical studies of bioconjugated ruthenium metal–ligand complexes incorporated in phospholipid membrane bilayers. *Inorg Chem* 2013;**52**:10835–45.
10. Sharmin A, Phung T, Ross, Rosenberg E, Atkins W. Site-specific and non-specific luminescence labeling of membrane scaffold protein (MSP) to study the dynamic properties of phospholipid nanodiscs, Abstracts, 69th Northwest Regional Meeting of the American Chemical Society, Missoula, MT; June 22–25 2014. NORM-205.
11. Ravera M, Sharmin A, Rosenberg E. Electrochemical studies of Ru(II) diimine bioconjugates. *Inorg Chim Acta* 2015;**429**:87–92.
12. Pohkrel S, Decato D, Rosenberg E, Ross JBA, Terwilliger M. Synthesis, structure, photophysical and electrochemical properties of Ru(TFA)(CO)(PPh3)2(L) (L = 2-phenylpyridine, 2-*p*-tolylpyridine) and Ru(CO)(PPhMe2)2(L)(L′) (L = TFA, H) (L′ = bipyridine, L′ = 4,4′-dimethylbipyridine) relationships between ancillary ligand structure and luminescent properties. *J Organometal Chem* 2017;**849-850**:306–14.
13. Abbott G, Brooks R, Rosenberg E, Terwilliger M, Ross JBA, Ichire OL. Surface bound ruthenium diimine organometallic complexes: luminescence and lifetimes. *Organometallics* 2014;**33**:2467–78.
14. Rosenberg E, Abbott G, Ross JBA, McVay R, Terwilliger M. Luminescence quenching and enhancement by adsorbed metal ions with ruthenium diimine complexes immobilized on silica polyamine composites. *Macromol Symp 2015* 2016;**364**:47–55.
15. Guo X, Li L, Castellano FN, Szmacinski H, Lakowicz JR. Synthesis and luminescence spectral characterization of long-lifetime lipid metal-ligand probes. *Anal Biochem* 1997;**244**:80–5.
16. Korade Z, Kenworthy AK. Lipid rafts cholesterol and the brain. *Neuropharmacology* 2008;**55**:1265–73.
17. Garino C, Ghiani S, Gobetto R, Nervi C, Salassa L, Ancarani V, et al. [Os(bpy)2(CO) enIA][OTf]$_2$ a novel sulfhydryl-specific metal-ligand complex. *Inorg Chem* 2005;**44**:3875.
18. Lakowicz JR. *Principles of fluorescence spectroscopy*. 3rd ed New York: Springer; 2006. p. 353–8.
19. Balazs GC, del Guerzo A, Schmehl RH, Russell H. Photophysical behavior and intramolecular energy transfer in Os(II) diimine complexes covalently linked to anthracene. *Photochem Photobiol Sci* 2005;**4**:89–94.
20. Kinosita Jr. K, Ikegami A, Kawato S. *Biophys J* 1982;**37**:461–4.
21. Minazzo AS, Darlington RC, Ross JBA. Loop dynamics of the extracellular domain of human tissue factor and activation of factor VIIA. *Biophys J* 2009;**96**:681–92.
22. Klauda JB, Roberts MF, Redfield AG, Brooks BR, Pastor RW. Rotation of lipids in membranes: molecular dynamics simulation, ^{31}P spin-lattice relaxation, and rigid-body dynamics. *Biophys J* 2008;**94**:3074–83.
23. Gidwani A, Holowka D, Baird B. Fluorescence anisotropy measurements of lipid order in plasma membranes and lipid rafts from RBL-2H3 mast cells. *Biochemistry* 2001;**40**:12422.
24. Morgan CR, Hebling CM, Rand KD, Stafford DW, Jorgenson JW, Engen JR. Conformational transitions in the membrane scaffold protein of phospholipid bilayer nanodiscs. *Mol Cell Proteom* 2011;**10**(9). M111.010876, 11.
25. Lab Chip 8 (2008) 1723–1728.

Chapter 13

Recent Development for Tetraplex DNA Organometallic Ligands

Shinobu Sato and Shigeori Takenaka
Department of Applied Chemistry, Kyushu Institute of Technology, Kitakyushu, Japan

13.1 INTRODUCTION

Deoxyribonucleic acid (DNA), with its double-stranded genetic structure, forms several kinds of high-ordered structures, including a tetraplex.[1] There are 370,000 potential tetraplex forming sequences in the human genome.[1,2] The G-rich repeating sequences of the DNA or RNA form the tetraplex structures through intra-molecular folding and the stacking of the G-quartet plane. The telomeric sequences located at the termini of the chromosomes and the promoter regions of the c-*Myc* or c-*Kit* genes are well-known for forming tetraplex structures, as shown in Fig. 13.1.

These tetraplex structure in Fig. 13.1 contributes to the regulation of transcription, replication, and translation.[3–6] Furthermore, this tetraplex structure is being increasingly recognized as an important motif that regulates several gene events, similar to the specific gene sequence TATA box (Fig. 13.2). When

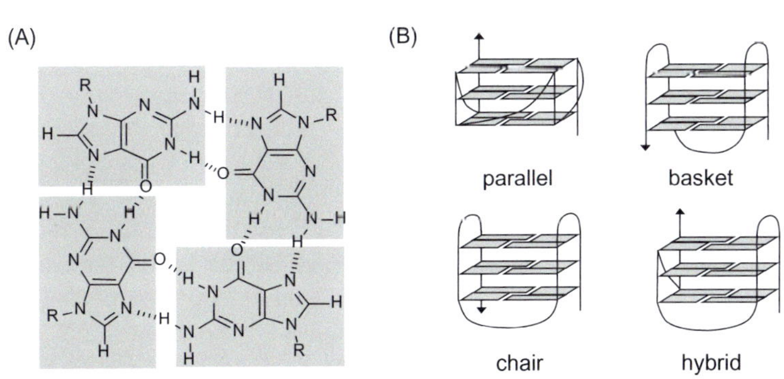

FIGURE 13.1 (A) Guanine (G)-quartet plane was stacked upon itself to form a tetraplex, and (B) the several kinds of tetraplex structures with different patterns of folding.

Advances in Bioorganometallic Chemistry. DOI: https://doi.org/10.1016/B978-0-12-814197-7.00013-3

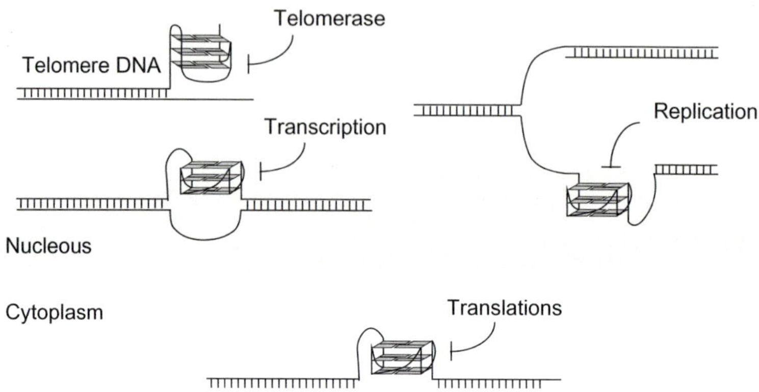

FIGURE 13.2 The tetraplex structure of DNA or RNA contributes to the processes of gene regulation, such as transcription, translation, replication, and telomere elongation.

some molecule stabilizes the tetraplex structure associated with the regulation of a specific gene, this molecule also impedes or completely stops the gene's regulation. For example, telomestatin, TMPyP4, or perylene diimide are known to bind strongly with the tetraplex structure of telomeric DNA, subsequently preventing the access of telomerase; TMPyP4 also binds to tetraplex DNA (Pu27), to inhibit the transcription of the c-*MYC* gene.[3] These ligands have been studied with the view to developing anticancer drugs based on the knowledge of the contribution of these tetraplex structures in cancer development.[7]

Many organometallic molecules bind DNA, as is represented by the anticancer drug cisplatin.[8] Some organometallic molecules have been shown to bind tetraplex DNA, promising a novel approach to diagnosing and treating cancer. In this chapter, we discuss the organometallic ligands of tetraplex DNA.

13.2 TETRAPLEX DNA-ORGANOMETALLIC LIGANDS AS ANTICANCER DRUGS

Table 13.1 shows the typical examples of the tetraplex DNA-organometallic ligands. The binding of metal-complex ligands with tetraplex DNA is determined by the following features of the metal-complex ligands: (1) the increased binding affinity of the metal cation; (2) control of stacking of the ligand's aromatic moiety with the G-quartet DNA structure to accommodate for the former's π−electron densities; and (3) retention of the plane of the aromatic part or the extended aromatic plane of the ligand through metal complex formation. These characteristics favor a stronger binding of the ligands with the G-quartet plane of the tetraplex DNA.

TMPyP4 and its complexes with several kinds of metal cation are well-known as tetraplex ligands.[9–12] The derivatives with Ni(II), Cu(II), Zn(II), Mn(III), or Au(III) have a higher binding affinity with tetraplex DNA than

TABLE 13.1 Typical Example of Tetraplex DNA Organometallic Ligands

Chemical Structure	Name or Abbreviation	Ref.
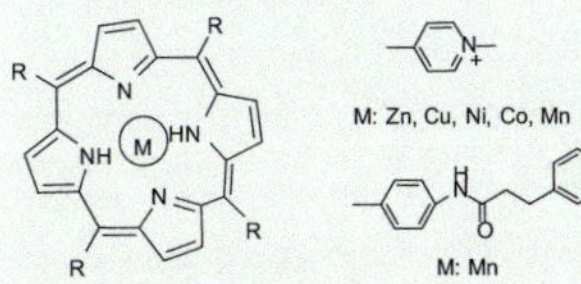	**TMPyP4** Derivativepentacationic Mg(III) porphyrin	9–12
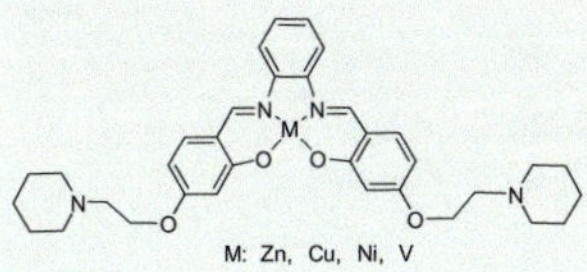 1 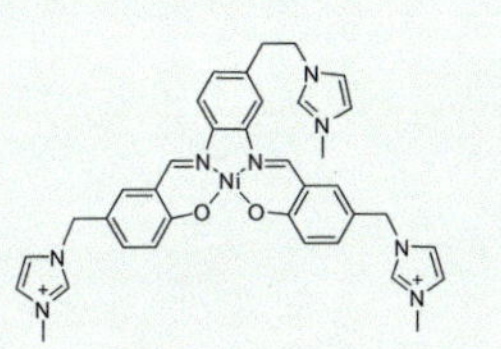 2	Salophen complexes	13,14
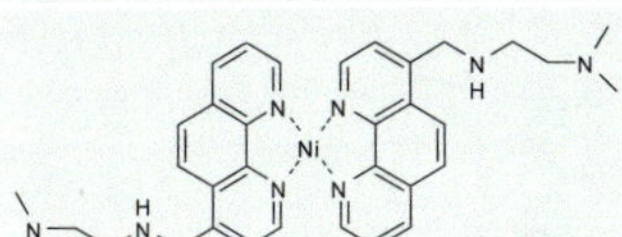	Ni(II) bis-phenanthroline Derivative **(K34)$_2$Ni(II)**	15
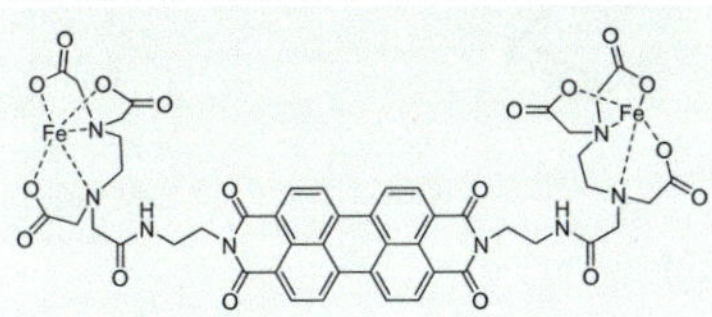	**Perylene diimide-EDTA-Fe**	16
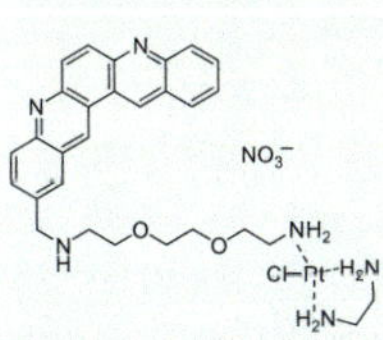	**Platinum Mono-para-quinacridine (Pt-MPQ)**	17,18
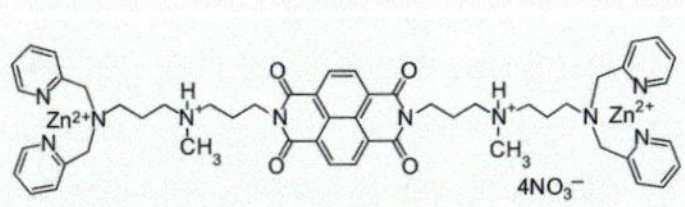	**NDI-Dpa-Zn**	19
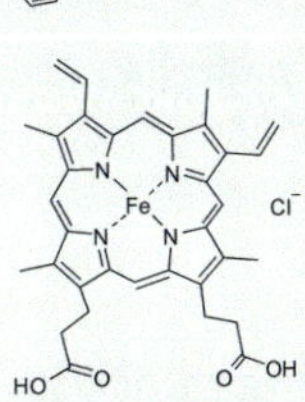	**Hemin**	20–24

(*Continued*)

TABLE 13.1 (Continued)

Chemical Structure	Name or Abbreviation	Ref.
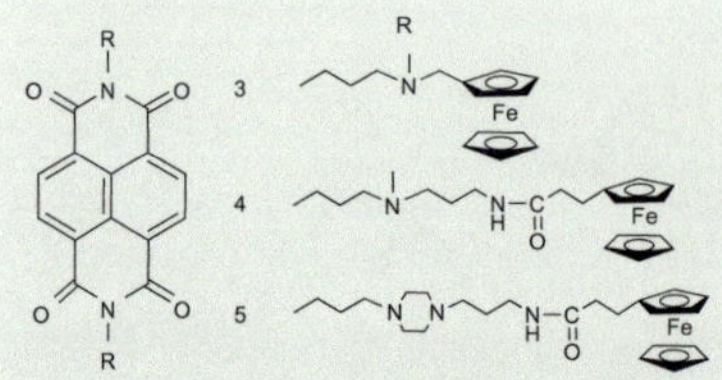	**Ferrocenylnaphthalene diimide, FND**	25–29
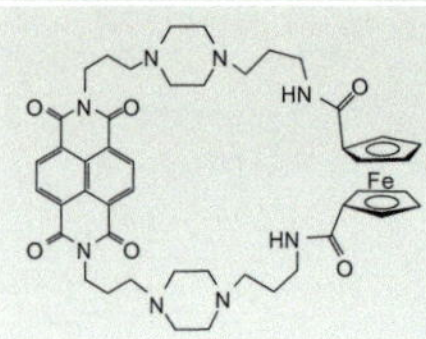	**Cyclicferrocenylnaphthalene diimide, cFND**	30

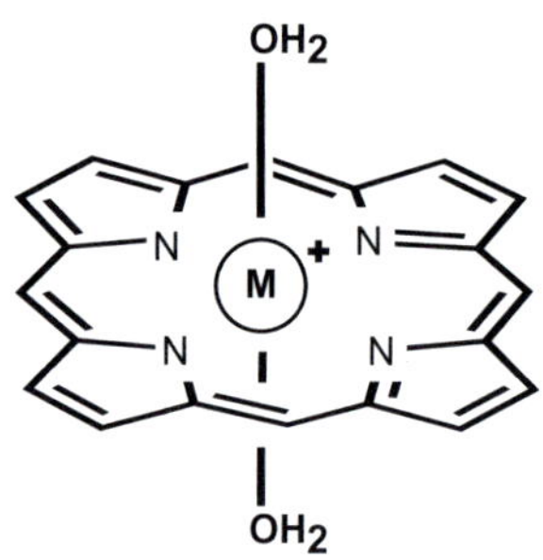

M:Mn(III), Co(III)

FIGURE 13.3 Porphyrin metal ligand coupled with axial water molecules.

that of H2-TMPyP4, a protonated one.[9–12] Although TMPyP4 has a divalent cation with $2H^+$, its metal complexes form tetravalent cations[9] and bind more stably with the tetraplex DNA due to the effect of characteristic (1) described above. However, the steric structure of the ligand is also an important factor for determining this binding. CuTMPyP4, which is the complex of Cu(II) with TMPyP4, has a similar plane as TMPyP4. On the contrary, the coordinated Zn(II) ion of the ZnTMPyP4 derivative is coupled with one water molecule and forms a tetrahedral structure, resulting in a binding affinity of $1.6 \times 10^7\ M^{-1}$, which is five times higher than the binding affinity of TMPyP4[10] in a corresponding tetraplex structure. Co(III) and Mn(III) of TMPyP4 coordinated two axial water molecules and forms an octahedral structure, as shown in Fig. 13.3.

In this case, the steric hindrance of the water molecules disturbs the $\pi-\pi$ stacking interaction with porphyrin and G-quartet planes, thereby decreasing their binding affinities.[11] However, pentacationic Mn(III) porphyrin showed a very high binding affinity with tetraplex DNA.[12] These findings indicate that the strong affinity of the four grooves of the tetraplex strand for the four linker chains introduced in the Mn(III) porphyrin is associated with the elimination of one axial water molecule. Furthermore, this porphyrin was reported to have a 10,000-times stronger preference for the tetraplex DNA over that for double-stranded DNA. Salophen complex **1** carrying Ni(II), Zn (II), Cu(II), or V(II) has also been reported to be a tetraplex DNA binder,[13] and Polycationic Ni(II)-Salophen **1** forms the most stable complex with tetraplex DNA.[14] Similar to TMPyP4, the Cu(II) complex of Salophen **1** couples with the axial water molecules, and the Zn(II) or V(II) complex of Salophen **1** forms a penta-coordinated structure, thereby destabilizing their complexes and rendering stacking with the G-quartet plane difficult. Tomas et al.[31] have reported a stable complex of Polycationic Ni(II)-Salophen **2**, carrying alkyl-imidazolium side-chains, with tetraplex DNA. They determined that this ligand's strong binding affinity is derived from the Van der Waals interaction of its imidazolium part with the unshared electron pairs of the guanine located in the groove of the G-quadruplex. Phenanthroline forms complexes with several metal cations and displays different binding modes, such as grove binder, intercalator, or insertor, to bind with double-stranded DNA.[31] Tris(phenanthroline) complexes display both $\Lambda-$ and Δ-enantiomers, although only the Δ-enantiomer is known to bind with the B-form of the DNA duplex.[31] Phenanthroline derivatives with Cu(II) or Ni(II) are also known to suitably bind DNA quadruplexes.[15] Interestingly, the Ni(II) complex formed by bis-phenanthroline derivative K34, $(K34)_2Ni(II)$, has been reported by Sissi et al. to show[15] different thermodynamic behaviors between the Hybrid-1 and Hybrid-2 of telomeric tetraplex DNA—the known polymorphisms of the human telomeric DNA. This constitutes a discriminatory ability of the Tris(phenanthroline) complex for the B-type and z-type DNA duplexes.

Non-organometallic tetraplex DNA binders have been connecting with metal complex as substituent as other type of tetraplex DNA organometallic ligand. Perylene diimide-EDTA · Fe(II) is categorized in this group.[32] It is known that perylene diimide has a large aromatic plane suitable for good stacking interactions with the G-tetrad,[16] and perylene diimide-EDTA · Fe(II) produces hydroxyl radicals upon reaction with reducing agents such as dithiothreitol (through Fenton's reaction), resulting in DNA cleaving. This reagent is expected to cause apoptosis of cancer cells, with specific cleavage of the overhang region of the telomeric DNA. This reagent was shown in vitro to bind to tetraplex DNA, with end-stacking at the 3'-position of the tetraplex DNA, and subsequent cleavage at this point. As a tetraplex DNA binder that binds both covalently and noncovalently, platinum mono-para-quinacridine, (Pt-MPQ), was synthesized by attaching cisplatin with quinacridine.[17] It is reported that the quinacridine plane of Pt-MPQ binds to the

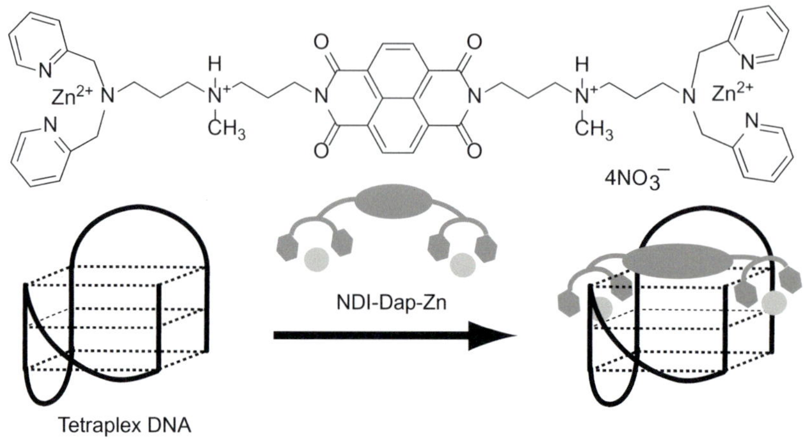

FIGURE 13.4 Expected structure of the NDI-Dpa-Zn complex with tetraplex DNA.

tetraplex DNA with end-stacking of its G-tetrad, while cisplatin is located at its opposite site, although both parts simultaneously contribute to the stable binding with the tetraplex DNA. It has been demonstrated experimentally that Pt-MPQ was localized at the termini of the chromosome and induced damage of the telomeric DNA.[18] This conjugation ensures the targeting of cisplatin for locating telomeric DNA. Naphthalene diimide is also utilized as a conjugated tetraplex binder that binds via a noncovalent/covalent binding mode. Naphthalene diimide binds to the tetraplex DNA through an electron-rich G-tetrad and an electron-deficient naphthalene diimide plane, and several types of naphthalene diimide derivatives, naphthalene diimide **3–5**, were reported in the context of strong binding with tetraplex DNA.[19,25–28,33–35] Synthesized naphthalene diimide carrying dipicolylamine, NDI-Dpa, forms stable complexes with Zn(II),[33] binding strongly with tetraplex DNA.[19] Since Dpa-Zn, the complex of dipicolylamine with Zn(II), is known to bind with phosphate anion, NDI-Dap-Zn should form stabile complexes with tetraplex DNA that has a more compact structure with a higher density than that of the phosphate anion (Fig. 13.4).

All of the reported tetraplex DNA binders carrying metal complexes are created to improve their binding affinities for tetraplex DNA and/or function as anticancer drugs.

13.3 BIOSENSING BASED ON TETRAPLEX DNA-ORGANOMETALLIC LIGAND

Tetraplex DNA-organometallic ligands have found application not only as anticancer drugs, but also as sensing ligands for diagnostic purposes. Telomerase is an enzyme that elongates or expands the telomeric DNA located at the termini of a chromosome. Since its activity is observed only in the cancer cell, telomerase is considered a good cancer marker.[36,37] To date, its activity has been detected by the telomerase repeat amplification protocol (TRAP),[38] however, this protocol is known to be affected by the extraneous substances contaminating biological samples.[39] Thus, several new methods to detect telomerase activity have been reported. Telomeric DNA, elongated with telomerase, forms the tetraplex DNA structure, and tetraplex DNA binders have been utilized to detect this elongated DNA in the form of the tetraplex structure. Hemin, which has a ferrous ion conjugated with porphyrin, binds to tetraplex DNA, and acts as a DNAzyme, carrying the peroxidase-like enzyme activity.[40] After the discovery of this system, several telomerase detection systems were published.[20] Willner and colleagues developed the telomerase detection system, which involves the hybridization of the elongated telomeric DNA with a telomerase containing a tagged probe carrying Hemin/G4 oligonucleotide (Fig. 13.5).[21]

The detection limit in this system was telomerase activity from 50 HeLa cell. This system, which relies on the sandwich method, allowed the researchers to enhance the detection limit of the BRCA1 oncogene using its DNA probe carrying this detection system.[22] The detection limit in this system was $1 \times 10^{-13}\,M^{-1}$ for the BRCA1 oncogene. This system also expanded the detection of the IgG1 antigen using Hemin/G4 as tag, as shown in Fig. 13.6.[23]

In this system, ferrocene and Hemin/G4 were combined though oligonucleotides. Enzymatic amplification of DNAzyme was achieved through ferrocene, and 0.1 pg/mL of IgG1 was detected with four-times amplification using this system, compared with the levels detected using only the ferrocene

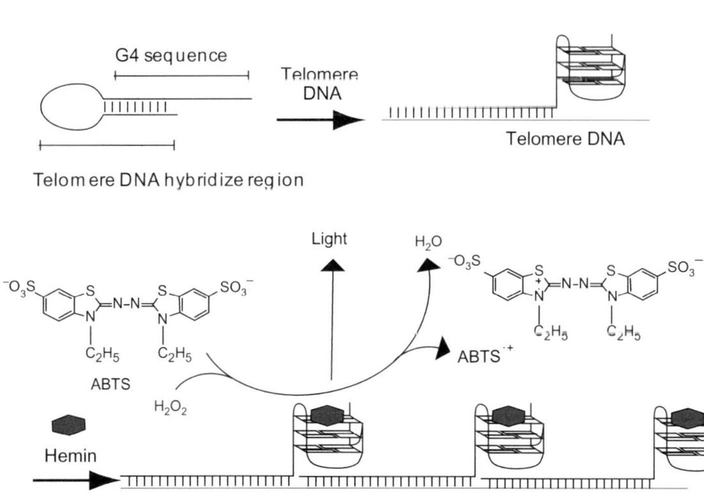

FIGURE 13.5 Telomerase detection using quadruplex/hemin.

system. This system had a very high preference for IgG1, without any response for the other antigens. Another interesting study involves the construction of allosteric DNAzyme by Ihara and colleagues. In this system, the G4 sequence was separated and the tetraplex structure bound by terpyridine and metal cation complexed in a ratio of 2:1, following the peroxidase activity of hemin.[24] This system promises to expand the detection of DNA or RNA through suitably designed oligonucleotide probes.

Hexaammine ruthenium (RuHex), is known to form complexes with DNA, where one ruthenium ligand binds to three phosphate anions to balance the charge of the complex. The amount of DNA on the electrode can be estimated from the coupled RuHex using the chronocoulometric method.[41] This was used in the electrochemical telomerase detection system (Fig. 13.7).

The TS-primer, which is substrate of telomerase, was immobilized on the electrode. This electrode was treated with telomerase and the elongated DNA was estimated with the increasing charge of RuHex.[34] Using this

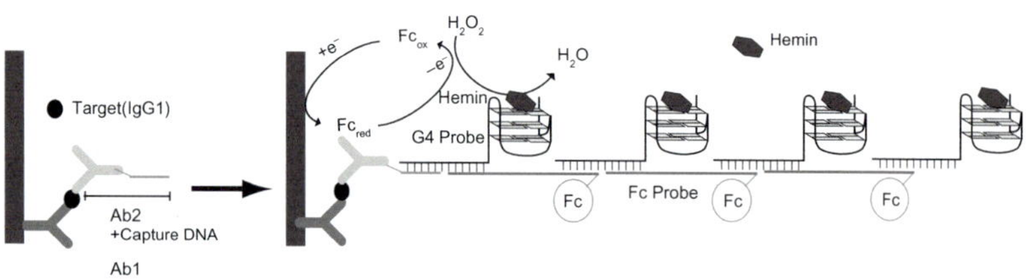

FIGURE 13.6 Antigen detection using Hemin/G4 and ferrocene conjugation.

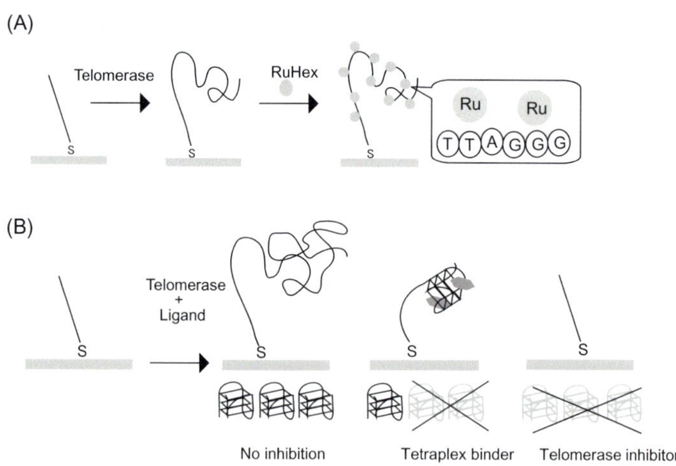

FIGURE 13.7 (A) Principle of electrochemical detection of telomerase activity based on RuHex and (B) the result of elongation patterns depending on three inhibition manners for screening molecules.

system, one can detect the telomerase activity from 10 HeLa cells. This system not only helps in diagnosing cancer, but also estimates how many nucleotides were elongated. Following the addition of the telomerase inhibitor in this system, observation of only $(TTAGGG)_4$ indicates inhibition with tetraplex DNA binding, whereas, no elongation indicates the direct inhibition of the telomerase enzyme. This finding shows that this system can be utilized for the screening of telomerase inhibitors and elucidate their mechanisms of inhibition.

Ferrocenylnaphthalene diimide (FND), 3–5 can stably bind to double stranded DNA through threading intercalation, and an FND-based hybridization assay was thus applied to detect target gene.[25]

FNDs bind to tetraplex DNA more strongly than they bind to duplex DNA, because of a good stacking interaction between the constituent electro-deficient naphthalene diimide plane and the electro rich G-tetrad plane.[26–30,35,42,43] A system for electrochemical telomerase detection (ECTA) was developed using FND **3** (Fig. 13.8).[26] In this system, the TS-primer-immobilized electrode is treated with telomerase, and the elongated TS-primer can detect FND in the electrolyte containing KCl, where the elongated DNA forms tetraplex DNA repeats on which the FND is concentrated, resulting in the increasing electrochemical signals. The electrochemical signal of telomerase activity was monitored by differential pulse voltammetry (DPV), which is simpler when compared to the chronocoulometric (CC) method. ECTA has been applied in diagnosing oral cancer.[27] Clear discrimination between cancer patients and healthy people was observed from tissue and saliva using this system, as this system can detect activity from 10 cells and a positive detection rate of 90% was obtained from clinical samples, resulting in its utilization for the screening of oral cancer.[27] This system is not affected by the presence of foreign substances in the clinical samples during electrochemical detection, and it can discriminate pre-cancer states.[28] FND suitable for ECTA was searched using binding studies and TRAP assays, as shown in Table 13.1.[29] These derivatives were categorized into

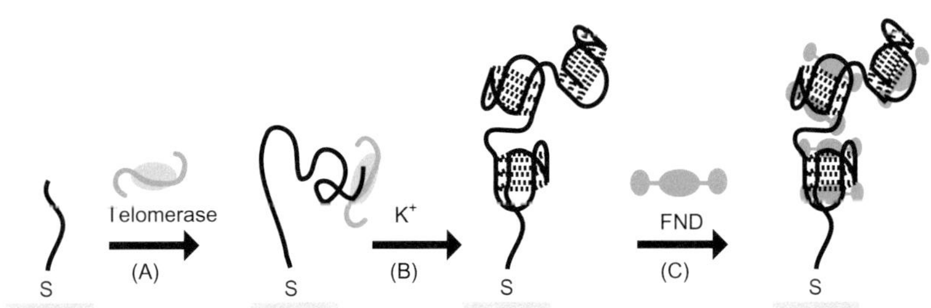

FIGURE 13.8 Principle of electrochemical telomerase activity detection (ECTA). (A) Telomerase elongation reaction, (B) tetraplex DNA formation, and (C) telomerase activity detection by FND.

three types: Inhibitors of telomerase only (FND**5**), inhibitors of telomerase and polymerase (FND**4**), or non-inhibitors of both enzymes (FND**3**). Cyclic naphthalene diimide derivatives have been reported to show high preference for tetraplex DNA.[35,42,43] Cyclic ferrocenylnaphthalene diimide shows higher preference for tetraplex DNA than that for double-stranded DNA.[30] As such, these derivatives should be established for cancer screening in the near future.

13.4 CONCLUSION

Recently, many tetraplex DNA organometallic ligands have been reported, with a goal to developing more effective anticancer drugs. The parallel tetraplex structure of DNA molecules is the next target for cancer therapy, such as that of the c-*Myc* oncogene. TMPyP4 has shown high binding affinity for the c-*Myc* DNA carrying the parallel tetraplex structure. It is important to inhibit the *Myc* gene-induced c-*Myc* protein expression, to inhibit the normal cell from turning cancerous. Since tetraplex DNA ligands bind specifically to the parallel structure of the c-*Myc* gene, these ligands may be targeted to act as anticancer drugs. At present, many researchers are focusing on stabilizing the specific nucleic acid tetraplex structures. The cancer markers described here have also been used for biosensing. Metal complexes have electrochemical and fluorescence characteristics, and they serve are suitable molecular probes. Here we show the application of organometallic ligands as clinical cancer probes using the example of FND. We highlight the potential applications of tetraplex DNA-organometallic ligands in cancer diagnosis and treatment in the future.

REFERENCES

1 Huppert JL, Balasubramanian S. Prevalence of quadruplexes in the human genome. *Nucleic Acid Res.* 2005;**33**(9):2908–16.

2. Todd AK, Johnston M, Neidle S. Highly prevalent putative quadruplex sequence motifs in human. *DNA, Nucleic Acid Res.* 2005;**33**(9):2901–7.

3. Siddiqui-Jain A, Grand CL, Bearss DJ, Hurley LH. Direct evidence for a G-quadruplex in a promoter region and its targeting with a small molecule to repress c-*MYC* transcription. *Proc. Natl. Acad. Sci. U.S.A.* 2002;**99**(18):11593–8.

4. Rhodes D, Lipps HJ. G-quadruplexes and their regulatory roles in biology. *Nucleic Acids Research* 2015;**43**(18):8627–37.

5. Valton AL, Prioleau MN. G-Quadruplexes in DNA replication: A problem or a necessity? *Trends Genet.* 2016;**32**(11):697–706.

6. Song J, Perreault JP, Topisirovic I, Richard S. RNA G-quadruplexes and their potential regulatory roles in translation. *Translation (Austin)* 2016;**4**:e1244031.

7. Monchaud D, Teulade-Fichou M-P. A hitchhiker's guide to G-quadruplex ligands. *Org. Biomol. Chem.* 2008;**6(4)**:627–36.

8. Liu H–K, Sadler PJ. Metal complexes as DNA intercalators. *Acc. Chem. Res.* 2011;**44**(5):349–59.
9. Bhattacharjee AJ, Ahluwalia K, Taylor S, Jin O, Nicoludis JM, Buscaglia R, et al. Induction of G-quadruplex DNA structure by Zn(II) 5,10,15,20-tetrakis (N-methyl-4-pyridyl)porphyrin. *Biochimie* 2011;**93**(8):1297–309.
10. Boschi E, Davis S, Taylor S, Butterworth A, Chirayath LA, Purohit V, et al. Interaction of a cationic porphyrin and its metal derivatives with G-quadruplex DNA. *J. Phys. Chem. B* 2016;**120**(50):12807–19.
11. Dejeu J, Lavergne T, Nora JD, Defrancq E, Pratviel G. Binding of metalloporphyrins to G-quadruplex DNA: The role of the central metal. *Inorg. Chim. Acta* 2016;**452**:98–103.
12. Dixon IM, Lopez F, Tejera AM, Estève J-P, Blasco MA, Pratviel G, et al. A G-quadruplex ligand with 10000-fold selectivity over duplex DNA. *J. Am. Chem. Soc.* 2007;**129**(6):1502–3.
13. Arola-Arnal A, Benet-Buchholz J, Neidle S, Vilar R. Effects of metal coordination geometry on stabilization of human telomeric quadruplex DNA by square-planar and square-pyramidal metal complexes. *Inorg. Chem.* 2008;**47**(24):11910–19.
14. Lecarme L, Prado E, De Rache A, Nicolau-Travers M-L, Bonnet R, van Der Heyden A, et al. Interaction of polycationic Ni(II)-salophen complexes with G-quadruplex DNA. *Inorg. Chem.* 2014;**53**:12519–31.
15. Musetti C, Krapcho AP, Palumbo M, Sissi C. Effect of G-quadruplex polymorphism on the recognition of telomeric DNA by a metal complex. *ProsOne* 2013;**8**:e5829.
16. Han H, Cliff CL, Hurley LH. A ccelerated assembly of G-quadruplex structures by a small molecule. *Biochemistry* 1999;**38**:6981–6.
17. Bertrand H, Bombard S, Monchaud D, Teulade-Fichou MP. A platinum–quinacridine hybrid as a G-quadruplex ligand. *J. Biol. Inorg. Chem.* 2007;**12**(7):1003–14.
18. Charif R, Granotier-Beckers C, Bertrand HC, Poupon J, Ségal-Bendirdjian E, Teulade-Fichou M-P, et al. Association of a platinum complex to a G-quadruplex ligand enhances telomere disruption. *Chem. Res. Toxicol.* 2017;**30**(8):1629–40.
19. Czerwinska I, Sato S, Takenaka S. Improving the affinity of naphthalene diimide ligand to telomeric DNA by incorporating Zn^{2+} ions into its dipicolylamine groups. *Bioorg. Med. Chem.* 2012;**20**:6416–22.
20. Kosman J, Juskowiak B. Peroxidase-mimicking DNAzymes for biosensing applications: A review. *Anal. Chim Acta* 2011;**707**(1-2):7–17.
21. Xiao Y, Pavlov V, Niazov T, Dishon A, Kotler M, Willner I. Catalytic beacons for the detection of DNA and telomerase activity. *J. Am. Chem. Soc.* 2004;**126**(24):7430–1.
22. Shimron S, Wang F, Orbach R, Willner I. Amplified detection of DNA through the enzyme-free autonomous assembly of hemin/G-quadruplex DNAzyme nanowires. *Anal. Chem.* 2012;**84**(2):1042–8.
23. Tang J, Hou L, Tang D, Zhang B, Zhou J, Chen G. Hemin/G-quadruplex-based DNAzyme concatamers as electrocatalysts and biolabels for amplified electrochemical immunosensing of IgG1. *Chem. Commun.* 2012;**48**(66):8180–2.
24. Ihara T, Ohura H, Shirahama C, Furuzono T, Shimada H, Matsuura H, et al. Metal ion-directed dynamic splicing of DNA through global conformational change by intramolecular complexation. *Nat. Commun.* 2015;**6**:6640.
25. Takenaka S, Yamashita K, Takagi M, Uto Y, Kondo H. DNA Sensing on a DNA probe-modified electrode using ferrocenylnaphthalene diimide as the electrochemically active ligand. *Anal. Chem.* 2000;**72**(6):1334–41.

26. S. Sato, H. Kondo, T. Nojima, S. Takenaka, Electrochemical Telomerase Assay with Ferrocenylnaphthalene Diimide as a Tetraplex DNA-Specific Binder, Anal. Chem. 77 (2005) 7304-7309.
27. Mori K, Sato S, Kodama M, Habu M, Takahashi O, Nishihara T, et al. Oral cancer diagnosis via a ferrocenylnaphthalene diimide–based electrochemical telomerase assay. *Clin. Chem.* 2013;**59**(1):289–95.
28. Hayakawa M, Sato S, Diala I, Kodama M, Tomoeda-Mori K, Haraguchi K, et al. Screening for oral cancer using electrochemical telomerase assay. *Electroanal.* 2016;**28**:503–7.
29. Sato S, Takenaka S. Ferrocenyl naphthalene diimides as tetraplex DNA binders. *J. Inorg. Biochem.* 2017;**167**:21–6.
30. Islam MM, Sato S, Shinozaki S, Takenaka S. Cyclic ferrocenylnaphthalene diimide derivative as a new class of G-quadruplex DNA binding ligand. *Bioorg. Med. Chem. Lett.* 2017;**27**(2):329–35.
31. Zeglis BM, Pierrea VC, Barton JK. Metallointercalators and metalloinsertors. *Chem. Commun.* 2007;**28**(44):4565–79.
32. Tuntiwechapikul W, Salazar M. Cleavage of telomeric G-quadruplex DNA with perylene-EDTA · Fe(II). *Biochemistry* 2001;**40** 43652-13658.
33. Watanabe S, Ohtsuka K, Sato S, Takenaka S. Discrimination of phosphorylated double stranded DNA by naphthalene diimide having zinc(II) dipicolylamine complexes. *Bioorg. Med. Chem.* 2011;**19**:1361–5.
34. Sato S, Takenaka S. PCR-free telomerase assay using chronocoulometry coupled with hexaammineruthenium(III) chloride. *Anal. Chem.* 2012;**84**(3):1772–5.
35. Islam MM, Fujii S, Sato S, Okauchi T, Takenaka S. A selective G-Quadruplex DNA-stabilizing ligand based on a cyclic naphthalene diimide derivative. *Molecules* 2015;**20**(6):10963–79.
36. Dhaene K, Van Marck E, Parwaresch R. Telomeres, telomerase and cancer: An up-date. *Virchows Arch* 2000;**437**(1):1–16.
37. Hiyama E, Hiyama K. Telomerase as tumor marker. *Cancer Lett.* 2003;**194**:221–33.
38. Kim NW, Piatyszek MA, Prowse KR, Harley CB, West MD, Ho PL, et al. Specific association of human telomerase activity with immortal cells and cancer. *Science* 1994;**266**:2011–15.
39. De Cian A, Cristofari G, Reichenbach P, De Lemos E, Monchaud D. Marie-paule teulade-fichou, reevaluation of telomerase inhibition by quadruplex ligands and their mechanisms of action. *Proc. Natl. Acad. Sci. U.S.A.* 2007;**104**(44):17347–52.
40. Travascio P, Li Y, Sen D. DNA-enhanced peroxidase activity of a DNA-aptamer-hemin complex. *Chem. Biol.* 1998;**5**(9):505–17.
41. Steel AB, Herne TM, Tarlov MJ. Electrochemical quantitation of DNA immobilized on gold. *Anal. Chem.* 1998;**70**(22):4670–7.
42. Esaki Y, Islam MM, Fujii S, Sato S, Takenaka S. Design of tetraplex specific ligands: Cyclic naphthalene diimide. *Chem. Commun.* 2014;**50**(45):5967–9.
43. Czerwinska I, Sato S, Juskowiak B, Takenaka S. Interactions of cyclic and non-cyclic naphthalene diimide derivatives with different nucleic acids. *Bioorg. Med. Chem.* 2014;**22**(9):2593–601.

Chapter 14

Organometallic Complexes for Biosensing

Toshihiro Ihara
Division of Materials Science and Chemistry, Faculty of Advanced Science and Technology, Kumamoto University, Kumamoto, Japan

14.1 INTRODUCTION

The involvement of metal ions in bioprocesses is one of the key aspects to understanding life.[1–3] Metal ions play essential roles as, for instance, the catalytic center of enzymes and the stabilizer of the specific structures of biopolymers. They also have importance in disease diagnosis and therapy. Several metal complexes are used as the contrast agents for SPECT (single photon emission computed tomography), PET (positron emission tomography), and MRI (magnetic resonance imaging). Currently, metallodrugs such as cisplatin are an integral part of pharmaceutical science.[4] A high number of molecular probes carrying metal ions have been developed based on coordination chemistry as research tools for use particularly in chemistry and biology. Each of these probes possesses unique physicochemical properties such as luminescence and electrochemical activity.

Various organometals have been developed, mainly as catalysts, and are used in research and industry. Remarkable progress in the theory and the techniques of organic chemistry would certainly have been impossible without the organometallic catalysts. Although naturally occurring organometals such as vitamin B_{12} (methylcobalamin) exist in our body, applications of exogenous organometals in biosciences are not so prevalent because of their limited choices. However, organometals themselves are biocompatible because they are generally inert to physiological conditions in aqueous media and susceptibility (reactivity) of naked metal ions is adequately controlled. In addition, the diversity in individual characters of metal ions is preserved and even expanded due to the variety of the kinds and combinations of the ligands. In particular, in the past two decades, organometals have been used as catalysts, luminescence and electrochemically-active molecular probes, and as drugs in bioanalytical, biochemical, and medicinal sciences.

Advances in Bioorganometallic Chemistry. DOI: https://doi.org/10.1016/B978-0-12-814197-7.00014-5

This chapter covers studies on organometallic complexes used mainly for analytical purposes in association with biorelated molecules. Relevant molecular recognitions and catalytic reactions on biopolymers are also involved because of their close connection with selectivity and signal amplification, respectively, in analytical applications. The research is first categorized by metal type and then subcategorized by purpose and binding targets.

14.2 IRON

Here, ferrocene (Fc) is mainly adopted as an Fe-containing organometallic. Fc is one of the typical organometals frequently used for various purposes in the biosciences. The stability in aqueous solution, accessibility of various chemical modifications, and well-defined reversible electrochemical nature of Fc have made it a very useful molecule for use in independent probes or ligands, or as reporter moieties in functional bioconjugates.[5]

14.2.1 DNA and RNA

14.2.1.1 Preparation of Fc-DNA Conjugates

A number of Fc derivatives, bioconjugates and small probe molecules have been developed for DNA/RNA analyses. Fc-DNA conjugates were used in most of the reports for DNA/RNA analyses. In 1994, Ihara et al., reported the first Fc-DNA conjugate,[6–11] which was prepared by postmodification of synthetic oligodeoxyribonucleic acids (ODNs); the activated ester of ferrocene carboxylic acid was coupled with the amino group modified on the end of an ODN to form the Fc-ODN conjugate (Fig. 14.1A).[7] Fc groups could also be introduced to an ODN backbone using a Fc amidite reagent in ODN synthesis by an automated DNA synthesizer (Fig. 14.1B).[12,13] Multiple modifications of Fc units are possible not only at the end, but also at any position in the middle of an ODN backbone. Enzymatic Fc modification of ODNs

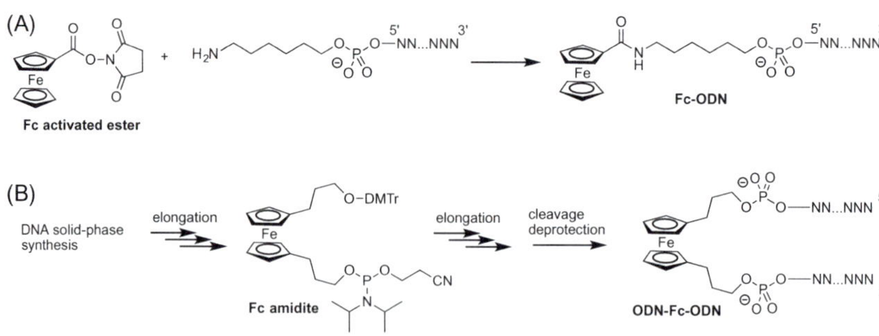

FIGURE 14.1 Two examples of ferrocene modification on chemically synthetic DNAs by (A) postmodification of activated ester of Fc carboxylic acid,[7] and (B) insertion of Fc amidite in automated solid-phase DNA synthesis.[13]

was also performed successfully. Using the triphosphates of Fc-modified nucleosides, Fc-nucleotides are incorporated into ODN strands by DNA polymerase.[14,15]

14.2.1.2 Gene Sensors Based on Target-Induced Probe Concentration

An electrochemical gene sensor prototype using a Fc-ODN conjugate, which had been applied as an electrochemical DNA probe for DNA/RNA analysis using a HPLC equipped with an electrochemical detector in their previous works,[6] was proposed by Ihara et al., in 1997.[8] In this gene sensor, Fc-ODN and an ODN anchored on a gold electrode formed a sandwich-type ternary tandem duplex with the target DNA to give a redox signal (Fig. 14.2A). The concentration level of Fc-ODN on the electrode depended even on the point mutation of the target sequences, making it possible to differentiate the targets by the current signals from voltammetry. The effect of the distance of the Fc moieties from the electrode in the sandwich complexes on their electrochemical signals was investigated by Li et al.,[16] and King et al.[15]

Signaling based on the sandwich-type format was improved by multiple Fc introduction to the electrode surface to amplify the signal. Pang et al.,

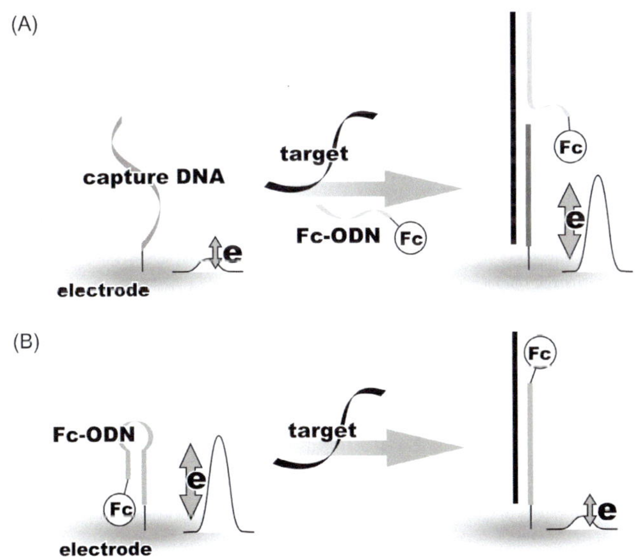

FIGURE 14.2 Typical two formats of electrochemical gene sensing using Fc-modified DNA probe. (A) Gene sensors based on target-induced probe concentration on the electrode surface.[8–10,17–20] (B) Gene sensors based on target-induced probe restructuring.[26,27] By using DNA/RNA aptamers as recognition interface, these detection principles could be extended to the sensing of relevant molecules such as proteins and metal ions.

reported PCR-free electrochemical DNA sensing.[17] They used Fc-modified Au nanoparticles as electrochemical labels, which enabled them to exclude PCR from the DNA analysis protocol. Yang et al., also amplified the signal by forming a Fc-ODN assembly on DNA concatemers grown on an electrode.[18] They succeeded in detecting concentrations of target DNA as low as 100 aM, even in complex samples. Umek et al., reported the successful diagnosis of HIV (human immunodeficiency virus) and HCV (hepatitis C virus) using a DNA probe with successive Fc modifications.[19] Block copolymers consisting of regions with DNA and Fc side chains were prepared as an electrochemical DNA probe by Mirkin et al. The cooperativity of the DNA side chains in hybridization made the selectivity of the probe very high.[20] Amplified redox signal was observed with multiple Fc units on the polymer probe.

Fc-ODN conjugates were obtained by target-induced enzymatic reactions in several reports. Jiang et al., prepared an effective Fc-ODN conjugate using target-templated gap ligation.[21] The obtained conjugate was subjected to duplex formation with the capture probe immobilized on the electrode to observe signals with high specificity. Only the target that displayed the specific nucleotide in the gap provided the Fc-ODN conjugate with an effective sequence. The stringent selectivity did not depend on that in hybridization, but came from that in ligase recognition of dNTPs. The same research group also used ligation to modify the electrode with Fc. They formed a stem-loop (hairpin) structured Fc-ODN on the electrode using the target as a template, and termed the system a "reverse molecular beacon."[22] Willner et al., succeeded in the detection of targeted cyclic virus DNA. Target-templated polymerization was performed from the primer DNA immobilized on the electrode. Fc-modified dUTPs were incorporated in the polymerase reaction to form complementary DNA carrying a number of Fc units on the electrode.[23] The redox-active replica was then coupled to the glucose oxidase (GOx)-mediated oxidation of glucose that provides bioelectrocatalytic amplification of the DNA detection process.

14.2.1.3 Gene Sensors Based on Target-Induced Probe Restructuring

The persistence lengths of single stranded (ss) and double stranded (ds) DNA are totally different to each other; the lengths of ss- and dsDNA are around 1 and 50 nm, respectively. That is, the duplex structure of DNA is much stiffer in bending than a single-stranded one. The global conformation of DNA could be designed through specific hybridization with partially complementary sequences (Fig. 14.2B).[24,25] Therefore, the distance of the Fc unit modified at the end of DNA, which is anchored onto the electrode at another end, from the electrode surface could be controlled through specific structure formation with the complementary target DNA/RNA.

Heeger et al., immobilized Fc-ODN on an Au electrode. The sequence of the Fc-ODN was designed to form a stem-loop structure in which a distal half sequence folded back to allocate a Fc moiety in close proximity to the electrode. The target DNA complementary to the loop region of Fc-DNA hybridized with it to form an extended duplex structure.[26] That is, the target distanced the Fc unit from the electrode to reduce the redox signal in a concentration-dependent manner. Grinstaff et al., prepared an ABA triblock PEG (polyethylene glycol)-ODN conjugate. The conjugate consisted of two ODN sequences (Fc- and thiol groups were modified on each of the ODNs ends) and an intervening PEG linker, and was immobilized on the electrode.[27] The sequences of the two ODN segments were designed to form a specific tandem duplex structure with the target, in which the Fc-end was located in close proximity to the electrode. They succeeded in reversing the signal response to the turn-on type.

In addition to this research, target-induced restructuring such as coil-duplex,[28–36] coil-triplex,[37] and bulge-quadruplex (G4)[38] were applied for the switching depending on DNA, RNA, and metal ions.

14.2.1.4 Immobilization-Free Systems for DNA/RNA Sensing

Most of the conventional electrochemical methods usually require one of the participants in the interaction of interest to be immobilized on electrodes. Generally, the procedures for modification of biomolecules on electrodes are time consuming and it is difficult to obtain reproducible results. The general electrochemical methods that work in homogeneous solutions would be beneficial to the development of convenient, low-cost, and sensitive sensors for biomolecules. To realize this, we need to develop general techniques that modulate the electrochemical signal by triggering a specific reaction.

Ihara et al., reported the modification of β-Cyclodextrin (βCyD) to the end of an ODN to prepare a DNA conjugate, βCyD-ODN.[39–41] They used βCyD-ODN with Fc-ODN as a pair of probes that work cooperatively for DNA sensing in homogeneous solutions. The electrochemical signal of Fc tethered to the end of one probe was significantly suppressed by the formation of an inclusion complex with the βCyD of the other probe on the DNA templates. βCyD could be regarded as a "quencher" for the Fc electrochemical signal (Fig. 14.3).[42] This method does not require electrode modification of one of the counterparts of the specific interaction of interest. This could be a common scheme for electrochemical sensing of susceptible biomolecules. Aoki et al., synthesized an electrochemical DNA probe bearing Fc and βCyD on each of the both ends respectively.[43] While an intramolecular inclusion complex, Fc/βCyD, formed in the absence of the target by circularization of the DNA probe, hybridization with the target DNA linearizes the structure to dissociate the complex. The hybridization was monitored in a homogeneous solution by observation of changes in redox behavior.

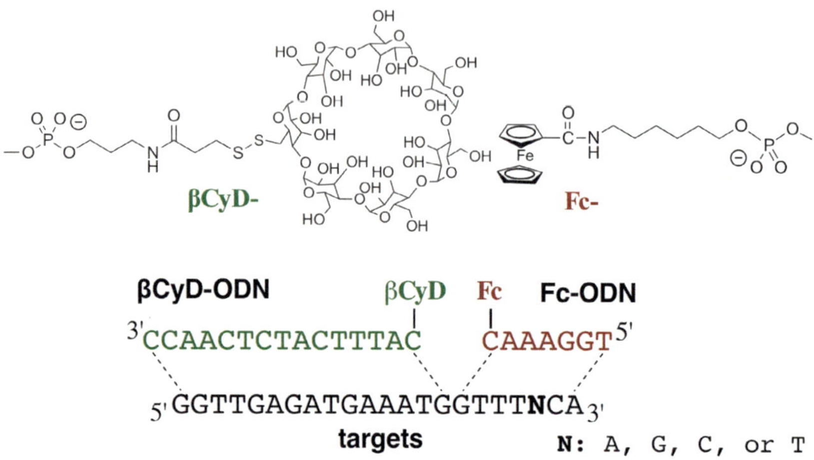

FIGURE 14.3 Immobilization-free system for electrochemical DNA/RNA sensing. The βCyD modified on the end of the probe functions as a "quencher" of electrochemical activity of Fc.[40]

Hsing et al., reported immobilization-free sensitive DNA detection using a Fc-PNA (peptide nucleic acid) conjugate.[44] While PNA is a neutral molecule, a PNA/DNA hybrid shows a negative charge. The electric current measured by an ITO (indium tin oxide) electrode was much reduced by hybridization due to the electrostatic repulsion between the negatively charged ITO surface and the PNA/DNA hybrid. In contrast, the current observed by the ITO coated with a positively charged polymer was significantly enhanced by the hybridization. The detection limits of the target DNA were 500 fmol and 40 fmol for each of the two electrodes, respectively.

14.2.1.5 Small Molecular Probes for DNA Analysis

Some small Fc derivatives used for DNA probing are shown in Fig. 14.4. Takenaka et al., have reported a series of systems of electrochemical DNA sensing using Fc-based small molecular probes.[45–48] Two hybrid molecules, adamantylnaphthalene diimide (Ad-NDI) and Fc-βCyD, were used as a dsDNA selective ligand and an electrochemical signal generator, respectively.[45] The capture DNA probes preimmobilized on the electrode were subjected to a duplex formation with the target DNAs in the presence of Ad-NDI. NDI moiety in Ad-NDI prefers to bind with dsDNA by threading intercalation. Therefore, dsDNA formed on the electrode was selectively labeled with multiple Ad groups. Then, by the addition of the signal generator, Fc-βCyD, to the electrode, intramolecularly accommodated Fc units were expelled from βCyDs by Ads displayed on dsDNA to form ternary complexes, dsDNA/Ad-NDI/Fc-βCyD, on the electrode. That is, Fc units exposed to the solution were concentrated on the electrode only in the presence of the complementary target DNA. In their successive works, the

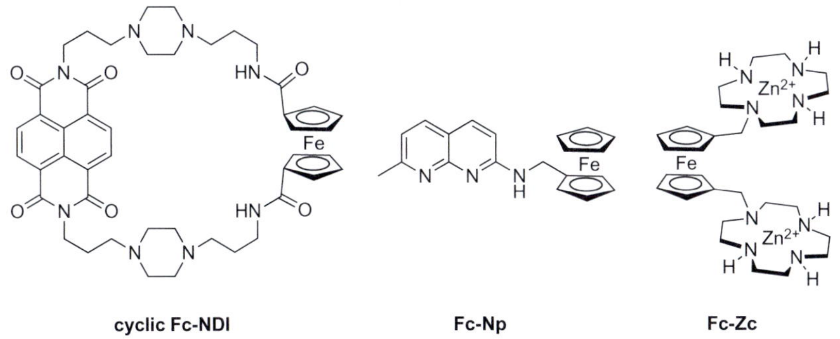

FIGURE 14.4 Ferrocene-based small electrochemical probes that are designed to recognize the specific structures of DNA such as G4, AP site, and T on dsDNA.[48–50]

ternary hybrid molecular probe, Fc-NDI-βCyD, was prepared and used for electrochemical labeling of dsDNA.[46] The intramolecular inclusion complex, Fc/βCyD, dissociated with NDI intercalation to dsDNA due to the steric requirement. The cyclic Fc-NDI probe worked as an electrochemically active probe for G4 DNA.[48] Steric hindrance of a Fc bridge covered one of the basal faces of NDI prevented Fc-NDI from intercalating into adjacent base pairs in dsDNA, making it a specific binder for G4 DNA. The electric currents based on Fc oxidation for these small probes were observed in a concentration-dependent manner of the target DNAs.

Teramae et al., prepared a hybrid molecular probe consisting of naphthyridine (Np) and Fc, Fc-Np. Np can form a hydrogen bond-mediated 1:1 complex with cytosine opposite the abasic (AP) site in a duplex.[49] Fc-Np showed a specific reduction in its fluorescent and electrochemical response by interaction with cytosine displayed in an AP site on dsDNA. The current decrease upon addition of the target DNA can be mainly ascribed to the decrease in the diffusion coefficient of Fc-Np. Bond et al., prepared the hybrid molecule, Zinc(II)-Cyclen (Zc) and Fc (Fc-Zc) and used it as a small electrochemical molecular probe for ds-DNA. Zc moiety can form a 1:1 complex with thymine.[50] Fc-Zc bound with dsDNA containing T to provide a redox signal. Coupling the redox chemistry of the bound Fc-Zc and dimethylamine provided an electrocatalytic pathway that increased the sensitivity of the assay and allowed the 100 fM target DNA sequence to be detected. Excellent selectivity against even single-base sequence mismatches was achieved.

14.2.1.6 *Electrochemiluminescence Using Fc as a Signal Modulator*

Landers et al., reported that the Fc molecule could quench the electrochemiluminescence (ECL) of the $Ru(bpy)_3^{2+}$ (bpy = 2,2′-bipyridine) effectively in an aqueous solution. ECL from the $Ru(bpy)_3^{2+}$ covalently attached at the end of the ODN was effectively quenched by the Fc tethered to the complementary ODN.[51] Fang et al., reported a controllable solid-state ECL film based on efficient and stable quenching of ECL from $Ru(bpy)_3^{2+}$ by oxidizing Fc at the electrode. Au nanoparticles (AuNP)/$Ru(bpy)_3^{2+}$ composite film was formed on the electrode. Stem-loop structured thiolated Fc-ODN was immobilized on the surface of the film. The target DNAs added to the film hybridized with Fc-ODN to form a duplex. The quenched ECL of the composite film was restored with an increase in the distance of the Fc from the film by linearization of the stem-loop structure.[52] Cui et al., succeeded in the detection of *rpoB* genes of Mycobacterium tuberculosis (MTB) in a homogeneous solution. ECL measurements were performed in a suspension of $Ru(bpy)_3^{2+}$ functionalized graphene oxide (Ru-GO). While GO absorbs ssDNA, the affinity of dsDNA to GO is very weak.[53] Initially the ECL from Ru-GO was quenched by preabsorbed Fc-ODN. Complementary target DNA hybridized with Fc-ODN to form a duplex and released Fc from Ru-GO, leading to recovery of ECL.[54] The ssDNA target can be detected in a range from 0.1 to 100 nM with a detection limit of 0.04 nM.

14.2.2 Protein

14.2.2.1 Protein Sensors Based on Target-Induced Probe Concentration

Enzymatic reactions of interest can be used for linking or release of Fc units to the electrode. The concentration or activity of the target enzymes could be determined from the change in redox signal from the Fc on the electrode. Lin et al., and Li et al., reported electrochemical approaches for determination of enzymes of interest using a Fc-peptide-modified Au electrode.[55,56] The peptides used here have the sequences recognized and cleaved by the enzymes. Therefore, the target enzymes are expected to release a Fc unit from the electrode and decrease the electric current signals. Lin et al., showed the detection of matrix metalloproteinases (MMPs), which have been proposed as important biomarkers of infection and inflammation.[55] Li et al., succeeded in the detection of caspase-3, which is a special proteinase to apoptosis. This method sensitively and specifically detected apoptotic cells with a signal decline of 85 percent.[56] Kraatz et al., reported the determination of protein kinase using a versatile strategy for biochemical, electrochemical, and immunoarray detection of phosphorylation. Protein kinases are the cellular proteins involved in the regulation of many cellular processes and have emerged as promising targets for the treatment of several diseases. They used Fc-ATP (5′-γ-Ferrocenyl-adenosine triphosphate) and the appropriate peptide immobilized on the electrode as substrates of a kinase reaction for bimodal analysis.[57] The Fc modified on the peptide was directly quantified by electrochemical techniques or recognized by anti-Fc-antibody and detected by a fluorescently-labeled second antibody.

In the following two works, Fc polymers were used to obtain amplified electrochemical signals. Tang et al., reported ultrasensitive immunoassay by DNA-based hybridization chain reaction (immune-HCR assay) using two hairpin DNAs bearing two Fc groups on both ends.[58] The assay protocol mainly involves the formation of the sandwiched immunocomplex between the immobilized capture antibodies on the magnetic beads and the secondary antibodies on the AuNPs conjugated with HCR initiator strands, the HCR reaction from DNA initiator strands on the AuNPs between the two hairpin DNA molecules, and electrochemical measurement of HCR product (Fc-array on dsDNAs). Leclerc et al. reported label-free electrochemical detection of protein based on a Fc-bearing cationic polythiophene and anti-target DNA aptamer. Combining with the PNA complementary to the aptamer, they succeeded in turn-on type sensitive detection of thrombin.[59]

Yu et al., showed the detection of a homodimer of platelet-derived growth factor B-chain (PDGF-BB) using proximity-dependent surface hybridization of the aptamer dimer. They used an ODN-modified Au electrode and PDGF aptamer with a 3′-Fc-labeled short ODN tail, which is complementary to the electrode-tethered ODN. Simultaneous recognition of PDGF-BB brings a pair of aptamer probes proximate, promoting the flanking Fc-labeled tail sequences to hybridize together with immobilized ODN strands, which triggers a redox current.[60] The absence of PDGF-BB only allows independent annealing of the tail sequences, disabling the hybridization of the short sequences due to a predesigned low melting temperature.

14.2.2.2 Protein Sensors Based on Target-Induced Probe Restructuring

Yu et al., immobilized a Fc-modified dsDNA on the electrode. One of the strands of the dsDNA carries the aptamer sequence for an early biomarker for lung cancer (CTAP III/NAP2) in a bulge region. The local structure of the bulged three-way junction of the immobilized dsDNA was designed so that the nucleobases stack better when CTAP III/NAP2 binds with the aptamer moiety. The structural change in dsDNA promoted the electron transfer from Fc to the electrode through nucleobase stacking.[61] Mao et al., and Ren et al., used a Fc-modified stem-loop DNAs immobilized on the electrode. Upon interaction with the targets, the distance between the Fc and the electrode changes with DNA restructuring. Thrombin[62] and prostate-specific antigen (PSA)[63] were determined, respectively.

Luong et al., and Gooding et al., immobilized Fc-statin and Fc-biotin on the electrode to detect HIV-1 proteinase[64] and anti-biotin IgG,[65] respectively. Electrochemical impedance spectroscopy (EIS) was used for determination of HIV-1 proteinase. Sensitive response to the proteinase enabled the application of the techniques for effective screening of HIV inhibitors. IgG measurement was performed using a molecular wire (Fc-biotin) immobilized on an electrode. A biotin (epitope model) was attached to the end of the wire through Fc linker. The binding of IgG with biotin changed the microenvironment of Fc in a large bulky protein structure, resulting in the physical blocking of Fc and/or the restriction of access of the counter ions required for Fc redox reactions. Conversely, the biotin added to IgG bound electrode caused a competition for the IgG, some of which are displaced from the electrode surface, with a concomitant increase in electrochemical signals.

Wu et al., proposed a universal strategy for aptamer-based nanopore sensing through host-guest interactions inside α-Hemolysin (αHL). They used a dsDNA consisting of a VEGF (vascular endothelial growth factor) aptamer and its complementary ODN conjugated with Fc, which was accommodated in cucurbit[7]uril (UB[7]).[66] dsDNA is too large to traverse through an αHL pore. However, the presence of VEGF may cause the aptamer/probe duplex to unwind. This process eventually frees the UB[7]/Fc-ODN probe, which generates characteristic current events when translocated through αHL under the transmembrane potential.

14.2.3 Cell

The nonconductive nature of the cell membrane has made it difficult to detect cells using electrochemical techniques. It greatly restricts the electrochemical communication between redox tags and target cells. Xu et al., proposed a cell-responsive electrochemical switch, which overcomes the limitation of cell membranes.[67] Mammalian cells, and in particular adherent cells, could be transfected by DNA molecules via the calcium phosphate coprecipitation (CPP) method. They showed that Fc moiety at the end of Fc-ODN immobilized on the electrode could penetrate cells together with DNA during cell transfection. After the occurrence of cell transfection, DNA molecules act as "cranes" that send the Fc redox tag from the electrode surface to the inside of the cells. This provides efficient electric signal change via the changes in distance between the Fc and the electrode surface and the electronic shielding effect of the cell membrane before and after cell transfection.

14.2.4 Other Biologically Active Substances

14.2.4.1 Molecular Sensors Based on Restructuring of Interface DNA

Electrochemical molecular sensors based on target-induced restructuring of Fc-ODNs modified on the electrode were reported by several research groups. DNA aptamers were used for ODN moieties in Fc-ODNs as the interfaces of target recognition. The conformational change in ODNs induced by specific target binding results in the translocation of Fc on the electrode. Fan et al., and Yu et al., reported ATP[68] and adenosine[69] sensors, Wang et al., and Gothelf et al., reported cocaine[70] and theophylline[71] sensors, respectively. Mayer et al., constructed a logic gate using a split ATP aptamer with multiple input signals such as ATP and adenosine deaminase.[72] A Fc-ODN carrying an L-histidine-dependent DNAzyme sequence was immobilized on a AuNP/graphene nanosheet composite by Li et al.[73] The Fc-ODN cleaved a complementary substrate sequence only in the presence of L-histidine, released the product to be a flexible single-stranded structure, and folded back to locate its Fc end close to the electrode. This system is regarded as an electrochemical L-histidine sensor with high selectivity.

Certain defined structures of DNA can be constructed by specific metal ion coordination to nucleobases. Potassium ions are essential to form a G4 structures of DNA/RNA. A potassium ion is accommodated in a pocket in between two planes of guanine quartets, in which eight carbonyl oxygens coordinate to a potassium to stabilize a whole G4 structure. O'Sullivan et al., immobilized Fc-ODN on an Au electrode. The G-rich sequence in the ODN moiety is expected to form a G4 structure only in the presence of potassium ions. They reported sensitive potassium ion sensing using EIS techniques.[74] Jiang et al., reported a mercury(II) sensor using T-rich Fc-ODN immobilized on an Au electrode. Two thymines coordinate to form a dimer, T-Hg^{2+}-T, which makes a metal ion-mediated stable dsDNA structure (M-DNA).[75] The distance between the Fc and the electrode changed with M-DNA formation.

14.2.4.2 Target Recognition Based on Host-Guest Chemistry and Selective Coordination

Various Fc derivatives have been designed by combining Fc with the functional molecules having specific affinity toward target molecules (Fig. 14.5). James et al., prepared the Fc derivative bearing boronic acid (Fc-BA) for sensing saccharides. Boronic acids are known to bind saccharides via covalent interactions in aqueous basic media. The most common interaction is with cis-1,2- or 1,3-diols of saccharides to form five- or six-membered rings respectively. The oxidation potential of the Fc became more anodic on binding with saccharides.[76] The interaction of the boronic acid and neighboring amine strengthened on saccharide binding, thereby reducing the electron density on the neighboring amine. This in turn destabilized the ferrocenium cation at higher saccharide concentrations, resulting in a more anodic ferrocene oxidation overpotential. Tucker et al., reported selective electrochemical sensing of acidic organic molecules using the Fc derivative bearing amidopyridine and crown ether, 18-Crown-6 (18C6), on each of the two cyclopentadienyl rings (Fc-Py18C6). The amidopyridine unit is well-established as a binder of the carboxylic acid motif through two complementary hydrogen bonds. 18C6 crown ether binds ammonium groups. By synergistic effect of these two functional groups, Fc-Py18C6 recognized L-phenylalanine and the oxidation potential of Fc shifted to the positive side.[77] Oxidation became

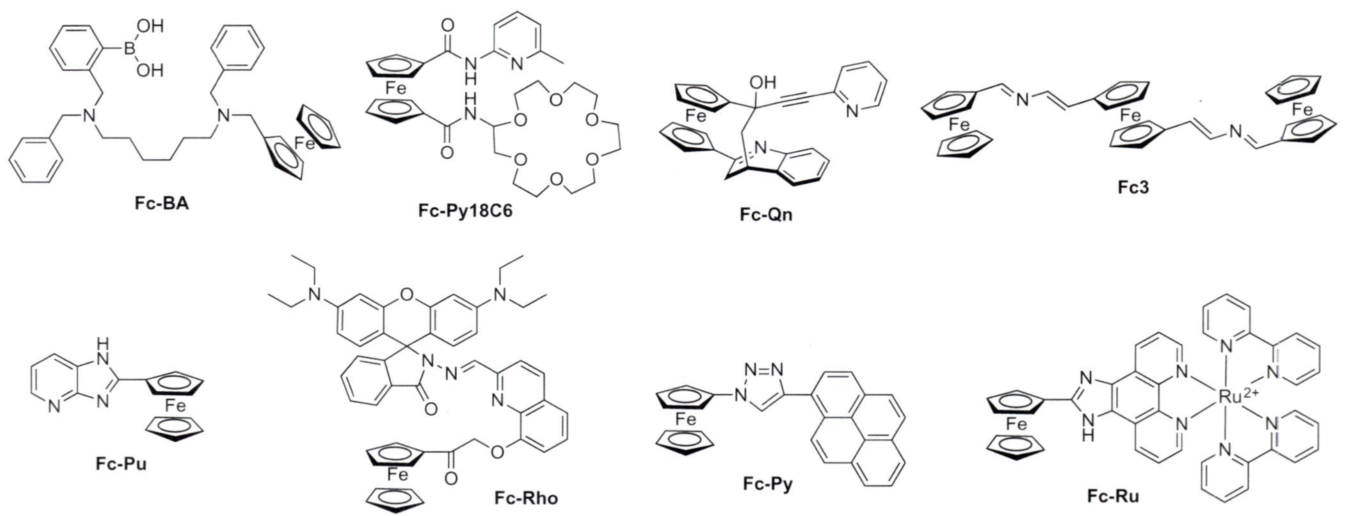

FIGURE 14.5 Ferrocene-based electrochemical molecular probes for saccharide, amino acid, metal ions, and anions. Some of them are bimodal probes that report the binding with their targets by the change in electric and spectroscopic signals.[76–84]

difficult upon complexation with a guest-to-host proton transfer reaction, since a positive charge is introduced onto the receptor in close proximity to the Fc.

Fc-based sensors for various cations have been reported. Molina et al., reported several sensors for metal ions such as Mg^{2+}, Ca^{2+}, Zn^{2+}, Ni^{2+} and Pb^{2+} (Fc-Qn, Fc-Pu, Fc3),[78–80] and Hung et al., for Hg^{2+} and Cr^{3+} (Fc-Rho).[81,82] Some of these are multimodal sensors; in addition to the cation dependent modulation in their redox behavior, they changed spectroscopic characters upon binding the target cations.[79–82] Anion sensors have also been designed in a similar manner. Molina et al., modified the Fc with pyrene (Fc-Py),[83] and ruthenium(II) polypyridyl complex (Fc-Ru)[84] for multimodal sensing for pyrophosphate and chloride, respectively. Beer et al., modified an Au electrode by a Fc derivative with thiolated linker chains tethered through an amide bond. The change in redox responses was observed with phosphate, halides and perrhenate.[85]

14.2.4.3 Activity Control of Redox Enzymes by Fc as an Electron Mediator

Zhou et al., constructed a mixed monolayer of ferrocenylalkanethiol (Fc-C_{11}) and encapsulated horseradish peroxidase (HRP) at an Au electrode for amperometric detection of H_2O_2 at trace levels. By tuning the alkanethiol chain lengths that tether the HRP and the Fc-C_{11} mediator, facile electron transfer between Fc and HRP could be achieved.[86] The electrode covered with Fc-C_{11} and HRP is shown to be capable of rapidly and reproducibly detecting H_2O_2 present in complex sample media. Willner et al., reported mercury(II) detection by using M-DNA formation based on T-Hg^{2+}-T bridging on an Au electrode. On the M-DNA scaffolds consisting of ODNs modified on the electrode, GOx-ODN and Fc-ODN, GOx and Fc were allocated at appropriate distances from the electrode so as to provide an effective electrocatalytic current fueled by glucose in the presence of Hg^{2+}.[87] They succeeded in ultrasensitive detection of Hg^{2+} in the Fc/GOx nanostructure. Zhan et al., constructed a sophisticated molecular system of photoswitchable bioelectrocatalysis using GOx/glucose. On/Off states were fabricated based on host-guest chemistry of azobenzene/βCyD at an interface electrode, where reversible immobilization and detachment of Fc-labeled βCyD polymer was controlled by photoirradiation to azobenzene anchored onto the electrode.[88]

14.3 IRIDIUM

Iridium is known to form various cyclometalated complexes, which have been widely used as catalysts and luminophores in industry and research. Here, the research on organoiridium complexes relevant to the biosciences (Fig. 14.6) is categorized by their applications.

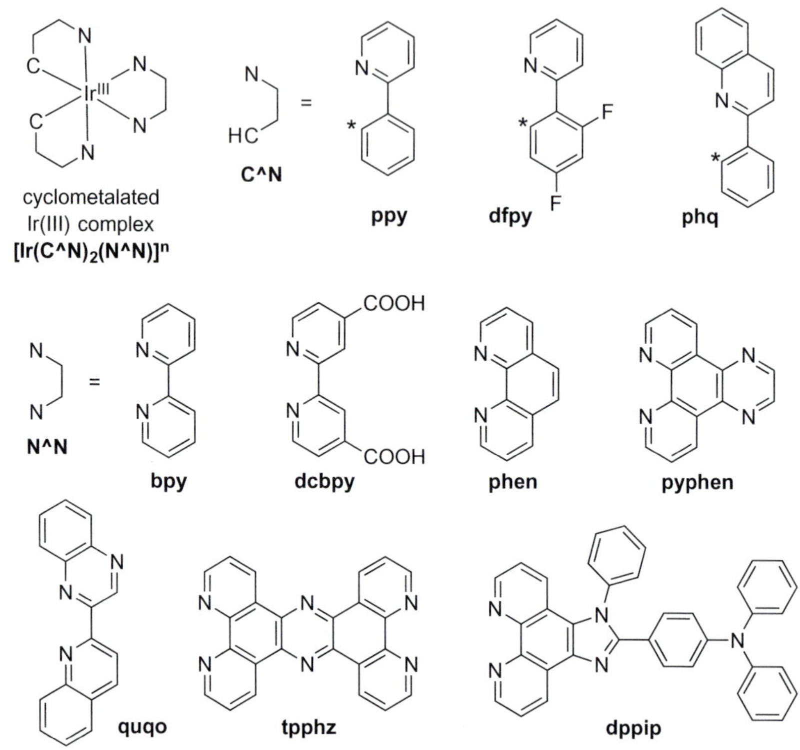

FIGURE 14.6 Typical phosphorescent cyclometalated Ir(III) complexes.[93–98] They show long-lived phosphorescence on the binding with their targets. Combined use with the techniques of time-gated signal acquisition made them very useful ion sensors or bioimaging agents. The asterisks (*) on C^N ligands show the carbons directly coordinated to Ir(III) in the complexes.

14.3.1 Luminescent Probes in Biology

Cyclometalated iridium(III) complexes show characteristic spectroscopic behavior. Wong et al., reported the turn-on probe, $[Ir(ppy)_2(solv)_2]^+$ (ppy = 2-phenylpyridine, solv = H_2O or CH_3CN), for histidine/histidine-rich protein. Up to 180-fold enhancement of luminescence was observed when it binds with the targets through ligand exchange.[89]

Phosphorescence signaling provides a valuable alternative to conventional bioimaging based on fluorescence. Long-lifetime phosphorescence emission can easily be discriminated from background noise caused by autofluorescence and scattered light. That is, the benefits of using phosphorescent molecules include improved sensitivity and capabilities for effective elimination of background signals by time-gated acquisition.[90–92] You et al., prepared a series of cyclometalated Ir(III) complexes having various C^N ligands, $[Ir(C^{\wedge}N)_2phen']^+$ (phen' = binary hybrid ligands between 1,10-phenanthroline and the chelators for target metals), with systematically varied electron densities and band-gap energies. They showed characteristic phosphorescence in the presence of biologically important transition-metal ions.[93] In particular, the phosphorescence zinc sensor could report the presence of intracellular zinc pools using confocal laser scanning microscopy and photoluminescence lifetime imaging microscopy techniques. It is likely that the complex binds to hydrophobic organelles such as the Golgi body and endoplasmic reticulum. Li et al., synthesized two cationic Ir(III) complexes, $[Ir(dfpy)_2(bpy)]^+$ and $[Ir(dfpy)_2(quqo)]^+$ (dfpy = 2-(2,4-difluorophenyl)pyridine, quqo = 2-(quinolin-2-yl)quinoxaline), with bright green and red phosphorescence and applied the complexes for living cell imaging. The complexes have significant features for their exclusive staining in cytoplasm, low cytotoxicity and reduced photobleaching, as well as cell membrane permeability, which make them promising candidates for the design of specific bioimaging agents.[94] Huang et al., reported a multifunctional phosphorescent Ir(III) complex, $[Ir(ppy)_2(tpphz)]^+$ (tpphz = tetrapyrido [3,2-a:2′,3′-c:3″,2″-h:2‴,3‴-j]phenazine), for specific nucleus staining and applied the complex to monitor the intranuclear oxygen level for hypoxia monitoring.[95] With the advantageous long emission lifetime of the complex, the influence of autofluorescence could be eliminated by using photoluminescence lifetime imaging and time-gated luminescence imaging techniques. Chao et al., reported cyclometalated Ir(III) complexes, $[Ir(ppy)_2(dppip)]^+$, (dppip = N,N-diphenyl-4-(1-phenyl-1H-imidazo [4,5-f][1,10]phenanthrolin-2-yl)aniline and derivatives), as AIE (aggregation-induced emission) phosphorescent probes. Owing to AIE, the Ir(III) complexes exhibited excellent photostability compared with MitoTracker Green. Moreover, they manifested satisfactory photostability in the mitochondrial physiological pH range. Using the complexes and LysoTracker Green, they succeeded to achieve real-time monitoring of mitophagy.[96]

Ma et al., reported a luminescent cocaine detection using a split DNA aptamer and a G4-selective Ir(III) complex, $[Ir(phq)_2(pyphen)]^+$ (phq = 2-phenylquinoline, pyphen = pyrazino[2,3-f][1,10]-phenanthroline). They split the cocaine aptamer into two parts and each of the split G4 DNAs was attached to the ends of the fragments of the split aptamer to prepare a pair of ODN fragments of half cocaine aptamer-half G4 DNA.[97] Cocaine was expected to facilitate the formation of a reconstructed aptamer structure due to the effect of its induced fit. The aptamer reconstruction would be concomitant with reconstitution of the G4 structure, which forms a luminous complex with the Ir(III) complex. A detection limit as low as 30 nM for cocaine was achieved by this sensing approach with a linear relationship between luminescence intensity and cocaine concentration established from 30 to 300 nM. Furthermore, this sensing approach could detect cocaine in diluted oral fluid. Zhang et al., reported the ECL from their cyclometalated Ir (III) complex, $[Ir(ppy)_2(dcbpy)]^+$ (dcbpy = 4,4′-dicarboxylic-2,2′-bipyridine). Combining use of the enzymatic reactions, they succeeded in ultrasensitive

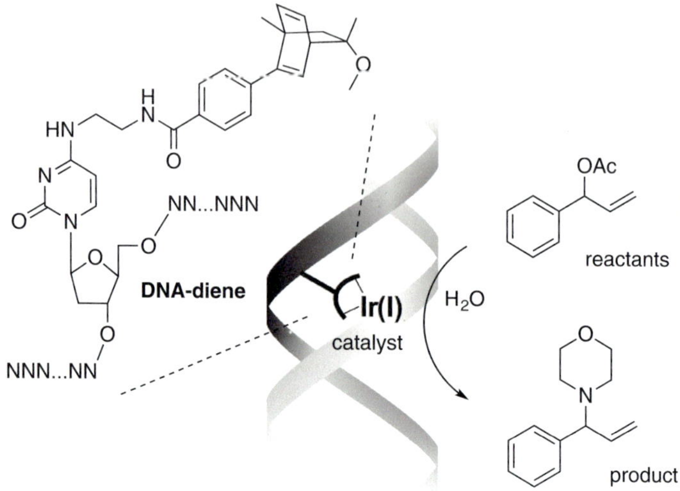

FIGURE 14.7 Allylic amination catalyzed by Ir(I) complexes containing the diene ligand tethered to a nucleobase of ODN.[100] DNA was expected to be a chiral source of the reaction product.

detection of Ramos cells (cancer cells) using the ECL signal from the Ir complexes tethered on AuNP.[98]

14.3.2 Catalyst on Biomolecules

Homogeneous catalysts such as transition metal complexes have been used for chemical reactions on biopolymers.[99] In these reports high stereoselectivities were obtained in Diels–Alder reactions, Michael additions, Friedel-Crafts alkylation, and fluorinations, with biopolymers as the sole source of chirality. Jäschke et al., demonstrated that the application of DNA-hybrid catalysis could be extended to organometallic chemistry beyond Lewis acid catalysis. They showed allylic aminations using the homogeneous hybrid catalyst, DNA–diene–Ir(I), in which diene ligands were covalently tethered to a nucleobase in DNA (Fig. 14.7). In the presence of the DNA-based diene, slightly higher activity was observed than with the Ir(I) complex of the free diene.[100] Complementary DNA and RNA strands were used to modulate the chiral microenvironment surrounding the DNA–diene–Ir(III) complex through hybridization. Although the ee% values observed remained low, these results clearly indicate chirality transfer from DNA to the iridium complex.

14.3.3 Donor or Acceptor of the Electron Transfer on dsDNA

Because of orbital overlap between the π systems of neighboring nucleobases, DNA can serve as a bridge in long-range electron transfer (ET) reactions. To improve the understanding of the chemical mechanisms underlying oxidative damage in DNA, as well as the biological factors affecting the prevalence, detection, and repair of such damage, it is necessary to study DNA-mediated charge transport.[101]

Barton et al., have studied ET on DNA using various transition metal complexes with polypyridyl ligands as an ET donor or acceptor. The cyclometalated Ir(III) complex, $[Ir(ppy)_2(dppz)]^+$ (dppz = dipyrido[2,3-a:2′,3′-c] phenazine), intercalates into DNA via the dppz ligand. The excited state of the Ir(III) complex has sufficient driving force to oxidize purines (G and A) and to reduce pyrimidine (C and T) nucleobases in dsDNA. This remarkable ability has enabled characterization of DNA-mediated ET and DNA-mediated hole transfer in identical sequence contexts, showing that both have a shallow distance dependence.[102–105]

14.4 MERCURY

The use of mercury in biosciences is very limited because the Hg(II) ion could be strongly coordinated by endocyclic nitrogens of nucleobases in DNA/RNA and thiol groups of cysteine residues in proteins, resulting in serious incompetence of related important bioprocesses. Organomercury complexes are stable under physiological conditions and, therefore, their coordination and reactivity are adequately controlled. Organomercury complexes have been applied in bioscience research, and such research concerning nucleic acid chemistry are reviewed in this section.

14.4.1 Hg(II)-Mediated Base Pairing

Metal ion-mediated base pairing could stabilize the duplex structure of M-DNA,[106–108] and therefore would be an attractive approach for targeting short nucleic acids such as miRNA. However, its therapeutic or analytical applications are limited due to the scarcity of relevant metal ions inside the cell. In addition, intracellular medium is not necessarily ideal for the complexation. To overcome these limitations it would be beneficial to replace one of the electronegative donor atoms with carbon, affording an organometallic metal–carbon (M–C) bond as an alternative for Watson–Crick base pairing. Generally, the strength of a single M–C bond is sufficient to effectively prevent dissociation of the complex.

Lönnberg et al., reported a mercuricytosine as a Janus nucleobase (Fig. 14.8A). The base-pairing properties of 5-mercuricytosine have been explored by NMR and UV melting experiments. For ODN duplexes, 5-mercuricytosine formed Hg(II)-mediated base pairs with T and G.[109] In the former case, the dsODN was as stable as the respective duplex comprising solely Watson–Crick base pairs. Based on detailed thermodynamic analysis

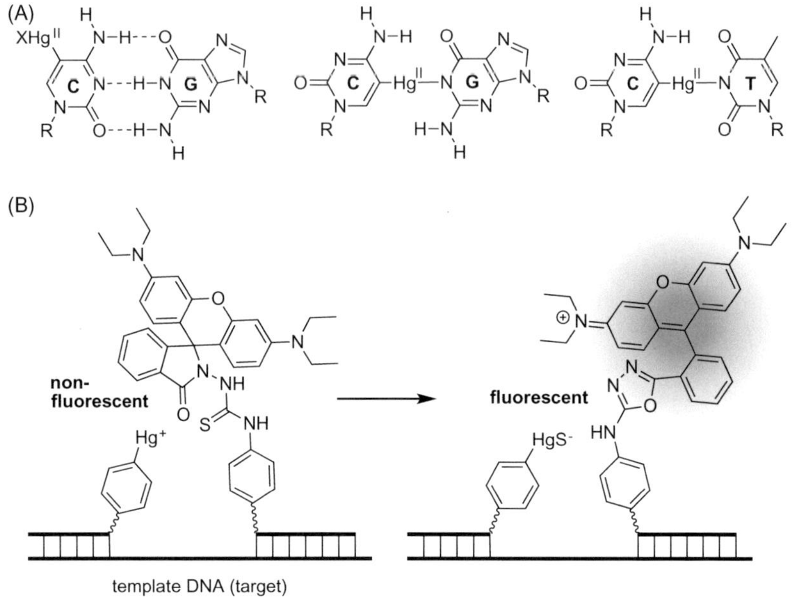

FIGURE 14.8 Organomercury complexes used in genome chemistry. (A) Base pairing of Janus nucleobase. 5-mercuricytosine formed Hg(II)-mediated base pairs with G and T.[109] (B) Template-directed activation of a fluorogen by bioorthogonal reaction with organomercury.[110] X denotes an exchangeable ligand such as Cl or H_2O.

of the melting curves, the stabilization by the Hg(II)-mediated base pairs was attributed to a lower entropic penalty of hybridization.

14.4.2 Organometallic Reaction for Bioanalysis

Fluorescent metal complexes may form by gathering complementary functionalities on the target molecule as a template. Labile coordination complexes, however, would preclude the application of such probes in a living cell. Organometallic molecules may provide stable, bioorthogonal reactant pairs that can be applied to templated fluorescence activation in cells.

Kool et al., reported organometallic activation of a fluorogen for templated nucleic acid detection. They prepared a pair of chemically modified ODN probes. Each of the probes carries a *p*-Mercuriobenzoate (HgBA) group and a fluorophore caged with a mercury-sensitive group (phenylthiosemicarbazide) on one of their ends. The sequences of the probes were designed to be complementary to the adjacent sites of the target DNA/RNA to form a tandem duplex with their auxiliary groups facing each other. In close proximity on the tandem duplex, HgBA uncages the counterpart to make a fluorescent structure (Fig. 14.8B).[110] The thiophilicity of mercury and the inertness of the Hg–C bond in aqueous conditions make HgBA a promising organometallic reactant for bioanalytical applications. This system could be useful as a fluorogenic reporter of nucleic acid sequences in solution, and it gave readable signals with single nucleotide discrimination.

14.5 PLATINUM

Cisplatin, *cis*-diamminedichloroplatinum(II), was the first inorganic antitumor drug used in clinics and, after more than 40 years, remains the most widely used and successful drug in cancer therapy.[4] Since then hundreds of platinum complexes have been designed not only as drugs but also as probes for use in bioscience research. In addition to Pt(II) complexes with N-donor ligands, organoplatinum complexes have been proposed based on their inert nature under cellular conditions and unique physicochemical properties (Fig. 14.9).

Several cyclometalated organoplatinum complexes have been prepared as DNA binders, which bind with dsDNA as intercalators or groove binders. Che et al., showed the specific blocking of CREB (cAMP response element binding protein)/DNA binding and succeeded in control of transcription by cyclometalated Pt(II) complex, $[Pt(pbpy)(dpt)]^+$ (pbpy = 6-phenyl-2,2′-bipyridyl, dpt = 2,4-di-amino-6-(4-pyridyl)-1,3,5-triazine) and $[Pt(pbpy)(icppy)]^+$ (icppy = 1-((4-isocyano-3,5-diisopropylphenyl)ethynyl)pyrene).[111,112] Square-planar complex structures strongly bind in the major-groove of dsDNA with CRE sequence to inhibit the binding of CREB. Ma et al., reported nanomolar Ag(I) ion sensing in aqueous solution. They used an ODN with a G4 structure

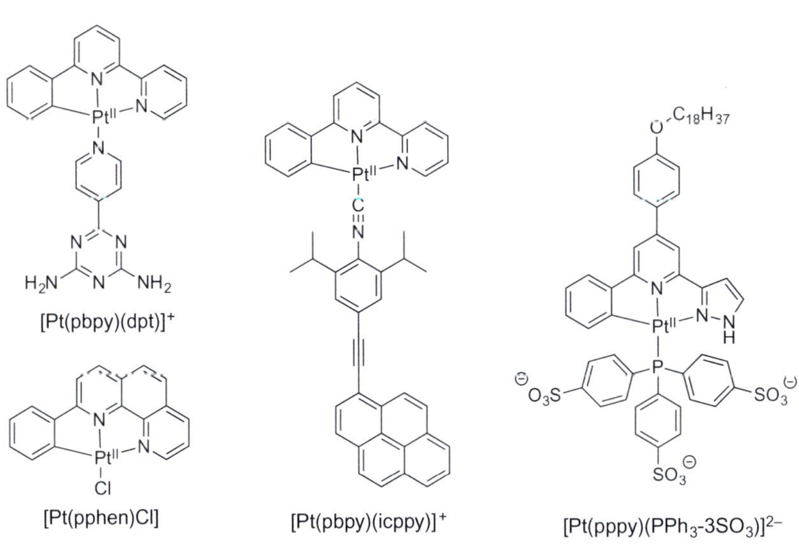

FIGURE 14.9 Cyclometalated Pt(II) complexes used for bioimaging.[111–114]

and the cyclometalated Pt(II) complex, [Pt(pphen)Cl] (pphen = 2-phenyl-1, 10-phenanthroline), as a silver ion binder and a phosphorescent probe, respectively.[113] Silver ion added to the G4 ODN induced the intramolecular transformation to the duplex structure. The Pt(II) complex intercalated to the duplex structure and concomitantly its phosphorescence increased. The time-gating techniques for obtaining long-lived phosphorescent signals from [Pt(pphen)Cl] enabled highly sensitive detection of Ag(I) ions.

Lam et al., reported two-photon plasma membrane imaging in living cells by an amphiphilic, water-soluble cyclometalated Pt(II) complex, $[Pt(pppy)(PPh_3\text{-}3SO_3)]^{2-}$ (pppy = amphiphilic derivative of 2-phenyl-6-(1H-pyrazol-3-yl)-pyridine, $PPh_3\text{-}3SO_3$ = *p*-Trisulfonated triphenylphosphine). In comparison to ordinary bioimaging techniques based on single-photon excitation, multiphoton imaging with near-infrared excitation offers a number of advantageous features including the reduction of phototoxicity to biological samples, the lowering of scattering, and the inherent capability of optical sectioning. By using the Pt(II) complex, they succeeded in the fast uptake of the complex by live cells, long resident lifetime in the plasma membrane, high two-photon induced luminescent intensity and low cytotoxicity.[114]

14.6 COBALT

Several metallocenes other than Fc have been studied in the biosciences. Metzler-Nolte et al., synthesized a cobaltocenium-PNA conjugate. Ferrocene and cobaltocenium groups are almost isostructural but differ in charge as well as redox properties.[115] They examined the electrochemical properties and thermal stabilities in interaction between the cobaltocenium-PNA and complementary DNA. The same research group reported a cobaltocenium-peptide conjugate that showed enhanced cellular uptake and directed nuclear delivery. The cobaltocenium group is exceedingly stable under physiological conditions. The cobaltocenium-NLS (nuclear localization signal) conjugate significantly accumulates in the nucleus of HepG2 cells.[116] Interestingly, the organometallic moiety in the conjugate is essential for active endocytosis and release of the conjugate into the cytoplasm.

14.7 CONCLUSION

This chapter covers molecular probes carrying organometals mostly used as reporter groups that produce unique signals that respond to biological targets. Fc is the most prevalent bioorganometal that is used for such purposes. The stability in aqueous solution, accessibility of various chemical modifications, and well-defined reversible electrochemical nature of Fc have made it a very useful molecule for employment as reporter moieties in bioconjugate probes. Cobaltocene is the isostructural molecule of Fc with different redox potential. Cyclometalated organoiridium complexes have been featured for their unique properties such as phosphorescence and catalytic activities. Phosphorescent nature is a very effective property for a reporter molecule. Employment of a time-resolved measurement technique provides high-quality luminescence with extremely low background. Generally, the Hg–C bond is stable in both air and moisture. Mercury(II) binds with cytosine to form an organometallic Janus nucleobase having double faces for allowing opponent nucleobases to be paired. In other work, the preference of Hg(II) to the soft atom was used for bioorthogonal chemical reactions to turn on the caged fluorophore on biomolecules. Platinum also forms cyclometalated organometallic complexes that specifically bind with ds- or G4 DNA and show intrinsic long-lived phosphorescence. Organometallic complexes encompass a wide variety of functions such as luminescence, electrochemical, catalytic, and pharmacological activities. This versatility would be very useful and promising for application in molecular biology, and the bioanalytical and biomedical sciences, which are growing and transforming rapidly.

REFERENCES

1. Thorp HH, Pecoraro VL, editors. *Mechanistic bioinorganic chemistry; Advances in chemistry series 246*. Washington DC: American Chemical Society; 1995.
2. Schmuck C, Wennemers H, editors. *Highlights in bioorganic chemistry, Methods and applications*. Weinheim: Wiley–VHC; 2004.
3. Sigel A, Sigel H, editors. *Metal ion in biological systems, Vol. 32, Interaction of metal ions with nucleotides, nucleic acids, and their constituents*. New York: Marcel Dekker, Inc; 1996.
4. Mjos KD, Orvig C. Metallodrugs in medicinal inorganic chemistry. *Chem Rev* 2014;**114**:4540–63.
5. van Staveren DR, Metzler-Nolte N. Bioorganometallic chemistry of ferrocene. *Chem Rev* 2004;**104**:5931–85.
6. Takenaka S, Uto Y, Kondo H, Ihara T, Takagi M. Electrochemically active DNA probe: Detection of target DNA sequence at femto mole level by high-performance liquid chromatography with electrochemical detection. *Anal Biochem* 1994;**218**:436–43.
7. Ihara T, Maruo Y, Takenaka S, Takagi M. Ferrocene–oligonucleotide conjugates for electrochemical probing of DNA. *Nucleic Acids Res.* 1996;**24**:4273–80.
8. Ihara T, Nakayama M, Murata M, Nakano K, Maeda M. Gene sensor using ferrocenyl oligonucleotide. *Chem Commun* 1997;1609–10.
9. Nakayama M, Ihara T, Nakano K, Maeda M. Gene sensor using ferrocene–oligonucleotide conjugate. *Talanta* 2002;**56**:857–66.
10. Ihara T, Nakayama M, Nakano K, Maeda M. *Electrochemical gene sensor in* Biomedical Diagnostic Science and Technology. New York: Marcel Dekker Inc; 2002. p. 227–41.
11. Nakano K, Kimura T, Kitamura Y, Ihara T, Ishimatsu R, Imato T. Potentiometric DNA sensing platform using redox-active DNA probe pair for sandwich-type dual hybridization at indicator electrode surface. *J Electroanal Chem* 2014;**720-721**:71–5.
12. Yu CJ, Wan Y, Yowanto H, Li J, Tao C, James MD, et al. Electronic detection of single-base mismatches in DNA with ferrocene-modified probes. *J Am Chem Soc* 2001;**123**:11155–61.
13. Ihara T, Sasahara D, Shimizu M, Jyo A. DNA conjugates bearing ferrocenyl group in backbone and their electrochemical behavior. *Supramol Chem* 2009;**21**:292–5.

14. Wlassoff WA, King GC. Ferrocene conjugates of dUTP for enzymatic redox rebelling of DNA. *Nucleic Acids Res* 2002;**30**:e58.
15. Giusto DA, Wlassoff WA, Giesebrecht S, Gooding JJ, King GC. Multipotential electrochemical detection of primer extension reactions on DNA self-assembled monolayers. *J Am Chem Soc* 2004;**126**:4120–1.
16. Zhang Y, Wang Y, Wang H, Jiang J-H, Shen G-L, Yu R-Q, et al. Electrochemical DNA biosensor based on the proximity-dependent surface hybridization assay. *Anal Chem* 2009;**81**:1982–7.
17. Wang J, Li J, Baca AJ, Hu J, Zhou F, Yan W, et al. Amplified voltammetric detection of DNA hybridization via oxidation of ferrocene caps on gold nanoparticle/streptavidin conjugates. *Anal Chem* 2003;**75**:3941–5.
18. Chen X, Lin Y-H, Li Juan, Lin L-S, Chen G-N, Yang H-H. A simple and ultrasensitive electrochemical DNA biosensor based on DNA concatamers. *Chem Commun* 2011;**47**:12116–18.
19. Umek RM, Lin SW, Vielmetter J, Terbrueggen RH, Irvine B, Yu CJ, et al. Electronic detection of nucleoside, A versatile platform for molecular diagnosis. *J Mol Diag* 2001;**3**:74–84.
20. Gibbs JM, Park S-J, Anderson DR, Watson JK, Mirkin CA, Nguyen ST. Polymer–DNA hybrids as electrochemical probe for the detection of DNA. *J Am Chem Soc* 2005;**127**:1170–8.
21. Huang Y, Zhang Y-L, Xu X, Jiang J-H, Shen G-L, Yu R-Q. Highly specific and sensitive electrochemical genotyping via gap ligation reaction and surface hybridization detection. *J Am Chem Soc* 2009;**131**:2478–80.
22. Wu Z-S, Jiang J-H, Shen G-L, Yu R-Q. Highly sensitive DNA detection and point mutation identification: An electrochemical approach based on the combined use of ligase and reverse molecular beacon. *Hum Mutat* 2007;**28**:630–7.
23. Patolsky F, Weizmann Y, Willner I. Redox-active nucleic-acid replica for the amplified bioelectrocatalytic detection of viral DNA. *J Am Chem Soc* 2002;**124**:770–2.
24. Tyagi S, Bratu DP, Kramer FR. Multicolor molecular beacons for allele discrimination. *Nat Biotechnol* 1998;**16**:49–53.
25. Liu J, Cao Z, Lu Y. Functional nucleic acid sensor. *Chem Rev* 2009;**109**:1948–98.
26. Fan C, Plaxco KW, Heeger AJ. Electrochemical interrogation of conformational change as a reagentless method for the sequence-specific detection of DNA. *Proc Natl Acad Sci USA* 2003;**100**:9134–7.
27. Immoos CE, Lee SJ, Grinstaff MW. DNA–PEG–DNA triblock macromolecules for reagentless DNA detection. *J Am Chem Soc* 2004;**126**:10814–15.
28. Anne A, Bouchardon A, Moiroux J. 3'-Ferrocene-labeled oligonucleotide chains end-tethered to gold electrode surface: Novel model systems for exploring flexibility of short DNA using cyclic voltammetry. *J Am Chem Soc* 2003;**125**:1112–13.
29. Inouye M, Ikeda R, Takase M, Tsuri T, Chiba J. Single-nucleotide polymorphism detection with "wire-like" DNA probes that display quasi "on-off" digital action. *Proc Natl Acad Sci USA* 2005;**102**:11606–10.
30. Ikeda R, Kobayashi S, Chiba J, Inouye M. Detection of mismatched duplexes by synchronizing the pulse potential frequency with the dynamics of ferrocene/isoquinoline conjugate-connected DNA probes immobilized onto electrodes. *Chem Eur J* 2009;**15**:4822–8.
31. Chiba J, Akaishi A, Ikeda R, Inouye M. Electrochemical detection of insertion/deletion mutation based on enhanced flexibility of bulge-containing duplexes on electrodes. *Chem Commun* 2010;**46**:7563–5.

32. Lereau M, Fournier-Wirth C, Mayen J, Farre C, Meyer A, Dugas V, et al. Development of innovative and versatile polythiol probes for use on ELOSA or electrochemical biosensors: Application in Hepatitis C virus genotyping. *Anal Chem* 2013;**85**:9204–12.
33. Anne A, Demaille C. Electron transport by molecular motion of redox-DNA strands: Unexpectedly slow rotational dynamics of 20-mer ds-DNA chains end-grafted onto surfaces via C_6 linkers. *J Am Chem Soc* 2008;**130**:9812–23.
34. Jenkins DM, Chami B, Kreuzer M, Presting G, Alvarez AM, Liaw Y. Hybridization probe for femtomolar quantification of selected nucleic acid sequences on a disposable electrode. *Anal Chem* 2006;**78**:2314–18.
35. Kang D, Zuo X, Yang R, Xia F, Plaxco KW, White RJ. Comparing the properties of electrochemical-based DNA sensors employing different redox tags. *Anal Chem* 2009;**81**:9109–13.
36. Yang W, Lai RY. A dual-signalling electrochemical DNA sensor based on target hybridization-induced change in DNA probe flexibility. *Chem Commun* 2012;**48**:8703–5.
37. Miao X, Guo X, Xiao Z, Ling L. Electrochemical molecular beacon biosensor for sequence-specific recognition of double-strandedDNA. *Biosens Bioelectron* 2014;**59**:54–7.
38. Ge B, Huang YC, Sen D, Yu H-Z. A robust electronic switch made of immobilized duplex/quadruplex DNA. *Angew Chem Int Ed* 2010;**49**:9965–7.
39. Kuzuya A, Ohnishi T, Wasano T, Nagaoka S, Sumaoka J, Ihara T, et al. Efficient guest inclusion by β-cyclodextrin attached to the ends of DNA oligomers upon hybridization to various DNA conjugates. *Bioconjugate Chem* 2009;**20**:1643–9.
40. Ihara T, Uemura A, Futamura A, Shimizu M, Baba N, Nishizawa S, et al. Cooperative DNA probing using a β-cyclodextrin–DNA conjugate and a nucleobase-specific fluorescent ligand. *J Am Chem Soc* 2009;**131**:1386–7.
41. Futamura A, Uemura A, Imoto T, Kitamura Y, Matsuura H, Wang C-X, et al. Rational design for cooperative recognition of specific nucleobases using β-cyclodextrin-modified DNAs and fluorescent ligands on DNA and RNA scaffolds. *Chem Eur J* 2013;**19**:10526–35.
42. Ihara T, Wasano T, Nakatake R, Arslan P, Futamura A, Jyo A. Electrochemical signal modulation in homogeneous solution using the formation of an inclusion complex between ferrocene and β-cyclodextrin on a DNA scaffold. *Chem Commun* 2011;**47**:12388–90.
43. Aoki H, Kitajima A, Tao H. Label-free and 'signal-on' DNA detection using a probe DNA terminated with ferrocene and β-cyclodextrin. *Supramol Chem* 2010;**22**:455–60.
44. Luo X, Lee TM-H, Hsing I-M. Immobilization-free sequence-specific electrochemical detection of DNA using ferrocene-labeled peptide nucleic acid. *Anal Chem* 2008;**80**:7341–6.
45. Sato S, Nojima T, Takenaka S. Electrochemical gene detection based on supramolecular complex formation by ferrocenyl–β-cyclodextrin and adamantylnaphthalene diimide bound to double strandedDNA. *J Organometal Chem* 2004;**689**:4722–8.
46. Watanabe S, Sato S, Ohtsuka K, Takenaka S. Electrochemical DNA analysis with a supramolecular assembly of naphthalene diimide, ferrocene, and β-cyclodextrin. *Anal Chem* 2011;**83**:7290–6.
47. Haraguchi K, Sato S, Habu M, Yada N, Hayakawa M, Takahashi O, et al. Oral cancer screening based on methylation frequency detection in hTERT gene using electrochemical hybridization assay via a multi-electrode chip coupled with ferrocenylnaphthalene diimide. *Electroanalysis* 2017;**29**:1596–601.
48. Islam MdM, Sato S, Shinozaki S, Takenaka S. Cyclic ferrocenylnaphthalene diimide derivative as a new class of G-quadruplex DNA binding ligand. *Bioorg Med Chem Lett* 2017;**27**:329–35.

49. Morita K, Sato Y, Seino T, Nishizawa S, Teramae N. Fluorescent and electrochemical detection of pyrimidine/purine transversion by ferrocenyl aminonaphthyridine derivative. *Chem Commun* 2008;**6**:266–8.
50. Shiddiky MJA, Torriero AAJ, Zeng Z, Spiccia L, Bond AM. Highly selective and sensitive DNA assay based on electrocatalytic oxidation of ferrocene bearing Zinc(II)cyclen complexes with diethylamine. *J Am Chem Soc* 2010;**132**:10053–63.
51. Cao W, Ferrance JP, Demas J, Landers JP. Quenching of the electrochemiluminescence of tris(2,2′-bipyridine)ruthenium(II) by ferrocene and its potential application to quantitative DNA detection. *J Am Chem Soc* 2006;**128**:7572–8.
52. Wang X, Yun W, Dong P, Zhou J, He P, Fang Y. A controllable solid-state $Ru(bpy)_3^{2+}$ electrochemiluminescence film based on conformation change of ferrocene-labeled DNA molecular beacon. *Langmuir* 2008;**24**:2200–5.
53. Kitamura Y, Miyahata T, Matsuura H, Hatakeyama K, Taniguchi T, Koinuma M, et al. Graphene oxide-based amplified fluorescence sensor for nucleic acid detection through target-catalyzed hairpin assembly. *Chem Lett* 2015;**44**:1353–5.
54. Li F, Yu Y, Li Q, Zhou M, Cui H. A homogeneous signal-on strategy for the detection of *rpoB* genes of mycobacterium tuberculosis based on electrochemiluminescent graphene oxide and ferrocene quenching. *Anal Chem* 2014;**86**:1608–13.
55. Liu G, Wang J, Wunschel DS, Lin Y. Electrochemical proteolytic beacon for detection of matrix metalloproteinase activities. *J Am Chem Soc* 2006;**128**:12382–3.
56. Xiao H, Liu L, Meng F, Huang J, Li G. Electrochemical approach to detect apoptosis. *Anal Chem* 2008;**80**:5272–5.
57. Martic S, Gabriel M, Turowec JP, Litchfield DW, Kraatz H-B. Versatile strategy for biochemical, electrochemical and immunoassay detection of protein phosphorylations. *J Am Chem Soc* 2012;**134**:17036–45.
58. Zhang B, Liu B, Tang D, Niessner R, Chen G, Knopp D. DNA-based hybridization chain reaction for amplified bioelectronics signal and ultrasensitive detection of proteins. *Anal Chem* 2012;**84**:5392–9.
59. Floch FL, Ho HA, Leclerc M. Label-free electrochemical detection of protein based on a ferrocene-bearing cationic polythiophene and aptamer. *Anal Chem* 2006;**78**:4727–31.
60. Zhang Y-L, Huang Y, Jiang J-H, Shen G-L, Yu R-Q. Electrochemical aptasensor based on proximity-dependent surface hybridization assay for single-step, reusable, sensitive protein detection. *J Am Chem Soc* 2007;**129**:15448–9.
61. Thomas JM, Chakraborty B, Sen D, Yu H-Z. Analyte-driven switching of DNA charge transport: De Novo creation of electronic sensors for an early lung cancer biomarker. *J Am Chem Soc* 2012;**134**:13823–33.
62. Lu Y, Zhang L, Yu P, Su L, Mao L. Aptamer-based electrochemical sensors with aptamer-complementary DNA oligonucleotides as probe. *Anal Chem* 2008;**80**:1883–90.
63. Ren K, Wu J, Yan F, Ju H. Ratiometric electrochemical proximity assay for sensitive one-step protein detection. *Sci Rep* 2014;**4**:4360.
64. Mahmoud KA, Luong HT. Impedance method for detecting HIV-1 protease and screening for its inhibitors using ferrocene-peptide conjugate/Au nanoparticle/single-walled carbon nanotube modified electrode. *Anal Chem* 2008;**80**:7056–62.
65. Liu G, Paddon-Row MN, Gooding JJ. Protein modulation of electrochemical signals: Application to immunobiosensing. *Chem Commun* 2008;3870–2.
66. Li T, Liu L, Li Y, Xie J, Wu H-C. A universal strategy for aptamer-based nanopore sensing through host-guest interactions inside α-hemolysin. *Angew Chem Int Ed* 2015;**54**: 7568–71.

67. Liu J, Zhou H, Xu J-J, Chen H-Y. An effective DNA-based electrochemical switch for reagentless detection of living cells. *Chem Commun* 2011;**47**:4388–90.
68. Zuo X, Song S, Zhang J, Pan D, Wang L, Fan C. A target-responsive electrochemical aptamer switch (TREAS) for reagentless detection of nanomolar ATP. *J Am Chem Soc* 2007;**129**:1042–3.
69. Wu Z-S, Guo M-M, Zhang S-B, Chen C-R, Jiang J-H, Shen G-L, et al. Reusable electrochemical sensing platform for highly sensitive detection of small molecules based on structure-switching signaling aptamer. *Anal Chem* 2007;**79**:2933–9.
70. Du Y, Chen C, Yin J, Li B, Zhou M, Dong S, et al. Solid-state probe based electrochemical aptasensor for cocaine: A potentially convenient, sensitive, repeatable, and integrated sensing platform for drugs. *Anal Chem* 2010;**82**:1556–63.
71. Ferapontova EE, Olsen EM, Gothelf KV. An RNA aptamer-based electrochemical biosensor for detection of theophylline in serum. *J Am Chem Soc* 2008;**130**:4256–8.
72. Feng L, Lyu Z, Offenhäusser A, Mayer D. Multi-level logic gate operation based on amplified aptasensor performance. *Angew Chem Int Ed* 2015;**54**:7693–7.
73. Liang J, Chen Z, Guo L, Li L. Electrochemical sensing of L-histidine based on structure-switching DNAzymes and gold nanoparticle-graphene nanosheet composites. *Chem Commun* 2011;**47**:5476–8.
74. Radi A-E, O'Sullivan CK. Aptamer conformational switch as sensitive electrochemical biosensor for potassium ion recognition. *Chem Commun* 2006;3432–4.
75. Liu S-J, Nie H-G, Jiang J-H, Shen G-L, Yu R-Q. Electrochemical sensor for mercury(II) based on conformational switch mediated by interstrand cooperative coordination. *Anal. Chem* 2009;**81**:5724–30.
76. Arimori S, Ushiroda S, Peter LM, Jenkins ATA, James TD. A modular electrochemical sensor for saccharides. *Chem Commun* 2002;2368–9.
77. Miyaji H, Gasser G, Green SJ, Molard Y, Strawbridge SM, Tucker JHR. Selective electrochemical sensing of acidic organic molecules via a novel guest-to-host proton transfer reaction. *Chem Commun* 2005;5355–7.
78. Tárraga A, Molina P, López JL, Velasco MD. Modulation of the metal recognition properties of a new type of azaferrocenophane-based chemosensor: Co-ordination chemistry towards Mg^{2+}, Ca^{2+}, Zn^{2+}, and Ni^{2+}. *Dalton Trans* 2004;1159–65.
79. Caballero A, Tárraga A, Velasco MD, Espinosa A, Molina P. Multifunctional linear trifer-rocene derivatives linked by oxidizable bridge: Optical, electronic, and cation sensing properties. *Org Lett* 2005;**7**:3171–4.
80. Zapata F, Caballero A, Espinosa A, Tárraga A, Molina P. Triple channel sensing of Pb(II) ions by a simple multiresponsive ferrocene receptor having a 1-deazapurine backbone. *Org Lett* 2008;**19**:41–4.
81. Yang H, Zhou Z, Huang K, Yu M, Li F, Yi T, et al. Multisignaling optical electrochemical sensor for Hg^{2+} based on a rhodamine derivative with a ferrocene unit. *Org Lett* 2007;**9**:4729–32.
82. Huang K, Yang H, Zhou Z, Yu M, Li F, Gao X, et al. Multisignal chemosensor for Cr^{3+} and its application in bioimaging. *Org Lett* 2008;**10**:2557–60.
83. Romero T, Caballero A, Tárraga A, Molina P. A click-generated triazoles tethered ferrocene–pyrene dyad for dual-mode recognition of the pyrophosphate anion. *Org Lett* 2009;**11**:3466–9.
84. Zapata F, Caballero A, Espinosa A, Tárraga A, Molina P. Cation coordination induced modulation of the anion sensing properties of a ferrocene–imidazophenanthroline dyad: Multichannel recognition from phosphate-related to chloride anion. *J Org Chem* 2008;**73**:4034–44.

85. Beer PD, Davis JJ, Drillsma-Milgrom DA, Szemes F. Anion recognition and redox sensing amplification by self-assembled monolayers of 1,1′-bis(alkyl-N-amido)ferrocene. *Chem Commun* 2002;1716–17.

86. Peng Y, Jiang D, Su L, Zhang L, Yan M, Du J, et al. Mixed monolayers of ferrocenylalkanethiol and encapsulated horseradish peroxidase for sensitive and durable electrochemical detection of hydrogen peroxide. *Anal Chem* 2009;**81**:9985–92.

87. Mor-Piperberg G, Tel-Vered R, Elbaz J, Willner I. Nanoengineered electrically contacted enzymes on DNA scaffolds: Functional assemblies for the selective analysis of Hg^{2+} ions. *J Am Chem Soc* 2010;**132**:6878–9.

88. Wan P, Xing Y, Chen Y, Chi L, Zhang X. Host-guest chemistry at interface for photoswitchable bioelectrocatalysis. *Chem Commun* 2011;**47**:5994–6.

89. Ma D-L, Wong W-L, Chung W-H, Chan F-Y, So P-K, Lai T-S, et al. A highly selective luminescent switch-on probe for histidine/histidine-rich proteins and its application in protein staining. *Angew Chem Int Ed* 2008;**47**:3735–9.

90. Kitamura Y, Ihara T, Tsujimura Y, Tazaki M, Jyo A. DNA-templated cooperative formation of the luminous lanthanide complex and its analytical application to gene detection. *Chem Lett* 2005;**34**:1606–7.

91. Kitamura Y, Ihara T, Tsujimura Y, Osawa Y, Tazaki M, Jyo A. Colorimetric allele typing through cooperative binding of DNA probes carrying a metal chelator for luminescent lanthanide ions. *Anal Biochem* 2006;**359**:259–61.

92. Kitamura Y, Ihara T, Tsujimura Y, Osawa Y, Sasahara D, Yamamoto M, et al. Template-directed formation of luminescent lanthanide complexes: Versatile tools for colorimetric identification of single nucleotide polymorphism. *J Inorg Biochem* 2008;**102**:1921–31.

93. You Y, Cho S, Nam W. Cyclometalated iridium(III) complexes for phosphorescence sensing of biological metal ions. *Inorg Chem* 2013;**53**:1804–15.

94. Yu M, Zhao Q, Shi L, Li F, Zhou Z, Yang H, et al. Cationic iridium(III) complexes for phosphorescence staining in the cytoplasm of living cells. *Chem Commun* 2008;2115–17.

95. Liu S, Liang H, Zhang KY, Zhao Q, Zhou X, Xu W, et al. A multifunctional phosphorescent iridium(III) complex for specific nucleus staining and hypoxia monitoring. *Chem Commun* 2015;**51**:7943–6.

96. Jin C, Liu J, Chen Y, Guan R, Ouyang C, Zhu Y, et al. Cyclometalated iridium(III) complexes as AIE phosphorescent probes for real-time monitoring of mitophagy in living cells. *Sci Rep* 2016;**6**:22039.

97. Ma D-L, Wang M, He B, Yang C, Wang W, Leung C-H. A luminescent cocaine detection platform using a split G-quadruplex-selective iridium(III) complex and a three-way DNA junction architecture. *ACS Appl Mater Interfaces* 2015;**7**:19060–7.

98. Li C, Lin J, Guo Y, Zhang S. A novel electrochemiluminescent reagent of cyclometalated iridium complex-based DNA biosensor and its application in cancer cell detection. *Chem Commun* 2011;**47**:4442–4.

99. Park S, Sugiyama H. DNA-based hybrid catalysts for asymmetric organic synthesis. *Angew Chem Int Ed* 2010;**49**:3870–8.

100. Fournier P, Fiammengo R, Jäschke A. Allylic amination by a DNA-diene-iridium(I) hybrid catalyst. *Angew Chem Int Ed* 2009;**48**:4426–9.

101. Barton JK, Olmon ED, Sontz PA. Metal complexes for DNA-mediated charge transport. *Coord Chem Rev* 2011;**255**:619–34.

102. Elias B, Genereux JC, Barton JK. Ping-Pong electron transfer through DNA. *Angew Chem Int Ed* 2008;**47**:9067–70.

103. Shao F, Elias B, Lu W, Barton JK. Synthesis and characterization of iridium(III) cyclometalated complexes with oligonucleotides: Insight into redox reactions with DNA. *Inorg Chem* 2007;**46**:10187–99.
104. Shao F, Barton JK. Long-range electron and hole transport through DNA with tethered cyclometalated iridium(III) complexes. *J Am Chem Soc* 2007;**129**:14733–8.
105. Olmon ED, Hill MG, Barton JK. Using metal complex reduced states to monitor the oxidation of DNA. *Inorg Chem* 2011;**50**:12034–44.
106. Scharf P, Müller J. Nucleic acids with metal-mediated base pairs and their applications. *ChemPlusChem* 2013;**78**:20–34.
107. Ihara T, Ishii T, Araki N, Wilson AW, Jyo A. Silver ion unusually stabilizes the structure of a parallel-motif DNA triplex. *J Am Chem Soc* 2009;**131**:3826–7.
108. Urata S, Miyahata T, Matsuura H, Kitamura Y, Ihara T. Alteration of DNAzyme activity by silver ion. *Chem Lett* 2014;**43**:1020–2.
109. Ukale D, Shinde VS, Lönnberg T. 5-Marcuricytosine: An organometallic Janus nucleobase. *Chem Eur J* 2016;**22**:7917–23.
110. Franzini RM, Kool ET. Organometallic activation of a fluorogen for templated nucleic acid detection. *Org Lett* 2008;**10**:2935–8.
111. Ma D-L, Che C-M. A bifunctional platinum(II) complex capable of intercalation and hydrogen-bonding interactions with DNA: Binding studies and cytotoxicity. *Chem Eur J* 2003;**9**:6133–44.
112. Wang P, Leung C-H, Ma D-L, Sun RW-Y, Yan S-C, Chen Q-S, et al. Specific blocking of CREB/DNA binding by cyclometalated platinum(II) complexes. *Angew Chem Int Ed* 2011;**50**:2554–8.
113. Man BY-W, Chan DS-H, Yang H, Ang S-W, Yang F, Yan S-C, et al. A selective G-quadruplex-based luminescent switch-on probe for the detection of nanomolar silver(I) ions in aqueous solution. *Chem Commun* 2010;**46**:8534–6.
114. Koo C-K, Wong K-L, Man CW-Y, Tam H-L, Tsao S-W, Cheah K-W, et al. Two-photon plasma membrane imaging in live cells by an amphiphilic, water-soluble cyclometalated platinum(II) complex. *Inorg Chem* 2009;**48**:7501–3.
115. Maurer A, Kraatz H-B, Metzler-Nolte N. Synthesis and electrochemical characterization of metallocene–PNA oligomers. *Eur J Inorg Chem* 2005;3207–10.
116. Noor F, Wüstholz A, Kinscherf R, Metzler-Nolte N. A cobaltocenium-peptide bioconjugate shows enhanced cellular uptake and directed nuclear delivery. *Angew Chem Int Ed* 2005;**44**:2429–32.

Part III

Metalloprotein, Hydrogenase, and Vitamin B12

Chapter 15

Artificially Created Metalloenzyme Consisting of an Organometallic Complex Immobilized to a Protein Matrix

Daniel F. Sauer[1], Takashi Matsuo[2], Akira Onoda[3], Jun Okuda[1] and Takashi Hayashi[3]

[1]Institute of Inorganic Chemistry, RWTH Aachen University, Aachen, Germany, [2]Division of Materials Science, Graduate School of Science and Technology, Nara Institute of Science and Technology (NAIST), Ikoma, Nara, Japan, [3]Department of Applied Chemistry, Osaka University, Suita, Japan

15.1 ARTIFICIAL METALLOENZYMES

15.1.1 Anchoring Approaches

Natural metalloenzymes that contain one or several transition metal ions such as Fe, Cu, Zn, Ni, and Co within a protein matrix are well known. These metalloenzymes are capable of catalyzing various classes of important reactions in biosynthesis and metabolism. These metal ions mainly act as Lewis acids or as redox-active sites.[1] Furthermore, certain enzymes have been employed in both laboratory reactions and industrial-scale reactions to synthesize valuable organic compounds. In one notable example, nitrile hydratase, which includes a Co(III) ion in the reactive site, has been used for industrial production of acrylamide.[2]

On the other hand, metal complexes containing precious metals such as Ru, Rh, or Pd have quite often been utilized as catalysts in production of many different compounds and drug precursors. Several research groups have investigated modification of such metal complexes to optimize not only their individual catalytic reactivities, but also their stereo- and regio-selectivities and substrate specificity.

Individual proteins with highly ordered structures including α-helices and β-sheets consisting of L-amino acids can each provide a unique chiral scaffold, which can potentially enable a metal-dependent reaction site to promote

Advances in Bioorganometallic Chemistry. DOI: https://doi.org/10.1016/B978-0-12-814197-7.00015-7

molecular transformations with high stereo- and regio-selectivities and substrate specificity. Therefore, we have been engaged in efforts to combine reactive transition metals or transition metal complexes with proteins in construction of protein-based artificial metalloenzymes. This strategy can provide custom-designed biocatalysts to address the needs of individual research projects. By utilization of metals which are not included in nature's toolbox (e.g., the platinum metals) we can expect to provide useful biocatalysts to promote reactions different from those seen in nature.

Design of protein-based artificial metalloenzymes uses strategies including: (1) site-directed mutations to construct metal ion coordination sites provided by side chains of amino acid residues; and (2) incorporation of a synthetic metal complex into a protein scaffold. The latter is a particularly promising strategy for constructing biohybrid catalysts designed to achieve various catalytic reactions.[3] These chemically engineered proteins with unnatural metal complexes are capable of extending the reaction scope of metalloenzymes and broadening the substrate specificity.

An important aspect to consider in construction of a biohybrid catalyst is the method used to incorporate a metal complex into a protein scaffold. Fig. 15.1 shows three methods for incorporating a synthetic metal complex (cofactor) through supramolecular, dative, or covalent anchoring methods.[3c] Supramolecular anchoring is commonly performed by providing a robust interaction between a cofactor and a cofactor-binding protein. This system was pioneered by Wilson and Whitesides in 1978, who developed a biohybrid catalyst for promotion of asymmetric hydrogenation of α-acetamidoacrylic acid.[4] An achiral rhodium-phosphine complex equipped with a biotin moiety was supramolecularly attached to avidin (Avi). This synthetic metalloprotein achieved enantioselectivity of 44% *ee* without any additional engineering of the protein. The supramolecular interaction between the biotin and Avi is one of the strongest supramolecular interactions found in nature ($K_d \approx 10^{-14}$ M),[5] enabling this system to be conveniently applied in construction of artificial metalloproteins with a variety of metal complexes. Ward et al., reported on the development of artificial metalloenzymes using

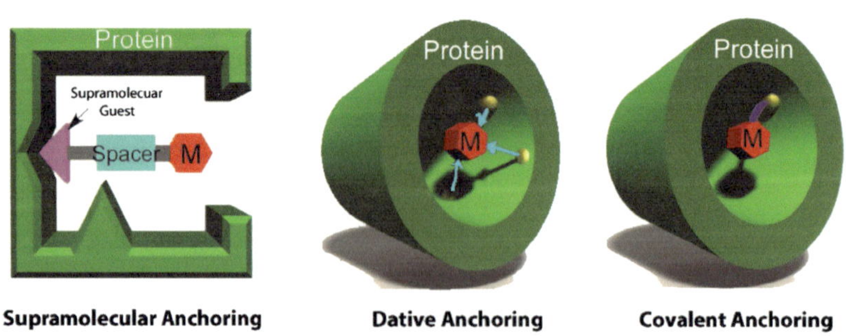

FIGURE 15.1 Illustration of three different anchoring methodologies.

this biotin-streptavidin (Sav) system in 2003 and increased the enantioselectivity of the reaction to 96% *ee* (*R*) with additional modifications of the protein.[6] The biotin-Sav platform has been further extended to other catalytic reactions.[7] In addition, artificial metalloproteins using other supramolecular interactions have been subsequently developed.[3a]

15.1.2 Artificial Metatheases

Formation and cleavage of carbon−carbon (C−C) bonds is one of the most important reaction classes in laboratory-scale organic chemical reactions and pharmaceutical processes, where several organometallic complexes including transition metal complexes have been employed as catalysts. This reaction class is rare in biological systems, with one exception being the alkyl group transfer reactions catalyzed by cobalamin-containing enzymes.[8] Thus, incorporation of a synthetic metal complex into a protein matrix to provide the reactivity required to mediate C−C bond formation/cleavage can be used to generate a biohybrid catalyst. In this context, olefin metathesis promoted by custom-designed biohybrid catalysts is expected to emerge as a challenging and useful strategy for providing C = C double bonds in lab-scale organic syntheses of important compounds such as pharmaceuticals, for example.[9]

Olefin metathesis, a key reaction promoting the rearrangement of olefins via C = C double bond formation/cleavage, requires a transition metal catalyst because of the high activation energy required for the [2 + 2] cycloaddition. The three types of olefin metathesis are shown in Scheme 15.1: (1) ring-opening metathesis polymerization (ROMP), (2) ring-closing metathesis (RCM), and (3) cross metathesis (CM). ROMP is usually promoted by the release of the ring-strain. RCM and CM are driven by the entropically-favored evolution of ethylene gas.

One of the best known and widely used catalysts is the Hoveyda-Grubbs (HG) type ruthenium carbene complex. In particular, the second generation

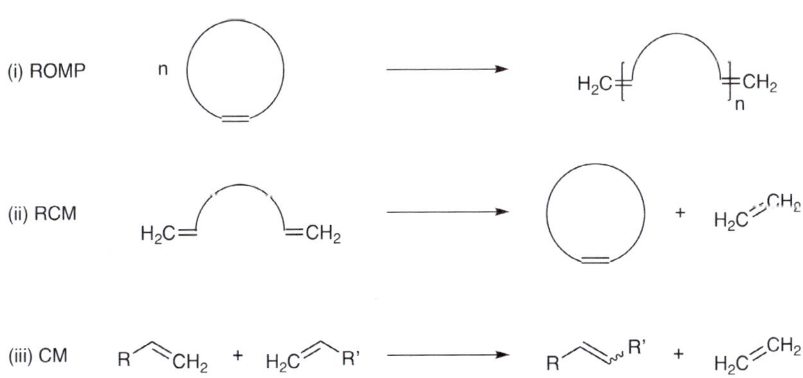

SCHEME 15.1 General types of olefin metathesis reactions.

HG catalyst which includes the *N*-heterocyclic carbene (NHC) ligand can work even in aqueous media with high catalytic activity. These findings have encouraged organometallic and inorganic chemists to anchor HG catalysts in protein matrices to control reactivity and stereoselectivity under mild reaction conditions in water, and incorporation of the HG catalysts into proteins to obtain biohybrid catalysts has been achieved during the past decade.[10] These biohybrid catalysts, which are known as "artificial metatheases", catalyze olefin metathesis reactions under mild conditions at neutral pH. Such olefin metathesis reactions are not catalyzed by any known natural enzymes. Therefore, this powerful C = C bond transformation can be considered as representing a "bioorthogonal reaction".

To date, all three catalyst anchoring methods shown in Fig. 15.1 have been used in creation of artificial metatheases. Supramolecular anchoring was performed by Ward et al.[11] An HG-type second generation metathesis catalyst was equipped with a biotin moiety at the periphery of an NHC ligand. The catalyst was incubated with Sav or Avi to generate the biohybrid conjugate and subsequently probed in the RCM reaction of *N,N*-diallyl-4-toluenesulfonamide in aqueous buffer solution. The best result achieved was a 95% conversion with Avi as a protein host with catalyst loading of 5 mol%.[11]

This combination using the HG-type catalyst and Avi was further developed in a system that performs RCM reactions within an intact whole-cell.[12] A mutant of the protein host Sav was expressed into the periplasm of *Escherichia coli* (*E. coli*) cells. The biotinylated catalyst was added and delivered to the Sav protein via diffusion through the outer membrane. As a test reaction, the RCM reaction of a water-soluble and charged diallylamine was performed with a per-cell turnover number (TON) $\approx$ 500,000. This is the first example of a metathesis biohybrid catalyst operating within an intact cell.[12]

In addition to supramolecular anchoring, Ward and coworkers established the dative anchoring approach for artificial metatheases using human carbonic anhydrase type II (hCAII) as a protein host. This metalloprotein contains a Zn^{2+} ion in the active site. Dative anchoring was therefore achieved by equipping the HG-type catalyst with an arylsulfonamide moiety, which is an inhibitor of hCAII that binds to the zinc center. In the RCM reaction of *N,N*-diallyl-4-toluenesulfonamide, this catalyst achieves a TON of 32, although it was stipulated that there may be some non-specific binding of the free catalyst on the protein surface.[13]

The covalent anchoring approach for the construction of artificial metatheases is the most prominent.[10] Hilvert and coworkers designed an artificial metathease based on *small heat shock protein 16.5* from *M. jannaschii* (MjHSP).[14] This protein forms a spherical capsid structure consisting of 24 subunits. The HG-type catalyst was covalently anchored through nucleophilic substitution at the bromoacetic acid unit attached to the catalyst. Cysteine C41 of MjHSP was quantitatively modified with the catalyst moiety. This artificial metathease catalyzes the RCM reaction of *N,N*-diallyl-4-toluenesulfonamide with the best activity being represented by a TON of 25 at pH = 2, although the capsid structure is disrupted under these reaction conditions.[14]

Klein Gebbink and coworkers demonstrated the utility of cutinase, a kind of lipases, as a protein host to accommodate the HG-type catalyst.[15] The HG-type catalyst was equipped with a phosphate-containing unit that binds to the active site and inhibits the intrinsic lipase activity. Nucleophilic attack of the Ser120-OH group of the lipase leads to formation of the covalent bond between the catalyst and the protein. The lipase containing the HG catalyst promotes the RCM reaction of *N,N*-diallyl-4-toluenesulfonamide in a 90% yield (catalyst loading: 5 mol%). Notably, a long spacer was required to expose the catalyst to the solution. Furthermore, this biohybrid catalyst was applied in the CM reaction of allylbenzene. The yield of the reaction with 5 mol% catalyst loading of the artificial metathease was 99%.[15]

15.2 β-BARREL PROTEINS AS SUPPORTS FOR ARTIFICIAL METALLOENZYMES

15.2.1 The Transmembrane Protein FhuA

A β-barrel motif offers a potentially robust second coordination sphere with respect to elevated temperatures, pH, and co-solvents.[16] An artificial metalloprotein designed to catalyze olefin metathesis was constructed using the transmembrane protein *Ferric Hydroxamate uptake protein component A* (FhuA) that is naturally located in the outer membrane of *E. coli*.[17] FhuA, an *E. coli* outer membrane protein with the function of an iron transporter, has a β-barrel structure consisting of 22 antiparallel β-strands. To generate sufficient space to harbor an organometallic catalyst, the cork domain responsible for mediating iron transport was deleted, yielding the FhuA Δ1-159 variant (amino acids from 1 to 159 are deleted relative to the wild-type protein). A cysteine residue was then introduced at position 545. We selected this position because computational studies suggest that this position is located in an area of the FhuA cavity which controls substrate flux. An artificial metathesis substrate is expected to approach this position within the cavity. This position has also been suggested to be a conformationally stable region surrounded by β-barrel structure. Furthermore, the N548V mutation was introduced to increase accessibility of the catalyst to position C545. In addition, E501 was replaced with Phe to avoid coordination of the Glu side chain to the metal catalyst which would deactivate the catalyst. Two specific TEV (Tobacco Etch Virus protease) cleavage sites were incorporated into loops 7 and 8 to allow the fragment containing the HG catalyst to be characterized by MALDI-TOF MS analysis, confirming the desired structure of the artificial metathease (FhuA Δ1-159_C545_V548_F501_tev; FhuA ΔCVF^{tev}) (Fig. 15.2).[17a,17b]

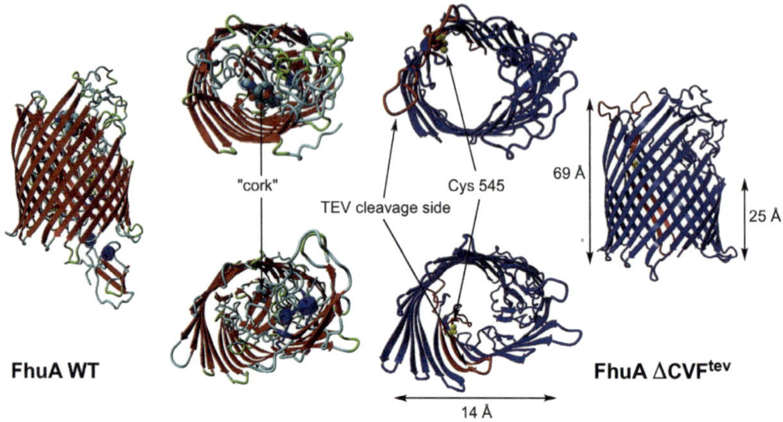

FIGURE 15.2 FhuA wildtype and FhuA ΔCVF^{tev}

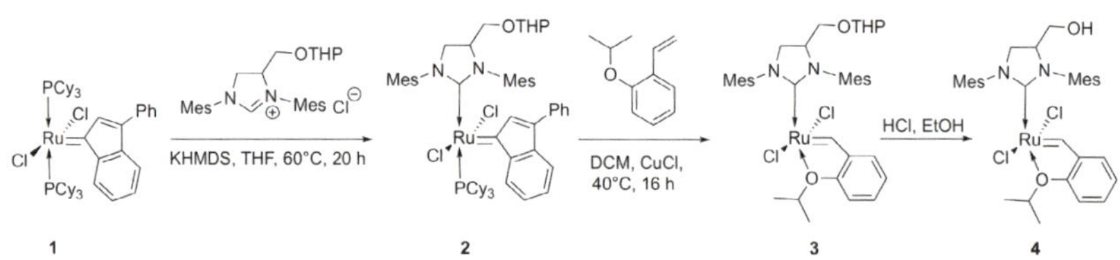

SCHEME 15.2 Synthesis of the HG-type catalyst **4**.

The cysteine residue is capable of covalently anchoring metal catalysts via an appropriate functional group such as maleimide.[18] To generate an HG-type catalyst **4** bearing a hydroxylmethyl group in the backbone of the NHC ligand, a straightforward synthetic scheme was applied (Scheme 15.2).[17a]

The main advantage of the second generation HG catalyst is its relatively high stability in the presence of air and water relative to the first generation Grubbs-type and Schrock-type catalysts.[19] Furthermore, the NHC ligand can be easily modified. The introduction of a hydroxyl function into the backbone of the NHC ligand provides the option to attach different anchoring units (Scheme 15.3).[17a,17c]

The hydroxyl group was used for esterification with acetyl halides containing either a bromoacetyl unit or a maleimide group with a spacer. Both anchoring groups allow covalent bond formation between the thiol group of the protein FhuA protein and the active site.[17a,17c] The bromoacetyl unit participates in a nucleophilic substitution mechanism and the maleimide group undergoes a thiol-ene reaction.[18] After the conjugation of the metal catalyst, the resulting metalloprotein was washed with dry THF to remove the unreacted catalyst.[17a,17c] The metalloproteins based on the membrane protein FhuA require detergents to be solubilized. Since the protein is partially unfolded in the presence of a detergent such as SDS, the detergent was

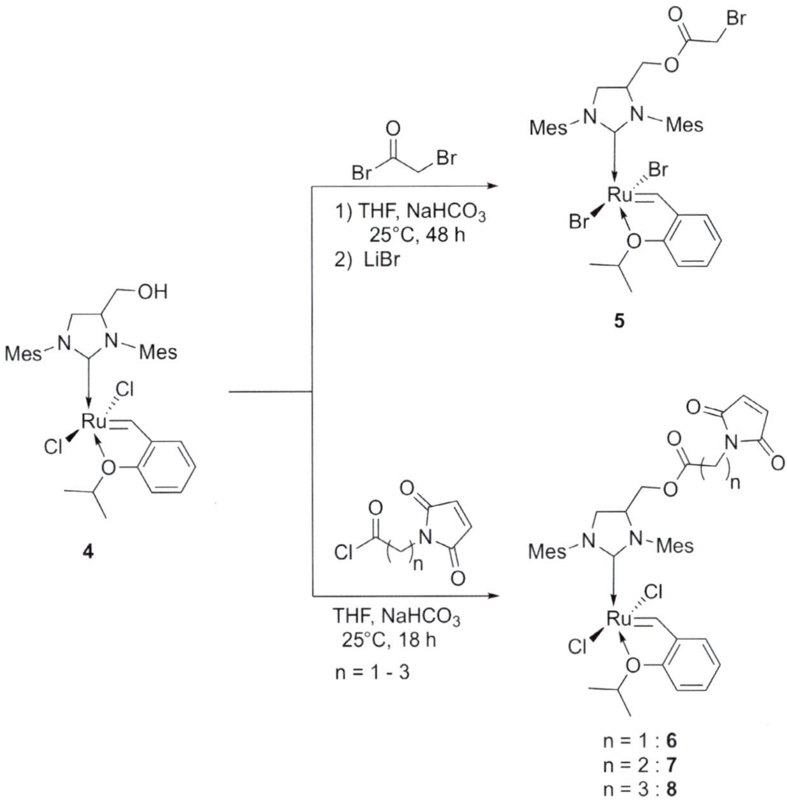

SCHEME 15.3 Hoveyda-Grubbs type catalysts with different anchoring units for attachment to FhuA.

removed by dialysis against PE-PEG to refold the protein as the natural β-barrel structure (Scheme 15.4).[17a,17c] Structural integrity was confirmed using CD spectroscopy and conjugation of the catalyst was determined in a fluorescence titration of the cysteine unit and by MALDI-TOF MS.

The partially unfolded biohybrid catalysts **6-8@FhuA*** and the refolded biohybrid catalysts **6-8@FhuA** demonstrate ROMP activity toward water-soluble ring-strained 7-oxanorbornene derivative **9**. However, the biohybrid conjugate of catalyst **5** with FhuA only has activity in the presence of SDS and is therefore not discussed (Table 15.1).[17a,17c]

The biohybrid catalysts **6-8@FhuA*** in SDS solution have the same activity as the protein-free catalyst **4** (Table 15.1, entries 1–4 compared with entries 5–8). After refolding, the activity decreases (Table 15.1, entries 9–12). This could be related to the steric demand of the renatured β-barrel structure. Additionally, the activity provided by the derivative with a shorter linker length, **6@FhuA** is higher (Table 15.1, entry 10 compared to entries 11–12). This is explained by a decrease in degrees of freedom of the

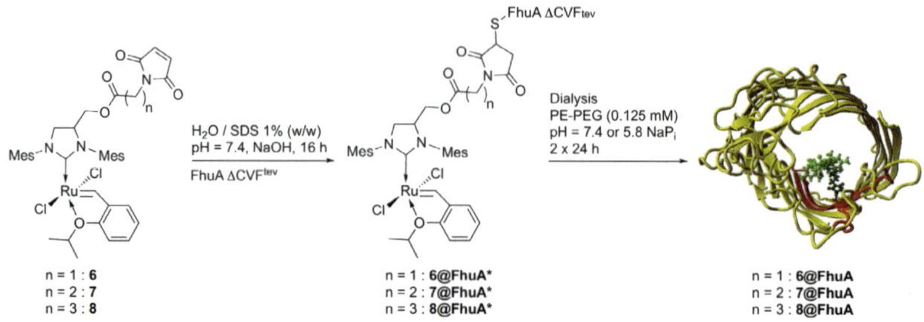

SCHEME 15.4 Anchoring and refolding of HG-type catalysts bearing the maleimide group to FhuA.

TABLE 15.1 ROMP of Substrate 9 Catalyzed by FhuA Based Biohybrid Catalysts

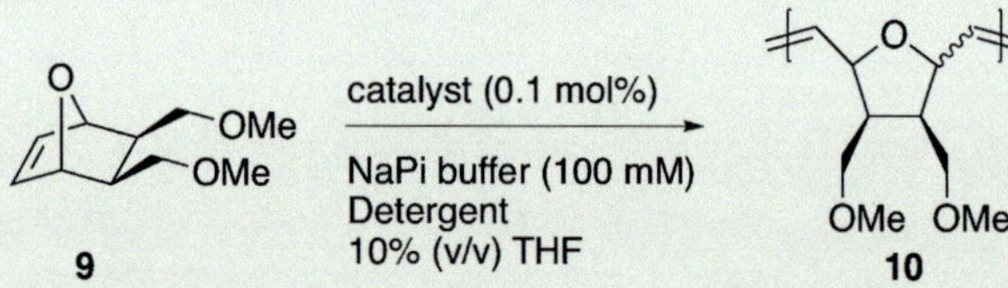

Entry[a]	Catalyst	Detergent	pH	Conv.[b] [%]	TON	*Cis/Trans*[b]
1	**4**	-	7.4	98	980	50/50
2	**4**	-	5.8	99	990	50/50
3	**4**	SDS[c]	5.8	99	990	60/40
4	**4**	PE-PEG[d]	5.8	99	990	50/50
5	**6@FhuA***	SDS[c]	7.4	90	900	60/40
6	**6@FhuA***	SDS[c]	5.8	99	990	61/39
7	**7@FhuA***	SDS[c]	5.8	99	990	60/40
8	**8@FhuA***	SDS[c]	5.8	99	990	60/40
9	**6@FhuA**	PE-PEG[d]	7.4	7	94	57/43
10	**6@FhuA**	PE-PEG[d]	5.8	41	555	58/42
11	**7@FhuA**	PE-PEG[d]	5.8	24	325	56/44
12	**8@FhuA**	PE-PEG[d]	5.8	37	365	56/44

[a]*Conditions: [**4**] = 0.1 mM which is determined with BCA assay and ThioGlo titration (approx. 0.09 mM)*
[b]*Determined by ^1H NMR spectroscopy in $CDCl_3$*
[c]*Containing 1% (w/w) SDS.*
[d]*[PE-PEG] = 0.125 mM.*

FIGURE 15.3 MD simulations of HG catalysts incorporated into FhuA. Left: Catalyst **8@FhuA**; Right: Catalyst **6@FhuA**. The numbers below the residues show their distances from the metal site in Å.

catalyst and, therefore, movement of the catalyst is restricted within the channel. Furthermore, the active metal center is moved away from potentially coordinating residues (Fig. 15.3).[17c]

15.2.2 Nitrobindin as a Protein Scaffold

Nitrobindin (NB) is a small β-barrel protein with a molecular weight of 19 kDa. Nitrobindin consists of 10 β-strands and naturally contains a heme unit as a prosthetic group.[20] After removal of the heme unit, the β-barrel structure is retained. This makes the protein a promising candidate to host a non-natural metal complex. Furthermore, nitrobindin provides a hydrophobic cavity. Previous examples reported by Hayashi et al.,[21] and Lewis et al.,[22] have demonstrated the suitability of nitrobindin as a protein scaffold for anchoring organometallic complexes. In the previous studies on reactions including phenylacetylene polymerization, the Diels-Alder reaction, hydrogen evolution, and epoxidation, the catalyst immobilized within the nitrobindin cavity was found to have increased activity compared to the protein-free catalyst under the same conditions.

An artificial metathease based on nitrobindin as a host[23] was synthesized using HG-type catalysts **6–8**. Covalent anchoring was performed within the cavity of two nitrobindin variants, which mainly differ in the size of the cavity volume. The so-called **NB4** mutant (mutations compared to wild-type: M75L/H76L/Q96C/M148L/H158L) has a cavity with a volume of 855 Å^3, whereas the **NB11** mutant (mutations compared to wild-type: M7AL/H76L/Q96C/M148L/H158A) has an increased cavity volume of 1161 Å^3. Only the larger cavity of **NB11** allows quantitative conjugation with the catalyst precursor bearing a longer maleimide linker. For the catalysts with shorter linkers, the coupling efficiency is dramatically reduced (Scheme 15.5).[23]

These biohybrid catalysts were characterized after isolation by various methods. CD spectroscopy confirmed their structural integrity. Metal content was determined by ICP-OES and revealed that only one metal per protein

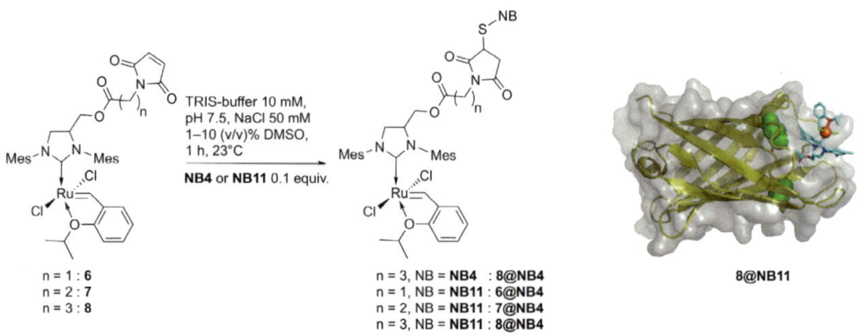

SCHEME 15.5 Coupling of HG-type catalysts **6–8** to **NB4** or **NB11**. Right figure: Illustration of the biohybrid conjugate **8@NB11**, where the red sphere represents the Ru atom of the HB catalyst.

TABLE 15.2 RCM with Nitrobindin-Based Metathease

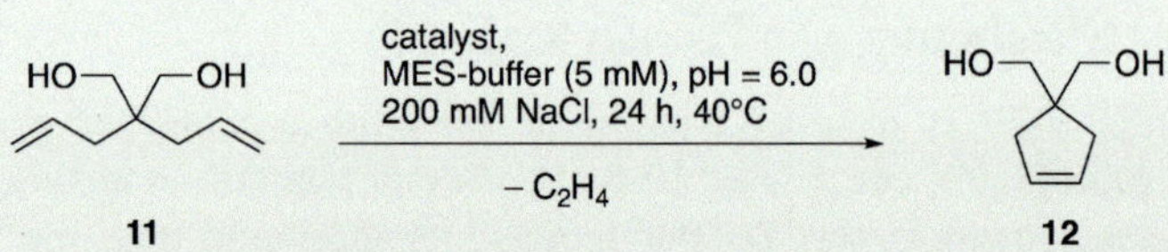

Entry	Catalyst [mol%]	Conv.[a] [%]	TON
1	**7@NB11** [1.0]	89	89
2	**8@NB11** [1.0]	>99	100
3	**8@NB11** [0.33]	57	178

[a]*Determined by* ^{1}*H NMR spectroscopy and GC analysis.*

molecule is present. In the UV-vis spectra, an additional absorption band for the HG-type catalyst was detected. ESI-MS showed the mass of the biohybrid conjugate, confirming the presence of covalent anchoring.[23]

These biohybrid catalysts were found to catalyze RCM and ROMP of water-soluble substrates. RCM reacts with 2,2-diallylpropane-1,3-diol (**11**) to give corresponding cyclopentene **12** (Table 15.2).[23]

Catalysts **7@NB11** and **8@NB11** have high activity toward RCM of substrate **11**. With a catalyst loading of 1.0 mol%, **7@NB11** achieves 89% conversion. In addition, **8@NB11** converts substrate **11** to **12** quantitatively. With decreased catalyst loading of 0.33 mol% under mild conditions (pH = 6.0, T = 40°C) the turnover number increases to 178 (Table 15.2, entry 3).

The ROMP reaction was also performed with a 7-oxanorbornene derivative **9** (Table 15.3).[23] With a catalyst loading of only 0.01 mol% the protein-free catalyst **4** demonstrates no activity in aqueous buffer solutions (Table 15.3, entry 1). After immobilization of the HG type catalyst **8** in the hydrophobic cavity of **NB4**, the biohybrid catalyst **8@NB4** showed ROMP activity with a TON of 1100 (Table 15.3, entry 2). Increasing the cavity size enables the incorporation of catalysts **6** and **7** with shorter linkers and the resultant biohybrid catalysts **6@NB11** and **7@NB11** have low to moderate activity, respectively (Table 15.3, entries 3–4). With catalyst **8@NB11**, a TON of 9700 was observed (Table 15.3, entry 5). Despite the high activity in the ROMP reaction, the corresponding polymer with M_n = 180.000 g/mol has a very narrow molecular weight distribution (PDI = 1.05). However, a change in regioselectivity (*cis/trans*) or tacticity is not observed.[23]

The increased activity compared to the artificial metathease based on the FhuA protein is related to the hydrophobic cavity as mentioned previously. Comparing the two cavities, the difference becomes obvious. In the case of FhuA, there are many amino acid residues that can potentially coordinate to the metal site. In the case of nitrobindin, the cavity is completely nonpolar and, therefore, coordination of amino acid side chains to the metal site should not occur (Fig. 15.4).

TABLE 15.3 **ROMP of Substrate 9 with Nitrobindin-Based Artificial Metathease**

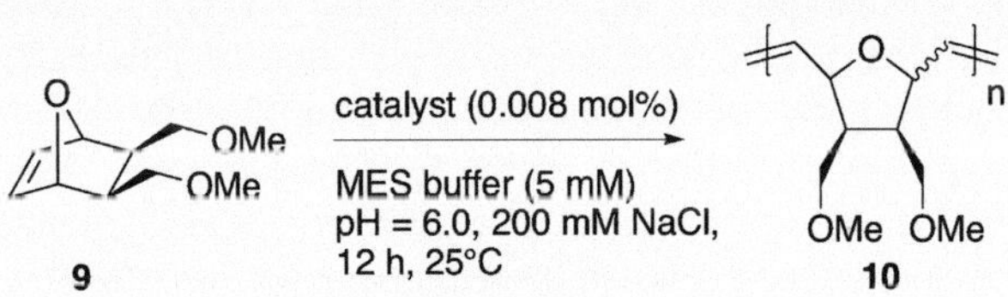

Entry	Catalyst	Conv.[a] [%]	TON	*cis/trans*[a]
1[b,c]	4	<5	n.d.	n.d.
2	8@NB4	10	1100	40/60
3	6@NB11	<5	n.d.	n.d.
4	7@NB11	18	2000	43/57
5	8@NB11	78	9700	43/57

[a]*Determined by* ^{1}*H NMR spectroscopy in* $CDCl_3$
[b]*Containing 10% (v/v) THF.*
[c]*Catalyst loading: 0.01 mol%.*

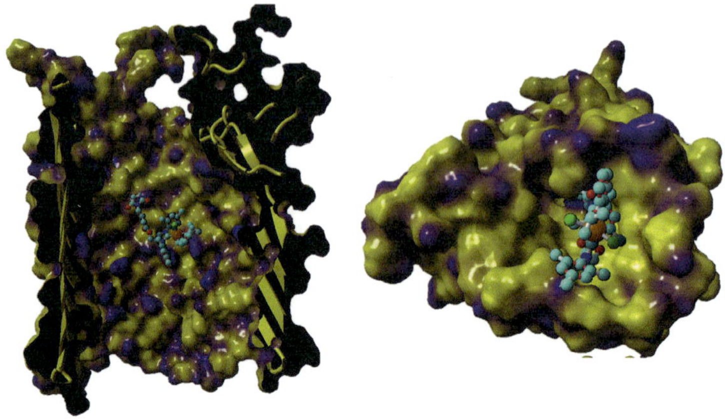

FIGURE 15.4 Left: Cross-section of FhuA containing **8**; Right: Nitrobindin based catalyst **8@NB11**. The Ru metal is shown as orange sphere. Yellow hydrophobic residues are colored in yellow, hydrophilic residues are colored in purple.

15.2.3 Conclusion on β-Barrel Proteins as Hosts for Artificial Metatheases

We demonstrated successful construction of artificial metalloproteins with two types of β-barrel proteins with olefin metathesis activity. Compared to other protein types, the β-barrel motif offers a robust and well-defined second coordination sphere. Several co-solvents can be conveniently applied and, therefore, the range of applicable substrates can be expanded. Furthermore, the channel structure provided by β-barrel proteins is an interesting feature, immobilizing the organometallic catalyst completely in the protein, even though specific advantages provided by this feature have not yet been identified. The combination of the β-barrel motif with the covalent anchoring approach provides the option to mutate several amino acid residues without denaturing the protein structure. This would provide a suitable non-coordinating protein environment to enhance the metathesis reaction within a biological system.

15.3 APPLICATION OF A PROTEIN CLEFT FOR OLEFIN METATHESIS WITHIN A PROTEIN MATRIX

15.3.1 Utility of a Protein Cleft

In the molecular design of metatheases based on covalent interactions between a protein scaffold and a metathesis center, the first concern is how to conjugate a Ru complex to a host protein. As demonstrated in Section 15.2, β-Barrel Proteins as Supports for Artificial Metalloenzymes, SN_2-type reactions of nucleophilically accessible functional groups, such as halogenoacetamide and maleimide toward reduced cysteine residues (−SH) is a popular and versatile method for introducing a synthetic molecule with a metal-based center. However, native proteins do not always have appropriate reduced cysteine residues. Even if reduced cysteine residues are introduced by protein engineering, host proteins which include several S−S linkages in their structures sometimes cannot be prepared with desired structures because protein misfolding may occur during the protein purification process. Although amino acid residues with moderate reactivities of side-chains such as His, Tyr, and Ser could provide appropriate conjugation points, the regio-selectivity of conjugation will often represent a significant problem. In this case, several modes of interaction (e.g., supramolecular interactions) are required in addition to the reactivities of functional groups at the conjugation point to enhance the chemo-selectivity of the protein modifications.

Another concern in the molecular design of metatheases is the location of the reaction center within a host protein. Since a catalytic process includes the three steps of (1) substrate access, (2) a chemical event at a reaction center, and (3) product release, we should optimize smooth entry of a substrate and release of a product as well as the intrinsic reactivity of a metal reaction center in the design of a catalyst. Biocatalysts based on protein scaffolds can provide various structural features such as chiral environments and highly charged surfaces. To fully utilize these characteristics, the selection of a location for a metal reaction center within a large macromolecule is an important aspect of designing molecular catalysts. In this context, the utilization of cleft within protein structures is expected to be useful in the design of biohybrid catalysts because the protein clefts tend to be relatively close to the protein structures but near protein surface, where both protein structural effects (e.g., steric hindrance and chiral environments) and bulk solvent effects (e.g., polarity) are expected.

As an example of metatheases with a metal reaction center in the protein cleft created by the combination of supramolecular and covalent interactions, this chapter will focus on a HG complex-containing protein with an α-chymotrypsin scaffold.[24]

15.3.2 Metathease with an α-Chymotrypsin Scaffold

15.3.2.1 Molecular Design and Construction

α-chymotrypsin from bovine pancreas, a commercially-available serine protease (*ca.* 25 kDa) from conventional vendors, has a protease catalytic site composed of Ser195, His57 and Asp102 in an arrangement known as a "catalytic triad".[25] The protein accommodates a peptidyl substrate in a cleft—indicated by an orange line in Fig. 15.5(A)—and recognizes the hydrophobic side-chains of Phe, Tyr, and Trp in the peptidyl substrate in a hydrophobic pocket—S1 pocket in Fig. 15.5(A)—at the catalytic site.[26] Other than the S1

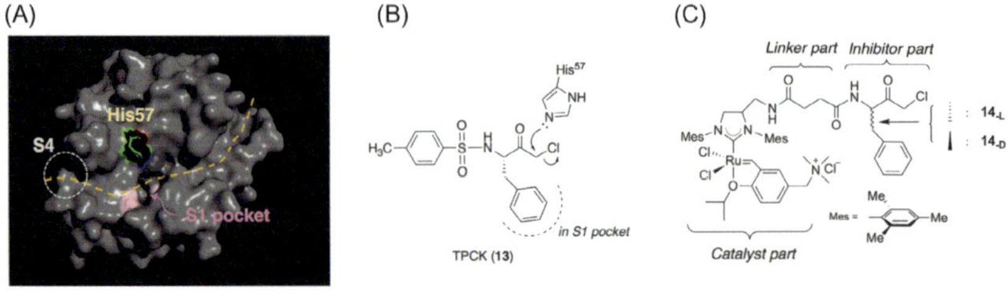

FIGURE 15.5 Structures of α-chymotrypsin ((A), PDB: 4CHA), TPCK (B) and compounds **14**$_{-L}$ and **14**$_{-D}$ (C). The protein cleft is indicated by an orange dotted line in figure (A). The side-chain of the L-phenylalanine moiety in TPCK is accommodated within the S1 pocket, inducing the nucleophilic attack of the imidazole of His57 to the chloromethyl ketone group in (B).

hydrophobic pocket, α-chymotrypsin has several sites (S2–S4 pockets) to accommodate amino acid residue units in a substrate molecule, and the S4 pocket is located at the entrance to the cleft.[27]

The protease activity of α-chymotrypsin is completely inhibited by *N*-tosyl-L-phenylalanyl chloromethyl ketone—TPCK, **13** in Fig. 15.5(B).[28] This is because the protein accommodates the side-chain of L-phenylalanine moiety within the S1 pocket, and the chloromethyl ketone group is forced toward His57, resulting in blockage of the histidine imidazole group—Fig. 15.5(B). According to these characteristics, an HG-type complex with a serine protease inhibitor moiety **14**$_{-L}$ and its stereoisomer, **14**$_{-D}$—Fig. 15.5 (C)—were designed. These compounds each contain three functional structures: (1) an inhibitor moiety, (2) a linker, and (3) a catalytic site. The catalytic site is prepared using the synthetic method for a Grubbs's water-soluble catalyst.[29] The length of the linker was determined based on the distance between the S1 and S4 pockets so that the catalytic center is situated in the vicinity of the S4 pocket. Furthermore, the His residue is used as a conjugate point because native α-chymotrypsin has five disulfide bonds in its structure and introduction of an additional Cys residue by protein engineering may not be reliable. Therefore, the construction of a metathease with compound **14** harnesses the intrinsic molecular recognition of a host protein in a combination of supramolecular interactions and chemical reactivity of functional groups at the conjugation site.

Scheme 15.6 demonstrates the synthetic route used in preparing of compound **14**$_{-L}$. The inhibitor moiety was synthesized using the scheme described for synthesis of an inhibitor for HIV-1 protease or renin.[30] This synthetic route avoids synthesis of the potentially dangerous diazomethane intermediate compounds employed in conventional syntheses of serine protease inhibitors. Another synthetic route subsequently reported for the chloromethyl ketone moiety may also be available.[31] The terminal carboxylic acid in the terminal of compound **19** was tethered to the amino group on the NHC ligand of HG-type complex **20** using a mixed anhydride method. Typical

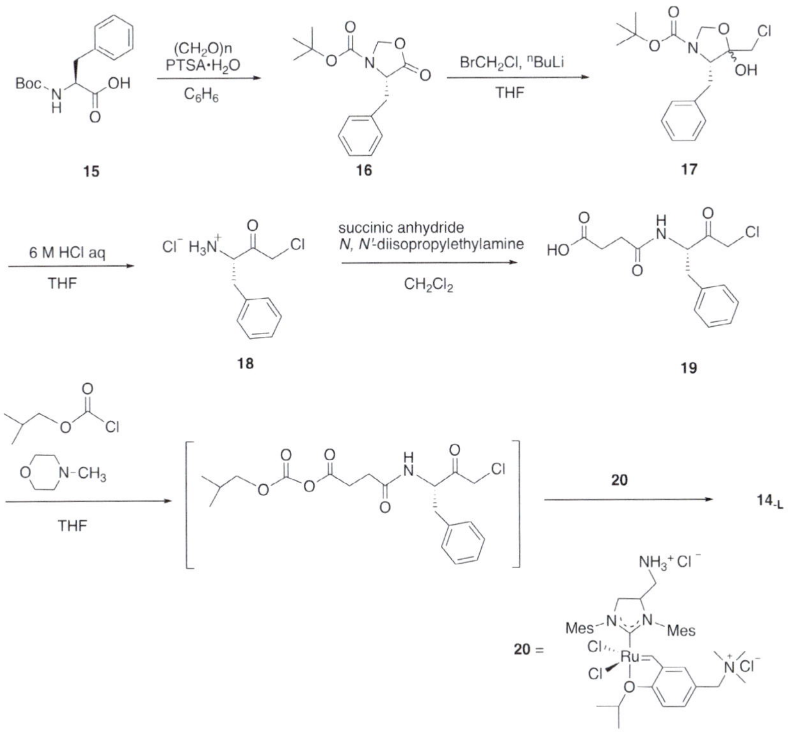

SCHEME 15.6 Synthesis of a serine protein inhibitor **14$_{\text{-L}}$** with an HG complex moiety **20**.

condensation reagents used for the amide bond formation (e.g., DCC, EDC, DMT-MM, etc.) cannot be used because they have a negative effect on the Ru complex moiety.

The introduction of **14$_{\text{-L}}$** into α-chymotrypsin can be attained by incubation of the protein and two equivalents of **14$_{\text{-L}}$** in HEPES buffer (pH = 7.0) in the presence of 100 mM KCl. After removal of excess **14$_{\text{-L}}$** by cation exchange column chromatography, the conjugation of **14$_{\text{-L}}$** is confirmable by ESI-MS and MALDI-MS. The treatment of α-chymotrypsin with **14$_{\text{-L}}$** abolishes its intrinsic protease activity, indicating that the His57 imidazole is regioselectively blocked by **14$_{\text{-L}}$**. Interestingly, the D-isomer **14$_{\text{-D}}$** (a stereoisomer of the inhibitor moiety) does not significantly affect the protease activity. A series of experimental results using **14$_{\text{-L}}$** and **14$_{\text{-D}}$** indicates that the host protein is capable of intrinsic molecular recognition.

The next concern is whether the catalyst moiety is subjected to position under protein structural effects. The CD spectrum of chymotrypsin with **14$_{\text{-L}}$** shows induced CD signals at 300–500 nm (Fig. 15.6), representing the MLCT region of the Ru complex center.[32] The appearance of induced CD signals is a good indication of conformational restriction and/or torsion at

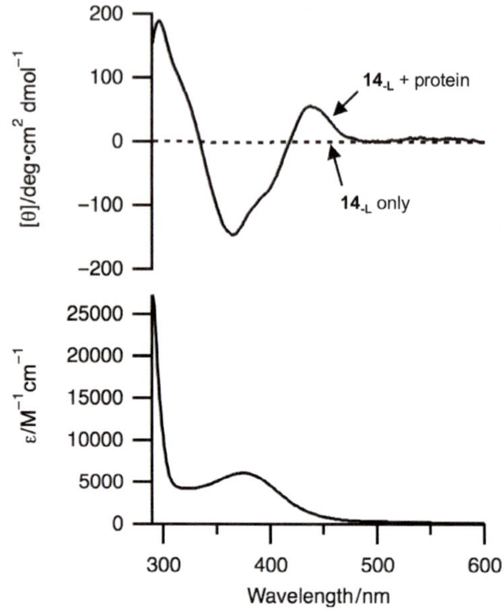

FIGURE 15.6 UV-vis (lower) and CD (upper) spectra of **14**-L-bound α-chymotrypsin. Without the protein, $\mathbf{14}_{\text{-L}}$ does not have CD signals in the expected region, proving that the appearance of induced CD signals in $\mathbf{14}_{\text{-L}}$-bound α-chymotrypsin originate from interactions between the Ru center and the protein core.

the Ru center induced by the protein scaffold. This observation indicates that the Ru center is located in the protein cleft and not outside of the protein core.

15.3.2.2 Ring-Closing Metathesis (RCM) Activity of the Chymotrypsin-Based Metathease

The chymotrypsin-based metathease has the higher RCM reactivity (Table 15.4) toward the glucose-pendant diallylamine compound **21** (water-soluble and electrostatically neutral) rather than positively charged diallylamine **22** and N-tosylated diallylamine (**23**, hydrophobic). The preference in substrates is ascribed to the charge state of the protein surface: The isoelectric point (p*I*) of α-chymotrypsin is ~8.5.[26] Because the protein surface is positively charged under the reaction conditions in the presence of 100 mM KCl, the cationic substrate **22** is blocked from approaching the Ru center by electrostatic repulsion. The reduction in catalytic activity for the hydrophobic substrate **23** is attributed to preferential hydration in the vicinity of the protein surface,[33] where a tight water molecule network formed around a highly charged biomacromolecule in the presence of an organic co-solvent may partially inhibit the access of the hydrophobic substrate to the Ru center within the protein matrix.

TABLE 15.4 Ring-Closing Metathesis (RCM) Activities of 14-L-Bound α-Chymotrypsin

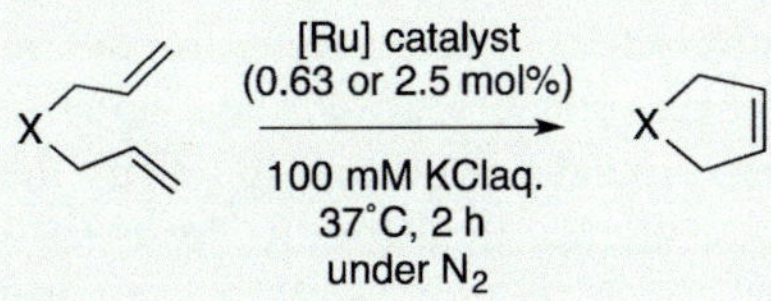

Entry	X =	[Ru] Catalyst	TON	Polarity of Substrate
1[a]	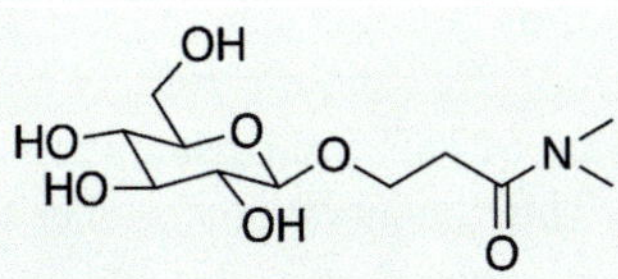	**14-L** in protein	20.3	hydrophilic, neutral
2[b,d]	**21**	**14-L** only	14.0	
3[b]	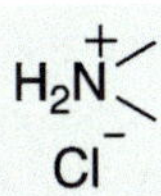	**14-L** in protein	N.R.[f]	hydrophilic, positively charged
4[b,e]	**22**	**14-L** only	< 2	
5[c,d]	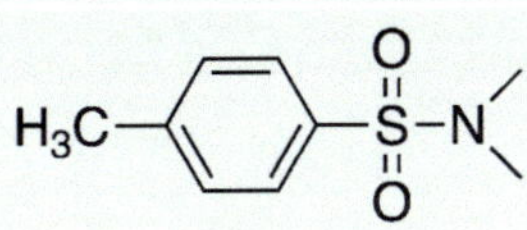	**14-L** in protein	4.1	hydrophobic
6[c,d]	**23**	**14-L** only	11.5	

[a] [Substrate] = 8.0 mM, [catalyst] = 50 μM (0.63 mol%); TON determined by ^1H NMR.
[b] [Substrate] = 8.0 mM, [catalyst] = 0.2 mM (2.5 mol%); TON determined by ^1H NMR.
[c] [Substrate] = 1.0 mM, [catalyst] = 25 μM (2.5 mol%); TON determined by HPLC.
[d] 10% DMSO (v/v) was added as co-solvent.
[e] 10% MeOH (v/v) was added as co-solvent.
[f] Product was not detected.

15.3.3 Importance of Medium Effects on RCM in Aqueous Solutions

Enzymatic reactions generally occur in aqueous media and, in most cases, additional components such as buffer and salt reagents are required in the reaction systems to maintain the pH of the solution and structures of host proteins. In metatheases with an organometallic ruthenium center, such additive reagents may cause the decrease in the reactivity of the Ru center. To conduct metathease-catalyzing reactions effectively, the effects of additives should be characterized.

The RCM activity of the chymotrypsin-based metathease was examined in unbuffered 100 mM KCl_{aq}. Furthermore, in the reaction mediated by metatheases with β-barrel protein scaffolds, the additive salts such as NaCl are required. It was found, interestingly, that the reactivity of the chymotrypsin-based metathease significantly decreases in HEPES or MOPS buffer solutions without KCl. In fact, the RCM activity of the water-soluble HG complex **20** significantly decreases in these buffer solutions.[34] Table 15.5 shows the bulk solvent effects on the catalytic activity of the chymotrypsin-based metathease.

As shown in Table 15.5, the addition of chloride salt (KCl) enhances the RCM activity (*cf.* entries 1 and 2). Even if the catalytic activity decreases in buffer solutions without KCl, the addition of KCl produces a recovery of the catalytic activity (*cf.* entries 3 and 4). This tendency proves that the effects on catalytic activity of the Ru catalyst in aqueous solutions involve a reversible process occurring at the Ru reaction center: The chloride anion inhibits deactivation of the Ru complex by ligand exchange with coordinating species which are present in aqueous solutions (such as H_2O, OH^-, nucleophilic buffer reagents, Scheme 15.7). Similar salt effects are not observed for nitrate salt, a weaker coordinative species (entry 5). Thus, it appears that

TABLE 15.5 Ring-Closing Metathesis (RCM) Activities of Ru Catalyst 20 in Aqueous Media

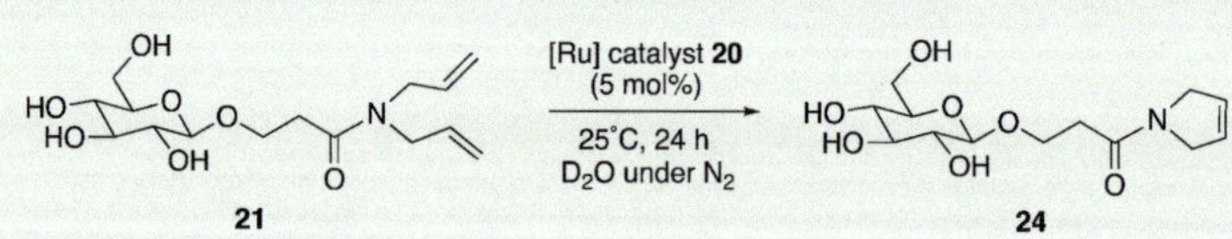

Entry	Additive	Yield(%)[a]
1	None	50
2	100 mM KCl	80
3	10 mM MES (pD = 6.4)	90
4	10 mM MES + 100 mM KCl (pD = 6.4)	40
5	100 mM KNO_3	47

[a]*Determined by ^1H NMR.*

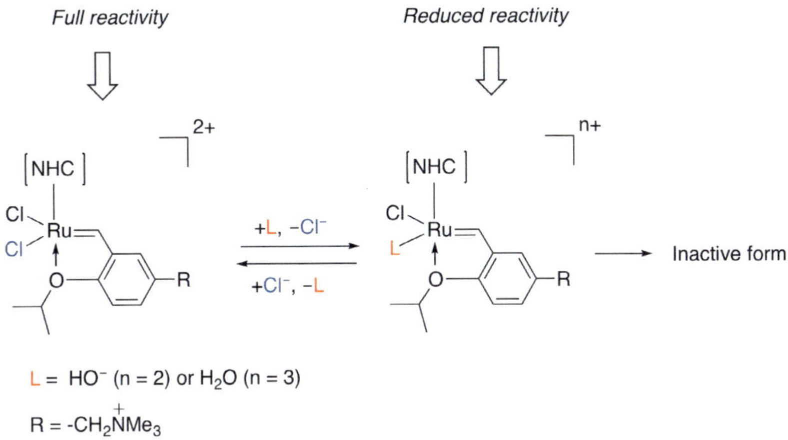

SCHEME 15.7 Mechanism of the chloride ligand effect on the reactivity of HG-type complexes.

coordination of the chlorido ligands to the Ru center precisely regulates the metathesis activity of HG-type catalysts, and the effect significantly appears in aqueous media (i.e., in the presence of a large excess of water molecules and hydroxide anions). In fact, the induction period at the initial stage of the reactions is shortened in $KCl_{aq.}$[34]: The coordination of the chlorido ligands facilitates the exchange of a Hoveyda-type ligand (2-alkoxybenzylidene moiety) with a substrate molecule. The effect of the chlorido ligands has also been observed in organic solvents,[35] which may be explained in terms of an electronic effect on the Ru center.

The mechanism described above has been found to operate under the neutral conditions. One of the characteristics required for effective olefin metathesis is high chemoselectivity toward alkene groups in the presence of several functional groups. Consequently, the experimental data shown in Table 15.5 gives us a good indication of the possibility that olefin metathesis is applicable as a biochemical tool in bioorthogonal reactions under mild conditions. Examples of olefin metathesis reactions on the surface of biomacromolecules have been reported.[36]

15.4 FINAL REMARKS AND PERSPECTIVES

All organisms are maintained by thousands of enzyme-catalyzed biochemical reactions over their entire lifetimes. To understand enzyme mechanisms and replicate their functions, chemists have, for example, attempted to mimic natural enzymes using small synthetic molecules such as chemically modified versions of cyclodextrin. Furthermore, the combination of synthetic molecules and proteins was recognized as a versatile strategy for constructing specifically designed biocatalysts or artificial enzymes because the merits of synthetic compounds and proteins provide complementary functions. In this context, combining a synthetic metal complex with an appropriate protein scaffold is a useful strategy for constructing artificial metalloenzymes which can be considered as biohybrid catalysts. Pioneering work was reported by Wilson and Whitesides in 1978.[4] This group constructed a hydrogenation catalyst where a biotin-attached Rh complex was immobilized to avidin protein. There were only limited efforts to construct of biohybrid catalysts in the ensuing years. However, over the course of the past decade, bioinorganic and organometallic chemists have accelerated research in this field with construction of new biohybrid catalysts which can promote various types of organic reactions such as reduction, oxidation, C–C bond formation, and hydration, among others.[3] In this chapter, we described olefin metathesis reactions supported by a Ru-containing HG complex, because this reaction has been frequently used in C = C bond formation/cleavage, as shown in Scheme 15.1. Our group then demonstrated this reaction in the hydrophobic cavity of a protein matrix having an immobilized Ru complex. These reactions are found to be controlled by the protein matrix and accelerated in aqueous media.[23] This finding highlights the advantages of organometallic complexes which provide chemoselectivity and stereoselectivity in the metathesis reactions. From this result, it is concluded that the combination of an organometallic species and a protein matrix is an effective strategy for constructing useful biocatalysts.

ACKNOWLEDGMENTS

This work was supported by the Deutsche Forschungsgemeinschaft (DFG) through the International Research Training Group Selectivity in Chemo- and Biocatalysis (SeleCa), the JSPS Japan-German Graduate Externship Program, and Grants-in-Aid for Scientific Research, KAKENHI Grant JP15H05804, JP23105530, JP25105738, and JP16H01029 from MEXT.

REFERENCES

1. Smith GM. *Biotechnology Set*. Wiley-VCH Verlag GmbH; 2008. p. 4–72.
2. Faber K, Fessner WD, Turner NJ. *Biocatalysis in Organic Synthesis*. Stuttgart: Thieme; 2015.
3. a) Schwizer F, Okamoto Y, Heinisch T, Gu Y, Pellizzoni MM, Lebrun V, Reuter R, Köhler V, Lewis JC, Ward TR. *Chem. Rev.* 2018;**118**:142–231. b) Lewis JC. *ACS Catal.* 2013;**3**:2954–75. c) Steinreiber J, Ward TR. *Coord. Chem. Rev.* 2008;**252**:751–66. d) Bos J, Roelfes G. *Curr. Opin. Chem. Biol.* 2014;**19**:135–43.
4. Wilson ME, Whitesides GM. *J. Am. Chem. Soc.* 1978;**100**:306–7.
5. Green NM. In: Anfinsen JTECB, Frederic MR, editors. *Adv. Protein Chem.*, Vol. 29. Academic Press; 1975. p. 85–133.
6. Collot J, Gradinaru J, Humbert N, Skander M, Zocchi A, Ward TR. *J. Am. Chem. Soc.* 2003;**125**:9030–1.

7. Ward TR. *Acc. Chem. Res.* 2010;**44**:47−57.
8. Kräutler B. *Metal-Carbon Bonds in Enzymes and Cofactors: Metal Ions in Life Sciences*, Vol. 6. The Royal Society of Chemistry; 2009. p. 1−51.
9. a) Deraedt C, d'Halluin M, Astruc D. *Eur. J. Inorg. Chem.* 2013;**2013**:4881−908. b) Higman CS, Lummiss JAM, Fogg DE. *Angew. Chem. Int. Ed.* 2016;**55**:3552−65.
10. Sauer DF, Gotzen S, Okuda J. *Org. Biomol. Chem.* 2016;**14**:9174−83.
11. Lo C, Ringenberg MR, Gnandt D, Wilson Y, Ward TR. *Chem. Commun.* 2011;**47**:12065−7.
12. Jeschek M, Reuter R, Heinisch T, Trindler C, Klehr J, Panke S, Ward TR. *Nature* 2016;**537**:661−5.
13. Zhao J, Kajetanowicz A, Ward TR. *Org. Biomol. Chem.* 2015;**13**:5652−5.
14. Mayer C, Gillingham DG, Ward TR, Hilvert D. *Chem. Commun.* 2011;**47**:12068−70.
15. Basauri-Molina M, Verhoeven DGA, van Schaik AJ, Kleijn H. R. J. M. Klein Gebbink. *Chem. Eur. J.* 2015;**21**:15676−85.
16. Tenne S-J, Schwaneberg U. *Int. J. Mol. Sci.* 2012;**13**:2459−71.
17. a) Philippart F, Arlt M, Gotzen S, Tenne S-J, Bocola M, Chen H-H, Zhu L, Schwaneberg U, Okuda J. *Chem. Eur. J.* 2013;**19**:13865−71. b) Zhu L, Arlt M, Liu H, Bocola M, Sauer DF, Gotzen S, Okuda T, Schwaneberg U. In: Boker A, van Rijn P, editors. *Bio-Synthetic Hybrid Materials and Bionanoparticles: A Biological Chemical Approach Towards Material Science*. The Royal Society of Chemistry; 2015. pp. 57−72. c) Sauer DF, Bocola M, Broglia C, Arlt M, Zhu L-L, Brocker M, Schwaneberg U, Okuda J. *Chem. Asian J.* 2015;**10**:177−82.
18. Hermanson GT. *Bioconjugate Techniques*. Third edition. Boston: Academic Press; 2013.
19. Garber SB, Kingsbury JS, Gray BL, Hoveyda AH. *J. Am. Chem. Soc.* 2000;**122**:8168−79.
20. Bianchetti CM, Blouin GC, Bitto E, Olson JS, Phillips GN. *Struct., Funct., Bioinf.* 2010;**78**:917−31.
21. a) Onoda A, Fukumoto K, Arlt M, Bocola M, Schwaneberg U, Hayashi T. *Chem. Commun.* 2012;**48**:9756−8. b) Fukumoto K, Onoda A, Mizohata E, Bocola M, Inoue T, Schwaneberg U, Hayashi T. *ChemCatChem* 2014;**6**:1229−35. c) Onoda A, Kihara Y, Fukumoto K, Sano Y, Hayashi T. *ACS Catal.* 2014;**4**:2645−8. d) Himiyama T, Sauer DF, Onoda A, Spaniol TP, Okuda J, Hayashi T. *J. Inorg. Biochem.* 2016;**158**:55−61.
22. Zhang C, Srivastava P, Ellis-Guardiola K, Lewis JC. *Tetrahedron* 2014;**70**:4245−9.
23. Sauer DF, Himiyama T, Tachikawa K, Fukumoto K, Onoda A, Mizohata E, Inoue T, Bocola M, Schwaneberg U, Hayashi T, Okuda J. *ACS Catal.* 2015;**5**:7519−22.
24. Matsuo T, Imai C, Yoshida T, Saito T, Hayashi T, Hirota S. *Chem Commun* 2012;**48**:1662−4.
25. Neurath H, Schwert GW. *Chem. Rev.* 1950;**46**:69−153.
26. a) Kraut J. In: Boyer PD, editor. *The Enzymes*, Vol. 3. Boston: Academic Press; 1971. p. 213−48. (b) Blow DM. In: Boyer) PD, editor. *The Enzymes*, Vol. 3. Boston: Academic Press; 1971. p. 185−212.
27. Tsukada H, Blow DM. *J. Mol. Biol.* 1985;**184**:703−11.
28. a) Ong EB, Shaw E, Schoellmann G. *J. Biol. Chem.* 1965;**240**:694−8. b) Schoellmann G, Shaw E. *Biochemistry* 1963;**2**:252−5. c) Schering CA, Zhong B, Woo JCG, Silverman RB. *Bioconjugate Chem.* 2004;**15**:673−6.
29. Jordan JP, Grubbs RH. *Angew. Chem. Int. Ed.* 2007;**46**:5152−5.
30. Onishi T, Hirose N, Nakano T, Nakazawa M, Izawa K. *Tetrahedron Lett.* 2001;**42**:5883−5.
31. Bing-Lei G, Shu-Tao M, Zhao-Peng L. *Lett. Org. Chem.* 2009;**6**:648−9.

32. a) Mingotaud AF, Mingotaud C, Moussa W. *J. Polym. Sci. Pol. Chem.* 2008;**46**:2833–44. b) Thiel V, Hendann M, Wannowius KJ, Plenio H. *J. Am. Chem. Soc.* 2012;**134**:1104–14.
33. a) Shimizu S, Smith DJ. *J. Chem. Phys.* 2004;**121**:1148–54. b) Timasheff SN. *Proc. Natl. Acad. Sci. USA* 2002;**99**:9721–6. c) Gekko K, Morikawa T. *J. Biochem.* 1981;**90**:39–50. (d) Gekko K, Timasheff SN. *Biochemistry* 1981;**20**:4667–76.
34. Matsuo T, Yoshida T, Fujii A, Kawahara K, Hirota S. *Organometallics* 2013;**32**:5313–19.
35. a) Krause JO, Nuyken O, Wurst K, Buchmeiser MR. *Chem. Eur. J.* 2004;**10**:777–84. (b) Tanaka K, Böhm VPW, Chadwick D, Roeper M, Braddock DC. *Organometallics* 2006;**25**:5696–8. c) Wappel J, Urbina-Blanco CA, Abbas M, Albering JH, Saf R, Nolan SP, Slugovc C. *Beilstein J. Org. Chem.* 2010;**6**:1091–8. (d) Zirngast M, Pump E, Leitgeb A, Albering JH, Slugovc C. *Chem. Commun.* 2011;**47**:2261–3.
36. (a) Binder JB, Raines RT. Olefin metathesis for chemical biology. *Curr Opin. Chem Biol* 2008;**12**(6):767–73. b) Chalker JM, Lin YA, Boutureira O, Davis BG. *Chem. Commun.* 2009;3714–16. (c) Bernardes GJL, Chalker JM, Errey JC, Davis BG. *J. Am. Chem. Soc.* 2008;**130**:5052–3.

Chapter 16

Tailoring Organometallic Complexes into Protein Scaffolds: Structures and Functions

Basudev Maity, Satoshi Abe and Takafumi Ueno
Department of Life Science and Technology, School of Life Science and Technology, Tokyo Institute of Technology, Yokohama, Japan

16.1 INTRODUCTION

Metals in suitable protein scaffolds are known to play important roles in biology. Several artificial metalloproteins and metalloenzymes have been developed by introducing synthetic metal complexes into protein environments which have important applications in catalysis, medicine, or material chemistry.[1,2] There is a growing interest in utilizing organometallic compounds in biology and medicine, since there exist several bioorganometallic species in nature such as methylcobalamine, cyanocobalamine, hydrogenases, the iron-molybdenum cofactor of nitrogenases, intermediate in Acetyl-CoA Synthase, etc.[3] Ferrocifen, ferroquine, etc., as representative of organometallic drugs, are known for promising biomedical applications in anticancer and antimalarial activities.[4,5] Similarly, metal-arene promoted transfer hydrogenation catalysis has been achieved in cancerous cells.[6,7] These examples suggest that organometallic compounds can be suitably designed for potential applications in biology, medicine, and catalysis, etc. However, the stability and solubility of organometallic compounds in an aqueous environment and toxicity limit their applications and, thus, remain as big challenges. One of the most convenient methods to resolve such issues is the utilization of suitable protein scaffolds which can accommodate the organometallic compounds through specific amino acid coordination or hydrophobic/hydrophilic interactions.[8,9] Organometallic compounds can be placed into the protein scaffold directly or through chemical modification and the binding sites can be modulated by protein engineering.[8] For example, ferrocene derivative has

Advances in Bioorganometallic Chemistry. DOI: https://doi.org/10.1016/B978-0-12-814197-7.00016-9

been covalently attached to the highly conserved Cys residue of apo-azurin and determined the X-ray crystal structure.[10] Similarly, the medicinally important half-sandwich organometallic Ru(*p*-cymene) complex specifically interacts with lysozyme protein and the corresponding coordination structure of the organometallic fragment has been determined by X-ray crystal structure analysis.[11] In the meantime, unique protein environments generated in 3D protein crystals are also promising for stabilizing organometallic compounds in the development of solid biohybrid materials. Porous lysozyme protein crystals stabilized with crosslink treatment are known for promoting catalytic reaction after immobilization with organometallic Ru(benzene) complex.[12] Protein scaffolds thus provide a powerful platform for stabilizing organometallic compounds in an aqueous environment and expand the scope of applications in medicine, biology, catalysis, and biomaterial science, etc. Until now, there are only a limited number of protein scaffolds such as ferritin cage, lysozyme, heme proteins, nitrobindin, etc., which have been structurally characterized with organometallic compounds and tested for specific functions like catalysis.[11,13,14] So, there is a necessity for further designing of new protein scaffold containing organometallic moieties with novel structures and new functions. In this context, the detail studies on the structure of organometallic compounds in a protein environment and their functions are significant for the growing field of bioorganometallic chemistry. This section focuses on describing the subsequent and significant research in the development of artificial organometalloproteins and metalloenzymes. We limited our scope to ferritin cage, protein needle, lysozyme crystal, and in vivo protein crystals (Fig. 16.1) which are the most promising proteins scaffolds developed and studied systematically. The design, X-ray crystal structure and functions of such artificial proteins are discussed in detail for future design and exploration in this field.

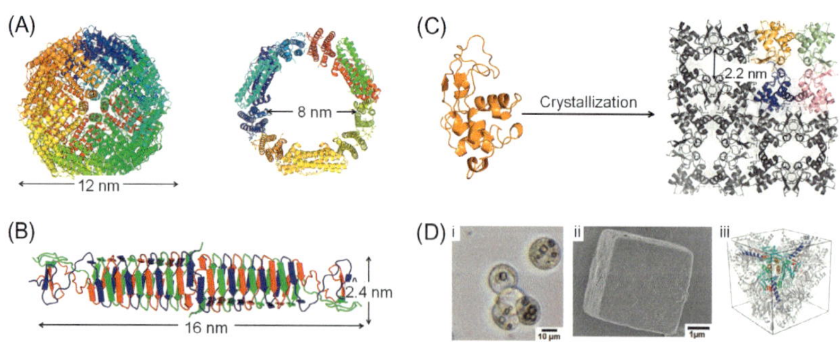

FIGURE 16.1 Examples of various protein scaffolds utilized for immobilizing organometallic compounds. (A) Ferritin protein and its internal cage structure. (B) The molecular structure of reconstructed β-helical protein needle from bacteriophage T4. (C) Lysozyme monomer and formation of porous structure in the tetragonal lattice. (D) Polyhedra crystals (PhC) formed in insect cells (i); SEM images of recombinant PhC (ii); and the assembling structure of polyhedrin monomer in a unit cell (iii).

16.2 STRUCTURE AND FUNCTIONS OF ORGANOMETALLIC CONJUGATES WITH FERRITIN PROTEIN CAGE

Ferritin is a 24-mer self-assembled protein cage which is naturally designed for iron storage.[15] The cage has an internal cavity with a diameter of 8 nm which is associated with several three-fold pores (Fig. 16.1A). The Fe^{2+} ions pass into the cage through these pores and form minerals. Other than iron, the ferritin protein cage is capable accommodating a number of metal ions and metal complexes including organometallic compounds and, thus, acts as a versatile building block for biotechnology, material science, and catalysis.[16] Encapsulation of organometallic compounds into the ferritin cage is of particular interest as it can stabilize these compounds inside the confined environment which is important for catalysis and drug delivery. Ueno and coworkers have incorporated functionalized organometallic ferrocene molecules into the ferritin protein cage similar to blue azurins and cytochrome $P450_{cam}$.[10,17,18]. Ferrocene derivatives non-covalently bind to the hydrophobic regions surrounding to the two-fold symmetric axis of the ferritin cage. The corresponding X-ray crystal structure without Cp rings has been determined.[18] The entrapped ferrocene derivatives inside the ferritin cage showed characteristic electrochemical redox behavior with an approximate shift of +8 mV due to hydrophobic protein environment compared to the analogous free ferrocene derivative. Similarly, piano stool Ru (arene) complexes, which are known for their potential applications in catalysis and medicine, can also be accommodated in protein scaffold. Organometallic Ru(*p*-cymene) complex has been immobilized on the interior surface of the ferritin cage.[19] The corresponding X-ray crystal structure analysis revealed that the Ru(*p*-cymene) moiety is coordinated by His49, His173, and Glu53 which are located on the interior surface of the protein cage (Fig. 16.2A). The presence of Ru(*p*-cymene) moiety at His49 suggests that the ferritin can stabilize the organometallic fragment without the disruption of the cage structure. There are two additional Ru binding sites without *p*-cymene unit which are located at Cys48 in the accumulation center and His114/Cys126 in the three-fold channel. Replacement of His49 by Ala did not alter the Ru coordination at the three-fold channel, although no binding site was observed at the accumulation center. In addition, a new Ru binding site was observed at His173 with low occupancy. Although these composites have no drug potency and catalytic activities, it is possible to arrange the coordination sites to improve the activities or to incorporate other organometallic compounds for suitable applications.

A catalytic polymerization reaction was achieved inside the confined space of ferritin cage after suitable immobilization of organometallic, Rh (nbd) [nbd: norbornadiene] complex on the interior surface of the protein cage (Fig. 16.2B, C).[20] It was found that a total of ~57 Rh atoms can be immobilized by reacting apo-ferritin with 100 equivalent of $[Rh(nbd)Cl]_2$.

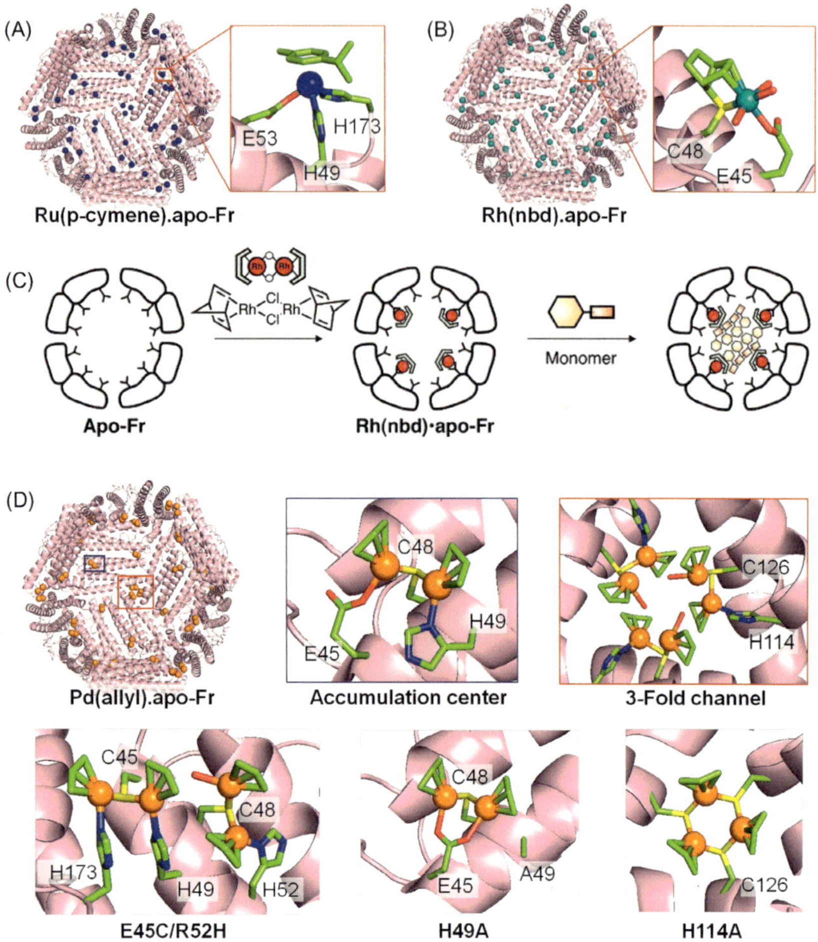

FIGURE 16.2 (A) Coordination structure of the organometallic Ru(*p*-cymene) moiety in the apo-ferritin cage and the positions of Ru atoms on the interior surface of the cage. (B) Coordination structure of the organometallic Rh(nbd) moiety and locations of the Rh atoms immobilized on the internal surface of the apo-ferritin cage. (C) Scheme for the incorporation of Rh(nbd) complex into the ferritin cage and catalytic polymerization of phenyl acetylene. *Reproduced with permission from ACS.*[20] (D) Coordination structures of the organometallic Pd(allyl) moiety in apo-ferritin cage and structure control in various mutants.

The resulting ferritin-Rh composite can promote catalytic polymerization of phenylacetylene. The confined space of the ferritin cage restricted the size and number of the polymer formed inside. Approximately two polymer chains consisting of 130 monomers in each chain were formed in a single cage. When a large number of monomers were used, a maximum of eight polymer chains per ferritin cage were formed. X-ray crystal structure analysis reveals the catalytic active sites. The Rh(nbd) moiety is bounded by

Cys48, Glu45 and three water molecules (Fig. 16.2B). Interestingly, the S^{γ} of Cys48 coordinates to both Rh and nbd ligand forming a bridge and stabilizes the nbd-Rh coordination. There are two additional Rh binding sites at His114 and His49 without the nbd ligand. Although there are in total three different Rh binding sites, all were not involved in the catalytic reactions.[21]

Organometallic Pd(allyl) complexes were immobilized on the internal surface of the ferritin cage and studied their mechanism of accumulation with a series of X-ray crystal structures (Fig. 16.2D).[13,22] In apo-ferritin, Pd (allyl) forms thiol bridged dinuclear structure at Cys48 and Cys126 which are supported by His49 and His114, respectively. No Pd(allyl) binding was observed at the three-fold channel as well as accumulation center when the respective Cys at positions 126 or 48 was replaced by Ala. The replacement of His by Ala at positions either 114 and 49 had little or no effect on Pd (allyl) coordination.[13] This signifies that Cys plays a crucial role in Pd(allyl) binding whereas nearby His acts as a supporting ligand. It is expected that the organometallic Pd(allyl) complexes enter into the cage through the three-fold pores. When His114 at the three-fold channel was replaced by Ala, an interesting thiol(Cys126) bridged trinuclear organometallic Pd(allyl) cluster was formed. As a result, the number of accumulated Pd complexes was reduced from 100 to 78 per cage (Fig. 16.2D).[13] Deletion of Cys126 at the three-fold channel significantly reduced the Pd(allyl) accumulation from 100 to 37 per cage which suggests that Cys126 accelerates the incorporation of Pd(allyl) complexes into the cage. It is possible to arrange the coordination structures of the organometallic Pd(allyl) fragments by changing suitable amino acid residues and similarly, new coordination sites can be created by introducing additional Cys. This is evident from the crystal structures of Pd(allyl) composite with various apo-ferritin mutants such as apo-E45C/48A-ferritin, apo-E45C/R52H-ferritin, apo-E45C/H49A/R52H-ferritin (Fig. 16.2D).[23] The organometallic Pd(allyl) fragments stabilized inside the ferritin cage were capable of catalyzing Suzuki-Maura cross-coupling reaction between phenyl boronic acid and 4-iodoaniline.

The turn over frequency(TOF) values were observed in the range of $800-4300\ h^{-1}$ and dependent on the number and coordination geometry of Pd(allyl) moiety. The highest TOF value of $4300\ h^{-1}$ was achieved when a new organometallic Pd(allyl) binding moiety was introduced in apo-E45C/R52H-ferritin mutant. The three-fold channel is important for the entry of both Pd(allyl) as well as substrates into the cage. Modification of the three-fold channel significantly reduced the catalytic activities with TOF values in the range of $800-900\ h^{-1}$. This represents that metal coordination in protein cage can be tuned to enhance the activity of catalytic reactions. Subsequently, two different organometallic IrCp* and Pd(allyl) complexes have been introduced into a single ferritin protein cage in order to construct a protein-based microcompartment (Fig. 16.3A).[24] Precise locations of both the metals inside the protein cage were determined using X-ray crystal

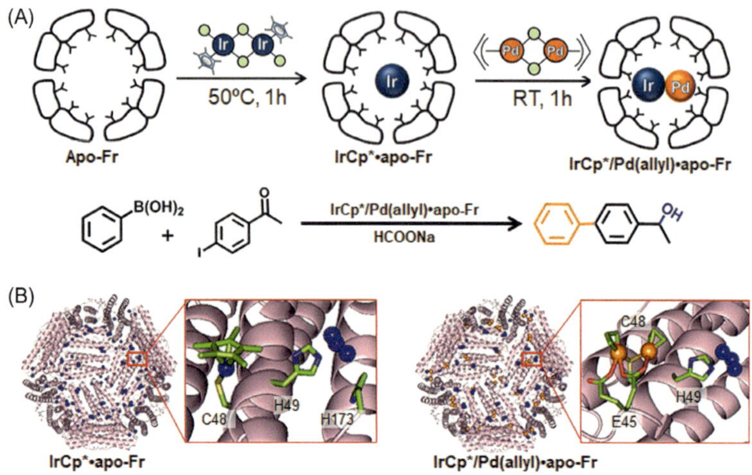

FIGURE 16.3 (A) Schematic representation for immobilizing two different organometallic complexes into a single ferritin cage and the corresponding catalytic cascade reactions promoted inside the cage. (B) Crystal structures of IrCp* ●apo-Fr and IrCp*/Pd(allyl)●apo-Fr showing the corresponding IrCp* and Pd(allyl) fragments.

structure analysis. The crystal structure shows two distinct binding positions for two metals and they do not compete for the same position. It was found that IrCp* prefers to bind to His whereas Pd(allyl) prefers Cys (Fig. 16.3B). Interestingly, both the metal complexes inside the ferritin cage were catalytically active to promote tandem reactions. The Pd(allyl) moiety catalyzed the C-C coupling reaction between 4-iodo acetophenone and phenyl boronic acid, whereas the IrCp* moiety catalyzed the hydrogenation of the carbonyl group present in four-iodo acetophenone. Although the overall catalytic performance is low, the concept can be utilized for promoting cascade reactions by suitable arrangement of the metal coordination sites as discussed previously for Pd(allyl) compound.

Carbon monoxide is an important signaling molecule in biology despite its extreme toxicity at high concentrations.[25,26] In nature, CO plays structural roles in hydrogenases. Besides its natural role, various CO-releasing molecules (CORMs) were incorporated into the protein scaffold for controlled release of CO in the cellular environment.[27,28] Subsequently, organometallic metal-carbonyl compounds were immobilized into designed ferritin cages for effective delivery of CO into the living cells and controlling cellular functions (Fig. 16.4).[29,30] The fixation of ruthenium carbonyl complex, $[Ru(CO)_3Cl_2]_2$, (CORM-2) was achieved by the reaction with apo-ferritin in a buffer solution and it was found that a total of 48 Ru atoms per cage were stabilized and immobilized inside the cage. The Ru atoms were bounded by the Cys and His residues as determined by X-ray crystal structure analysis. Organometallic *cis*-$Ru(CO)_2$ fragment was observed at Cys48 coordination

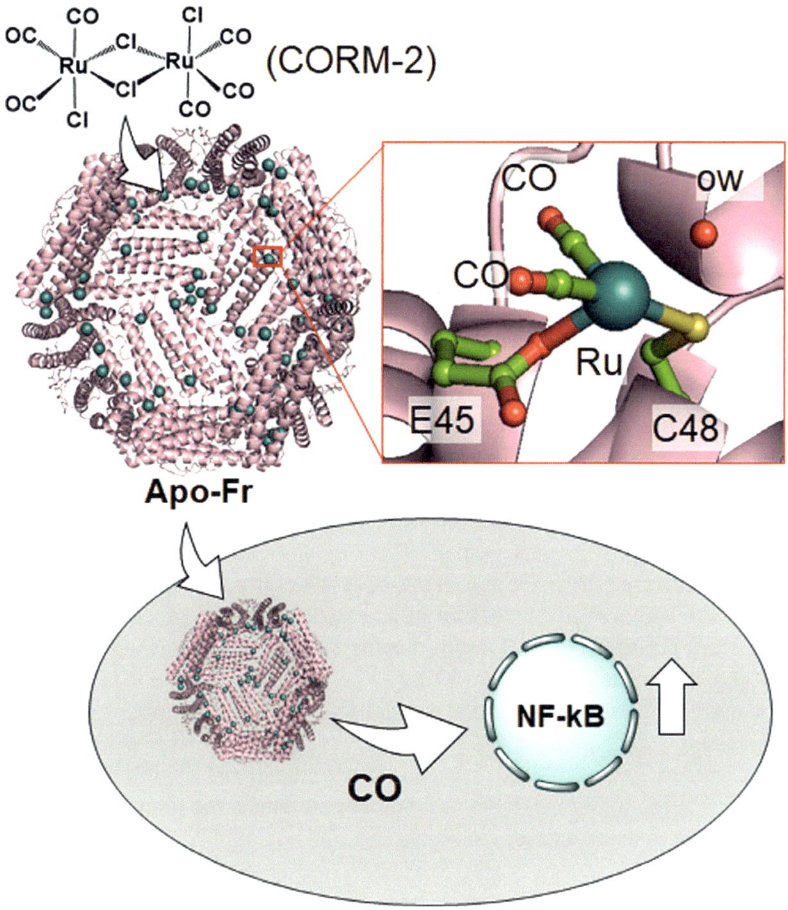

FIGURE 16.4 Immobilization of the CORM-2, ($[Ru(CO)_3Cl_2]_2$) in the apo-ferritin cage with the crystal structure, cellular uptake in the HEK293/kB-Fluc cell and activation of NF-κB by released CO gas. Crystal structure showing the coordination of organometallic *cis*-$Ru(CO)_2$ moiety and locations of the Ru atoms on the interior surface of the cage.

site (Fig. 16.4). However, the CO ligands were not observed in other binding positions at His114 and His132. To modulate the CO coordination, new mutants such as apo-E45C/C48A-ferritin and apo-R52C-ferritin were designed. Both the mutants maintained thiol bridged dinuclear Ru structure at Cys45 and Cys48, respectively. Although no electron density was observed for CO ligand, the presence of CO was confirmed from the ATR-IR stretching frequencies at $\sim 2040\ cm^{-1}$ and $\sim 1960\ cm^{-1}$ which correspond to the *cis*-$Ru(CO)_2$ unit. This suggests that due to disorder or low occupancy, the ancillary CO ligands were not observed from the crystal structures. All the three ferritin-RuCO composites were capable of releasing

CO with a release rate which was ~18-times slower than the commercial CORM-3 and thus becomes more convenient for cellular applications. It is known that CO gas activates the nuclear factor kB (NF-κB) cooperatively with tumor necrosis factor-α (TNF-α).[31] Ferritin-RuCO composites were found to activate NF-κB with ~10 fold higher efficiency compared to the commercial CORM-3 while remaining nontoxic to cells. In addition, ferritin-RuCO had ~4-times higher uptake efficiency than the commercial CORM-3, which is significant for stabilization and delivery of organometallic complexes into living cells. In the next step, they regulated the CO release using light stimulus by constructing photoactivable ferritin-MnCO composite.[30] The CORM-1, $Mn_2(CO)_{10}$ which is known to release CO in presence of light, was reacted with the designed apo-ferritin mutant (R52C) in the buffer solution to construct photoactive ferritin-MnCO conjugates.[30,32] From X-ray crystal structure, it was found that a Mn atom was stabilized and bounded by both Cys48 and Cys52 without the CO ligands. The typical CO stretching frequencies at 2028 cm^{-1} and 1917 cm^{-1}suggested the presence of CO ligands which were not observed from the crystal structure. Another Mn binding was observed near Asp38. The organometallic ferritin-MnCO construct can release CO inside the living cells when irradiated with visible light of 456 nm which was evident from the appearance of green fluorescence from a CO probe (COP-1). Despite having similar cellular uptake efficiency, only the organometallic ferritin-MnCO was capable to activate NF-κB after light irradiation, whereas the commercial CORM-1 having the larger amount of released CO remained inactive. This suggests that the organometallic MnCO moiety gets stabilized in protein environment and exhibit unique properties inside the living cells. So, the examples stipulate the potential of ferritin protein cage as a carrier tool for organometallic compounds into living cells.

16.3 ORGANOMETALLIC CONJUGATES WITH PROTEIN NEEDLE

Protein needle (PN) is a reconstructed 16 nm long head-to-head dimer assembly of the triple-stranded β-helix fusion protein of gene product 5 $(gp5)_3$ from bacteriophage T4 with the fold-on domain (Fig. 16.1A, 16.5A).[34,35] The original role of gp5 is to puncture the outer cell membrane of host bacteria during infection. The structure of the engineered protein needle, $[(gp5)_3\beta f]_2$ is highly robust and has excellent stability over a wide range of pH and temperature. Subsequently, it has been recognized as a molecular template for the delivery of various external molecules including DNA, proteins, etc., into the cells.[36] Chemical modification at Lys or Cys with succinimidyl ester or maleimide conjugation is an important strategy for arranging various functionalities such as metal complexes and organic molecules on the surface of the needle. Ueno and coworkers have precisely arranged the organometallic ReCO complexes in combination with Ru(bpy) complexes on the protein needle surface to promote the photocatalytic reduction of CO_2

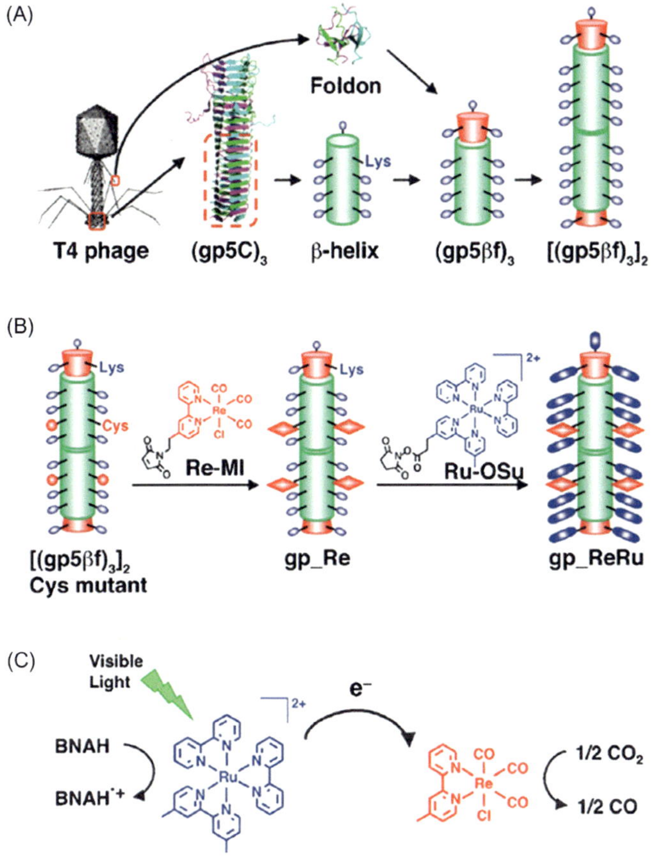

FIGURE 16.5 (A) Construction of protein needle from bacteriophage T4. (B) Schematic representation of the arrangement of organometallic $Re(bpy)(CO)_3Cl$ and $Ru(bpy)_3$ complexes on the surface of protein needle by chemical modification. (C) Visible light induced photocatalytic reduction of CO_2 to CO. *Reproduced with permission from RSC*[33].

(Fig. B,C).[33] Lys at position 41 was replaced by Cys which was then fixed with organometallic $Re(bpy)(CO)_3Cl$ complex through maleimide conjugation. The other Lys residues on the surface of protein needle were conjugated with $Ru(bpy)_3$ complex through succinimidyl ester. The $C_\alpha-C_\alpha$ distance between Cys41 and nearest Lys25 is 11 Å. It was found that 6 out of 6 Cys and 22 out of 30 Lys were functionalized with $Re(bpy)(CO)_3Cl$ and Ru $(bpy)_3$ complexes, respectively. In the presence of visible light, electron transfer occurred from photo-activated Ru(bpy) moiety to ReCO moiety in the presence of 1-benzyl-1,4-dihydronicotinamide (BNAH), which reduced CO_2 to CO (Fig. C). Due to the close proximity of Ru(bpy) and Re(CO) on the PN surface, the photocatalytic activity was found to be 3.3 times higher than that of the mixture of PN-ReCO and analogous Ru(bpy) complex. The photocatalytic efficiency was dependent on the higher flexibility of Re moiety as the activity was reduced to almost half when the Re(bpy)$(CO)_3$Cl complex was conjugated through to Cys57 or Cys69. This suggests that catalytic reactions can be cooperatively promoted by suitable arrangement of different metal complexes on the protein needle surface. In another example, CO-releasing protein needle was constructed by immobilizing organometallic ruthenium carbonyl complexes using His-tag triad at the C-terminus of the protein needle.[37] A total of 6 Ru atoms per needle were determined from quantitative analysis and the IR bands corresponding to *cis*-$Ru(CO)_2$ fragments were observed at 2031 cm^{-1} and 1955 cm^{-1}. As expected, the resulting protein needle-RuCO conjugate had significantly higher uptake efficiency compared to free metal complex or other carriers such as ferritin cage. The CO delivered by the organometallic bioconjugate was capable of activating the transcriptional nuclear factor-kappa B (NF-κB) and induced the expression of its target genes such as HO1, NQO1, and IL6, via ROS (reactive oxygen species) mediated pathway. So, these examples suggest that the protein needle is a useful and promising molecular template for arranging various organometallic functionalities for catalytic reactions as well as efficient delivery into the cells.

16.4 POROUS LYSOZYME CRYSTALS FOR FUNCTIONALIZING ORGANOMETALLIC COMPLEXES

Protein crystals are considered as densely packed monomeric protein assembly which has been recognized as a scaffold for nanomaterials preparation and catalysis.[38] One such example is hen egg white lysozyme (HEWL) protein which crystallizes in three different polymorphic forms such as tetragonal, orthorhombic and monoclinic depending on crystallizing conditions (Fig. 16.1C). Orthorhombic and tetragonal lattice structures have the largest solvent channels with the diameter of 2.6 and 2.2 nm, respectively. The internal surface of such solvent channels can be immobilized with metal ions or complexes by coordination of amino acid residues for preparation of inorganic materials and catalytic reactions. For example, platinum nanoparticles were prepared using dye encapsulated porous lysozyme crystals which showed photocatalytic hydrogen production.[39] Organometallic metal carbonyl complexes were immobilized and stabilized in lysozyme protein and their precise locations were determined with X-ray crystal structure.[40–44] Recently, lysozyme crystals have been stabilized by crosslink treatment in 1%glutaraldehyde solution. Such crosslinked crystals maintain their crystallinity and have high mechanical strength as well as stability which can be used under harsh conditions. Crosslinked porous lysozyme crystals were utilized for immobilizing the organometallic ruthenium carbonyl complex in order to construct artificial CO-releasing extracellular scaffold (Fig. 16.6A).[45] The fixation of the RuCO complex

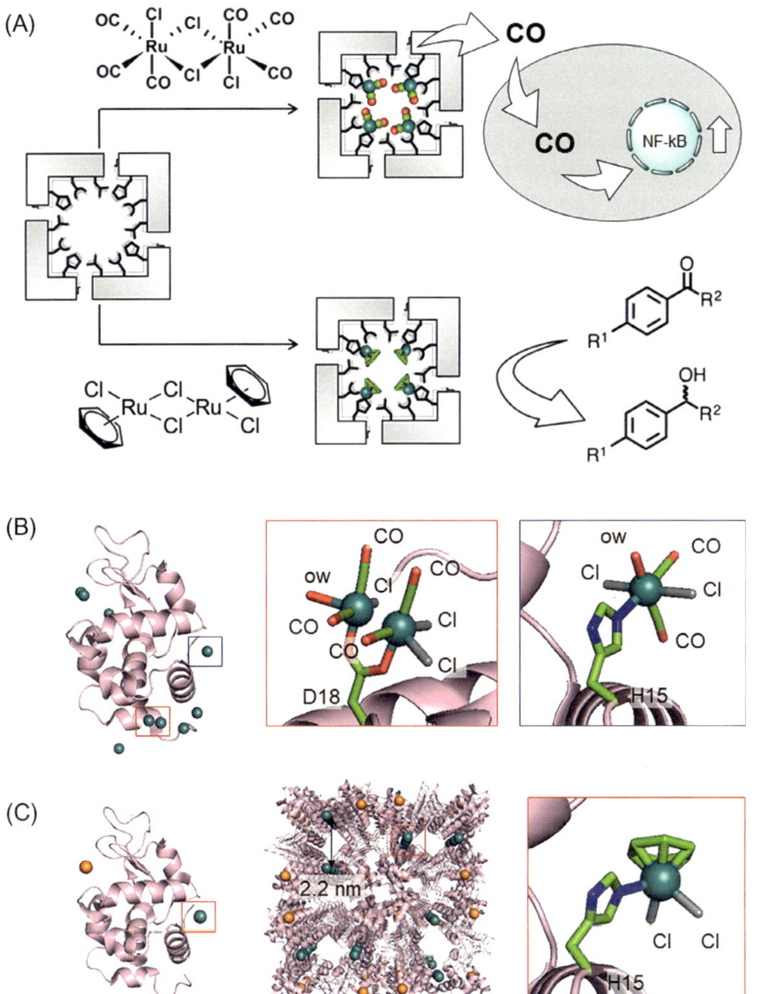

FIGURE 16.6 Structure and functions of porous lysozyme crystal. (A) Schematic representation for the immobilization organometallic[$Ru(CO)_3Cl_2]_2$ and $[Ru(benzene)Cl_2]_2$ compounds into the solvent channel of porous lysozyme crystal and their functions in NF-κB activation in HEK293/ kB-Fluc cell and asymmetric catalysis, respectively. (B) Crystal structure showing the positions of Ru atoms (blue sphere) in lysozyme monomer and coordination of organometallic *cis*-$Ru(CO)_2$ moieties. (C) Crystal structure of the lysozyme monomer showing the locations of Ru atoms (blue and orange spheres), porous solvent channels in the tetragonal lattice and coordination structure of the $Ru(benzene)Cl_2$ moiety which is exposed towards the large solvent channel.

was achieved by soaking the crystals in the CORM-2 solution. From X-ray crystal structure analysis, it was found that there are total of six Ru binding sites present in lysozyme. However, the electron densities for organometallic Ru-CO moiety was observed only at binding positions His15 and

Asp18. An octahedral Ru coordination structure was observed at His15 with two CO ligands at *cis* positions, two Cl atoms and one water molecule (Fig. 16.6b). The characteristic CO stretching frequencies of this CO coordination were observed at 2055 cm^{-1} and 1981 cm^{-1}. Another Ru binding site was observed at Asp18 where two Ru was coordinated by two carboxylic oxygen atoms with half occupancies suggesting that they do not exist simultaneously. The observed IR stretching frequencies of this CO coordination are at 2025 $cm^{-1}$and 1940 cm^{-1}. The CO ligands of other Ru binding sites at Asp52, Asp101, Asp119, and Leu129 were not observed due to low occupancy. X-ray crystal structure analysis after the CO release during myoglobin (Mb) assay showed two interesting facts. First, the RuCO coordination at His15 remained unaltered suggesting no CO release occurred from this site. The RuCO moiety at His15 was stabilized by steric hindrance and H-bonding interactions with surrounding amino acid residues which prevent the ligand exchange reaction with CO. Second, the disappearance of CO ligands from Ru coordination Asp18 which strongly represent the active site. This was also reflected from the disappearance of IR bands at 2025 $cm^{-1}$and 1940 cm^{-1}. The CO release from this lysozyme crystal was found to have 10 times slower rate than the commercial CORM-2 and effectively induced the NF-κB activity similar to organometallic ferritin-RuCO or protein needle-RuCO conjugates when treated with cells.

Similar to the artificial CO-releasing extracellular scaffold, the crosslinked porous lysozyme crystals were immobilized with organometallic Ru(benzene) Cl_2 complex for promoting transfer hydrogenation reaction inside the solvent channel (Fig. 16.6A,C). X-ray crystal structures of both orthorhombic and tetragonal lattice have a common Ru binding site at His15 which is exposed towards the largest solvent channel (Figure 16.6C). The Ru(benzene) coordination at His15 is similar to the Ru(p-cymene)-enzyme complex and consists of two labile Cl atoms which are necessary for the catalytic reaction.[11] The lysozyme-Ru(benzene) crystals catalyzed a number of acetophenone derivatives to the corresponding alcohol. Tetragonal lysozyme crystals gave the highest conversion up to 53% for propiophenone whereas orthogonal crystal gave the highest enantioselectivity up to 36% (S) for isobutyrophenone. It is expected that the enantioselectivity was influenced by Asp7 in the tetragonal lattice and Arg14 in the orthogonal lattice. Interestingly, the catalytic lysozyme crystals were reused and the activity was retained up to 70% after the 10th cycle of reactions which make it an important scaffold for heterogeneous catalysis.

16.5 ORGANOMETALLIC BIOCONJUGATES WITH IN VIVO PROTEIN CRYSTAL

Protein crystals can be prepared in vivo in contrast to the conventional methods. Crystallization of proteins in living cells are of particular interest as it skips the protein purification and crystallization steps. It is also possible to prepare the crystals in large scale using this method. One such example is Polyhedra crystals (PhCs) produced by polyhedrin proteins in baculovirus-infected insect cells (Fig. 16.1D). PhCs can be utilized for the development of solid biohybrid materials by conjugating organometallic compounds. Ruthenium carbonyl complexes were immobilized in the PhCs similar to the lysozyme crystals in order to construct extracellular scaffold for releasing CO into living cells.[46] It was found that the incorporation of His-tag to the C-terminus of the polyhedrin monomer significantly increases the amount of immobilized RuCO. The Ru complexes were fully dispersed in the crystal which was evidenced from STEM-EDX. The hybrid composite can release CO with a release rate which is four-fold greater than the commercial CORM-2 and activates the NF-kB. Subsequently, organometallic MnCO complex has been immobilized into the PhC in order to develop photoactivable CO-releasing scaffold.[47] Similar to organometallic RuCO composite, the His-tag at C-terminus significantly enhances the MnCO uptake compared to WT polyhedrin (Fig. 16.7A). X-ray crystal structure analysis showed only the protein chain without any metal. This indicates that the MnCO binding sites are highly flexible in the PhC. In the presence of visible light irradiation, the MnCO-polyhedra crystal effectively released CO and the characteristic CO stretching frequencies at 1921 cm^{-1} and 2028 cm^{-1} disappeared. The total amount of released CO per Mn was found to be 2.9 compared to 1.9 for WT polyhedrin. The photoinduced CO release from the engineered PhCs was capable to activate the nuclear factor kappa B (NF-κB). So, these two examples suggest that the fusion of His-tag at the C-terminus of the polyhedrin effectively enhances the metal immobilization which was reflected in their performance in NF-κB activation by suppressing the cytotoxicity of the precursor metal complexes. In another example, the molecular structure of organometallic Pd(allyl) complex immobilized in the PhC has been determined by microfocus X-ray diffraction.[48] The immobilization of the Pd(allyl) complexes was performed by soaking the crystals in an aqueous acetonitrile solution containing Pd complexes. In WTPhC, there are total 5 Pd binding sites involving either Cys or His (Fig. 16.7B). The ancillary allyl moiety was observed for Pd with full occupancy at His238 and Cys211. At the three-fold symmetric center in the unit cell, a trinuclear Pd structure supported by Cys204 from three different monomers was observed. Deletion of the residues 192–194 in the L4 loop (ΔL4PhC) had no effect on the lattice structure, however, the total Pd binding sites were reduced to four (Fig. 16.7C). The structure of Pd(allyl)•ΔL4PhC shows highly flexible regions located at the interface of polyhedrin trimers where the appeared electron densities of several amino acid residues were too weak to determine the structure. Interestingly, two new Pd binding sites were observed at His183 and His223 in the structure of Pd(allyl)•ΔL4PhC. Due to the presence of highly flexible regions, the mutant crystal is expected to promote chemical reactions after Pd(allyl) immobilization.

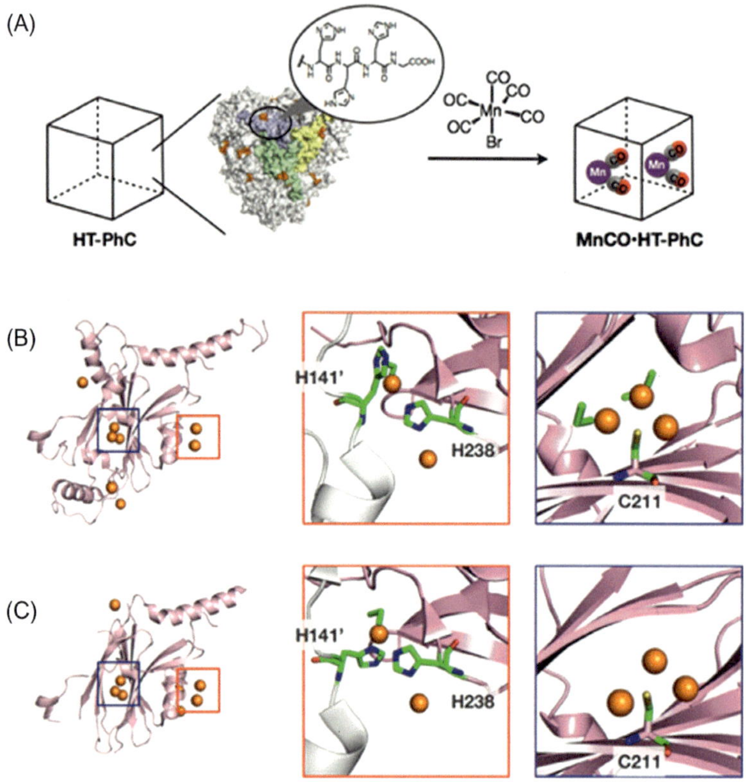

FIGURE 16.7 (A) Schematic representation of the preparation of photoactivable CO-releasing scaffold by immobilization of MnCO into His-tag(HT)-PhC. (B, C) Crystal structures of in vivo polyhedral crystals functionalized with organometallic Pd(allyl) complex. (B) Structure of WTPhM showing the location of Pd atoms and selected coordination of Pd(allyl) binding. (C) Monomeric structure of Pd(allyl)•ΔL4PhC showing the location of Pd atoms and selected coordination of Pd(allyl) binding.

16.6 STABILIZATION OF ORGANOMETALLIC FUNCTIONALITIES IN OTHER PROTEIN SCAFFOLDS

In addition to these protein templates, there are several examples where the organometallic functionalities were incorporated into suitable protein scaffolds and determined their X-ray crystal structures. For example, organometallic gold N-heterocyclic carbene compound was reacted with thaumatin and the corresponding X-ray crystal structure showed the coordination of Au (NHC) fragment to distinct protein sites at lysine side chain and N-terminal tail.[49] Organometallic rhodium complex, Rh(Cp)(COD) was covalently conjugated inside the scaffold of an apo-nitrobindin mutant which showed *trans*-polymerization of phenylacetylene.[14]

16.7 CONCLUSIONS

This chapter describes the structure and functions of various artificial metalloproteins containing organometallic fragments which were stabilized in specific cavities. Such organometalloproteins were successfully applied in catalysis and biology. The efficiency of such artificial proteins can be enhanced by arranging the coordination sites of the active metal center. This strongly supports the fact that protein scaffolds can act as macromolecular ligands for stabilizing organometallic fragments in an aqueous medium, even in the cellular environment, and their structures can be tuned by protein engineering. This will significantly influence the biomedical and cellular applications of organometallic compounds which are relatively less explored due to their air sensitivity and aqueous stability. The precise structure determination of metalloproteins with organometallics compounds will enhance the understanding of their interactions with biomolecules. In addition, the structural diversity of organometallic compounds can be explored with biomolecules to mimic natural functions and promoting enzymatic reactions. Although several organometallic compounds such as metal-carbonyl, metal-arene, etc., form a stable complex with protein support, the challenges remain to fix more and more sensitive organometallics into a protein scaffold and to study their reactivity. In future, it is expected that more sophisticated metal–carbon bonds will be prepared by a suitable design of the protein environment which will give unique functionality and expand the diversity of organometalloproteins.

ACKNOWLEDGMENT

Parts of this work were supported by JSPS KAKENHI grant Nos. JP16H04177, JP16K13095, JP16H00755, and JP17H05358 to T.U. and JP16K17931, JP16H00827 and JP17H05872 to S.A.

REFERENCES

1. Yu F, Cangelosi VM, Zastrow ML, Tegoni M, Plegaria JS, Tebo AG, et al. Protein design: toward functional metalloenzymes.. *Chem. Rev.* 2014;**114**(7):3495–578.
2. Thomas CM, Ward TR. Design of artificial metalloenzymes. *Appl. Organomet. Chem.* 2005;**19**(1):35–9.
3. Hartinger CG, Dyson PJ. Bioorganometallic chemistry from teaching paradigms to medicinal applications. *Chem. Soc. Rev.* 2009;**38**(2):391–401.
4. Patra M, Gasser G. The medicinal chemistry of ferrocene and its derivatives. *Nat. Rev. Chem.* 2017;**1**:0066.
5. Gasser G, Ott I, Metzler-Nolte N. Organometallic anticancer compounds. *J. Med. Chem.* 2011;**54**(1):3–25.
6. Soldevila-Barreda JJ, Romero-Canelón I, Habtemariam A, Sadler PJ. Transfer hydrogenation catalysis in cells as a new approach to anticancer drug design. *Nat. Commun.* 2015;**6**:6582.

7. Coverdale JPC, Romero-Canelón I, Sanchez-Cano C, Clarkson GJ, Habtemariam A, Wills M, et al. Asymmetric transfer hydrogenation by synthetic catalysts in cancer cells. *Nat. Chem.* 2018;**10**(3):347–54.
8. Herrick RS, Ziegler CJ, Leeper TC. Structure and function in organometallic•protein complexes. *J. Organomet. Chem.* 2014;**751**(1):90–110.
9. Onoda A, Hayashi T, Salmain M. *Bioorganometallic Chemistry*. Wiley-VCH Verlag GmbH & Co. KGaA; 2014. p. 305.
10. Hwang HJ, Carey JR, Brower ET, Gengenbach AJ, Abramite JA, Lu Y. Blue ferrocenium azurin: an organometalloprotein with tunable redox properties. *J. Am. Chem. Soc.* 2005;**127**(44):15356–7.
11. McNae IW, Fishburne K, Habtemariam A, Hunter TM, Melchart M, Wang F, et al. Half-sandwich arene ruthenium(II)-enzyme complex. *Chem. Commun* 2004;(16):1786–7.
12. Tabe H, Abe S, Hikage T, Kitagawa S, Ueno T. Porous protein crystals as catalytic vessels for organometallic complexes. *Chem. Asian. J.* 2014;**9**(5):1373–8.
13. Abe S, Niemeyer J, Abe M, Takezawa Y, Ueno T, Hikage T, et al. Control of the coordination structure of organometallic palladium complexes in an apo-ferritin cage. *J. Am. Chem. Soc.* 2008;**130**(32):10512–14.
14. Onoda A, Fukumoto K, Arlt M, Bocola M, Schwaneberg U, Hayashi T. A rhodium complex-linked ß-barrel protein as a hybrid biocatalyst for phenylacetylene polymerization. *Chem. Commun.* 2012;**48**(78):9756–8.
15. Honarmand Ebrahimi K, Hagedoorn P-L, Hagen WR. Unity in the biochemistry of the iron-storage proteins ferritin and bacterioferritin. *Chem. Rev.* 2015;**115**(1):295–326.
16. Jutz G, van Rijn P, Santos Miranda B, Böker A. Ferritin: A versatile building block for bionanotechnology. *Chem. Rev.* 2015;**115**(4):1653–701.
17. Di Gleria K, Nickerson DP, Hill HAO, Wong L-L, Fülöp VJ. Enantioselective cyclopropanation of allylic alcohols with dioxaborolane ligands: scope and synthetic applications. *J. Am. Chem. Soc.* 1998;**120**(46):11943–52.
18. Niemeyer J, Abe S, Hikage T, Ueno T, Erker G, Watanabe Y. Noncovalent insertion of ferrocenes into the proteinshell of apo-ferritin. *Chem. Commun.* 2008;6519–21.
19. Takezawa Y, Bockmann P, Sugi N, Wang Z, Abe S, Murakami T, et al. Incorporation of organometallic Ru complexes into apo-ferritin cage. *Dalton Trans.* 2011;**40**(10):2190–5.
20. Abe S, Hirata K, Ueno T, Morino K, Shimizu N, Yaamoto M, et al. Polymerization of phenylacetylene by rhodium complexes within a discrete space of apo-ferritin. *J. Am. Chem. Soc.* 2009;**131**(20):6958–70.
21. Ke Z, Abe S, Ueno T, Morokuma K. Catalytic mechanism in artificial metalloenzyme: QM/MM study of phenylacetylene polymerization by rhodium complex encapsulated in apo-ferritin. *J. Am. Chem. Soc.* 2012;**134**(37):15418–29.
22. Abe S, Hikage T, Watanabe Y, Kitagawa S, Ueno T. Mechanism of accumulation and incorporation of organometallic Pd complexes into the protein nanocage of apo-ferritin. *Inorg. Chem.* 2010;**49**(15):6967–73.
23. Wang Z, Takezawa Y, Aoyagi H, Abe S, Hikage T, Watanabe Y, et al. Definite coordination arrangement of organometallic palladium complexes accumulated on the designed interior surface of apo-ferritin. *Chem. Commun.* 2011;**47**(1):170–2.
24. Maity B, Fukumori K, Abe S, Ueno T. Immobilization of two organometallic complexes into a single cage to construct protein-based microcompartments. *Chem. Commun.* 2016;**52**(31):5463–6.
25. Mann BE, Jaouen G, Metzler-Nolte N. *Medicinal Organometallic Chemistry. Topics in Organometallic Chemistry*, 32. Springer; 2010.

26. Motterlini R, Otterbein LE. The therapeutic potential of carbon monoxide. *Nat. Rev. Drug Discovery* 2010;**9**(9):728–43.
27. Inaba H, Fujita K, Ueno T. Design of biomaterials for intracellular delivery of carbon monoxide. *Biomater. Sci.* 2015;**3**(11):1423–38.
28. Kautz AC, Kunz PC, Janiak C. CO-releasing molecule (CORM) conjugate systems. *Dalton Trans.* 2016;**45**(45):18045–63.
29. Fujita K, Tanaka Y, Sho T, Ozeki S, Abe S, Hikage T, et al. Intracellular CO release from composite of ferritin and ruthenium carbonyl complexes. *J. Am. Chem. Soc.* 2014;**136**(48):16902–8.
30. Fujita K, Tanaka Y, Abe S, Ueno T. A photoactive carbon-monoxide-releasing protein cage for dose-regulated delivery in living cells. *Angew. Chem. Int. Ed.* 2016;**55**(3):1056–60.
31. Kim HS, Loughran PA, Rao J, Billiar TR, Zuckerbraun BS. Carbon monoxide activates NF-kB via ROS generation and Akt pathways to protect against cell death of hepatocytes. *Am. J. Physiol. Gastrointest. Liver Physiol.* 2008;**295**(1):G146–52.
32. Motterlini R, Clark JE, Foresti R, Sarathchandra P, Mann BE, Green C. Carbon monoxide-releasing molecules: characterization of biochemical and vascular activities. *Circ. Res.* 2002;**90**(2):e17–24.
33. Yokoi N, Miura Y, Huang C-Y, Takatani N, Inaba H, Koshiyama T, et al. Dual modification of a triple-stranded ß-helix nanotube with Ru and Re metal complexes to promote photocatalytic reduction of CO_2. *Chem. Commun.* 2011;**47**(7):2074–6.
34. Kanamaru S, Leiman PG, Kostyuchenko VA, Chipman PR, Mesyanzhinov VV, Arisaka F, et al. Structure of the cell-puncturing device of bacteriophage T4. *Nature* 2002;**415**:553–7.
35. Yokoi N, Inaba H, Terauchi M, Stieg AZ, Sanghamitra NJM, Koshiyama T, et al. Construction of robust bio-nanotubes using the controlled self-assembly of component proteins of bacteriophage T4. *Small* 2010;**6**(17):1873–9.
36. Inaba H, Sanghamitra NJM, Fukai T, Matsumoto T, Nishijo K, Kanamaru S, et al. Intracellular protein delivery system with protein needle–GFP construct. *Chem. Lett.* 2014;**43**(9):1505–7.
37. Inaba H, Sanghamitra NJM, Fujita K, Sho T, Kuchimaru T, Kitagawa S, et al. A metal carbonyl-protein needle composite designed for intracellular CO delivery to modulate NF-κB activity. *Mol. Biosyst.* 2015;**11**(11):3111–18.
38. Abe S, Maity B, Ueno T. Functionalization of protein crystals with metal ions, complexes and nanoparticles. *Curr. Opin. Chem. Biol.* 2018;**43**:68–76.
39. H. Tabe, H. Takahashi, T. Shimoi, S. Abe, T. Ueno, Y. Yamada, Photocatalytic hydrogen evolution systems constructed in cross-linked porous protein crystals, *Appl. Catal. B* 2018. http://doi.org/10.1016/j.apcatb.2018.01.046
40. Binkley SL, Ziegler CJ, Herrick RS, Rowlett RS. Specific derivatization of lysozyme in aqueous solution with $Re(CO)_3(H_2O)_3^+$. *Chem. Commun.* 2010;**46**(8):1203–5.
41. Razavet M, Artero V, Cavazza C, Oudart Y, Lebrun C, Fontecilla-Camps JC, et al. Tricarbonylmanganese(I)-lysozyme complex: A structurally characterized organometallic protein. *Chem. Commun.* 2007;(27):2805–7.
42. Seixas JD, Mukhopadhyay A, Santos-Silva T, Otterbein LE, Gallo DJ, Rodrigues SS, et al. Characterization of a versatile organometallic pro-drug (CORM) for experimental CO based therapeutics. *Dalton Trans.* 2013;**42**(17):5985–98.
43. Binkley SL, Leeper TC, Rowlett RS, Herrick RS, Ziegler CJ. $Re(CO)_3(H_2O)_3^+$ binding to lysozyme: structure and reactivity. *Metallomics* 2011;**3**(9):909–16.

44. Pontillo N, Ferraro G, Messori L, Tamasi G, Merlino A. Ru-Based CO releasing molecules with azole ligands: interaction with proteins and the CO release mechanism disclosed by X-ray crystallography. *Dalton Trans.* 2017;**46**(29):9621–9.
45. Tabe H, Fujita K, Abe S, Tsujimoto M, Kuchimaru T, Kizaka-Kondoh S, et al. Preparation of a cross-linked porous protein crystal containing Ru carbonyl complexes as a CO-releasing extracellular scaffold. *Inorg. Chem.* 2015;**54**(1):215–20.
46. Tabe H, Shimoi T, Fujita K, Abe S, Ijiri H, Tsujimoto M, et al. Design of a CO-releasing extracellular scaffold using in vivo protein crystals. *Chem. Lett.* 2015;**44**(3):342–4.
47. Tabe H, Shimoi T, Boudes M, Abe S, Coulibaly F, Kitagawa S, et al. Photoactivatable CO release from engineered protein crystals to modulate NF-κB activation. *Chem. Commun.* 2016;**52**(24):4545–8.
48. Abe S, Atsumi K, Yamashita K, Hirata K, Mori H, Ueno T. Structure of in cell protein crystals containing organometallic complexes. *PCCP* 2018;**20**(5):2989–4548.
49. Ferraro G, Gabbiani C, Merlino A. First crystal structure for a gold carbene-protein adduct. *Bioconjugate Chem.* 2016;**27**(7):1584–7.

Chapter 17

[FeFe]-Hydrogenases Models

Catherine Elleouet, François Y. Pétillon and Philippe Schollhammer
Université de Bretagne Occidentale, Brest; CNRS, UMR 6521, Chimie, Electrochimie Moléculaires et Chimie Analytique, CS 93837, Brest Cedex 3, France

17.1 INTRODUCTION

[FeFe]- and [NiFe]-hydrogenases are metalloenzymes able to catalyze the reversible H^+/H_2 conversion with an efficiency comparable to that of platinum.[1–4] A third type of hydrogenases, called [Fe]-hydrogenases, was structurally described subsequently,[5] but is not involved in the consumption or catalytic production of dihydrogen according to the needs of microorganisms in the nature. Twenty years ago, the structural X-ray determinations of [FeFe]-hydrogenases[6,7] revealed that the active site of these enzymes, named the H-cluster, contains an organometallic subsite, $[2Fe]_H$, covalently linked by a cysteine bridge to a $[Fe_4S_4]$ ferredoxin cluster (Fig. 17.1). Each iron atom in the bimetallic subunit is bonded to two terminal ligands, CO and CN^-, a third carbonyl is in a bridging position.[2,8–10] Several studies afforded evidence that the bridging dithiolate is functionalized with an amine group (NH).[11–13]

During the past two decades, well-known organometallic synthetic routes[14] have been used for the design of diiron dithiolate carbonyl compounds $[Fe_2(CO)_{6-x}(L)_x(\mu\text{-dithiolate})]$ mimicking, mainly, the Fe^IFe^I (H_{red}) and Fe^IFe^{II} (H_{ox}) catalytically active states of [FeFe]-hydrogenases. The modeling, at a molecular level, of the H-cluster requires the introduction of a combination, more or less sophisticated, of three functional components: a redox functionality (Fe_4S_4 cluster or surrogates) as an electron-reservoir, a base-relay in the second sphere of coordination acting as a proton-shuttle, and an open coordination site (the rotated state at the distal iron atom (Fe_d) in the subunit) where the H^+/H_2 conversion occurs.

Some aspects of our contribution to this topic are presented herein. They concern the conception of diiron molecules in order to understand how to control the geometry of such mimics, but also the studies of the protonation processes of these molecules to know the becoming of protons and the location of protonation sites as well as electrochemical investigations to describe the redox properties of these models in the presence or not of substrates, such as H^+ or H_2.[15–17]

Advances in Bioorganometallic Chemistry. DOI: https://doi.org/10.1016/B978-0-12-814197-7.00017-0

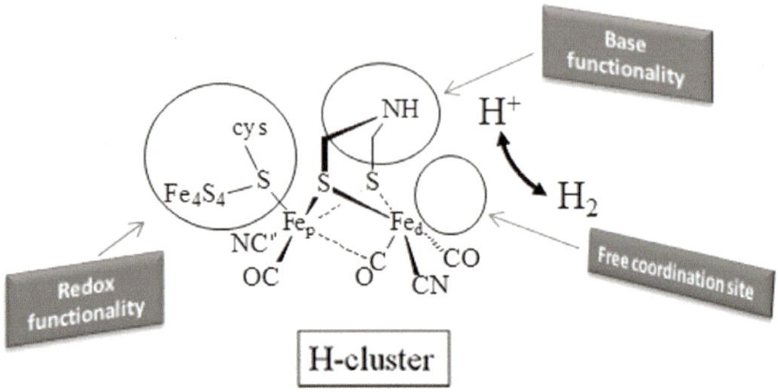

FIGURE 17.1 The active site (H-cluster) of [FeFe]-hydrogenases.

17.2 STRUCTURE AND PROTONATION OF UNSYMMETRICAL Fe^IFe^I MODELS

Theoretical studies have suggested that, in diiron hexacarbonyl dithiolate precursors, an unsymmetrical substitution of two carbonyl ligands by better σ-donor ligands at one metal center may stabilize Fe^IFe^I complexes with a geometry having an inverted conformation.[18] As the use of monodentate ligands does not allow the readily substitution of two carbonyls at the same metal center, the introduction of bidentate ligands has been considered, leading to unsymmetrically disubstituted complexes of general formula $[Fe_2(CO)_4(\kappa^2\text{-LL})(\mu\text{-dithiolate})]$.[19–26] Propanedithiolate (pdt) complexes $[Fe_2(CO)_4(\kappa^2\text{-LL})(\mu\text{-pdt})]$ were obtained with various chelating ligands (κ^2-LL), such as diphosphine (dppe = 1,2-bis(diphenylphosphino) ethane; dmpe = 1,2-bis(dimethylphosphino)ethane), phenanthroline, biscarbene (I_{Me}-CH_2-I_{Me}; I_{Me} = 1-methylimidazol-2-ylidene) and phosphine-N-heterocyclic-carbene (I_{Me}-$(CH_2)_2$-PPh_2; Ph = phenyl). They retain an eclipsed structure, both in solution and solid state (Fig. 17.2).

17.2.1 Protonation of Unsymmetrically Substituted Fe^IFe^I Models

The studies of their protonation in solution, using HBF_4 as acid, by ^1H, ^{13}C, and ^{31}P NMR spectroscopies at low temperature (below −40°C) demonstrated the formation of transient protonated species featuring a terminal hydride suggesting, as expected by theoretical calculations, that an intermediate with an inverted conformation is stabilized during the protonation process.[21,22] These terminal hydride species isomerize at room temperature, affording thermodynamic bridging hydride complexes (Fig. 17.3).

Above −40°C, the terminal hydride species $[Fe_2(CO)_4(\kappa^2\text{-dppe})(\mu\text{-pdt})(t\text{-H})]^+$ was not stable enough to study its electrocatalytic behavior toward proton reduction, but the electrochemical study by cyclic voltammetry (CV) of the

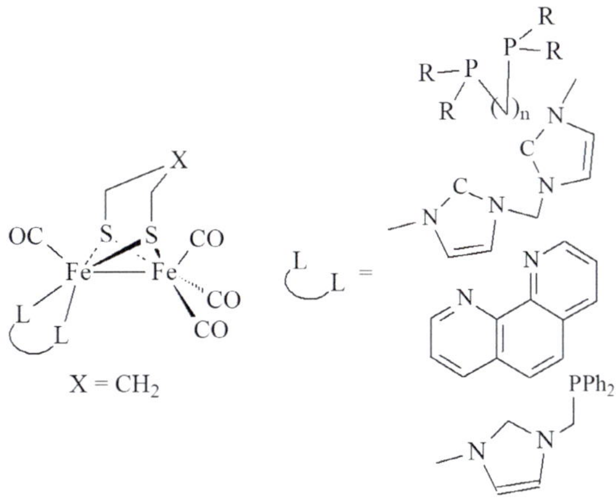

FIGURE 17.2 Unsymmetrically disubstitued diiron complexes.

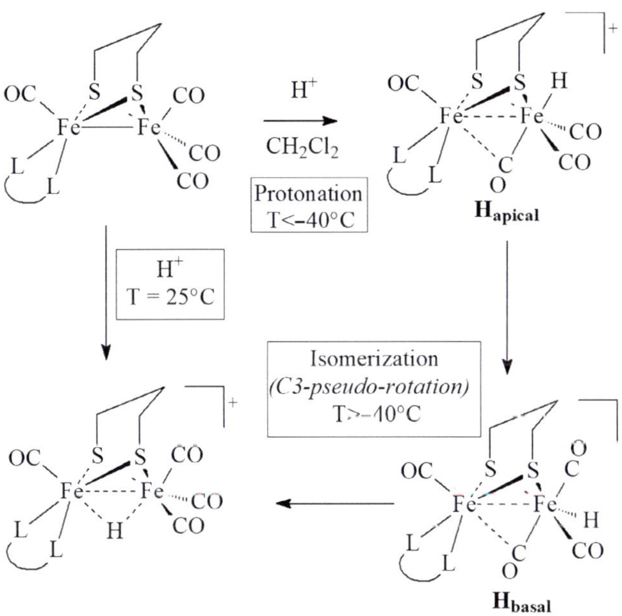

FIGURE 17.3 Protonation of complexes $[Fe_2(CO)_4(\kappa^2\text{-LL})(\mu\text{-pdt})]$ at 25°C and below −40°C.

bridging hydride isomer $[Fe_2(CO)_4(\kappa^2\text{-dppe})(\mu\text{-pdt})(\mu\text{-H})]^+$ revealed that this latter species behaves as a HER-catalyst (Hydrogen Evolution Reaction) at mild potential, ~ −1.3 V vs $[Fe(\eta^5\text{-}C_5H_5)_2]^+/[Fe(\eta^5\text{-}C_5H_5)_2]$ (Fc^+/Fc), according to the mechanism proposed in Fig. 17.4. This pathway involves the neutral reduced bridging hydride species as the first intermediate.[27]

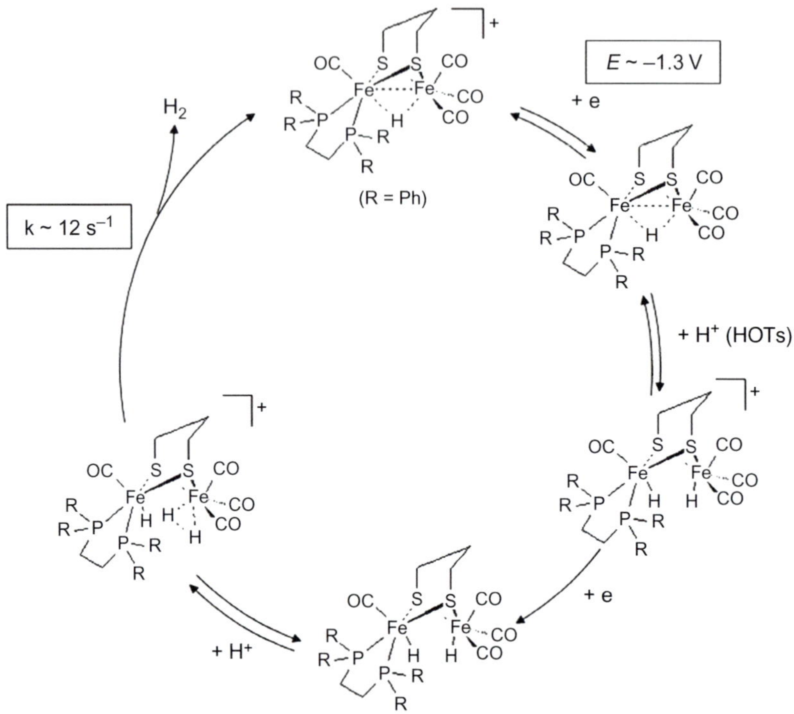

FIGURE 17.4 Proposed mechanism of proton reduction, from tosylic acid (HOTs, $pK_a^{MeCN} = 8.7$),[28] catalyzed by $[Fe_2(CO)_4(\kappa^2\text{-dppe})(\mu\text{-pdt})(\mu\text{-H})]^+$ in MeCN.

17.2.2 Fe^IFe^I Models with Inverted Conformation

Another approach to promote an inversion of the local square-pyramidal geometry at one iron atom is to crowd the dithiolate bridge.[29,30] Among the hundreds of model complexes of the H_{red} state of [FeFe]-hydrogenases that have been described,[8,14–16,31–33] only two examples have, in the solid state, a conformation structurally close to that of the active site of [FeFe]-hydrogenases (Fig. 17.5A).[34,35] The X-ray diffraction analysis of these complexes, $[Fe_2(CO)_4(\kappa^2\text{-dmpe})(\mu\text{-adt}^{Bn})]$ ($adt^{Bn} = \{SCH_2\}_2NCH_2C_6H_5$)[34] (Fig. 17.5B) and $[Fe_2(CO)_4(\kappa^2\text{-dppv})(\mu\text{-}\{SCH_2\}_2C(CH_2CH_3)_2)]$ (dppv = 1,2-bis(diphenylphosphino)ethylene)[35] (Fig. 17.5C), revealed the inverted geometry of one iron atoms. This conformation is stabilized by the presence of a hindered dithiolate bridge, an agostic bond and a dissymmetrically disubstituted diiron center. Density functional theory (DFT) calculations performed for the complex $[Fe_2(CO)_4(\kappa^2\text{-dmpe})(\mu\text{-adt}^{Bn})]$ confirmed the importance of the agostic interaction between the methylene proton of the benzyl and the adjacent iron atom to stabilize this peculiar conformation.[34] A third example having a semi-inverted conformation in the solid state has been also described

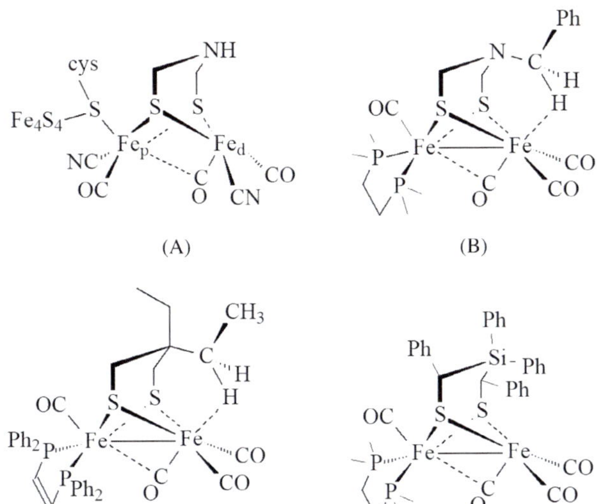

FIGURE 17.5 Rotated conformation in the H-cluster (A), and mimics (B-D).

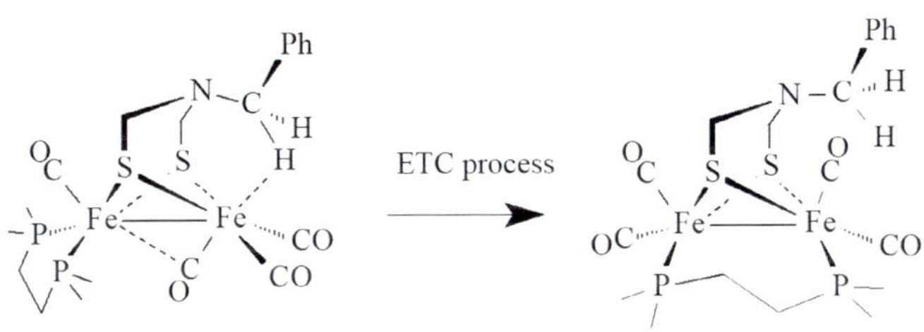

FIGURE 17.6 ETC isomerization of [$Fe_2(CO)_4(\kappa^2$-dmpe)(μ-adt^{Bn})].

(Fig. 17.5D).[36] In contrast to the two previous complexes, the compound [$Fe_2(CO)_4(\kappa^2$-dmpe)(μ-{SCHPh}$_2$SiPh$_2$)] has no agostic interaction, as indicated by the Fe---H distance (3.184 Å).[36,37] On the basis of IR studies, it has been proposed that the rotated conformation of [$Fe_2(CO)_4(\kappa^2$-dmpe) (μ-adt^{Bn})] is slithgly persistent in solution. In addition, the replacement of the benzyl group (Bn) by an isopropyl (i-Pr) one, as well as dmpe by dppe, results in the loss of the inverted conformation in the solid state; an eclipsed conformation (versus the Fe-Fe axis) of the two fragments {$Fe(CO)_3$} and {$Fe(CO)P_2$} is adopted in the complexes [$Fe_2(CO)_4(\kappa^2$-dmpe)(μ-adt$^{i\text{-}Pr}$)] (adt$^{i\text{-}Pr}$ = {SCH_2}$_2$NCH(CH_3)$_2$) and [$Fe_2(CO)_4(\kappa^2$-dppe)(μ-adt^{Bn})].

Chelated disubstituted compounds [$Fe_2(CO)_4(\kappa^2$-diphosphine) {μ-(SCH_2)$_2$NR}] (R = Bn, i-Pr, $CH_2CH_2OCH_3$) isomerize into bridging diphosphine compounds [$Fe_2(CO)_4$(μ-diphosphine)(μ-{SCH_2}$_2$NR)] through a reductive ETC mechanism (ETC = electron transfer chain) (Fig. 17.6).[38,39]

The comparison of the structure in solid state of the chelated and bridging isomers $[Fe_2(CO)_4(dmpe)\{\mu\text{-}(SCH_2)_2NBn\}]$ revealed that the inverted conformation is lost when dmpe adopts a symmetrical bridging position. The symmetrization (κ^2-dmpe)→(μ-dmpe) hinders the presence of the agostic Fe-H(CH) interaction stabilizing the rotated geometry.[39]

17.2.3 Models with Amine-Functionalized-Diphosphines

The use of amine-functionalized-diphosphines, as bases-relay in the second coordination sphere of unsymmetrically substituted diiron complexes[40–48] allows to study and to control the intra- and intermolecular proton circulation around the bimetallic site, as well as to speed up the protonation and/or to make it easier. Indeed, numerous studies have demonstrated the interest of introducing into mononuclear complexes of iron, nickel or cobalt, a diphosphine functionalized by one or two amine functions [$(Ph_2PCH_2)_2NR$ (PN^RP), $(PhPCH_2N^RCH_2)_2$ ($P_2^{Ph}N_2^R$, R = Ph, Bn, C_6H_4X)] for the activation (heterolytic activation) or production (hydride-proton interaction) of H_2. In particular, Ni^{II}-bis(diphosphine) complexes have been extensively studied.[49–52] The presence of the amine in the second coordination sphere increases the catalytic activity of these complexes toward the proton reduction because of the metal-base proximity and the stabilization of NH species.[53] On the basis of these works, diiron complexes having such amine functionalized-diphosphines (PN^RP or $P_2^{Ph}N_2^R$) were synthesized in order to introduce a non-coordinating basic function in the second coordination sphere (Fig. 17.7) of the bimetallic site and to evaluate their effect on the HER activity.

The study of these complexes has highlighted the role of the amine as a proton-relay when it is positioned near the diiron site, as it is in the biological process, able to capture a proton in the medium and to promote its transfer onto the bimetallic center according to a tautomerization process (Fig. 17.8). The process is controlled by the nature of the group R carried by the amine as well as the electronic behavior of the dithiolate bridge.[44] Unlike the process involved with the H-cluster, the proton is transferred into a bridging position.[48]

The various reduction processes observed experimentally and analyzed by DFT are rationalized in Fig. 17.9. The most interesting thermodynamically

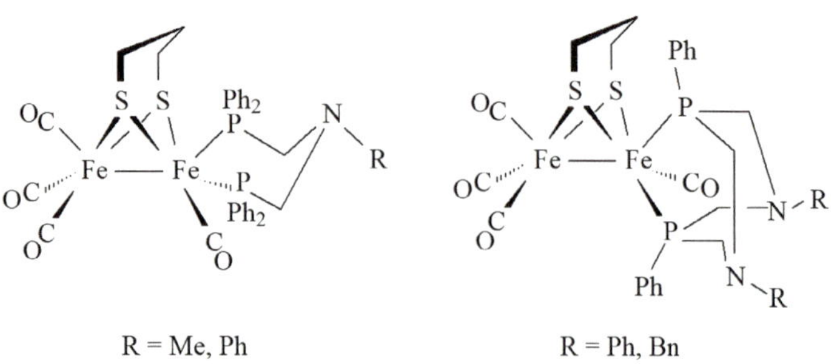

FIGURE 17.7 Diiron complexes with amine functionalized-diphosphines (PN^RP or $P_2^{Ph}N_2^R$).

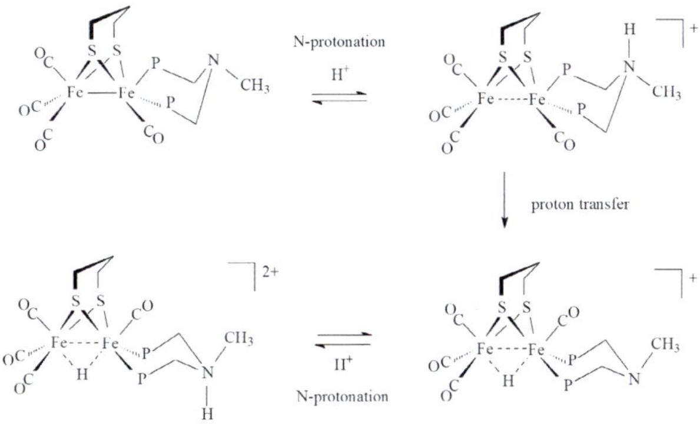

FIGURE 17.8 Protonation process of the complex $[Fe_2(CO)_4(\kappa^2\text{-}PN^{Me}P)(\mu\text{-}pdt)]$ ($PN^{Me}P = (Ph_2PCH_2)_2NCH_3$, phenyl on phosphorus atoms are omitted for clarity).

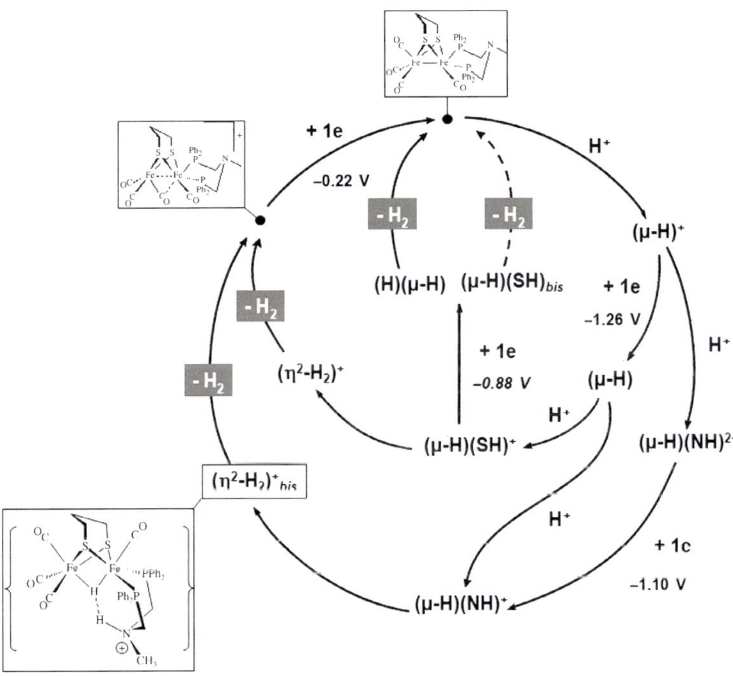

FIGURE 17.9 Proposed catalytic proton reduction mechanism for $[Fe_2(CO)_4(\kappa^2\text{-}PN^{Me}P)(\mu\text{-}pdt)]$ (potentials are referenced to the standard potential of Fc^+/Fc, the value in italic has been calculated.

reaction sequence is that corresponding to the outer cycle, comprising two successive protonations followed by a reduction step and a proton migration leading to the formation of a dihydrogen ligand coordinated in a η^2-mode, and then to the releasing of H_2. The second electron transfer occurs at the protonation/reduction level $\{2H^+/1e\}$, at a low negative potential (−0.22 V vs Fc^+/Fc). Theoretical calculations have shown that this catalytic cycle (outer circle) is unfortunately inefficient for kinetic reasons (proton migration and slow hydrogen release), whereas the lack of efficiency of the second catalytic process has a thermodynamic origin, in particular because of the formation of stable {$2H^+/2e$} species ((H)(μ-H) and (μ-H)$(SH)_{bis}$ stages in Fig. 17.9).

17.3 OXIDATION OF UNSYMMETRICALLY SUBSTITUTED Fe^IFe^I MODELS

The oxidative processes (one- or two-electron oxidation) involving diiron models have been less explored than the electrocatalytic proton reduction with Fe^IFe^I complexes. Fe^IFe^{II} and $Fe^{II}Fe^{II}$ species serve as models for H_{ox}, H_{ox}^{CO} and H_{ox}^{air} states of [FeFe]-H_2ases active site and the knowledge of their reactivity may help to understand how to generate an efficient-free coordination site at a diiron framework. Moreover, efficient activation of H_2 at non-noble catalysts is still a challenging perspective.

The theoretical study of Fe^IFe^I complexes $[Fe_2(CO)_4(\kappa^2\text{-LL})(\mu\text{-pdt})]$ (LL = dppe or I_{Me}-CH_2-I_{Me}), synthesized in the laboratory, suggests that the one-electron oxidation of these compounds leads initially to eclipsed Fe^IFe^{II} cationic species, which in a second stage evolves into a more stable inverted form whose formation could be observed by IR.[54] DFT calculations also showed the existence of two different isomers with an inverted conformation, at the iron atom carrying the bidendate ligand for the first, and at the iron atom bearing the three carbonyls for the second. The most stable isomer corresponds to the species exhibiting the pyramidal inversion on the {Fe(CO) (κ^2-LL)} fragment (Fig. 17.10). As expected, electrochemical studies carried out on these compounds showed that the nature of the chelating ligand affects the redox potentials of the various complexes. Indeed, the bis-carbene ligand (I_{Me}-CH_2-I_{Me}) being more electron-donor than dppe, the complex

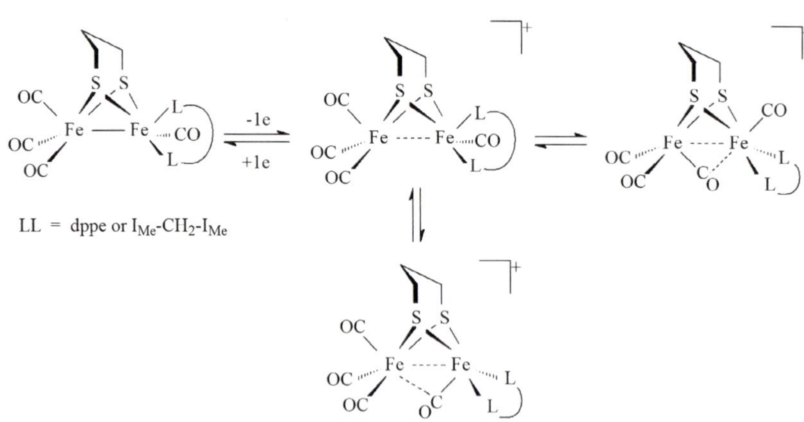

FIGURE 17.10 Oxidation of complexes $[Fe_2(CO)_4(\kappa^2\text{-LL})(\mu\text{-pdt})]$.

$[Fe_2(CO)_4(\kappa^2\text{-}I_{Me}\text{-}CH_2\text{-}I_{Me})(\mu\text{-pdt})]$ is oxidized at a more negative potential than the related dppe derivative.[54]

17.3.1 Oxidation in the Presence of CO and Other Substrates (MeCN, $P(OMe)_3$)

In the case of mono-oxidized forms of the complexes $[Fe_2(CO)_4(\kappa^2\text{-LL})(\mu\text{-pdt})]$ (LL = dppe or $I_{Me}\text{-}CH_2\text{-}I_{Me}$), the cationic adduct $[Fe_2(CO)_4(\kappa^2\text{-dppe})(\mu\text{-CO})(\mu\text{-pdt})]^+$ is detected during the electrochemical oxidation of $[Fe_2(CO)_4(\kappa^2\text{-dppe})(\mu\text{-pdt})]$ under a CO atmosphere, while the analogous bis-carbene adduct $[Fe_2(CO)_4(\kappa^2\text{-}I_{Me}\text{-}CH_2\text{-}I_{Me})(\mu\text{-CO})(\mu\text{-pdt})]^+$ is not observed in cyclic voltammetry.[55] DFT results indicate that their free energies and the related activation barriers to form them are similar, but do not explain this difference of reactivity. In addition, it was evidenced that the electrochemical oxidation of the complex $[Fe_2(CO)_4(\kappa^2\text{-bipy})(\mu\text{-pdt})]$ (bipy = 2,2′-bipyridine) is more reversible under CO than under inert atmosphere suggesting that the oxidation step can also result in the loss of a carbonyl.[56] This loss is not observed for $[Fe_2(CO)_4(\kappa^2\text{-dppe})(\mu\text{-pdt})]$ and $[Fe_2(CO)_4(\kappa^2\text{-}I_{Me}\text{-}CH_2\text{-}I_{Me})(\mu\text{-pdt})]$.[54,55] The reactivity of the electrogenerated oxidized parent species $[Fe_2(CO)_4(\kappa^2\text{-LL})(\mu\text{-pdt})]^+$ (LL = dppe or $I_{Me}\text{-}CH_2\text{-}I_{Me}$) toward different substrates, such as MeCN and $P(OMe)_3$, has also been described.[55] These studies revealed the formation of adducts in which a MeCN or $P(OMe)_3$ ligand coordinates to the metal center. Although the DFT study predicts that, in the absence of substrate, the most stable isomer corresponds to the species exhibiting the pyramidal inversion at the $\{Fe(CO)(\kappa^2\text{-LL})\}$ fragment (Fig. 17.10), the coordination of the incoming substrate MeCN or $P(OMe)_3$ takes place at the $\{Fe(CO)_3\}$ fragment (Fig. 17.11).

In these cationic complexes, the nature of the chelating ligand (LL) controls the reactivity of the adjacent iron $\{Fe(CO)_3\}$ group. Indeed, the replacement of dppe by bis-carbene induces a decrease of the reactivity toward MeCN. A difference in the reactivity of the mono-cations $[Fe_2(CO)_3(\kappa^2\text{-LL})(\mu\text{-CO})(\mu\text{-pdt})]^+$ (LL = dppe or $I_{Me}\text{-}CH_2\text{-}I_{Me}$) is also observed when $P(OMe)_3$ is used. The $[Fe_2(CO)_2\{P(OMe)_3\}_2(\kappa^2\text{-}I_{Me}\text{-}CH_2\text{-}I_{Me})(\mu\text{-CO})(\mu\text{-pdt})]^{2+}$ cation is obtained via a disproportionation reaction while this reaction is not favored when the cation $[Fe_2(CO)_3(\kappa^2\text{-dppe})(\mu\text{-CO})(\mu\text{-pdt})]^+$ reacts with $P(OMe)_3$, the major products in this latter case are the derivatives $[Fe_2(CO)_{4\text{-}x}\{P(OMe)_3\}_x(\kappa^2\text{-dppe})(\mu\text{-pdt})]^+$ ($x = 1$ or 2) (Fig. 17.12). The electrochemical experiments and theoretical calculations reveal the complexity of the mechanisms involving these diiron models. They suggest that the difference of lability of CO in cationic intermediates, due to the nature of the chelating ligand, induces the formation of very different products. In the case of the dppe complex, the short lifetime of the intermediate would annihilate the competitive disproportionation reaction.

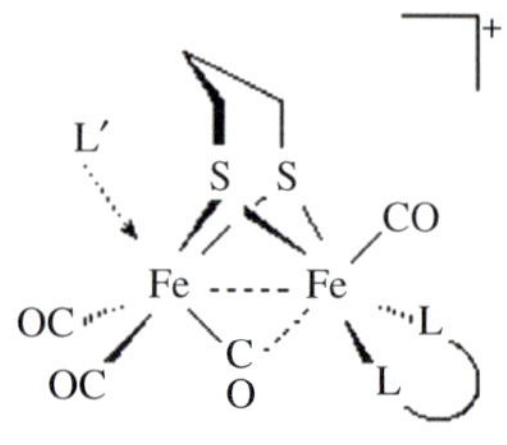

LL = dppe or I_{Me}-CH_2-I_{Me} ; L′ = MeCN, $P(OMe)_3$

FIGURE 17.11 Coordination of an incoming substrate L' at $[Fe_2(CO)_4(\kappa^2\text{-LL})(\mu\text{-pdt})]^+$.

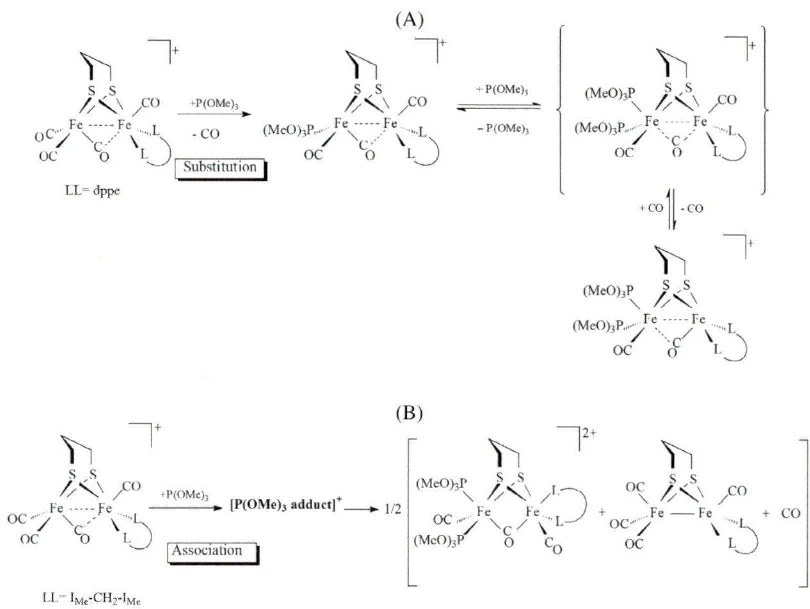

FIGURE 17.12 Reactivity of $[Fe_2(CO)_3(\kappa^2\text{-LL})(\mu\text{-CO})(\mu\text{-pdt})]^+$ (LL = dppe (a); I_{Me}-CH_2-I_{Me} (b)) with $P(OMe)_3$.

17.3.2 $Fe^{II}Fe^{II}$ Complexes as Models of H_{ox}^{air} State

It has been previously noted that the reaction of $[Fe_2(CO)_4(\kappa^2\text{-}I_{Me}\text{-}CH_2\text{-}I_{Me})(\mu\text{-CO})(\mu\text{-pdt})]^+$ with the substrate $P(OMe)_3$ leads to $[Fe_2(CO)_2\{(P(OMe))_3\}_2(\kappa^2\text{-}I_{Me}\text{-}CH_2\text{-}I_{Me})(\mu\text{-CO})(\mu\text{-pdt})]^{2+}$ (Fig. 17.13). This $Fe^{II}Fe^{II}$ complex, with six terminal ligands and a bridging CO, is a model of the inactive form H_{ox}^{air} of [FeFe]-hydrogenases. Cyclic voltammetry revealed that it is reduced according to a ECE mechanism (with C corresponding to the rapid loss of a $P(OMe)_3$) to give a product which is re-oxidized in $[Fe_2(CO)_2\{P(OMe)_3\}_2(\kappa^2\text{-}I_{Me}\text{-}CH_2\text{-}I_{Me})(\mu\text{-CO})(\mu\text{-pdt})]^{2+}$, but involving a different reaction pathway (EEC mechanism with C corresponding to the addition of a $P(OMe)_3$).[57] Theoretical calculations and CV simulations suggest the formation of $Fe^{II}Fe^{II}$ intermediates, stabilized by an agostic

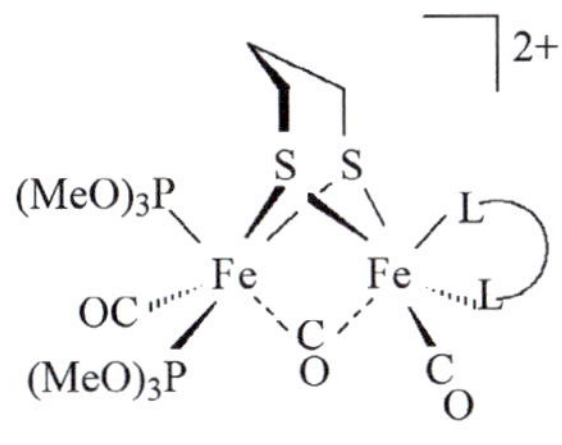

LL= I_{Me}-CH_2-I_{Me}

FIGURE 17.13 Geometry of the dication $[Fe_2(CO)_2\{P(OMe)_3\}_2(\kappa^2\text{-}I_{Me}\text{-}CH_2\text{-}I_{Me})(\mu\text{-}CO)(\mu\text{-pdt})]^{2+}$.

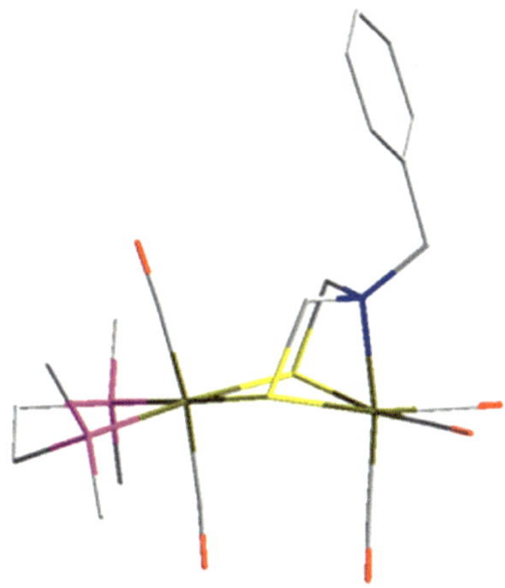

FIGURE 17.14 Stick drawing of the molecular structure of the cation $[Fe_2(CO)_5(\kappa^2\text{-dmpe})(\mu\text{-adt}^{Bn})]^{2+}$.

interaction between the non-central CH_2 group of the dithiolate bridge and one iron atom, in the reoxidation process.

17.3.3 $Fe^{II}Fe^{II}$ Complexes with an Azadithiolate Bridge

When the dithiolate bridge has an amine function, important changes in the oxidative behavior of the $Fe^{I}Fe^{I}$ complexes are observed with respect to the pdt analogs. The bielectronic oxidation of some of these complexes in the presence of substrates such as CO leads to the coordination of the amine function at one iron atom (Fig. 17.14).[58] Few examples of this type of complex have been described until now.[59,60]

The comparison of the electrochemical properties of the two complexes $[Fe_2(CO)_4(\kappa^2\text{-dmpe})(\mu\text{-dithiolate})]$ (dithiolate = adt^{Bn}, pdt) and the rationalization of the redox processes by DFT calculations by modelizing the oxidized parent species reveal the striking influence of an amine functionality in the dithiolate bridge on the oxidative properties of diiron complexes $[Fe_2(CO)_4(\kappa^2\text{-chelate})(\mu\text{-dithiolate})]$.[58] Particularly, the role of stabilizing interactions in elusive cationic species proposed as intermediates during the oxidation of these $Fe^{I}Fe^{I}$ derivatives has been evidenced (Fig. 17.15).

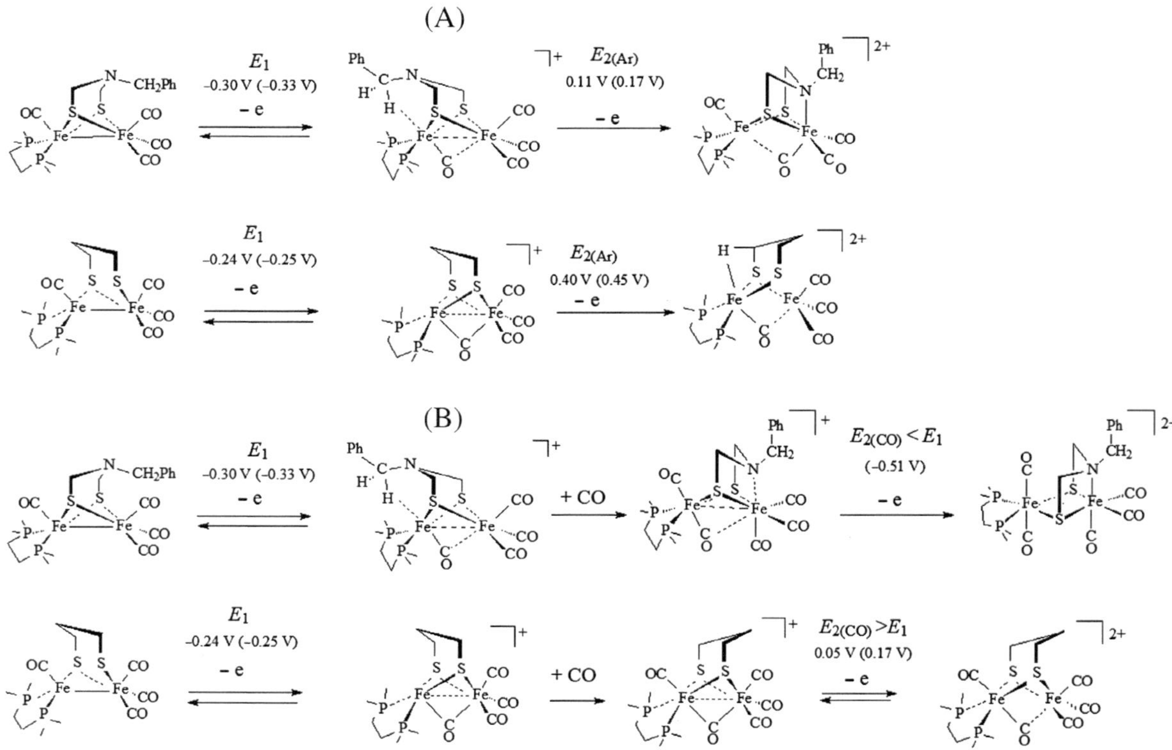

FIGURE 17.15 Proposed mechanism for the oxidation of $[Fe_2(CO)_4(\kappa^2\text{-dmpe})(\mu\text{-adt}^{Bn})]$ and $[Fe_2(CO)_4(\kappa^2\text{-dmpe})(\mu\text{-pdt})]$ under inert atmosphere (A) or under CO (B). Potentials are quoted against the (Fc^+/Fc) couple. Calculated values are in brackets.

Indeed, the different properties observed when the adt^{Bn} bridge is replaced by pdt is due to the presence, or not, of interactions between the dithiolate bridge and the diiron site (agostic Fe...HC interaction versus Fe...N interaction) depending on the redox stages, $Fe^{I}Fe^{II}$ or $Fe^{II}Fe^{II}$, of the complexes.

It has to be noted that the coordination of the amine function of the azadithiolate bridge at one iron atom has never been described in the inactive forms of the H-cluster and, therefore, does not seem to be a cause of deactivation of [FeFe]-hydrogenases. Theoretical calculations suggest that the protein environment of the enzyme would prevent the coordination of the amine.[61]

17.4 CONCLUSION

The activity of the H-cluster of [FeFe]-hydrogenases is governed by fine redox changes that allow its remarkable catalytic efficiency toward H^+/H_2 conversion. Despite numerous works, a deep understanding of redox properties of models is still required to have an intimate knowledge of the steps controlling the activity of such diiron auxiliaries to be able to reproduce an efficient functioning without a proteic matrix and for having a hydrogenases-like activity. Whereas the reduction of H^+ into H_2 has been widely studied, the oxidative processes involving such diiron mimics have been less explored despite the challenging perspective that would be to activate efficiently H_2 at such a bimetallic site.

ACKNOWLEDGMENTS

We warmly thank Dr. J. Talarmin (Université de Bretagne Occidentale), who is now retired, for his invaluable contribution to the works presented here, our colleagues from the research group in Brest who have contributed to this topic, our colleagues from the University of Milano-Bicocca who have performed the DFT calculations, and the PhD students who appeared in our references.

REFERENCES

1. Lubitz W, Ogata H, Rüdiger O, Reijerse E. Hydrogenases. *Chem Rev* 2014;**114**(8):4081–148. Available from: https://doi.org/10.1021/cr4005814.
2. Lubitz W, Reijerse E, van Gastel M. [NiFe] and [FeFe] Hydrogenases studied by advanced magnetic resonance techniques. *Chem Rev* 2007;**107**(10):4331–65. Available from: https://doi.org/10.1021/cr050186q.
3. Frey M. Hydrogenases: hydrogen-activating enzymes. *ChemBioChem* 2002;**3**:153–60 doi:10.1002/1439-7633(20020301)3:2/3 < 153::AID-CBIC153 > 3.0.CO;2-B.
4. Weigand W, Schollhammer P, editors. *Bioinspired catalysis: Metal-sulfur complexes*. Weinheim: Wiley-VCH-Verl; 2015.
5. Hiromoto T, Ataka K, Pilak O, Vogt S, Stagni MS, Meyer-Klaucke W, et al. The crystal structure of C176A mutated [Fe]-hydrogenase suggests an acyl-iron ligation in the active site iron complex. *FEBS Lett* 2009;**583**(3):585–90. Available from: https://doi.org/10.1016/j.febslet.2009.01.017.

6. Peters JW. X-ray crystal structure of the Fe-only hydrogenase (CpI) from *Clostridium pasteurianum* to 1.8 angstrom resolution. *Science* 1998;**282**(5395):1853–8. Available from: https://doi.org/10.1126/science.282.5395.1853.
7. Nicolet Y, Piras C, Legrand P, Hatchikian CE, Fontecilla-Camps JC. *Desulfovibrio desulfuricans* iron hydrogenase: the structure shows unusual coordination to an active site Fe binuclear center. *Structure* 1999;**7**:13–23. Available from: https://doi.org/10.1016/S0969-2126(99)80005-7.
8. Tard C, Pickett CJ. Structural and functional analogues of the active sites of the [Fe]-, [NiFe]-, and [FeFe]-hydrogenases. *Chem Rev* 2009;**109**(6):2245–74. Available from: https://doi.org/10.1021/cr800542q.
9. Fontecilla-Camps JC, Volbeda A, Cavazza C, Nicolet Y. Structure/function relationships of [NiFe]- and [FeFe]-hydrogenases. *Chem Rev* 2007;**107**(10):4273–303. Available from: https://doi.org/10.1021/cr050195z.
10. Vincent KA, Parkin A, Armstrong FA. Investigating and exploiting the electrocatalytic properties of hydrogenases. *Chem Rev* 2007;**107**(10):4366–413. Available from: https://doi.org/10.1021/cr050191u.
11. Nicolet Y, de Lacey AL, Vernède X, Fernandez VM, Hatchikian EC, Fontecilla-Camps JC. Crystallographic and FTIR spectroscopic evidence of changes in Fe coordination upon reduction of the active site of the Fe-only hydrogenase from *Desulfovibrio desulfuricans. J Am Chem Soc* 2001;**123**:1596–601. Available from: https://doi.org/10.1021/ja0020963.
12. Silakov A, Wenk B, Reijerse E, Lubitz W. ^{14}N HYSCORE investigation of the H-cluster of [FeFe] hydrogenase: Evidence for a nitrogen in the dithiol bridge. *Phys Chem Chem Phys* 2009;**11**:6592–9. Available from: https://doi.org/10.1039/B905841A.
13. Berggren G, Adamska A, Lambertz C, Simmons TR, Esselborn J, Atta M, et al. Biomimetic assembly and activation of [FeFe]-hydrogenases. *Nature* 2013;**499**:66–9. Available from: https://doi.org/10.1038/nature12239.
14. Li Y, Rauchfuss TB. Synthesis of diiron(I) dithiolato carbonyl complexes. *Chem Rev* 2016;**116**(12):7043–77. Available from: https://doi.org/10.1021/acs.chemrev.5b00669.
15. Capon J-F, Gloaguen F, Pétillon FY, Schollhammer P, Talarmin J. Electron and proton transfers at diiron dithiolate sites relevant to the catalysis of proton reduction by the [FeFe]-hydrogenases. *Coord Chem Rev* 2009;**253**:1476–94. Available from: https://doi.org/10.1016/j.ccr.2008.10.020.
16. Capon J-F, Gloaguen F, Pétillon FY, Schollhammer P, Talarmin J. Organometallic diiron complex chemistry related to the $[2Fe]_H$ subsite of $[FeFe]H_2$ase. *Eur J Inorg Chem* 2008;4671–81. Available from: https://doi.org/10.1002/ejic.200800717.
17. Capon J-F, Gloaguen F, Pétillon FY, Schollhammer P, Talarmin J. On the electrochemistry of diiron dithiolate complexes related to the active site of the $[FeFe]H_2$ase. *C R Chim* 2008;**11**:842–51. Available from: https://doi.org/10.1016/j.crci.2008.03.007.
18. Tye JW, Darensbourg MY, Hall MB. De Novo design of synthetic di-iron(I) complexes as structural models of the reduced form of iron – iron hydrogenase. *Inorg Chem* 2006;**45**(4):1552–9. Available from: https://doi.org/10.1021/ic051231f.
19. Duan L, Wang M, Li P, Na Y, Wang N, Sun L. Carbene–pyridine chelating 2Fe2S hydrogenase model complexes as highly active catalysts for the electrochemical reduction of protons from weak acid (HOAc). *Dalton Trans* 2007;(13):1277–83. Available from: https://doi.org/10.1039/B616645H.
20. Morvan D, Capon J-F, Gloaguen F, Le Goff A, Marchivie M, Michaud F, et al. N-Heterocyclic carbene ligands in nonsymmetric diiron models of hydrogenase active sites. *Organometallics* 2007;**26**:2042–52. Available from: https://doi.org/10.1021/om061173l.

21. Ezzaher S, Capon J-F, Gloaguen F, Pétillon FY, Schollhammer P, Talarmin J, et al. Evidence for the formation of terminal hydrides by protonation of an asymmetric iron hydrogenase active site mimic. *Inorg Chem* 2007;**46**(9):3426–8. Available from: https://doi.org/10.1021/ic0703124.
22. Ezzaher S, Capon J-F, Gloaguen F, Kervarec N, Pétillon FY, Pichon R, et al. Diiron chelate complexes relevant to the active site of the iron-only hydrogenase. *C R Chim* 2008;**11**:906–14. Available from: https://doi.org/10.1016/j.crci.2008.04.002.
23. Orain P-Y, Capon J-F, Kervarec N, Gloaguen F, Pétillon F, Pichon R, et al. Use of 1,10-phenanthroline in diiron dithiolate derivatives related to the [Fe–Fe] hydrogenase active site. *Dalton Trans* 2007;(34):3754–6. Available from: https://doi.org/10.1039/B709287C.
24. Adam FI, Hogarth G, Richards I. Models of the iron-only hydrogenase: Reactions of [$Fe_2(CO)_6(\mu$-pdt)] with small bite-angle diphosphines yielding bridge and chelate diphosphine complexes [$Fe_2(CO)_4$(diphosphine)(μ-pdt)]. *J Organomet Chem* 2007;**692**:3957–68. Available from: https://doi.org/10.1016/j.jorganchem.2007.05.050.
25. Adam FI, Hogarth G, Richards I, Sanchez BE. Models of the iron-only hydrogenase: Structural studies of chelating diphosphine complexes [$Fe_2(CO)_4(\mu$-pdt)(κ^2 P,P′-diphosphine)]. *Dalton Trans* 2007;2495–8. Available from: https://doi.org/10.1039/B706123B.
26. Morvan D, Capon J-F, Gloaguen F, Pétillon FY, Schollhammer P, Talarmin J, et al. Modeling [FeFe] hydrogenase: Synthesis and protonation of a diiron dithiolate complex containing a phosphine-N-heterocyclic-carbene ligand. *J Organomet Chem* 2009;**694**:2801–7. Available from: https://doi.org/10.1016/j.jorganchem.2009.01.018.
27. Ezzaher S, Capon J-F, Dumontet N, Gloaguen F, Pétillon FY, Schollhammer P, et al. Electrochemical study of the role of a H-bridged, unsymmetrically disubstituted diiron complex in proton reduction catalysis. *J Electroanal Chem* 2009;**626**:161–70. Available from: https://doi.org/10.1016/j.jelechem.2008.12.005.
28. Felton GAN, Glass RS, Lichtenberger DL, Evans DH. Iron-only hydrogenase mimics. thermodynamic aspects of the use of electrochemistry to evaluate catalytic efficiency for hydrogen generation. *Inorg Chem* 2006;**45**(23):9181–4. Available from: https://doi.org/10.1021/ic060984e.
29. Singleton ML, Bhuvanesh N, Reibenspies JH, Darensbourg MY. Synthetic support of De Novo design: Sterically bulky [FeFe]-hydrogenase models. *Angew Chem Int Ed* 2008;**47**(49):9492–5. Available from: https://doi.org/10.1002/anie.200803939.
30. Hsieh C-H, Erdem ÖF, Harman SD, Singleton ML, Reijerse E, Lubitz W, et al. Structural and spectroscopic features of mixed valent $Fe^{II}Fe^{I}$ complexes and factors related to the rotated configuration of diiron hydrogenase. *J Am Chem Soc* 2012;**134**(31):13089–102. Available from: https://doi.org/10.1021/ja304866r.
31. Gloaguen F, Rauchfuss TB. Small molecule mimics of hydrogenases: Hydrides and redox. *Chem Soc Rev* 2009;**38**:100–8. Available from: https://doi.org/10.1039/B801796B.
32. Felton GAN, Mebi CA, Petro BJ, Vannucci AK, Evans DH, Glass RS, et al. Review of electrochemical studies of complexes containing the Fe_2S_2 core characteristic of [FeFe]-hydrogenases including catalysis by these complexes of the reduction of acids to form dihydrogen. *J Organomet Chem* 2009;**694**:2681–99. Available from: https://doi.org/10.1016/j.jorganchem.2009.03.017.
33. Tschierlei S, Ott S, Lomoth R. Spectroscopically characterized intermediates of catalytic H_2 formation by [FeFe] hydrogenase models. *Energy Environ Sci* 2011;**4**:2340–52. Available from: https://doi.org/10.1039/C0EE00708K.

34. Munery S, Capon J-F, De Gioia L, Elleouet C, Greco C, Pétillon FY, et al. New Fe^{I}-Fe^{I} complex featuring a rotated conformation related to the $[2Fe]_H$ subsite of [Fe-Fe] hydrogenase. *Chem Eur J* 2013;**19**(46):15458–61. Available from: https://doi.org/10.1002/chem.201303316.
35. Wang W, Rauchfuss TB, Moore CE, Rheingold AL, De Gioia L, Zampella G. Crystallographic characterization of a fully rotated, basic diiron dithiolate: Model for the H_{red} state?. *Chem Eur J* 2013;**19**(46):15476–9. Available from: https://doi.org/10.1002/chem.201303351.
36. Goy R, Bertini L, Elleouet C, Görls H, Zampella G, Talarmin J, et al. A sterically stabilized Fe^{I}–Fe^{I} semi-rotated conformation of [FeFe] hydrogenase subsite model. *Dalton Trans* 2015;**44**(4):1690–9. Available from: https://doi.org/10.1039/C4DT03223C.
37. Brookhart M, Green MLH, Parkin G. Agostic interactions in transition metal compounds. *Proc Natl Acad Sci* 2007;**104**(17):6908–14. Available from: https://doi.org/10.1073/pnas.0610747104.
38. Ezzaher S, Capon J-F, Gloaguen F, Pétillon FY, Schollhammer P, Talarmin J. Electron-transfer-catalyzed rearrangement of unsymmetrically substituted diiron dithiolate complexes related to the active site of the [FeFe]-hydrogenases. *Inorg Chem* 2007;**46**(23):9863–72. Available from: https://doi.org/10.1021/ic701327w.
39. De Gioia L, Elleouet C, Munery S, Pétillon FY, Schollhammer P, Talarmin J, et al. Reductive behavior of $[Fe_2(CO)_4(\kappa^2\text{-dmpe})\{\mu\text{-}(SCH_2)_2NBn\}]$: Effect of symmetrization on the rotated conformation in Fe^{I}-Fe^{I} models of $[2Fe]_H$ subsite of [Fe-Fe]-H_2ases. *Eur J Inorg Chem* 2014;3456–61. Available from: https://doi.org/10.1002/ejic.201402335.
40. Ezzaher S, Capon J-F, Gloaguen F, Pétillon FY, Schollhammer P, Talarmin J, et al. Influence of a pendant amine in the second coordination sphere on proton transfer at a dissymmetrically disubstituted diiron system related to the $[2Fe]_H$ subsite of $[FeFe]H_2$ase. *Inorg Chem* 2009;**48**:2–4. Available from: https://doi.org/10.1021/ic801369u.
41. Wang N, Wang M, Zhang T, Li P, Liu J, Sun L. A proton–hydride diiron complex with a base-containing diphosphine ligand relevant to the [FeFe]-hydrogenase active site. *Chem Commun* 2008;(44):5800–2. Available from: https://doi.org/10.1039/B811352A.
42. Wang N, Wang M, Liu J, Jin K, Chen L, Sun L. Preparation, facile deprotonation, and rapid H/D exchange of the μ-hydride diiron model complexes of the [FeFe]-hydrogenase containing a pendant amine in a chelating diphosphine ligand. *Inorg Chem* 2009;**48**(24):11551–8. Available from: https://doi.org/10.1021/ic901154m.
43. Lounissi S, Capon J-F, Gloaguen F, Matoussi F, Pétillon FY, Schollhammer P, et al. Reactivity of $[Fe_2(CO)_6(\mu\text{-}S_2)]$ toward a base-containing diphosphine $(Ph_2PCH_2)_2NCH_3$: Formation of diiron carbonyl compounds having polydentate heterofunctionalized phosphine (PNS = $Ph_2PCH_2N(CH_3)CH_2S$) and bidentate thiophosphinito (Ph_2PS = PS) bridges. *Organometallics* 2010;**29**:1296–301. Available from: https://doi.org/10.1021/om9011054.
44. Ezzaher S, Gogoll A, Bruhn C, Ott S. Directing protonation in [FeFe] hydrogenase active site models by modifications in their second coordination sphere. *Chem Commun* 2010;**46**(31):5775–7. Available from: https://doi.org/10.1039/C0CC00724B.
45. Wang Y, Wang M, Sun L, Ahlquist MSG. Pendant amine bases speed up proton transfers to metals by splitting the barriers. *Chem Commun* 2012;**48**(37):4450–2. Available from: https://doi.org/10.1039/C2CC00044J.
46. Wang Y, Ahlquist MSG. Mechanistic studies on proton transfer in a [FeFe] hydrogenase mimic complex. *Dalton Trans* 2013;**42**(21):7816–22. Available from: https://doi.org/10.1039/C3DT33087G.
47. Lounissi S, Capon J-F, Gloaguen F, Matoussi F, Pétillon FY, Schollhammer P, et al. Diiron species containing a cyclic $P_2^{Ph}N_2^{Ph}$ diphosphine related to the $[FeFe]H_2$ases active

site. *Chem Commun* 2011;**47**(3):878–80. Available from: https://doi.org/10.1039/C0CC03295F.

48. Lounissi S, Zampella G, Capon J-F, De Gioia L, Matoussi F, Mahfoudhi S, et al. Electrochemical and theoretical investigations of the role of the appended base on the reduction of protons by $[Fe_2(CO)_4(\kappa^2\text{-PNPR})(\mu\text{-S}(CH_2))_3S]$ (PNP R = $\{Ph_2PCH_2\}_2NR$, R = Me, Ph). *Chem Eur J* 2012;**18**(35):11123–38. Available from: https://doi.org/10.1002/chem.201201087.
49. Ginovska-Pangovska B, Dutta A, Reback ML, Linehan JC, Shaw WJ. Beyond the active site: The impact of the outer coordination sphere on electrocatalysts for hydrogen production and oxidation. *Acc Chem Res* 2014;**47**(8):2621–30. Available from: https://doi.org/10.1021/ar5001742.
50. DuBois DL. Development of molecular electrocatalysts for energy storage. *Inorg Chem* 2014;**53**(8):3935–60. Available from: https://doi.org/10.1021/ic4026969.
51. Rakowski DuBois M, DuBois DL. The roles of the first and second coordination spheres in the design of molecular catalysts for H_2 production and oxidation. *Chem Soc Rev* 2009;**38**(1):62–72. Available from: https://doi.org/10.1039/B801197B.
52. Rakowski DuBois M, DuBois DL. The role of pendant bases in molecular catalysts for H_2 oxidation and production. *C R Chim* 2008;**11**:805–17. Available from: https://doi.org/10.1016/j.crci.2008.01.019.
53. Wilson AD, Shoemaker RK, Miedaner A, Muckerman JT, DuBois DL, Rakowski DuBois M. Nature of hydrogen interactions with Ni(II) complexes containing cyclic phosphine ligands with pendant nitrogen bases. *Proc Natl Acad Sci* 2007;**104**(17):6951–6. Available from: https://doi.org/10.1073/pnas.0608928104.
54. Chouffai D, Zampella G, Capon J-F, De Gioia L, Gloaguen F, Pétillon FY, et al. Oxidatively induced reactivity of $[Fe_2(CO)_4(\kappa^2\text{-dppe})(\mu\text{-pdt})]$: An electrochemical and theoretical study of the structure change and ligand binding processes. *Inorg Chem* 2011;**50**(24):12575–85. Available from: https://doi.org/10.1021/ic201601q.
55. Chouffai D, Zampella G, Capon J-F, De Gioia L, Le Goff A, Pétillon FY, et al. Electrochemical and theoretical studies of the impact of the chelating ligand on the reactivity of $[Fe_2(CO)_4(\kappa^2\text{-LL})(\mu\text{-pdt})]^{1+}$ complexes with different substrates (LL = $I_{Me}\text{-}CH_2\text{-}I_{Me}$, dppe; I_{Me} = 1-methylimidazol-2-ylidene). *Organometallics* 2012;**31**:1082–91. Available from: https://doi.org/10.1021/om201143p.
56. Roy S, Groy TL, Jones AK. Biomimetic model for [FeFe]-hydrogenase: Asymmetrically disubstituted diiron complex with a redox-active 2,2′-bipyridyl ligand. *Dalton Trans* 2013;**42**(11):3843–53. Available from: https://doi.org/10.1039/C2DT32457A.
57. Chouffai D, Capon J-F, De Gioia L, Pétillon FY, Schollhammer P, Talarmin J, et al. A diferrous dithiolate as a model of the elusive H_{ox}^{inact} state of the [FeFe] hydrogenases: An electrochemical and theoretical dissection of its redox chemistry. *Inorg Chem* 2015;**54**(1):299–311. Available from: https://doi.org/10.1021/ic5024746.
58. Arrigoni F, Mohamed Bouh S, De Gioia L, Elleouet C, Pétillon FY, Schollhammer P, et al. Influence of the dithiolate bridge on the oxidative processes of diiron models related to the active site of [FeFe] hydrogenases. *Chem Eur J* 2017;**23**(18):4364–72. Available from: https://doi.org/10.1002/chem.201605060.
59. Olsen MT, Rauchfuss TB, Wilson SR. Role of the azadithiolate cofactor in models for [FeFe]-hydrogenase: Novel structures and catalytic implications. *J Am Chem Soc* 2010;**132**(50):17733–40. Available from: https://doi.org/10.1021/ja103998v.

60. Zheng D, Wang M, Chen L, Wang N, Sun L. Redox reactions of [FeFe]-hydrogenase models containing an internal amine and a pendant phosphine. *Inorg Chem* 2014;**53** (3):1555–61. Available from: https://doi.org/10.1021/ic4025519.
61. Miyake T, Bruschi M, Cosentino U, Baffert C, Fourmond V, Léger C, et al. Does the environment around the H-cluster allow coordination of the pendant amine to the catalytic iron center in [FeFe] hydrogenases? Answers from theory. *J Biol Inorg Chem* 2013;**18** (6):693–700. Available from: https://doi.org/10.1007/s00775-013-1014-4.

Chapter 18

Structural Insights into the Protective Mechanisms against O_2 in the [NiFe]-Hydrogenases

Yasuhito Shomura
Institute of Quantum Beam Science, Graduate School of Science and Engineering, Ibaraki University, Hitachi, Japan

18.1 INTRODUCTION

Biological cleavage of the H−H bond is challenging because of its high bond-dissociation energy and nonpolar property. Hydrogenases are enzymes that reversibly catalyze the decomposition of dihydrogen into protons and electrons, and their physiological functions are mainly divided into two groups.[1] The first is the elimination of excess amount of the reducing equivalents by the synthesis of H_2 (termed H_2-evolution), where two protons are utilized as electron acceptors. The second is the decomposition of H_2 as an energy source (termed H_2-uptake). Hydrogenases consist of three structurally and evolutionally different types, [Fe]-,[2,3] [FeFe]-,[4−6] and [NiFe]-hydrogenases,[5−7] named after the metal composition of the active site. While the three types independently emerged and evolved, they share a common structural characteristic in that the Fe−CO organometal compound is included as a component of the active site (Fig. 18.1). The [Fe]-hydrogenase is a mononuclear non-heme iron enzyme that utilizes methenyltetrahydromethanopterin (mthenyl-H_4MPT^+) as another substrate.[2] The reaction catalyzed by the enzyme is limited to the reversible hydride transfer from H_2 to mthenyl-H_4MPT^+, and the orthologs are found only in some methanogenic archaea. In contrast to the [Fe]-hydrogenases, the [FeFe]- and [NiFe]-hydrogenases widely occur in bacteria and archaea,[1] and the reaction catalyzed by the enzymes is a reversible conversion of H_2 into two protons and electrons, leading the ability of enzymatic transformations between chemical and electrical energies. Common structural features of the active sites of the [FeFe]- and [NiFe]-hydrogenases are the presence of CO and CN^- ligands for the Fe atom and cysteinyl thiolate ligands for two metals. In general, [FeFe]-hydrogenases show higher turnover numbers as large as 10000 s^{-1} for

Advances in Bioorganometallic Chemistry. DOI: https://doi.org/10.1016/B978-0-12-814197-7.00018-2

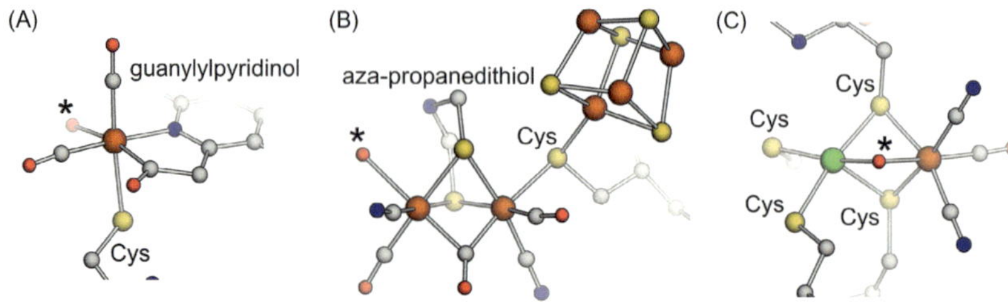

FIGURE 18.1 Structures of the active site: (A) of [Fe]-hydrogenase (PDB ID: 3F47), (B) of [FeFe]-hydrogenase (PDB ID: 3C8Y), and (C) of [NiFe]-hydrogenase (PDB ID: 1YRQ). Hydrogen atoms are omitted in all figures. Chemical species of the ligands denoted by asterisks are mobile and differ by the states. Atom colors in all figures: gray = C, blue = N, red = O, yellow = S, brown = Fe, green = Ni.

H_2-evolution[8] than typical [NiFe]-hydrogenases that show the turnover numbers for H_2-evolution and H_2-uptake of $500\ s^{-1}$ and $2000\ s^{-1}$, respectively.[9] Another remarkable difference between the two types of hydrogenases is the resistance against O_2. While most [NiFe]-hydrogenases are stable under aerobic environments, [FeFe]-hydrogenases promptly lose catalytic activities due to the reaction with O_2 at the active sites.[10–12] It has been proposed that the difference in the O_2-resistances between [NiFe]- and [FeFe]-hydrogenases is derived from the reactivity with reactive oxygen species (ROS) produced by the partial reduction of O_2. As will be described below, a subgroup of [NiFe]-hydrogenases that showed a high O_2-tolerant enzymatic property suppresses the formation of ROS by the four-electron reduction of O_2 to produce nonhazardous molecule H_2O.

The minimum catalytic unit of the [NiFe]-hydrogenases consists of a 50–70 kDa large subunit with the Ni–Fe cluster as the active site for H_2-oxidation/H^+-reduction, and 20–30 kDa small subunit with one to three Fe–S clusters for the electron transfer between the Ni–Fe cluster and its extrinsic electron acceptor/donor. The Ni–Fe cluster is buried deep inside the heterodimer and, therefore, the hydrogen-bond network as the proton pathway and gas channel as the H_2-pathway connecting from outside the protein to the Ni–Fe cluster are also equipped as well as the electron transfer pathway. The physiological function of the [NiFe]-hydrogenases largely depends on the cellular localization and the binding partner with the catalytic unit.[13] Based on the functions and phylogenetic aspects, the [NiFe]-hydrogenases are classified into four groups.[5,7] The most divergent and widely found in nature are group 1 [NiFe]-hydrogenases that include so called standard [NiFe]-hydrogenases (standard H_2ases), membrane-bound [NiFe]-hydrogenases (MBHs), [NiFeSe]-hydrogenases, and the subsequently identified actinobacteria type. Most of the members of group 1 function as H_2-uptake enzymes in the periplasmic space. Some MBHs form a complex with the integral membrane cytochrome *b* subunit that accepts two electrons derived from the H_2-decomposition by the hydrogenase catalytic unit and reduces quinone species in the inner membrane. In subsequent studies, MBHs have been further subdivided into several variants including Hyd-1 types (or 6C-hydrogenases)[14] that are distinguished from others by the comparison of the amino acid sequences and show high O_2-tolerant enzymatic properties.[15] Although Hyd-1 is originally a specific isozyme name in *Escherichia coli* that also contains Hyd-2, Hyd-3, and Hyd-4, Hyd-1 will be used hereafter in this chapter to refer to orthologs from other organisms. The standard H_2ase is the subgroup that has been most extensively studied in the earlier investigations, and are mainly found in sulfate-reducing bacteria. They are also known as representatives of O_2-intolerant (O_2-sensitive) [NiFe]-hydrogenases. It has long been known that the standard H_2ases take two major inactive forms, Ni-A and Ni-B, by exposure to O_2. Both inactive states have common features in that they contain Ni(III) and an oxygenous bridging ligand[16] (at the position denoted by asterisk in Fig. 18.1C). The chemical species of the ligand of Ni-B has been assigned to hydroxide, whereas hydroperoxide[17,18] and hydroxide[19] have been proposed for Ni-A. The reversible inhibition of the standard H_2ases by oxidation has been explained by the formation of Ni-A that takes a few hours for reactivation, whereas Ni-B is readily reactivated within a few seconds.[20] While Ni-A is dominantly formed when the Ni–Fe active site reacts with O_2 under electron deficient conditions, Ni-B is prone to be produced in electron-rich conditions and is also formed by the electrochemical oxidation in the absence of O_2.[21]

18.2 O_2-TOLERANCE MECHANISM OF HYD-1

Progress in electrochemical techniques, which enable the measurements of the H_2-uptake activities of hydrogenases in the presence of O_2, first delineated the unusual O_2-tolerance properties of Hyd-1.[22,23] Further electrochemical kinetic and thermodynamic analyses have suggested that the O_2-tolerance property of Hyd-1 is ascribable to the ability of the four-electron reduction of O_2,[24,25] and that the inactive state electrochemically formed by oxidation under anaerobic conditions[26] is indistinguishable from Ni-B. The subsequent spectroscopic studies on Hyd-1 from *Ralstonia eutropha* H16 (*Re*MBH)[27] and *Aquifex aeolicus* (*Aa*Hase 1)[28] have further demonstrated that the proximal Fe–S cluster in Hyd-1 has an unusual property in that it takes three oxidation states in the range of physiological redox potentials. They have also presumed that the formation of the superoxidized state of the proximal Fe–S cluster would be attributed to the two unique supernumerary cysteine residues of Hyd-1. The [4Fe–4S] clusters including those of the standard H_2ases have been classified into either ferredoxin or HiPIP types based on their structure and redox potential. The ferredoxin and HiPIP types generally exhibit redox potentials ranging from −417 to −280 mV for 2+/1+ transition and from +93 to +458 mV for 3+/2+ transition, respectively.[29]

The structural insights into the unique property of the proximal Fe–S cluster of Hyd-1 have been reported by our group for Hyd-1 from *Hydrogenovibrio marinus* (*Hm*MBH)[30] and others for *Re*MBH[31] and Hyd-1 from *Escherichia coli* (*Ec*Hyd-1).[32] Overall structures of the catalytic heterodimers of the three orthologs of Hyd-1 are very similar to those of the previously reported standard H_2ases except that the catalytic heterodimeric units of *Hm*MBH and *Ec*Hyd-1 form a dimer of protomers in the crystals, which has later been proven to be a physiological oligomerization state by the X-ray crystal structure analysis of the *Ec*Hyd-1 complex including the cytochrome *b* subunit.[33] On the other hand, it has also later been reported that the *Re*MBH complex takes a trimer of protomers in the membrane-embedded physiological state.[34] As predicted, the structure of the proximal Fe–S cluster of Hyd-1 was significantly different from those of the standard H_2ases that have a conventional four-cysteine coordinated [4Fe–4S] cluster with the cubane form (Fig. 18.2A). The crystal structure analyses of the three orthologs of Hyd-1 have revealed that the proximal Fe–S cluster of Hyd-1 consists of four iron and three sulfide ions with the coordination of six cysteine residues (Fig. 18.2B). Furthermore, we have, for the first time, reported that the unprecedented [4Fe–3S] cluster shows a redox-dependent conformational change, which is critical to the four-electron reduction of O_2 at the [NiFe]-active site,[30] whereas it has been well known that the crystal structures of the cubane-type [4Fe–4S] clusters in the oxidized and reduced states are basically indistinguishable at normal resolutions.[35] The H_2-reduced states of the [4Fe–3S] clusters from three organisms share identical molecular geometry, where the two supernumerary cysteinyl thiolates coordinate iron atoms as the terminal and μ^2-Bridging ligands, replacing the sulfide ion as the μ^3-Bridging ligand for three Fe atoms in the [4Fe–4S] cluster (Fig. 18.2A and B). Upon oxidation of the crystals by ferricyanide, however, the iron atom labeled Fe4 in Fig. 18.2B moved toward the main-chain amide nitrogen to form the coordination bond, resulting in the bond cleavage between Fe4 and a sulfide ligand (Fig. 18.2C). Under air-oxidized conditions, a merge of the two structures has been observed in *Hm*MBH, suggesting that the two states are in equilibrium under the weaker oxidative conditions than in the treatment with ferricyanide. The redox-dependent conformational change of the proximal [4Fe–3S] cluster has been subsequently reported for *Ec*Hyd-1, where one critical difference from our structure analysis has been pointed out in the ferricyanide-oxidized state.[36] According to the report by Volbeda et al., the structure of the [4Fe–3S] cluster in the ferricyanide-oxidized state is also the equilibrium of two structures, where Fe4 and the nearby glutamic acid residue slightly come close to each other to form a coordination bond in one of the two states (Fig. 18.2C). The same disorder has also been detected in our structural data, but in our original paper the double conformation of Glu82 and Fe4 had been interpreted as large anisotropic displacement parameters of constituting atoms, which has

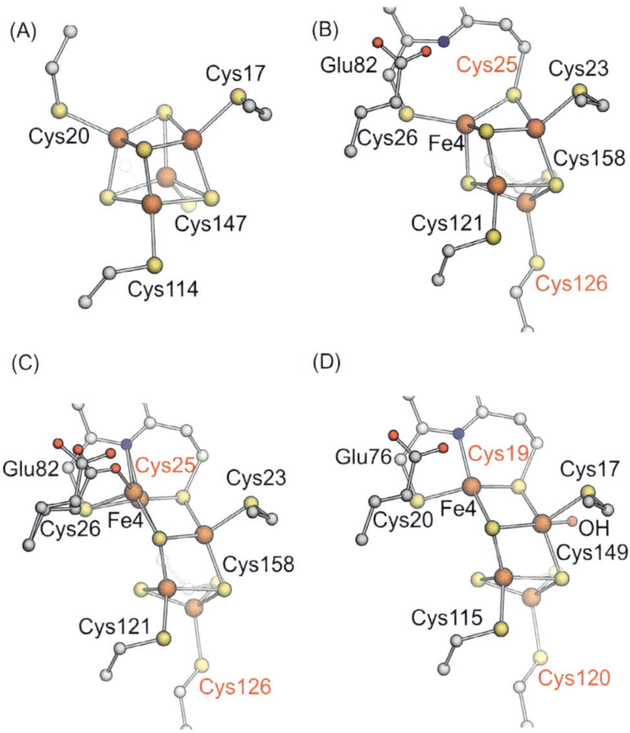

FIGURE 18.2 Proximal Fe–S clusters of the group 1 [NiFe]-hydrogenase: (A) of standard H_2ase (PDB ID: 1YRQ), (B) of *Hm*MBH in the H_2-reduced state (PDB ID: 3AYX), (C) of *Hm*MBH in the ferricyanide-oxidized state (PDB ID: 5Y34), and (D) of *Re*MBH in the oxidized state (PDB ID: 4IUB). Red labels are for the supernumerary cysteines characteristic of Hyd-1. In (C), double conformations of Glu82 and Fe4 are shown.

later been corrected, and the original PDB ID of 3AYZ has been superseded by 5Y34.[37] On the other hand, the movement of Glu82 has not been confirmed in the air-oxidized state of *Hm*MBH, while it occurred in the air-oxidized (as-purified) state of *Ec*Hyd-1. Subsequent reports on the structure analysis on *Re*MBH have also demonstrated its redox-dependent conformational changes of the [4Fe–3S] cluster similar to those observed in *Hm*MBH and *Ec*Hyd-1 with one additional conformational difference between reduced and oxidized states.[38,39] As shown in Fig. 18.2D, an iron ion in the oxidized state harbors monooxygen species with the distance of 1.8 Å, whose occupancies significantly vary by their crystal conditions. The chemical species has been identified as hydroxide based on resonance Raman spectroscopy experiments,[38] which may lead an assumption that the ligand functions as a stabilizer for the superoxidized state of the Fe–S cluster, but the details of the derivation and mechanism for the formation of the ligand remains unclear.

It has now widely been accepted that the unusual redox property of the proximal Fe–S cluster of Hyd-1 is attributed to the unique asymmetric structure of the [4Fe–3S] cluster. As we first proposed, deprotonation of the main-chain amide nitrogen to form the coordination bond with Fe4 is a key factor contributing to the stabilization of the superoxidized $[4Fe-3S]^{5+}$ state, leading to the donating ability of two electrons and one proton at once for the reduction of O_2 to H_2O at the Ni–Fe active site. As proposed by Volbeda et al., the glutamic acid residue close to Fe4 also plays a key role as the first proton acceptor and in the stabilization of superoxidized state by the coordination to Fe4.

18.3 REDOX-DEPENDENT CONFORMATIONAL CHANGE IN THE NI–FE ACTIVE SITE OF THE NAD^+-REDUCING SOLUBLE [NIFE]-HYDROGENASE

Since the first crystal structure report in 1995,[40] a number of the structural data on the group 1 [NiFe]-hydrogenases have been deposited in the Protein Data Bank (~60 structures in March 2018), but it was not until 2014 that the first crystal structure of [NiFe]-hydrogenase other than group 1 was reported.[41] Up to now, the crystal structures of the group 3 [NiFe]-hydrogenases from three organisms have been reported, but none for the group 2 and 4 was known. In contrast to the members of the group 1, the group 3 [NiFe]-hydrogenases are localized in the cytoplasm in the soluble form and, thus, also called cytosolic bidirectional [NiFe]-hydrogenases, which consist of four subgroups: the F_{420}-reducing type and heterodisulfide reductase-linked type from methanogenic archaea, and the NADP-coupled and NAD-coupled types from eubacteria and cyanobacteria. Among these, the NAD^+-reducing soluble [NiFe]-hydrogenase (SH) belongs to the NAD-coupled type and have been mainly found in hydrogen-oxidizing bacteria and photosynthetic bacteria, which can be further divided into two types based on the requirement of the peripheral subunit termed HoxE.[42,43] The common components of SH include the HoxHY unit as the minimum catalytic unit of the [NiFe]-hydrogenases and the HoxFU diaphorase unit for the reduction/oxidation of NAD. Based on the amino acid sequence similarity of each subunit with the corresponding subunit of NADH:Quinone oxidoreductase (complex I), it has been predicted that the active sites of the two units are connected by the electron transfer pathway consisting of the several Fe–S clusters.[44] It has long been known that SH reversibly catalyzes both the H_2-driven $NAD(P)^+$ reduction and H_2-synthesis upon NAD(P)H oxidation *in vitro*,[45] but the precise physiological function of SH remains unknown and may differ by the organisms.[43,46–48]

To investigate the molecular evolutionary aspects of SH and complex I, we have subsequently determined the first crystal structure of SH from *Hydrogenophilus thermoluteolus* TH-1 (*Ht*SH).[49] The advantages of using

*Ht*SH for the crystallization experiments were its remarkably higher stability of the intact heterotetrameric complex under atmospheric environments compared to other orthologs, and the absence of peripheral subunits, which would complicate the stoichiometry of the purified sample. In addition, it has recently been demonstrated that *Ht*SH is a moderate O_2-tolerant hydrogenase, leaving ~50% of the maximal H_2-driven NAD^+-reducing ability even at 2%(v/v) O_2.[48]

Unexpectedly, the coordination geometry of the Ni–Fe active site of *Ht*SH in the air-oxidized state was significantly different from any of those reported thus far (Fig. 18.3A). The conventional Ni–Fe clusters in the oxidized states include the coordination by two cysteinyl thiolates as the terminal ligands for Ni, two additional cysteinyl thiolates as the bridging ligands between Ni and Fe, three diatomic ligands for Fe, and the mobile bridging OH^- ligand (Fig. 18.1C). On the other hand, the active site of *Ht*SH in the air-oxidized state consists of the bidentate coordination by the glutamyl carboxy group for Ni, one terminal cysteinyl thiolate ligand for Ni, three bridging cysteinyl thiolate ligands between Ni and Fe, and three diatomic ligands for Fe. As a result, both Ni and Fe show the octahedral coordination

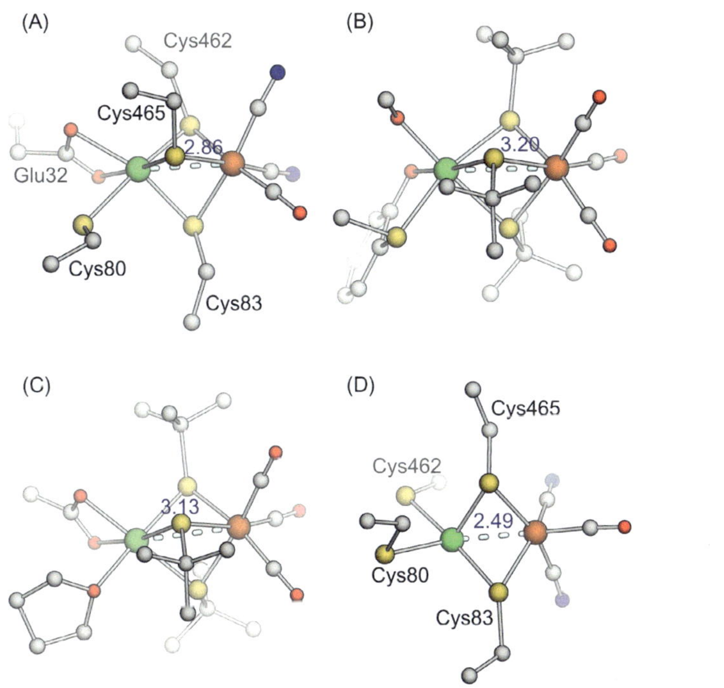

FIGURE 18.3 Ni–Fe active sites: (A) of *Ht*SH in the air-oxidized state (PDB ID: 5XF9), (D) of *Ht*SH in the H_2-reduced state (PDB ID: 5XFA); structurally related model complexes—CSD ID: 673707 (B) and 673709 (C). Blue numbers indicate the distance between Ni and Fe connected by dotted lines in angstroms (Å).

geometry without the mobile OH^- ligand, preventing the metal ions from being directly attacked by O_2 which would cause the production of ROS and oxidative damage in the active site. At the same time, it should be noted that the Ni–Fe active site has no coordination site for the binding of H_2 or H^-. While the structure of the Ni–Fe cluster is unprecedented in known metalloproteins, two model complexes with similar coordination geometries have previously been reported (Fig. 18.3B and C).[50] The two compounds are devoid of H_2-uptake/evolution activities and the distances between Ni and Fe are longer by ~0.3 Å than that of *Ht*SH in the air-oxidized state. In order to determine the crystal structure of *Ht*SH in the enzymatically active state, the crystals of *Ht*SH obtained under the atmospheric conditions have been soaked into the anaerobic buffer containing 1 mM oxidized benzyl viologen (BV) and saturated H_2 at 0.1 M Pa for 24 hours at 293 K, which is defined as the H_2-reduced state. The X-ray structure analysis has revealed that the Ni–Fe active site of *Ht*SH in the H_2-reduced state takes the same coordination geometry as those of the group 1 [NiFe]-hydrogenases in the H_2-reduced state (Fig. 18.3D). In addition to the coordination manner of the four cysteinyl thiolates, the disappearance of the OH^- ligand and the distance between Ni and Fe shortened to ~2.5 Å are also characteristic features of the Ni–Fe active sites in the H_2-reduced state.[37]

The structure comparison around the active site between the air-oxidized and H_2-reduced states have revealed that the rearrangement of the Ni–Fe cluster would be originally derived from the flip of the side-chain of Arg58 that is in close proximity to the proximal [4Fe–4S] cluster in the H_2-reduced state (Figs. 18.4A and B). It is assumed that Arg58 flips away from the Fe–S cluster by the electrostatic repulsion when the Fe–S cluster is in the oxidized state long-term. The flip of Arg58 would cause the displacement of Glu56, which liberates Ser464 and induces the reorientation of the short α helix accommodating Ser464 and Cys465, producing the steric hindrance

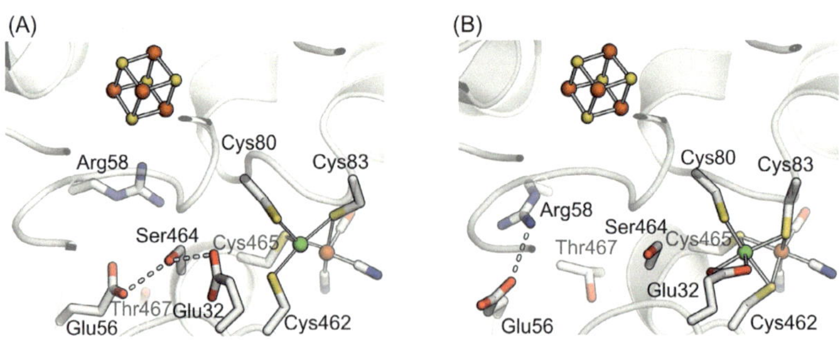

FIGURE 18.4 Comparison of the structures around the Ni–Fe active site and proximal Fe–S cluster in between (A) H_2-reduced state and (B) air-oxidized state. Dotted lines represent hydrogen bonds. Side-chains of the residues crucial for the redox-dependent rearrangement of the Ni–Fe cluster are shown with stick models.

between the α helix and Glu32. As the result of the approximation of Glu32 and Cy462 to Ni and Fe, the rearrangement of the coordination geometry occurs. According to the model, the major oxidation state of the proximal [4Fe−4S] cluster determines the coordination geometry of the Ni−Fe cluster, resulting in the protection of the active site from damages caused by O_2 at higher oxidation environments. When preparing the crystals of *Ht*SH in the H_2-reduced state in our experiment, it is assumed that BV was slowly reduced by H_2, which should be enough to reactivate a small amount of enzyme, and they produced the reducing environment by their enzymatic activities. It remains unknown whether the redox-dependent conformational change of the Ni−Fe active site is involved in the O_2-tolerance property of SH, whereas it should be noted that Glu56 (glutamine is also found in some orthologs) and Ser464 are conserved only among orthologs of SH and that nonpolar residues such like methionine or leucine for the former residue and less bulky and nonpolar residues such like alanine or glycine for the latter residue are found in other [NiFe]-hydrogenases.[49]

18.4 CONCLUDING REMARKS

Structural studies on the O_2-tolerant [NiFe]-hydrogenase have demonstrated the differences between O_2-tolerant and the O_2-sensitive hydrogenases, and given insights into the understanding of the defending mechanisms against O_2. In most cases, the determinants of the properties are not found in the active site itself, but in the protein environment around the active site. The mechanisms have been mainly divided into two types. The one involves the physical elimination of O_2 from the active site as in the case of *Ht*SH. The other mechanism is more active and related to the four-electron reduction of O_2 to convert into harmless H_2O, which is also shown by taking *Hm*MBH as an example.

Our interest is also focused on other O_2-tolerant [NiFe]-hydrogenases, whose structure and molecular mechanism have not been elucidated. For instance, the Hyd-2 type [NiFe]-hydrogenase from *Citrobacter* sp. S-77 (S77Hyd2), which has shown the O_2-tolerant H_2-uptake activity and the remarkably high stability and the activity on the electrode.[51,52] The amino acid sequence analysis of S77Hyd2 has detected neither supernumerary cysteines for the proximal [4Fe−3S] cluster nor the conserved serine and glutamic acid residues among orthologs of SH. Therefore, a new structural feature related to the enzymatic properties of S77Hyd2 would be expected to occur. Another interest in the structure−function relationship may be in the structure analysis of the group 4 [NiFe]-hydrogenase, which is exceptionally O_2-unresistant and unstable under atmospheric environment, as in the case with the [FeFe]-hydrogenases. These structural insights would help to improve the stability of the hydrogenases in nature and to design stable molecular catalysts aimed at the industrial applications.

ACKNOWLEDGMENTS

The author thanks all the collaborators who were involved in the studies on *Hm*MBH and *Ht*SH. Financial support from Japan Science and Technology Agency CREST (grant JPMJCR12M1) and Specially Promoted Research (grant 26000008) is gratefully acknowledged.

REFERENCES

1. Vignais PM, Billoud B. Occurrence, classification, and biological function of hydrogenases: An overview. *Chem. Rev* 2007;**107**(10):4206–72.
2. Shima S, Pilak O, Vogt S, Schick M, Stagni MS, Meyer-Klaucke W, et al. The crystal structure of [Fe]-hydrogenase reveals the geometry of the active site. *Science (80-.).* 2008;**321**(5888):572–5.
3. Tamura H, Salomone-Stagni M, Fujishiro T, Warkentin E, Meyer-Klaucke W, Ermler U, et al. Crystal structures of [Fe]-hydrogenase in complex with inhibitory isocyanides: Implications for the H_2 -activation site. *Angew. Chemie Int. Ed* 2013;**52**(37):9656–9.
4. Mulder DW, Shepard EM, Meuser JE, Joshi N, King PW, Posewitz MC, et al. Insights into [FeFe]-hydrogenase structure, mechanism, and maturation. *Structure* 2011;**19**(8):1038–52.
5. Peters JW, Schut GJ, Boyd ES, Mulder DW, Shepard EM, King PW, et al. [FeFe]- and [NiFe]-hydrogenase diversity, mechanism, and maturation. *Biochim. Biophys. Acta - Mol. Cell Res* 2015;**1853**(6):1350–69.
6. Lubitz W, Ogata H, Rüdiger O, Reijerse E. Hydrogenases. *Chem. Rev.* 2014;**114**(8):4081–148.
7. Greening C, Biswas A, Carere CR, Jackson CJ, Taylor MC, Stott MB, et al. Genomic and metagenomic surveys of hydrogenase distribution indicate H2 is a widely utilised energy source for microbial growth and survival. *ISME J* 2016;**10**(3):761–77.
8. Hatchikian EC, Forget N, Fernandez VM, Williams R, Cammack R. Further characterization of the [Fe]-hydrogenase from desulfovibrio desulfuricans ATCC 7757. *Eur. J. Biochem* 1992;**209**(1):357–65.
9. Fauque G, Peck HD, Moura JJG, Huynh BH, Berlier Y, DerVartanian DV, et al. The three classes of hydrogenases from sulfate-reducing bacteria of the genus desulfovibrio. *FEMS Microbiol. Lett* 1988;**54**(4):299–344.
10. Stripp ST, Goldet G, Brandmayr C, Sanganas O, Vincent KA, Haumann M, et al. How oxygen attacks [FeFe] hydrogenases from photosynthetic organisms. *Proc. Natl. Acad. Sci. U. S. A* 2009;**106**(41):17331–6.
11. Goldet G, Brandmayr C, Stripp ST, Happe T, Cavazza C, Fontecilla-Camps JC, et al. Electrochemical kinetic investigations of the reactions of [FeFe]-hydrogenases with carbon monoxide and oxygen: Comparing the importance of gas tunnels and active-site electronic/redox effects. *J. Am. Chem. Soc* 2009;**131**(41):14979–89.
12. Lambertz C, Leidel N, Havelius KGV, Noth J, Chernev P, Winkler M, et al. O2 reactions at the six-iron active site (H-Cluster) in [FeFe]-hydrogenase. *J. Biol. Chem* 2011;**286**(47):40614–23.
13. Vignais PM, Billoud B, Meyer J. Classification and phylogeny of hydrogenases. *FEMS Microbiology Reviews* 2001;455–501.
14. Pandelia M-E, Lubitz W, Nitschke W. Evolution and diversification of group 1 [NiFe] hydrogenases. Is there a phylogenetic marker for O2-tolerance? *Biochim. Biophys. Acta - Bioenerg* 2012;**1817**(9):1565–75.

15. Goris T, Wait AF, Saggu M, Fritsch J, Heidary N, Stein M, et al. A unique iron-sulfur cluster is crucial for oxygen tolerance of a [NiFe]-hydrogenase. *Nat. Chem. Biol* 2011;**7**(5):310–18.
16. Volbeda A, Martin L, Barbier E, Gutiérrez-Sanz O, De Lacey AL, Liebgott PP, et al. Crystallographic studies of [NiFe]-hydrogenase mutants: Towards consensus structures for the elusive unready oxidized states. *J. Biol. Inorg. Chem* 2015;**20**(1):11–22.
17. Ogata H, Hirota S, Nakahara A, Komori H, Shibata N, Kato T, et al. Activation process of [NiFe] hydrogenase elucidated by high-resolution X-ray analyses: Conversion of the ready to the unready state. *Structure* 2005;**13**(11):1635–42.
18. Volbeda A, Martin L, Cavazza C, Matho M, Faber BW, Roseboom W, et al. Structural differences between the ready and unready oxidized states of [NiFe] hydrogenases. *J. Biol. Inorg. Chem* 2005;**10**(3):239–49.
19. Volbeda A, Martin L, Liebgott P-P, De Lacey AL, Fontecilla-Camps JC. [NiFe]-Hydrogenases revisited: Nickel–carboxamido bond formation in a variant with accrued O 2 -tolerance and a tentative re-interpretation of Ni-SI states. *Metallomics* 2015;**7**(4):710–18.
20. Fernandez VM, Hatchikian EC, Cammack R. Properties and reactivation of two different deactivated forms of desulfovibrio gigas hydrogenase. *Biochim. Biophys. Acta - Protein Struct. Mol. Enzymol* 1985;**832**(1):69–79.
21. Abou-Hamdan A, Ceccaldi P, Lebrette H, Gutiérrez-Sanz O, Richaud P, Cournac L, et al. A threonine stabilizes the NiC and NiR catalytic intermediates of [NiFe]-hydrogenase. *J. Biol. Chem* 2015;**290**(13):8550–8.
22. Vincent Ka, Cracknell Ja, Lenz O, Zebger I, Friedrich B, Armstrong Fa. Electrocatalytic hydrogen oxidation by an enzyme at high carbon monoxide or oxygen levels. *Proc. Natl. Acad. Sci. U. S. A* 2005;**102**(47):16951–4.
23. Vincent KA, Parkin A, Armstrong FA. Investigating and exploiting the electrocatalytic properties of hydrogenases. *Chem. Rev* 2007;**107**(10):4366–413.
24. Armstrong FA, Belsey NA, Cracknell JA, Goldet G, Parkin A, Reisner E, et al. Dynamic electrochemical investigations of hydrogen oxidation and production by enzymes and implications for future technology. *Chem. Soc. Rev* 2009;**38**(1):36–51.
25. Cracknell Ja, Wait AF, Lenz O, Friedrich B, Armstrong FaA. Kinetic and thermodynamic understanding of O2 tolerance in [NiFe]-hydrogenases. *Proc. Natl. Acad. Sci. U. S. A* 2009;**106**(49):20681–6.
26. Jones AK, Lamle SE, Pershad HR, Vincent KA, Albracht SPJ, Armstrong FA. Enzyme electrokinetics: Electrochemical studies of the anaerobic interconversions between active and inactive states of allochromatium vinosum [NiFe]-hydrogenase. *J. Am. Chem. Soc* 2003;**125**(28):8505–14.
27. Saggu M, Zebger I, Ludwig M, Lenz O, Friedrich B, Hildebrandt P, et al. Spectroscopic insights into the oxygen-tolerant membrane-associated [NiFe] hydrogenase of ralstonia eutropha H16. *J. Biol. Chem* 2009;**284**(24):16264–76.
28. Pandelia M-E, Nitschke W, Infossi P, Giudici-Orticoni M-T, Bill E, Lubitz W. Characterization of a unique [FeS] cluster in the electron transfer chain of the oxygen tolerant [NiFe] hydrogenase from aquifex aeolicus. *Proc. Natl. Acad. Sci. U. S. A* 2011;**108**(15):6097–102.
29. Bertini I, Gray H, Steifel E, Valentine JS. *B*iological Inorganic Chemistry: Structure and Reactivity. University Science Book; 2006.
30. Shomura Y, Yoon K-S, Nishihara H, Higuchi Y. Structural basis for a [4Fe-3S] cluster in the oxygen-tolerant membrane-bound [NiFe]-hydrogenase. *Nature* 2011;**479**:253–6.

31. Fritsch J, Scheerer P, Frielingsdorf S, Kroschinsky S, Friedrich B, Lenz O, et al. The crystal structure of an oxygen-tolerant hydrogenase uncovers a novel iron-sulphur centre. *Nature* 2011;**479**:249–52.
32. Volbeda A, Amara P, Darnault C, Mouesca J, Parkin A, Roessler MM, et al. Supporting information X-ray crystallographic and computational studies of the O 2 -tolerant [NiFe] -hydrogenase 1 from escherichia coli. *Proc. Natl. Acad. Sci. U. S. A* 2012;**109** (14):5305–10.
33. Volbeda A, Darnault C, Parkin A, Sargent F, Armstrong FA, Fontecilla-Camps JC. Crystal structure of the O2-tolerant membrane-bound hydrogenase 1 from escherichia coli in complex with its cognate cytochrome B. *Structure* 2013;**21**(1):184–90.
34. Frielingsdorf S, Schubert T, Pohlmann A, Lenz O, Friedrich B. A Trimeric supercomplex of the oxygen-tolerant membrane-bound [NiFe]-hydrogenase from ralstonia eutropha H16. *Biochemistry* 2011;**50**(50):10836–43.
35. Ohno H, Takeda K, Niwa S, Tsujinaka T, Hanazono Y, Hirano Y, et al. Crystallographic characterization of the high-potential iron-sulfur protein in the oxidized state at 0.8 Å resolution. *PLoS One* 2017;**12**(5):e0178183.
36. Volbeda A, Amara P, Darnault C, Mouesca J-M, Parkin A, Roessler MM, et al. X-ray crystallographic and computational studies of the O2-tolerant [NiFe]-hydrogenase 1 from escherichia coli. *Proc. Natl. Acad. Sci* 2012;**109**(14):5305–10.
37. Shomura Y, Higuchi Y. Structural aspects of [NiFe]-hydrogenases. *Rev. Inorg. Chem.* 2013;**33**(4):173–92.
38. Frielingsdorf S, Fritsch J, Schmidt A, Hammer M, Löwenstein J, Siebert E, et al. Reversible [4Fe-3S] cluster morphing in an O2-tolerant [NiFe] hydrogenase. *Nat. Chem. Biol.* 2014;**10**(5):378–85.
39. Kalms J, Schmidt A, Frielingsdorf S, Utesch T, Gotthard G, von Stetten D, et al. Tracking the route of molecular oxygen in O2-tolerant membrane-bound [NiFe] hydrogenase. *Proc. Natl. Acad. Sci. U. S. A* 2018;**115**(10):E2229–37.
40. Volbeda A, Charon MH, Piras C, Hatchikian EC, Frey M, Fontecillacamps JC. Crystal-structure of the nickel-iron hydrogenase from desulfovibrio-gigas. *Nature* 1995;**373** (6515):580–7.
41. Vitt S, Ma K, Warkentin E, Moll J, Pierik AJ, Shima S, et al. The F420-reducing [NiFe]-hydrogenase complex from methanothermobacter marburgensis, the first X-ray structure of a group 3 family member. *J. Mol. Biol.* 2014;**426**(15):2813–26.
42. Schmitz O, Boison G, Salzmann H, Bothe H, Schütz K, Wang S, et al. HoxE—a subunit specific for the pentameric bidirectional hydrogenase complex (HoxEFUYH) of cyanobacteria. *Biochim. Biophys. Acta - Bioenerg* 2002;**1554**(1–2):66–74.
43. Maróti J, Farkas A, Nagy IK, Maróti G, Kondorosi E, Rákhely G, et al. A second soluble hox-type NiFe enzyme completes the hydrogenase set in thiocapsa roseopersicina BBS. *Appl. Environ. Microbiol* 2010;**76**(15):5113–23.
44. Böhm R, Sauter M, Böck A. Nucleotide sequence and expression of an operon in escherichia coli coding for formate hydrogenylase components. *Mol. Microbiol* 1990;**4** (2):231–43.
45. Schneider K, Schlegel HG. Purification and properties of soluble hydrogenase from alcaligenes eutrophus H 16. *Biochim. Biophys. Acta - Enzymol* 1976;**452**(1):66–80.
46. Eckert C, Boehm M, Carrieri D, Yu J, Dubini A, Nixon PJ, et al. Genetic analysis of the hox hydrogenase in the cyanobacterium synechocystis Sp. PCC 6803 reveals subunit roles in association, assembly, maturation, and function. *J. Biol. Chem* 2012;**287**(52):43502–15.

47. Gutekunst K, Chen X, Schreiber K, Kaspar U, Makam S, Appel J. The bidirectional NiFe-hydrogenase in synechocystis Sp. PCC 6803 is reduced by flavodoxin and ferredoxin and is essential under mixotrophic, nitrate-limiting conditions. *J. Biol. Chem* 2014;**289**(4):1930–7.
48. Preissler J, Wahlefeld S, Lorent C, Teutloff C, Horch M, Lauterbach L, et al. Enzymatic and spectroscopic properties of a thermostable [NiFe]-hydrogenase performing H2-driven NAD + -reduction in the presence of O2. *Biochim. Biophys. Acta - Bioenerg* 2018;**1859**(1):8–18.
49. Shomura Y, Taketa M, Nakashima H, Tai H, Nakagawa H, Ikeda Y, et al. Structural Basis of the Redox Switches in the NAD $^+$ -Reducing Soluble [NiFe]-Hydrogenase. *Science (80-.)* 2017;**357**(6354):928–32.
50. Ohki Y, Yasumura K, Kuge K, Tanino S, Ando M, Li Z, et al. Complexes relevant to the active site of [NiFe] hydrogenase. *Proc. Natl. Acad. Sci. U. S. A.* 2008;**105**(22):7652–7.
51. Eguchi S, Yoon K-S, Ogo S. O2-stable membrane-bound [NiFe]hydrogenase from a newly isolated citrobacter Sp. S-77. *J. Biosci. Bioeng* 2012;**114**(5):479–84.
52. Matsumoto T, Eguchi S, Nakai H, Hibino T, Yoon K-S, Ogo S. [NiFe]hydrogenase from *Citrobacter* Sp. S-77 surpasses platinum as an electrode for H_2 oxidation reaction. *Angew. Chemie Int. Ed* 2014;**53**(34):8895–8.

Chapter 19

Bioorganometallic B_{12} as Versatile Catalyst for Green Organic Synthesis

Hisashi Shimakoshi and Yoshio Hisaeda
Department of Chemistry and Biochemistry, Graduate School of Engineering, Kyushu University, Fukuoka, Japan

19.1 INTRODUCTION

Bioorganometallic compounds are unique metal complexes having a metal-organic σ-bond present in biological systems.[1] The unique physicochemical properties of these organometallic complexes prompted chemists to explore novel multidisciplinary applications of these compounds in metallo-drug, bioimaging, bioanalysis, and bioinspired catalyses.[2,3] Among the bioorganometallic compounds, coenzyme B_{12} is a well-known example present in various B_{12}-dependent mutases (Fig. 19.1).[4–8] The reduction of ribonucleotides to 2′-deoxyribonucleotides during DNA biosynthesis is also catalyzed by adenosylcobalamin in a ribonucleotide reductase (Fig. 19.1B).[9] Adenosylcobalamin is an alkylated cobalt complex having a cobalt−carbon σ-bond in which the adenosyl group is bound to the cobalt ion as an axial ligand. One of the unique properties of these compounds is the facile formation of carbon-centered radical species by homolysis of the cobalt−carbon bond. The B_{12}-inspired catalytic reactions have been developed using this property of the alkylated cobalt complexes.[4,10,11] The alkylated cobalt complex is generally formed by the reaction of the Co(I) state of the complex with electrophiles such as an organic halide. Therefore, the reductive formation of the Co(I) species followed by formal oxidative addition of the organic halide to form the alkylated cobalt complex are key steps for the development of the B_{12} catalyzed reaction. In an early studies, using chemical reductants, such as sodium borohydride, zinc, amalgam, dithiothreitol, and titanium (III) citrate, were dominant for the reduction of the cobalt complex to form Co(I) species. Recently, various kinds of photosensitizers have been utilized for the reductive activation of a metal complex.[12,13] The significant

Advances in Bioorganometallic Chemistry. DOI: https://doi.org/10.1016/B978-0-12-814197-7.00019-4

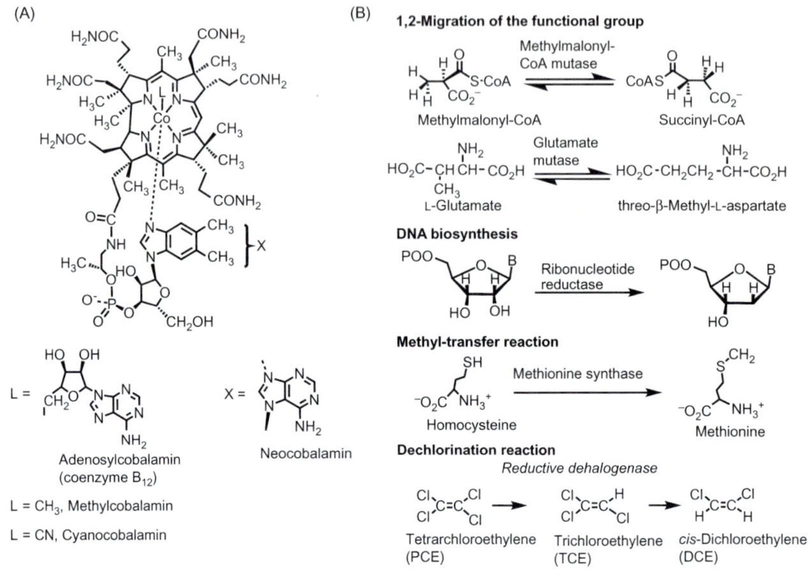

FIGURE 19.1 (A) Structure of B_{12} derivatives. (B) Examples of B_{12}-dependent enzyme reactions.

advantages of the photoredox system are reducing chemical waste using a clean and abundant light source and a simple and facilitated reaction system.

Electrolysis of the cobalt complex is another approach to form Co(I) species. Electroorganic synthesis is recognized as a green process in synthetic organic chemistry since the redox process of a substrate occurs due to an electric current without any chemical reagent and waste after the reaction.[14–16] The reducing or oxidizing powers of an electrode are easily controlled by the applied potential and current density. Electroorganic syntheses mediated by the B_{12} complexes have also been developed based on these backgrounds.[17] Various kinds of B_{12}-inspired reactions with a bioorganometallic character occurred by not only a photochemical method, but also an electrochemical procedure. In this chapter, recent advances and the potential applications of the bioorganometallic B_{12} as a catalyst for various organic syntheses are highlighted.

19.2 PHOTO-DRIVEN B_{12}-INSPIRED REACTIONS

As the natural vitamin B_{12}, cyanocobalamin (Fig. 19.1A), has a low solubility in conventional organic solvents, the amphiphilic vitamin B_{12} derivative with a high solubility in an organic solvent was developed for the practical application of the vitamin B_{12} derivative in organic synthesis. Among the amphiphilic vitamin B_{12} derivatives, heptamethyl cobyrinate perchlorate

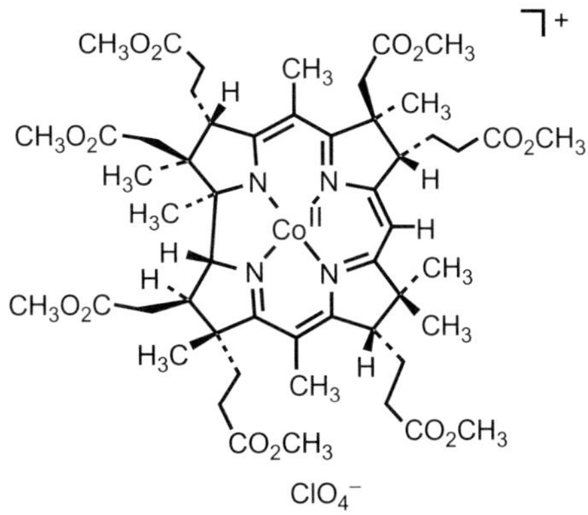

FIGURE 19.2 Structure of B_{12} derivatives, heptamethyl cobyrinate perchlorate.

with peripheral methyl ester groups instead of natural amide groups has an excellent solubility in various organic solvents (Fig. 19.2).[18] Various B_{12}-mimetic and -inspired reactions were developed using the heptamethyl cobyrinate as a catalyst.[19–21]

19.2.1 B_{12}-Photosensitizer Combination System

A photosensitizer supports the light-induced molecular transformation by electron or energy transfer processes using its abundant light absorbance and redox property.[12,13] Among them, the Ru(II) trisbipyridine complex, [Ru(II)(bpy)$_3$]Cl$_2$ (where bpy is 2,2′-bipyridine), is widely known as an excellent photoredox compound and has been utilized as a photosensitizer for the reductive activation of the B_{12} derivative to form Co(I) species. As the [Ru(II)(bpy)$_3$]Cl$_2$ is activated by visible light irradiation, various molecular transformations were conducted by the B_{12} derivatives with the [Ru(II)(bpy)$_3$]Cl$_2$ such as carbon dioxide and water reductions[22] and dechlorination of 1,1,1-trichloro-2,2-bis(4-chlorophenyl)ethane (DDT) in the presence of a sacrificial reductant as the electron source (Fig. 19.3).[23,24] Since DDT is characterized by a pronounced insecticidal property and has been used worldwide for the past decades despite its known hazardous effects on human health and wild life,[25] the efficient and effective dechlorination of DDT is strongly required. It is noted that the chlorine atom in DDT was removed as the nontoxic chloride anion in the B_{12} catalyzed reaction. The combined use of a photosensitizer and cobalamin derivative as the binuclear Ru^{II}–Co^{II} complex for the photoredox catalytic system was also reported (Fig. 19.4).[26] Advantages of such kinds of photosensitizers are the easy way to expand their molecular design for control of the redox potential and light absorption. Based on these viewpoints, cyclometalated iridium(III) complexes have been rapidly developing due to their superior photophysical and

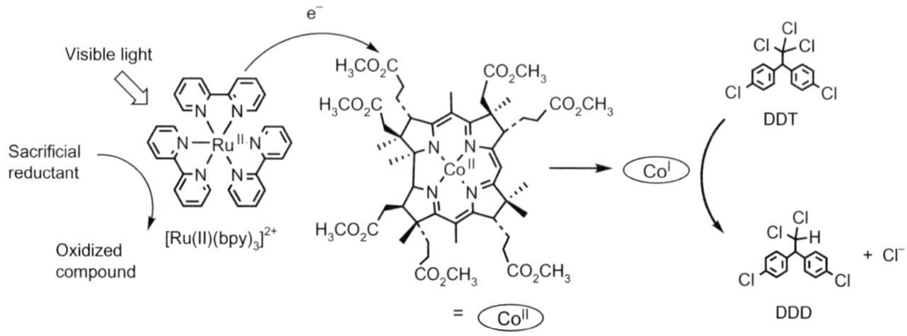

FIGURE 19.3 Photoredox catalytic system of the B_{12} catalyst with the photosensitizer [Ru(II)(bpy)$_3$]Cl_2.

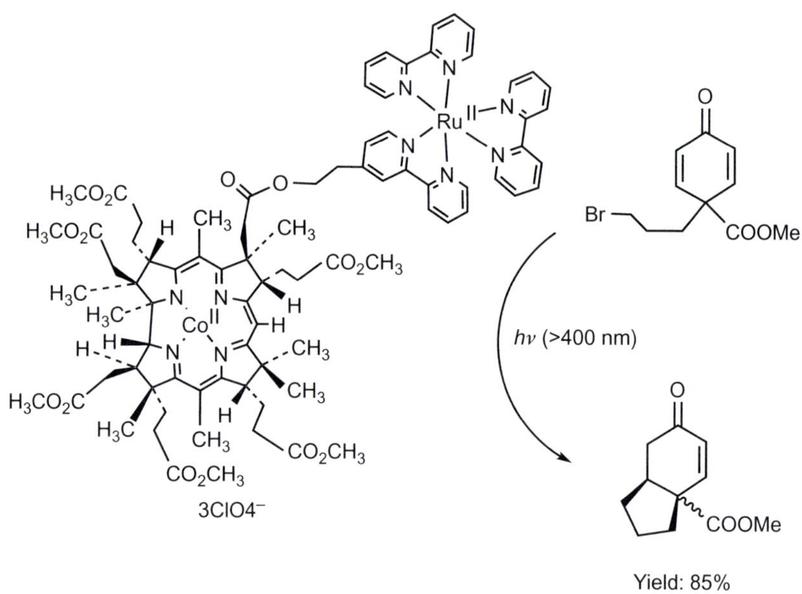

FIGURE 19.4 Photoredox catalytic reaction catalyzed by the binuclear Ru(bpy)$_3$–B_{12} system.

photochemical properties.[27] Especially, the robust organometallic structure in the cyclometalated iridium(III) complexes is the focus in organic synthesis for practical applications. Consequently, they have been utilized as photoredox catalysts in organic synthesis as well as serving as photosensitizers. The combined use of the cyclometalated iridium(III) photosensitizer and B_{12} catalyst achieved a significantly high efficiency for the DDT dechlorination reaction, and the turnover numbers for the catalytic reaction were over 10,000 and 80,000 based on the iridium(III) photosensitizer and B_{12} catalyst, respectively (Fig. 19.5).[28]

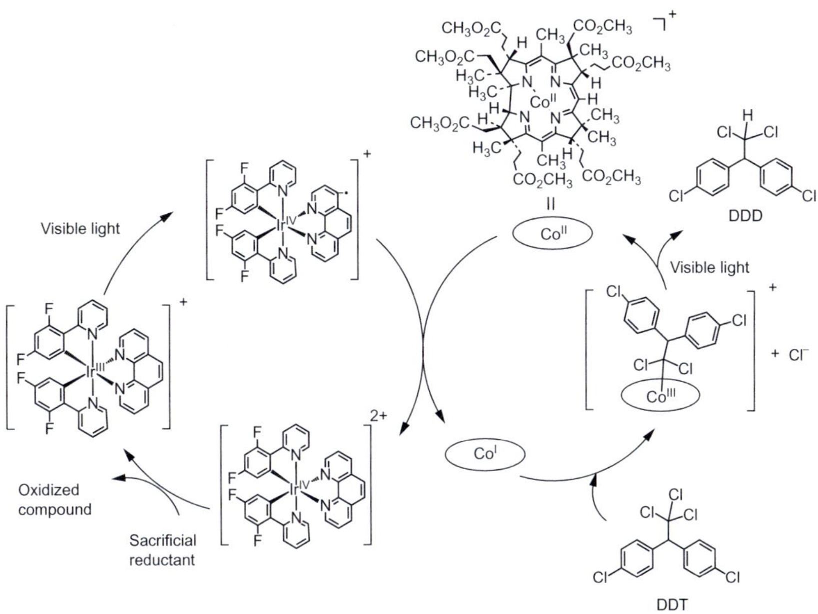

FIGURE 19.5 Photoredox catalytic system of the B_{12} catalyst with the photosensitizer cyclometalated iridium(III) complex.

To enhance the photocatalytic reaction, both the photosensitizer and B_{12} catalyst were incorporated into the same polymer unit for the proximity effect (Fig. 19.6).[29,30] The hybrid polymer was synthesized by the radical polymerization of a B_{12} derivative and a Ru(II) complex having styrene moieties in each of the peripheral positions. Thus, the prepared hybrid polymer showed an over 10 times higher reactivity for the catalytic reaction than that of the nonpolymer system just by mixing $[Ru(bpy)_3]Cl_2$ and the B_{12} complex in the dehalogenation reaction. The UV–vis spectral change for the Co(I) formation of the B_{12} complex under visible light irradiation showed that the intra-polymer electron transfer from the photosensitizer to the B_{12} complex should be prior to that of the nonpolymer system.[29] Another approach is using the reduced graphene oxide (rGO) as an electron relay platform during the catalytic reaction. An anionic B_{12} derivative was immobilized on the surface of the poly(ionic liquid)-modified rGO by electrostatic attraction and $\pi-\pi$ stacking attraction (Fig. 19.7).[31] The hybrid catalyst showed an enhanced photocatalytic activity in the presence of $[Ru(II)(bpy)_3]Cl_2$ for the DDT dechlorination with complete conversion and the total yield for the didechlorinated products, 1-chloro-2,2-bis(4-chlorophenyl)ethylene (DDMU), (*E*/*Z*)-2,3-dichloro-1,1,4,4-tetrakis (4-chlorophenyl)but-2-ene (TTDB), and 1-chloro-2,2-bis(4-chlorophenyl)ethane (DDMS) were over 60% after 1 h of visible light irradiation.

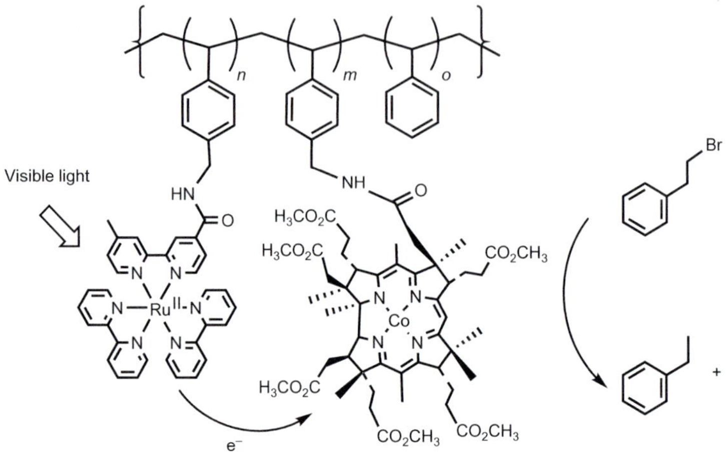

FIGURE 19.6 Photoredox catalytic reaction by the polymer-supported B_{12} catalyst and photosensitizer.

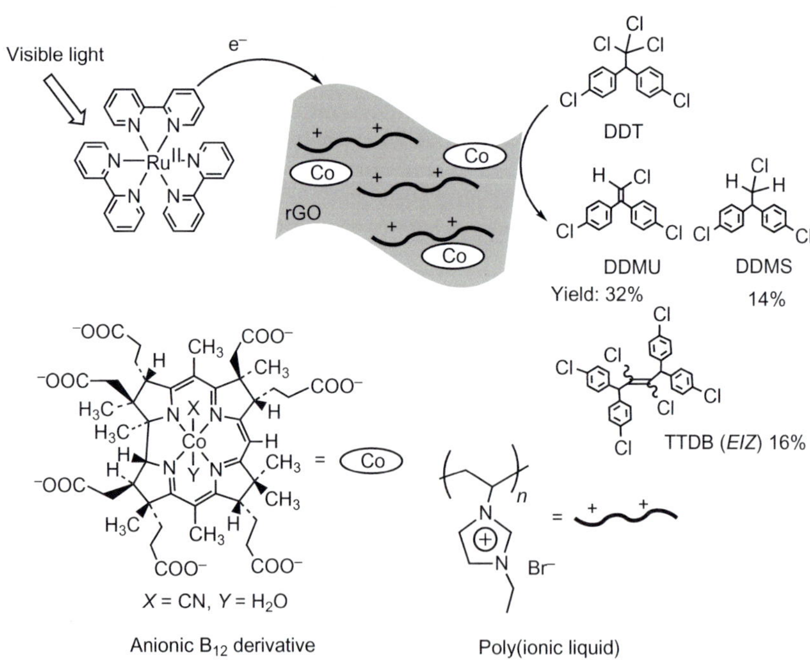

FIGURE 19.7 Photoredox catalytic system of the anionic B_{12} complex–poly(ionic liquid)/rGO hybrid catalyst with the photosensitizer $[Ru(II)(bpy)_3]Cl_2$.

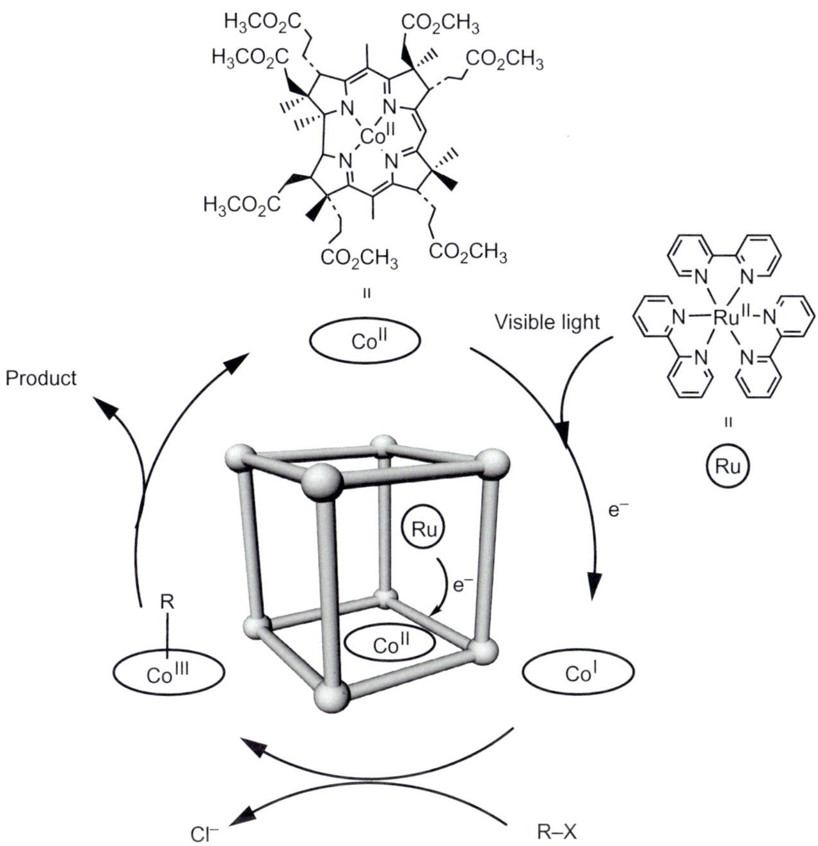

FIGURE 19.8 Photoredox catalytic system of the B_{12} catalyst with the photosensitizer [Ru(II)(bpy)$_3$]Cl$_2$ in the MOF.

Furthermore, this photosensitizing B_{12} catalytic system was developed as a heterogeneous system. The cationic $[Ru(II)(bpy)_3]^{2+}$ and B_{12}, heptamethyl cobyrinate, are immobilized to the metal-organic framework (MOF), B_{12}-Ru@MOF, by a cation exchange reaction (Fig. 19.8).[32] The B_{12}-Ru@MOF hybrid material worked as a heterogeneous catalyst for the DDT dechlorination and 1,2-migration reactions. Moreover, B_{12}-Ru@MOF can be recovered by filtration and subsequently reused several times and maintain its photocatalytic activity.

19.2.2 B_{12}–TiO_2 Hybrid Catalyst System

A semiconductor is a functional material, which is activated by light irradiation corresponding to its band gap energy. The band gap generation of a hole and an excited electron is utilized for various purposes. Among the semiconductors, the application of titanium oxide (TiO_2) has increased in various areas, such as solar cell devices, antibiotic and environmental applications, surface self-cleaning, and organic synthesis due to its chemical and physical stabilities and inexpensive price as well as its semiconductor property.[33–35] The TiO_2 mainly has three crystal forms (anatase, rutile, and brookite) and generates electron–hole pairs during band gap excitation (3.0–3.2 eV based on their crystal forms) by UV light irradiation. As the conduction band electron ($e_{CB}-$) for TiO_2 has an E_{red} of c. −0.5 V and c. −0.3 V versus normal hydrogen electrode (NHE) (in pH 7 aqueous solution) in the anatase and rutile, respectively, these are strong enough to reduce the B_{12} complex to form the Co(I) species. Using the reducing ability of the conduction band electron of TiO_2, the light-driven B_{12}–TiO_2 hybrid catalyst system was constructed (Fig. 19.9). The unique catalysis of the B_{12}–TiO_2 hybrid was developed for the B_{12}-inspired reactions as well as the B_{12}-mimetic reactions.

By the catalysis of B_{12}–TiO_2, the dechlorination of DDT or 1,1-dichloro-2,2-bis(4-chlorophenyl)ethane (DDD) occurred by UV light irradiation under anaerobic conditions to form various dechlorinated compounds in ethanol (Scheme 19.1A).[36] The B_{12}–TiO_2 hybrid catalyst was also effectively used for the radical reactions.[37] UV light irradiation of a diethyl 2-bromomethyl-2-phenylmalonate produced the phenyl migrated product, diethyl benzylmalonate, catalyzed by the B_{12}–TiO_2 (Scheme 19.1B). The B_{12}–TiO_2 catalyst was also used for the ring-expansion reactions of the alicyclic ketones (Scheme 19.1C). This is a kind of isomerization with 1,2-migration of the

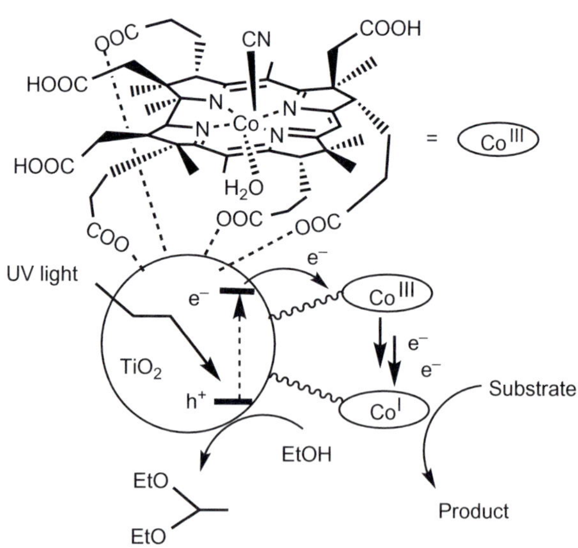

FIGURE 19.9 Structure of the B_{12}–TiO_2 hybrid catalyst.

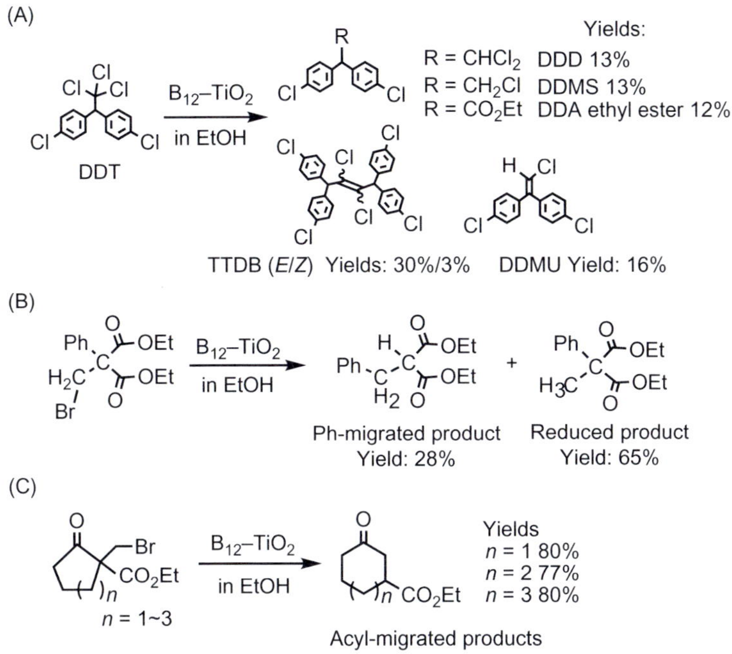

SCHEME 19.1 Molecular transformations catalyzed by the B_{12}-TiO_2.

acyl group. Ring expanded products (6-, 7-, and 8-membered rings) were obtained in moderate to good yields (77%–80%) and the turnover numbers (TONs) of the B_{12} catalyst in the B_{12}–TiO_2 were c. 200.[37,38]

For the development of the heterogeneous B_{12}–TiO_2 system, the B_{12}–TiO_2 powder was immobilized on a glass plate for practical use (Fig. 19.10).[39] The hybrid glass showed catalysis for the dechlorination of tetrachloroethylene (PCE) to trichloroethylene (TCE) by UV light irradiation in MeOH under N_2. After the reaction, the glass plate was removed from the solution and the products and catalyst were easily separated.

An oxygen switch for the catalysis of the B_{12}–TiO_2 hybrid catalyst has been developed. Generally, the B_{12} mediated reactions are carried out under anaerobic conditions to avoid the auto-oxidation of the labile Co(I) species. Furthermore, the carbon-centered radical formed by cleavage of the cobalt–carbon bond of the alkylated complex easily reacts with oxygen, which inhibits the 1,2-migration reaction during the B_{12}-mimic reactions. Take advantage of this point, oxygen-incorporated products were formed from the trichlorinated organic compounds by the B_{12}–TiO_2 hybrid catalyst.[40] Esters were formed from benzotrichloride under aerobic conditions in good yields, while dichlorostilbenes (*E* and *Z* forms) were formed under anaerobic conditions (Fig. 19.11). For the oxygen-incorporated reaction,

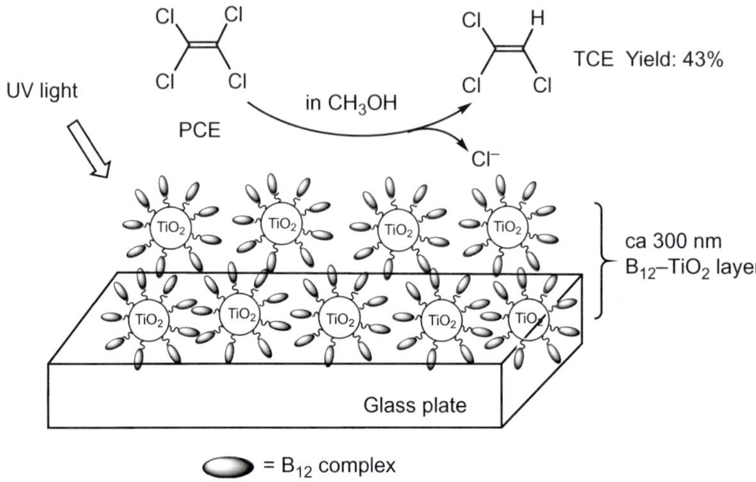

FIGURE 19.10 Dechlorination of PCE by the B_{12}–TiO_2 glass plate.

benzoyl chloride was thought to be an intermediate. The benzoyl chloride should react with the solvent alcohol to form an ester. When the reaction was applied to other trichlorinated compounds, such as DDT, the yield of the ester decreased to 56%, but with a high selectivity (87%) (Fig. 19.11C).

The photochemically produced reactive Co(I) species have been utilized for other organic syntheses. The cob(I)alamin catalyzed reduction of an α,β-unsaturated ethyl ester in aqueous acetic acid has been reported (Scheme 19.2).[41] The vitamin B_{12} catalyzed reductive dimerizations of arylalkenes in the presence of Ti(III)citrate or Zn reductant were also reported (Scheme 19.3).[42] The facile and efficient formation of the Co(I) species in the B_{12}–TiO_2 system by UV light irradiation prompted the development of these kinds of reactions. The B_{12}–TiO_2 catalyst mediates the reductions of conjugated alkenes, such as the styrene derivatives and alkylacrylates, by UV light irradiation under N_2 (Scheme 19.4).[43] During the reaction, the photochemically produced Co(I) species of the B_{12} derivative could react with a proton to form the cobalt–hydrogen complex (Co–H complex), then the Co–H complex was considered to catalyze the alkene reduction. During the alkene reduction by the B_{12}–TiO_2, a unique α-substituent dependence of the styrene derivatives reduction was observed. α-Substituted styrenes afforded a radical coupling product, hydrogenated product, and C–C bonded product by changing the α-substituent from H or CH_3 to COOH, and Br (Scheme 19.5). For the development of the substrate scope, α-trifluoromethyl styrene, having a trifluoromethyl group on the C=C bond, produced the *gem*-difluoroolefin by C–F bond cleavage in the CF_3 group.[44] The *gem*-difluoroolefin is a useful synthetic intermediate for a variety of fluorinated compounds.

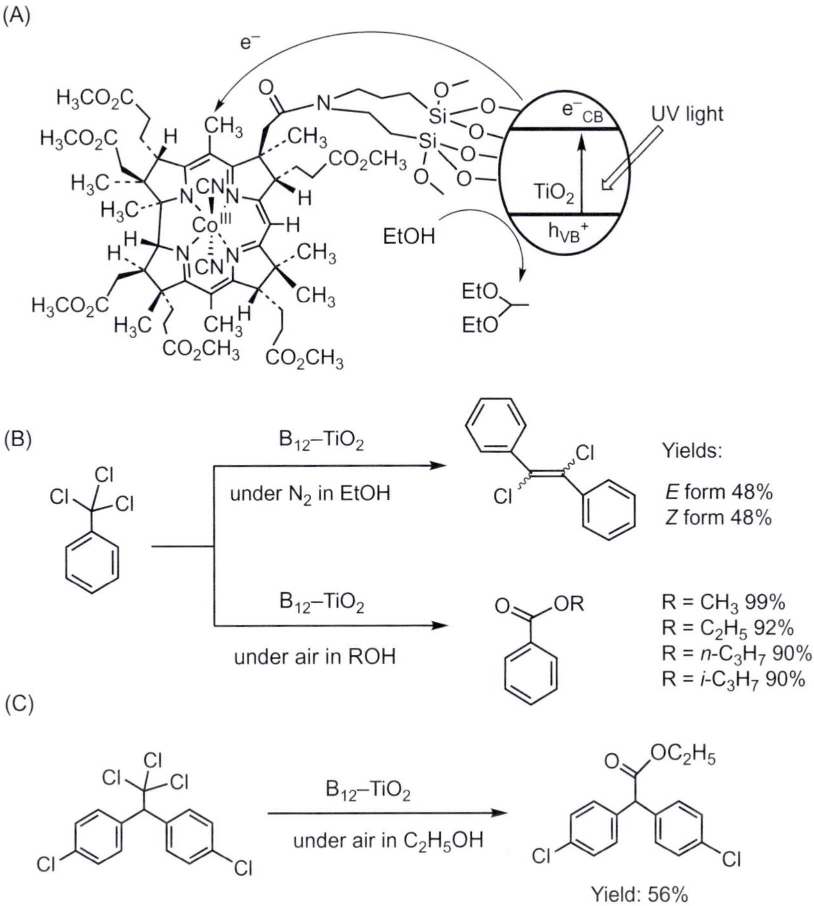

FIGURE 19.11 (A) Structure of covalently bound B_{12}–TiO_2. (B) Oxygen-controlled dechlorination of benzotrichloride by the B_{12}–TiO_2. (C) DDT esterification by the B_{12}–TiO_2 under air.

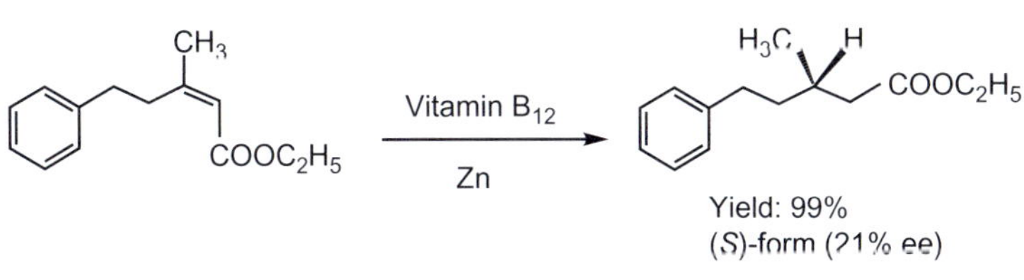

SCHEME 19.2 Reduction of an α, β-unsaturated ethyl ester catalyzed by the cob(I)alamin.

The B_{12}–TiO_2 hybrid catalyst also mediates proton reduction to form hydrogen. The recent growth of studies on hydrogen evolution by the cobalt complex comes from the high reactivity of the Co(I) complex toward a proton to form the Co–H complex as the intermediate.[45] UV light irradiation of the aqueous EDTA•2Na solution produced hydrogen in the presence of the

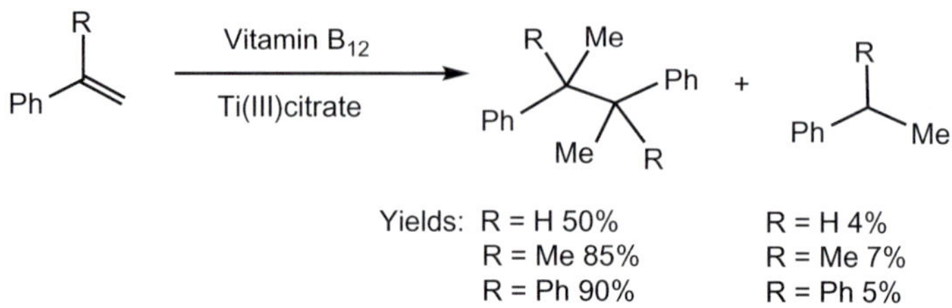

SCHEME 19.3 Reductive dimerizations of arylalkenes catalyzed by the vitamin B_{12}.

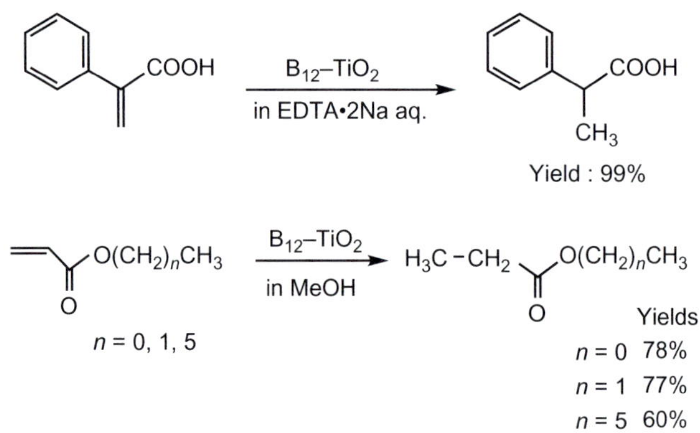

SCHEME 19.4 Reductions of conjugated alkenes catalyzed by the B_{12}-TiO_2.

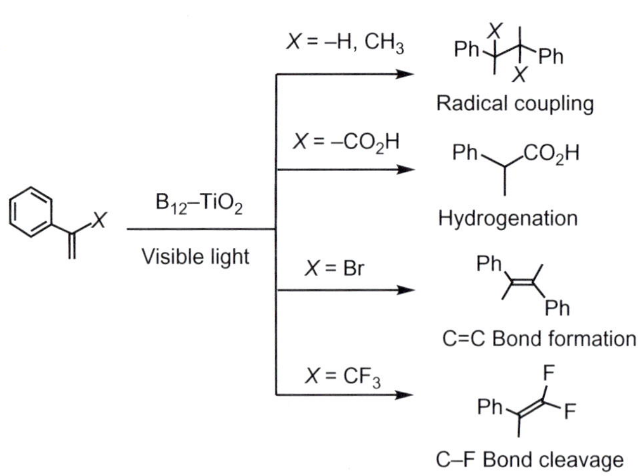

SCHEME 19.5 Photocatalytic reductions of styrene derivatives catalyzed by the B_{12}-TiO_2.

B_{12}–TiO_2.[43] The hydrogen production efficiency of the hybrid catalyst using the anatase-type TiO_2 was superior to that of the rutile TiO_2 since the conduction band electron is more negative for anatase TiO_2 (E_{red} = c. −0.5 V versus NHE in pH 7 aqueous solution) than that of the rutile (E_{red} = c. −0.3 V versus NHE in pH 7 aqueous solution) for the B_{12} reduction. An initial turnover frequency for the hydrogen evolution was 2.1 h^{-1} and 0.47 h^{-1} for the anatase TiO_2 and the rutile TiO_2, respectively. The turnover number based on the B_{12} complex was estimated to be about 1 per hour in B_{12}–TiO_2 (anatase) for a 10 hour reaction and this value was quite low compared to that of the alkene reduction which was 100 per hour. When the reaction was carried out in the presence of alkene substrates, the hydrogen evolution was inhibited. This result implied that the Co–H complex is the plausible intermediate for both reactions.

19.3 ELECTROCHEMICALLY DRIVEN B_{12} CATALYTIC REACTIONS

The cobalt complex of the B_{12} derivatives is electrochemically active, and the Co(I) species should be formed by an electrode process. The electrochemical behaviors of vitamin B_{12} and its derivatives have been studied for a long time and applications for electroorganic synthesis have been developed.[17,46] Since indirect electrolysis using a mediator facilitates the redox process of the substrate to lower the applied potential resulting in a reduced energy cost and achieving an efficient molecular transformation, vitamin B_{12} and its derivatives were utilized as the mediator for the electroorganic reactions.

19.3.1 B_{12} Mediated Electrochemical Dehalogenation of Organic Halides Pollutants

As the monoanionic corrin ring can stabilize the low oxidation state of the cobalt atom in the cobalamin, the Co(II)/Co(I) redox couple appeared at a relatively positive potential around −0.4 V ~ −0.6 V versus Ag/AgCl in a nonaqueous solvent.[46] The in situ generated Co(I) species of the vitamin B_{12} derivative, heptamethyl cobyrinate, in a nonaqueous electrolyte solution reacts with organic halides by dehalogenation via an alkylated complex. The electrolysis of DDT catalyzed by heptamethyl cobyrinate perchlorate was conducted at −1.4 V versus Ag/AgCl in N,N-dimethylformamide (DMF) under anaerobic conditions.[47] Monodechlorinated products, such as DDD and 1,1-dichloro-2,2-bis(4-chlorophenyl)ethylene (DDE), and didechlorinated products, such as DDMU and TTDB (*E/Z*), were produced (Fig. 19.12). The electrolysis was also performed in a conductive ionic liquid, 1-butyl-3-methylimidazolium tetrafluoroborate, without the use of the DMF electrolyte solution. The catalyst solution of an ionic liquid was recycled and reused for

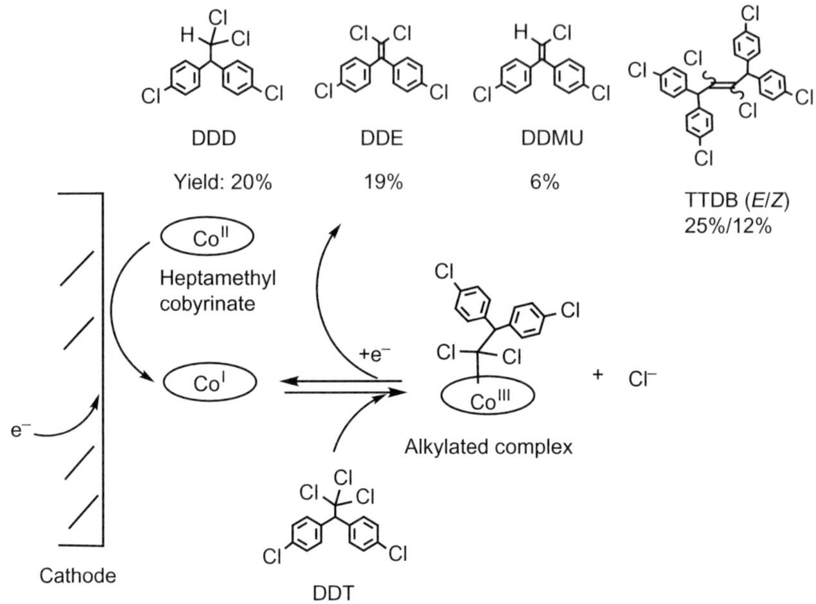

FIGURE 19.12 Electrochemical dechlorination of DDT catalyzed by the B_{12} catalyst.

the electrolysis of DDT without any loss of catalysis efficiency due to the high stability and solubility of the catalyst in the ionic liquid. It should be noted that more dechlorinated products, such as DDMS, 1,1-bis(4-chlorophenyl)ethylene (DDNU) and 1,1-bis(4-chlorophenyl)ethane (DDO), were obtained in an ionic liquid in addition to DDD, DDE, DDMU, and TTDB (*E*/*Z*) (Fig. 19.13).[48] The enhanced reactivity of the B_{12} derivative–ionic liquid system over the B_{12} derivative–DMF system could be explained by the Hughes–Ingold predictions of the solvent polarity effects on the reaction rates. The advantages of an ionic liquid, such as high conductivity, very low volatility, and immiscible in an organic solvent combined with such a solvent effect, could provide a new methodology for electroorganic synthesis.

The reductive electrolysis of an organic halide under aerobic conditions catalyzed by heptamethyl cobyrinate perchlorate was also reported.[49] Though the labile Co(I) species of heptamethyl cobyrinate should be lost during the reaction by auto-oxidation in air, the aerobic electrolysis of trichlorinated organic compounds by the B_{12} catalyst afforded unique products. The electrolysis of benzotrichloride under nitrogen afforded 1% of the 1,1,2,2-tetrachloro-1,2-diphenylethane and 15% and 72% of the 1,2-dichlorostilibenes, *E* and *Z* forms, respectively. On the other hand, the electrolysis of benzotrichloride at -0.9 V versus Ag/AgCl in ethanol in air provided ethylbenzoate in 56% yield with a 92% selectivity using heptamethyl cobyrinate perchlorate as the catalyst (Scheme 19.6). The oxygen-controlled

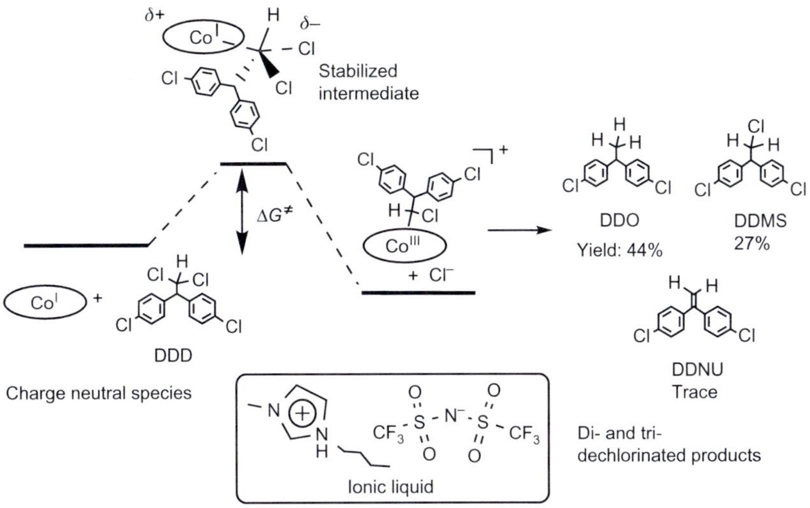

FIGURE 19.13 Enhanced reactivity of the B_{12} catalyst in ionic liquid. Schematic energy diagram for the B_{12} complex catalyzed dechlorination reaction was shown.

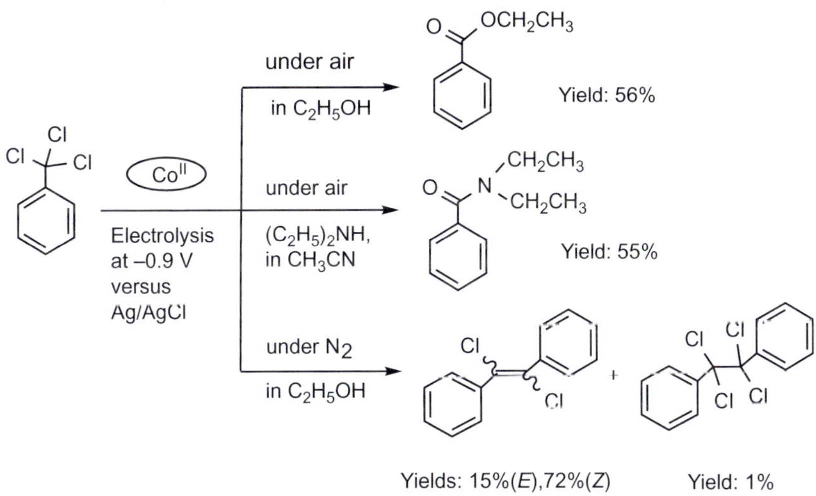

SCHEME 19.6 Electrolyses of benzotrichloride by the B_{12} complex under aerobic or anaerobic conditions.

electrolysis of trichlorinated organic compounds for product distributions was achieved. The substrate radical should react with oxygen to form the oxyl-radical, then changes to benzoyl chloride by disproportionation. Thus benzoyl chloride was deduced as an intermediate for the ester formation. This method, the aerobic electrochemical reduction, was also applied to amide synthesis, and moderate yields of the amides were obtained in the presence of amines due to the high reactivity of the amines to benzoyl chloride.

19.3.2 Bioinspired Electrochemical Reaction of Alkenes and Alkyne by the B_{12} Derivative

B_{12}-inspired reactions for the alkene and alkyne reductions also electrochemically proceeded. During the reaction, the electrochemically produced Co(I) species of the B_{12} catalyst reacts with a proton to form the cobalt–hydrogen complex (Co(III)–H complex), then the Co(III)–H complex was considered to catalyze the alkene and alkyne reductions. Generally, the direct reduction of alkenes requires an extremely negative potential over −2 V versus saturated calomel electrode (SCE) in an organic solvent, while the alkene reduction may occur at a more positive potential in the presence of the B_{12} derivative since the Co(I) species of the B_{12} complex can be formed at the potential and reacts with a proton to form the Co(III)–H complex as a reactive intermediate for the alkene reduction (Fig. 19.14).[50] The electrochemical reductive coupling and hydrogenation of alkenes catalyzed by heptamethyl cobyrinate perchlorate were conducted in the presence of an acid. The electrolysis was also applied to alkynes, such as phenylacetylene and diphenylacetylene (Scheme 19.7). The spin-trapping experiment using 5,5-dimethylpyrroline N-oxide (DMPO) provided the DMPO-H electron spin resonance (ESR) signal during the electrolysis. It should be expected that the Co(III)–H complex has a hydrogen radical character and the attack of the hydrogen radical on the C=C bond of the alkene initiates the reaction.

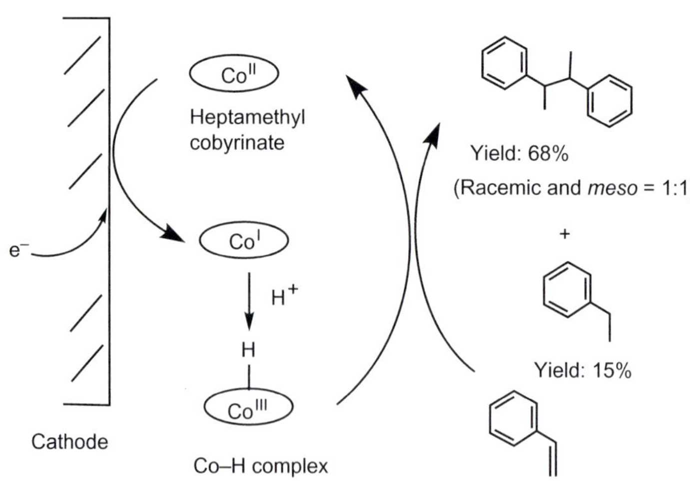

FIGURE 19.14 Electrolysis of styrene catalyzed by the B_{12} catalyst in the presence of acid.

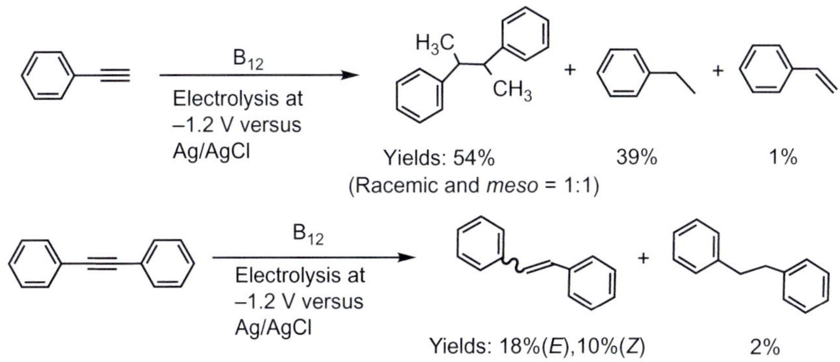

SCHEME 19.7 Electrolysis of alkynes by the B_{12} complex in the presence of an acid.

The bioinspired electrolysis of the alkene and alkyne by the B_{12} model complex proceeds under mild conditions, for example, at room temperature and normal pressure without use of the hydrogen gas and a precious metal catalyst, such as platinum or ruthenium and the reductive coupling and hydrogenation of the alkenes proceeded at a positive applied potential compared to that of the direct reduction of the alkenes. The electrolysis was applied to a variety of alkenes and the conjugated alkenes to form moderate yields of the products. The B_{12} derivative showed a good catalysis for the reaction, and the Co—H species of the B_{12} complex was proposed as an intermediate for the reaction.

19.4 CONCLUSION AND SCOPE

Bioorganometallic B_{12} compounds are useful intermediates for organic synthesis. Applications of the bioinspired intermediate, the Co—H complex, as well as biomimetic intermediate, the Co—C complex, have been developed for photochemical and electrochemical catalytic systems. Biomimetic isomerization and dechlorination efficiently proceeded. As for the bioinspired reaction, hydrogenation and hydrogen evolution were catalyzed by the B_{12} complex. If the suitable photosensitizer is selected for the photochemical system, a low energy visible light responsible catalytic system could be constructed, and utilization of clean and abundant sunlight for molecular transformation is realized. In the electrochemical system, the combination of the B_{12} complex with recent progress in electrochemistry, such as a flow microreactor and light assisted electrolysis, will offer a new area for electroorganic synthesis.[51] Furthermore, as the vitamin B_{12} derivatives have several chiral centers in the corrin ring, the asymmetric reaction by the vitamin B_{12} derivatives may become the next target. The progress of bioorganometallic B_{12} compounds in organic synthesis will expand to many fields.

ACKNOWLEDGMENTS

This study was partially supported by KAKENHI from the Japan Society for the Promotion of Science (JSPS) and a Grant from the Iwatani Naoji Foundation.

REFERENCES

1. Hillard EA, Jaouen G. Bioorganometallics: future trends in drug discovery. *Organometallics* 2011;**30**:20–7.
2. Patra M, Gasser G, Metzler-Nolte N. Small organometallic compounds as antibacterial agents. *Dalton Trans* 2012;**41**:6350–8.
3. Zhang P, Sadler PJ. Advances in the design of organometallic anticancer complexes. *J Organomet Chem* 2017;**839**:5–14.
4. Brown KL. Chemistry and enzymology of vitamin B_{12}. *Chem Rev* 2005;**105**:2075–149.
5. Gruber K, Puffer B, Kräutler B. Vitamin B_{12}-derivatives-enzyme cofactors and ligands of proteins and nucleic acids. *Chem Soc Rev* 2011;**40**:4346–63.
6. Buckel W, Golding BT. Radicals enzymes in anaerobes. *Annu Rev Microbiol* 2006;**60**:27–49.
7. Toraya T. Radical catalysis in coenzyme B_{12}-dependent isomerization (eliminating) reactions. *Chem Rev* 2003;**103**:2095–127.
8. Bridwell-Rabb J, Drennan CL. Vitamin B_{12} in the spotlight again. *Curr Opin Chem Biol* 2017;**37**:63–70.
9. Licht S, Gerfen GJ, Stubbe J. Thiyl radicals in ribonucleotide reductases. *Science* 1996;**271**:477–81.
10. Giedyk M, Goliszewska K, Gryko D. Vitamin B_{12} catalyzed reactions. *Chem Soc Rev* 2015;**44**:3391–404.
11. Shimakoshi H, Hisaeda Y. A hybrid catalyst for light-driven green molecular transformations. *ChemPlusChem* 2017;**82**:18–29.
12. Skubi KL, Blum TR, Yoon TP. Dual catalysis strategies in photochemical synthesis. *Chem Rev* 2016;**116**:10035–74.
13. Fabry DC, Rueping M. Merging visible light photoredox catalysis with metal catalyzed C–H activations: on the role of oxygen and superoxide ions as oxidants. *Acc Chem Res* 2016;**49**:1969–79.
14. Francke R, Little RD. Redox catalysis in organic electrosynthesis: basic principles and recent developments. *Chem Soc Rev* 2014;**43**:2492–521.
15. Yoshida J, Kataoka K, Horcajada R, Nagaki A. Modern strategies in electroorganic synthesis. *Chem Rev* 2008;**108**:2265–99.
16. Yan M, Kawamata Y, Baran PS. Synthetic organic electrochemical methods since 2000: on the verge of a renaissance. *Chem Rev* 2017;**117**:13230–319.
17. Shimakoshi H, Hisaeda Y. Electrochemistry and catalytic properties of vitamin B_{12} derivatives in nonaqueous media. *Curr Opin Electrochem* 2018;**8**:24–30.
18. Murakami Y, Hisaeda Y, Kajihara A. Hydrophobic vitamin B_{12}. I. Preparation and axial ligation behavior of hydrophobic vitamin B_{12r}. *Bull Chem Soc Jpn* 1983;**56**:3642–6.
19. Murakami Y, Hisaeda Y, Ozaki T, Tashiro T, Ohno T, Tani Y, et al. Hydrophobic vitamin B_{12}. V. Electrochemical carbon-skeleton rearrangement as catalyzed by hydrophobic vitamin B_{12}: reaction mechanism and migratory aptitude of functional groups. *Bull Chem Soc Jpn* 1987;**60**:311–24.

20. Ociepa M, Baka O, Narodowiec J, Gryko D. Light-driven vitamin B_{12}-catalyzed generation of acyl radicals from 2-*S*-pyridyl thioesters. *Adv Synth Catal* 2017;**359**:3560–5.
21. Sun F, Darbre T. The Co(I) induced methymalonyl–succinyl rearrangement in a model for the coenzyme B_{12} dependent methymalonyl-CoA mutase. *Org Biomol Chem* 2003;**1**:3154–9.
22. Ziessel R, Hawecker J, Lehn J-M. Photogeneration of carbon monoxide and of hydrogen via simultaneous photochemical reduction of carbon dioxide and water by visible-light irradiation of organic solutions containing tris(2,2′-bipyridine)ruthenium(II) and cobalt(II) species as homogeneous catalysts. *Helv Chim Acta* 1986;**69**:1065–84.
23. Shimakoshi H, Tokunaga M, Baba T, Hisaeda Y. Photochemical dechlorination of DDT catalyzed by a hydrophobic vitamin B_{12} and a photosensitizer under irradiation with visible light. *Chem Commun* 2004;1806–7.
24. Shimakoshi H, Kudo S, Hisaeda Y. Recycled catalysis of a hydrophobic vitamin B_{12} in a ionic liquid. *Chem Lett* 2005;**34**:1096–7.
25. Kiflom WG, Wandiga SO, Kamau GN. Status of organochlorine (DDT) pollutants and steps toward electrocatalytic reductions. *Pure Appl Chem* 2001;**73**:1907–16.
26. Steiger B, Eichenberger E, Walder L. Ru(II)$(bpy)_3$-Cob(II)yrinate -ein binuklearer photoredox-photokatalysator. Synthese und anwendung in der organischen chemie. *Chimia* 1991;**45**:32–7.
27. Singh A, Teegardin K, Kelly M, Prasad KS, Krishnan S, Weaver JD. Facile synthesis and complete characterization of homoleptic and heteroleptic cyclometalated iridium(III) complexes for photocatalysis. *J Organomet Chem* 2015;**776**:51–9.
28. Tian H, Shimakoshi H, Park G, Kim S, You Y, Hisaeda Y. Photocatalytic function of B_{12} complex with cyclometalated iridium(III) complex as photosensitizer under visible light irradiation. *Dalton Trans* 2018;**47**:675–83.
29. Shimakoshi H, Nishi M, Tanaka A, Chikama K, Hisaeda Y. Photocatalytic function of a polymer-supported B_{12} complex with a ruthenium trisbipyridine photosensitizer. *Chem Commun* 2011;**47**:6548–50.
30. Zhang W, Shimakoshi H, Houfuku N, Song X, Hisaeda Y. A polymerized ionic liquid-supported B_{12} catalyst with a ruthenium trisbipyridine photosensitizer for photocatalytic dechlorination in ionic liquid. *Dalton Trans* 2014;**43**:13972–8.
31. Sun Y, Zhang W, Tong J, Zhang Y, Wu S, Liu D, et al. Significant enhancement of visible light photocatalytic activity of the hybrid B_{12}–PIL/rGO in the presence of $Ru(bpy)_3^{2+}$ for DDT dehalogenation. *RSC Adv* 2017;**7**:19197–204.
32. Xu J, Shimakoshi H, Hisaeda Y. Development of metal-organic framework (MOF)–B_{12} system as new bio-inspired heterogeneous catalyst. *J Organomet Chem* 2015;**782**:89–95.
33. Ma Y, Wang X, Jia Y, Chen X, Han H, Li C. Titanium dioxide-based nanomaterials for photocatalytic fuel generations. *Chem Rev* 2014;**114**:9987–10043.
34. Sang L, Zhao Y, Burda C. TiO_2 nanoparticles as functional building blocks. *Chem Rev* 2014;**114**:9283–318.
35. Ohtani B. Photocatalysis A to Z—what we know and what we do not know in a scientific sense. *J Photochem Photobiol C: Photochem Rev* 2010;**11**:157.
36. Shimakoshi H, Sakumori E, Kaneko K, Hisaeda Y. B_{12}–TiO_2 hybrid catalyst for dehalogenation of organic halides. *Chem Lett* 2009;**38**:468–9.
37. Shimakoshi H, Abiru M, Izumi S, Hisaeda Y. Green molecular transformation by a B_{12}–TiO_2 hybrid catalyst as an alternative to tributyltin hydride. *Chem Commun* 2009;6427–9.

38. Izumi S, Shimakoshi H, Abe M, Hisaeda Y. Photo-induced ring-expansion reactions mediated by $B_{12}-TiO_2$ hybrid catalyst. *Dalton Trans* 2010;**39**:3302–7.
39. Shimakoshi H, Abiru M, Kuroiwa K, Kimizuka N, Watanabe M, Hisaeda Y. Preparation and reactivity of vitamin $B_{12}-TiO_2$ hybrid catalyst immobilized on a glass plate. *Bull Chem Soc Jpn* 2010;**83**:170–2.
40. Shimakoshi H, Hisaeda Y. Oxygen-controlled catalysis by vitamin $B_{12}-TiO_2$: formation of esters and amides from trichlorinated organic compounds by photoirradiation. *Angew Chem Int Ed* 2015;**54**:15439–43.
41. Fischli A, Suss D. Cob(I)alamin as catalyst. 5. Communication [1]. Enantioselective reduction of α,β-unsaturated carbonyl derivatives. *Helv Chim Acta* 1979;**62**:2361–73.
42. Shey J, McGinley CM, McCauley KM, Dearth AS, Young BT, van der Donk WA. Mechanistic investigation of a novel vitamin B_{12}-catalyzed carbon–carbon bond forming reaction, the reductive dimerization of arylalkenes. *J Org Chem* 2002;**67**:837–46.
43. Shimakoshi H, Hisaeda Y. $B_{12}-TiO_2$ hybrid catalyst for light-driven hydrogen production and hydrogenation of C–C multiple bonds. *ChemPlusChem* 2014;**79**:1250–3.
44. Tian H, Shimakoshi H, Imamura K, Shiota Y, Yoshizawa K, Hisaeda Y. Photocatalytic alkene reduction by $B_{12}-TiO_2$ hybrid catalyst coupled with C–F bond cleavage for *gem*-difluoroolefin synthesis. *Chem Commun* 2017;**53**:9478–81.
45. Artero V, Chavarot-Kerlidou M, Fontecave M. Splitting water with cobalt. *Angew Chem Int Ed* 2011;**50**:7238–66.
46. Murakami Y, Hisaeda Y, Kajihara A, Ohno T. Hydrophobic vitamin B_{12}. II. Coordination geometry and redox behavior of heptamethyl cobyrinate in nonaqueous media. *Bull Chem Soc Jpn* 1984;**57**:405–11.
47. Shimakoshi H, Tokunaga M, Hisaeda Y. Hydrophobic vitamin B_{12}. Part 19: Electroorganic reaction of DDT mediated by hydrophobic vitamin B_{12}. *Dalton Trans* 2004;878–82.
48. Jabbar MA, Shimakoshi H, Hisaeda Y. Enhanced reactivity of hydrophobic vitamin B_{12} towards the dechlorination of DDT in ionic liquid. *Chem Commun* 2007;1653–5.
49. Shimakoshi H, Luo Z, Inaba T, Hisaeda Y. Electrolysis of trichloromethylated organic compounds under aerobic conditions catalyzed by B_{12} model complex for ester and amide formations. *Dalton Trans* 2016;**45**:10173–80.
50. Shimakoshi H, Luo Z, Tomita K, Hisaeda Y. Cathodic reductive coupling and hydrogenation of alkenes and alkynes catalyzed by the B_{12} model complex. *J Organomet Chem* 2017;**839**:71–7.
51. Atobe M. Organic electrosynthesis in flow microreactor. *Curr Opin Electrochem* 2017;**2**:1–6.

Chapter 20

Biological Organometallic Chemistry of Vitamin B_{12}-Derivatives

Bernhard Kräutler
Institute of Organic Chemistry and Centre of Molecular Biosciences University of Innsbruck, Innsbruck, Austria

20.1 INTRODUCTION

The importance of organometallic chemistry in life processes was first revealed in the early 1960s, when coenzyme B_{12} was identified as the organometallic vitamin B_{12} derivative 5′-deoxyadenosyl-cobalamin (**AdoCbl**).[1] Its discovery followed the one of the related red cobalt corrin vitamin B_{12} or cyanocobalamin (**CNCbl**),[1–3] an exceptionally complex member of the natural tetrapyrroles,[4] prepared by total synthesis by Eschenmoser, Woodward, and their teams.[5,6]

B_{12}-coenzymes are, perhaps, nature's physiologically most relevant organometallic cofactors.[7–10] They are required in the metabolism of a broad range of organisms, including humans. Remarkably, only some microorganisms have the capacity to biosynthesize B_{12} and other natural corrinoids.[11] All other B_{12}-dependent organisms rely on B_{12}-derivatives as their vitamins.[12] Thus, their functioning metabolism depends on the uptake and cellular import of useful B_{12}-derivatives,[13] their metabolic transformation to relevant B_{12}-cofactors,[14] as well as the catalysis by B_{12}-dependent enzymes.[10,15–17] Interestingly, some physiologic effects of B_{12} in humans are still rather puzzling.[18]

Vitamin B_{12} (cyanocobalamin, **CNCbl**) is the most important commercially available vitamin form of the naturally occurring B_{12}-derivatives, but has no physiological function itself.[14] The "inorganic" B_{12}-derivatives aquocobalamin (**H_2OCbl^+**) and hydroxocobalamin (**HOCbl**) are also used as B_{12}-vitamers. However, the organometallic analogues coenzyme B_{12} (**AdoCbl**) and methylcobalamin (**MeCbl**) are the physiologically directly

Advances in Bioorganometallic Chemistry. DOI: https://doi.org/10.1016/B978-0-12-814197-7.00020-0

relevant cobalamins (Cbl), and help to catalyze exceptional enzymatic reactions by engaging the reactivity of their (Co-C)-bond.[10,12,15,19–21]

During the past 50 years, remarkable scientific advances in the B_{12}-field have been made that solved some earlier major "B_{12}-mysteries". Much of it is described in earlier[22,23] and in more recent[7,8,12,24] books in the B_{12}-field.

20.2 BASIC STRUCTURES OF B_{12}-DERIVATIVES

Vitamin B_{12} (**CNCbl**) and coenzyme B_{12} (**AdoCbl**) are classified as "complete" corrinoids, in which a pseudo-nucleotide function is attached via the f-side chain to the corrin moiety.[19] The "incomplete" corrinoid cobyric acid (Cby) is the corrinoid moiety of **CNCbl**, but lacks a nucleotide "loop." In the likewise "incomplete" cobinamides (Cbi) only the first linker section of the nucleotide "loop," (R)-isopropanolamine, extends from the f-side chain (see Fig. 20.1).

20.2.1 "Incomplete" B_{12}-Derivatives

The crystalline "incomplete" corrinoid cyano,aquo-cobyric acid (**CN,H_2O-Cby**) played historically important roles, as its X-ray investigation led to the

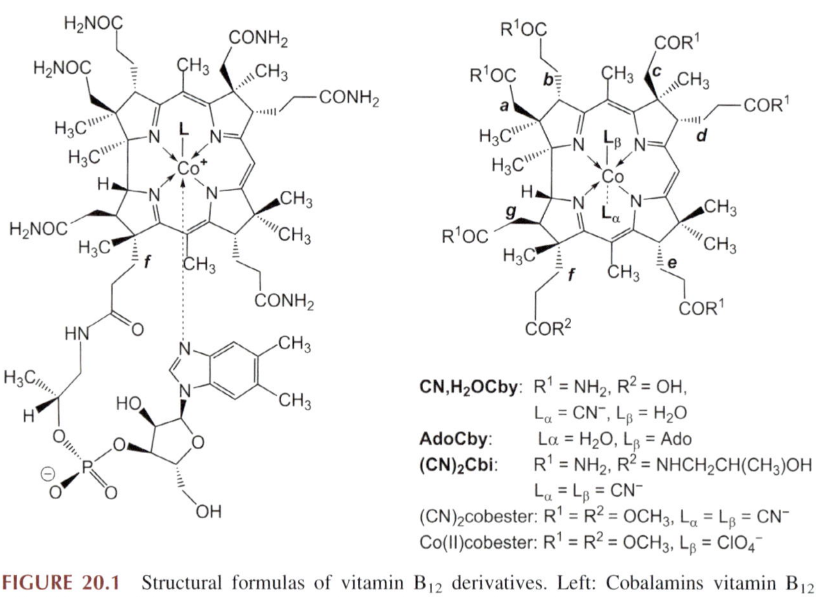

FIGURE 20.1 Structural formulas of vitamin B_{12} derivatives. Left: Cobalamins vitamin B_{12} (**CNCbl**, L = CN), coenzyme B_{12} (**AdoCbl**, L = 5′-deoxy-5′-adenosyl), methylcobalamin (**MeCbl**, L = CH_3), cob(II)alamin (**Cbl^{II}**, L = e^-), hydroxo-Cbl (HOCbl, L = OH), ethylphenyl-Cbl (**PhCbl**, L = phenyl), 4-ethylphenyl-Cbl (**EtPhCbl**, L = 4-ethylphenyl), 2-phenyl-ethynyl-Cbl (**PhEtyCbl**, L = 2-phenylethynyl), 2(2,4-difluorophenyl]-ethynyl-Cbl (**F2PhEtyCbl**, L = 2 (2,4-difluorophenyl]-ethynyl); right: Some important "incomplete" corrinoids.

discovery of the structure of the corrin ligand (see[1,25]). **CN,H_2O-Cby** also represented the direct corrinoid target in the total synthesis of vitamin B_{12}.[5,6] Adenosylcobyrate (**AdoCby**), an often inferred common organometallic intermediate of the B_{12}-biosynthesis,[11] has been characterized subsequently and was also used in an efficient chemical synthesis of **AdoCbl**.[26] The C13-epimer of **CN,H_2O-Cby** crystallized as a dimer, in which the f-carboxylate function of one molecule coordinated to the cobalt-center of the other.[27] The lipophilic heptamethylcobyrinate "cobester" was prepared as a B_{12}-model compound[5,28,29] and its crystal structure was analyzed.[30] The first detailed insights into the structure of a paramagnetic Co(II)-corrin were gained from the crystal structure of heptamethyl-cob(II)yrinate perchlorate ("cob(II)ester"), which confirmed the expected five-coordinate Co(II)-center.[31]

20.2.2 "Complete" B_{12}-Derivatives

"Complete" corrinoids, such as **CNCbl** and **AdoCbl**, are conjugates of natural cobyrates with the unusual, B_{12}-typical α-nucleotide functions.[9,10,19] The crystal structure of vitamin B_{12} (**CNCbl**) revealed the unique 3D-architecture of the mutually interacting nucleotide loop and corrin moieties.[1,25,32,33] **CNCbl** is a 5′,6′-dimethylbenzimidazolyl-cobamide or cobalamin (Cbl), in which 5,6-dimethylbenzimidazole (DMB) coordinates the cobalt ion from the "lower" axial (or α) side.[34] It is more specifically named cyanocobalamin (**CNCbl**), since a cyanide ligand is bound at the "upper" axial coordination site (or β-face)[34]). Crystal structures of a range of Cbls and other "complete" corrinoids have been solved.[10,32,33] The 3D-structure of the asymmetric Cbls in their cobalt-coordinated "base-on"-form represents a unique and topologically chiral molecule.[10]

Purinyl-cobamides occur in various micro-organisms, such as pseudovitamin B_{12} (a 7″-adeninylcobamide) and factor A (a 7″ [2 methyl]adeninylcobamide), a second important class of "complete" corrinoids[35–37] in which, e.g., adenine or 2-methyl-adenine, respectively, replace the DMB nucleotide function of the Cbls (see Fig. 20.2).[38–40] In other, semisynthetic Cbas, an imidazole, as in Co$_β$-cyano-imidazolylcobamide, substitutes for the DMB-base.[41] *Nor*-pseudovitamin B_{12} (Co$_β$-cyano-7″-adeninyl-176-norcobamide) is the first natural 176-*nor*-cobamide discovered lacking the methyl group C176 of the cobamide moiety.[42] This modification destabilizes the "base-on" form,[43] important in some B_{12}-dependent dehalogenases[44,45] (see below).

Extensive crystallographic studies of coenzyme B_{12} (**AdoCbl**) revealed both axial bonds [(Co−C) (2.030 Å) and (Co−N) (2.237 Å)] to be relatively long.[32,33] The crystal structure of methylcobalamin (**MeCbl**), the "other" biologically important organocobalamin, showed two axial bonds shorter than the ones in **AdoCbl**.[32] Structural information on cob(II)alamin (**Cbl^{II}**)

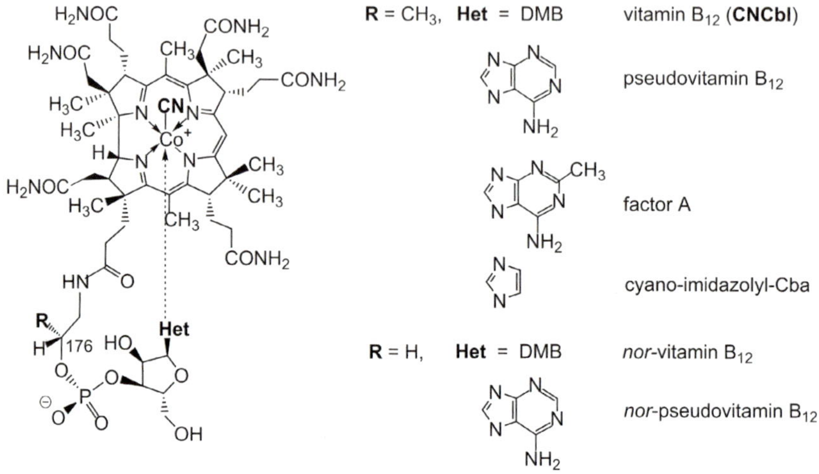

FIGURE 20.2 Structural formulas of some natural "complete" corrinoids differing by their nucleotide moiety: Cobamides (**R** = CH_3) and 176-*nor*-cobamides (**R** = H).

has also been particularly important,[46] as it is the product of (Co–C)-bond homolysis of coenzyme B_{12} (**AdoCbl**), and occurs during the catalytic cycle of coenzyme B_{12}-dependent enzymes. Crystal structures of the oxygen sensitive **Cbl^{II}** and of **AdoCbl** have revealed very similar corrin moieties.[46] These observations suggested that in AdoCbl-dependent enzymes, the enigmatic protein-induced activation of the bound AdoCbl toward homolysis of its (Co–C)-bond would largely come about by way of a protein induced separation of the homolysis fragments, made possible by strong binding of the separated components.[46] Hence, the "stretched" homologue of coenzyme B_{12}, "homocoenzyme B_{12}" (Co_β-(5′-deoxy-5′-adenosylmethyl)-cob(III)alamin), was suggested to function as a covalent structural mimic of the hypothetical enzyme bound "activated" state of the B_{12}-cofactor.[47] The distance between the cobalt center and C5′ of the homoadenosine moiety of "homocoenzyme B_{12}" was increased to 2.99 Å. This is roughly the same distance, as the one found between the corrin-bound cobalt center and C5′ in one of the two "activated" forms of coenzyme B_{12} in the crystal structure of glutamate mutase.[47]

The crystalline "inorganic" B_{12}-derivative aquocobalamin perchlorate (**H_2OCbl^+** ClO_4^-) contains a very short axial (Co_α–N)-bond.[33] The structures of vinylcobalamin and of *cis*-chlorovinylcobalamin[48] were the first examples of organocobalamins with sp^2-hybridized carbon ligands. The crystal structures of the first phenylcobalamins[49,50] and of the newly available phenylalkynylcobalamins,[51–54] revealed the existence of very short organometallic bonds between aromatic sp^2- or sp-carbons and cobalt.

20.2.3 The "Base-on/Base-off" Switch of Cobalamins

The nucleotide moiety of cobalamins (Cbls) features an exceptional α-configuration, which allows for the stable intramolecular coordination of the heterocyclic base to the "lower" α-axial coordination site of the corrin-bound cobalt center.[1,25,32,33,55] The DMB-nucleotide function undergoes relatively strong intramolecular cobalt-coordination, with little build-up of strain.[28,43] Hence, the DMB-base can either be cobalt-coordinated ("base-on" form) or it may be de-coordinated, generating the "base-off" form (see Fig. 20.3).[10,55] The existence in two forms ("base-off" or "base-on") not only leads to a strong restructuring, but it also modifies the reactivity of B_{12}-derivatives in biologically relevant organometallic reactions. First of all, the coordinating DMB-base in Cbls may stabilize the (organo)-B_{12} derivative significantly in their "base-on" form.[9,34,56] The coordinating DMB-nucleotide function also steers the face-selectivity at the corrin-bound cobalt center.[39] By coordinating to the "lower" face, it may direct alkylation (and other ligation) reactions (in cobalamins) to the "upper" (or β-face).

Cobalamins and related "complete" corrinoids are, thus, natural "molecular switches".[10,55] A complete "base-on" to "base-off" switch results from protonation of the nucleotide base and de-coordination from the corrin-bound cobalt-ion. The associated acidity of the protonated "base-off" form (as expressed by its pK_a) reflects quantitatively the strength of the intramolecular DMB-coordination. The (protonated) "base-off" form is readily

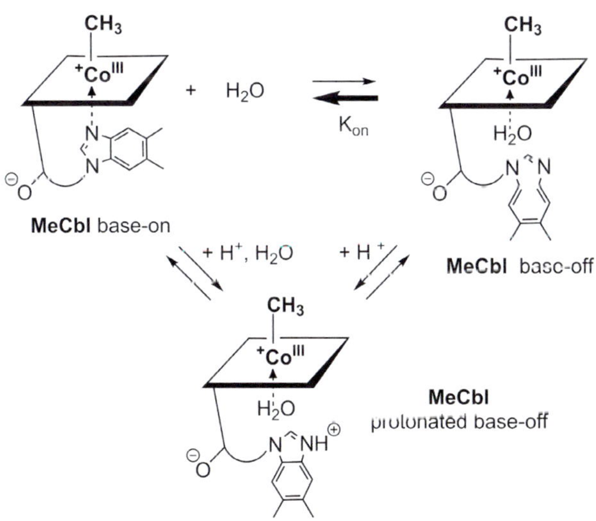

FIGURE 20.3 **MeCbl** as a "molecular switch."[55] The DMB-base of **MeCbl** is cobalt-coordinated in the "base-on"-form or de-coordinated in the less stable "base-off"-form ($K_{on} = 93$ at 25°C[9]). Protonation of the DMB-base of **MeCbl** furnishes the stable protonated "base-off" form.

accessible in **AdoCbl** and **MeCbl**, with a $pK_a = 2.9$.[9,55] On the other hand, the proton-assisted de-coordination is inhibited by strong cobalt-coordination, as is the case for **CNCbl** [9] and alkynyl-Cbls, with a pK_a near 0.[52]

The structure of Cbls is an effective determinant for the selective and tight binding by B_{12}-binding proteins.[57] This is crucial for discriminating Cbls from various other natural B_{12}-derivatives by the B_{12}-uptake and transport system of mammals, which recognizes and binds its Cbl-load in the "base-on" form.[13,58] "Base-on" and "base-off" forms of the "complete" corrinoids may (or may not) be structured correctly for binding by specific B_{12}-apoenzymes (see below). Hence, a protein environment may be pre-structured to bind and switch the bound B_{12}-cofactors from "base-on" to "base-off",[59,60] or vice versa.[57,61]

20.3 ORGANOMETALLIC AND REDOX-CHEMISTRY OF B_{12}-DERIVATIVES

20.3.1 Biological formation and cleavage of the (Co−C)-bond in B_{12}-derivatives

Formation and cleavage of the (Co−C)-bond are key to the chemistry of organometallic B_{12}-cofactors and are essential steps in the reactions catalyzed by B_{12}-dependent enzymes.[9,10,15–17,39,62–64] In solution, the typical cleavage and formation of the (Co−C)-bond mediates between two of the three basic oxidation levels of the corrin-bound cobalt center (see Figs. 20.4 and 20.5).[9,39,56] However, the cobalt-corrin remains on the single oxidation level Co(III) in the specific cases of the proton- or electrophile-induced substitution reactions. Proton-induced heterolytic cleavage of the Co(III)-corrin **AdoCbl** occurs slowly at low pH in aqueous solution and furnishes **H_2OCbl^+**.[65] Likewise, depending upon the substitution pattern at the phenyl group, (phenyl)-alkynyl-Cbls are cleaved in acidic medium by proton-induced heterolysis of the (Co-C)-bond and resulting in the formation of

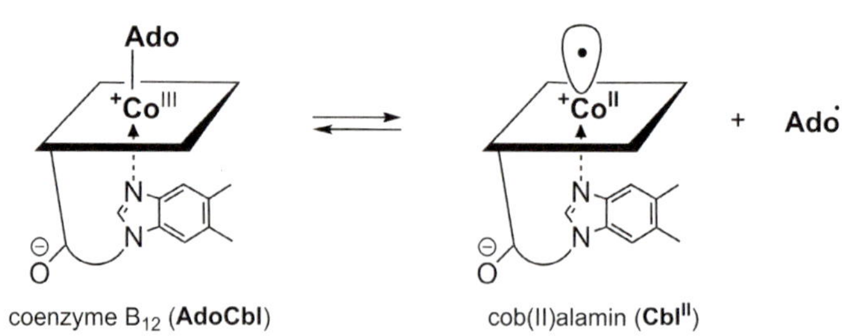

FIGURE 20.4 The (Co-C)-bond of coenzyme B_{12} (**AdoCbl**) cleaves by thermally reversible homolysis, furnishing the "radical trap" cob(II)alamin (**Cbl^{II}**) and the 5′-desoxy-5′adenosyl radical (**Ado•**) reversibly.

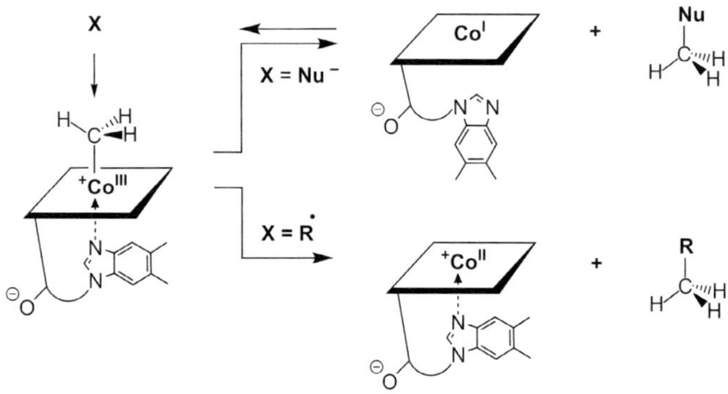

FIGURE 20.5 Biologically important modes of co-substrate induced cleavage of the (Co-C)-bond of methylcorrinoids, such as **MeCbl**. Abstraction of the cobalt-bound methyl group are induced by nucleophiles (**Nu$^-$**) or by radicals (**R$^\bullet$**).

(substituted phenyl)ethine and **H_2OCbl^+**.[51] In contrast, proton-induced cleavage of the (Co–C)-bond of **MeCbl** is not documented. The cobalt-bound methyl group of **MeCbl** may also be readily abstracted by polarizable electrophilic metal ions, such as by Hg^{+2} ions. It is an environmentally important path to the poisonous Hg-CH_3 ion.[66,67]

The three main organometallic reaction paths that have been found to be relevant in known enzymes also involve a change of the (formal) oxidation level of cobalt:

i. The homolytic mode – essential for the cofactor role of coenzyme B_{12} (**AdoCbl**):

$$5'\text{-adenosyl-Co(III)-corrin} \rightleftarrows \text{Co(II)-corrin} + 5'\text{-adenosyl radical}$$

ii. The nucleophile induced, heterolytic mode - typical of the reactivity of **MeCbl**:

$$\text{methyl-Co(III)-corrin} + \text{nucleophile} \rightleftarrows \text{Co(I)-corrin} + \text{methylating agent}$$

iii. The radical abstraction mode involving methyl-corrinoids, such as **MeCbl**[68]:

$$\text{methyl-Co(III)-corrin} + \text{radical} \rightarrow \text{Co(II)-corrin} + \text{methylated radical}$$

20.3.1.1 The Homolytic Mode

Coenzyme B_{12} (**AdoCbl**) undergoes selective thermal homolysis of its organometallic bond readily (see Fig. 20.4), and it has been considered a "reversible carrier of an alkyl radical" (or a reversibly functioning "radical source" [64]). Indeed, this homolytic mode of the cleavage of the (Co–C)-bond of **AdoCbl** is particularly important for its cofactor role (see below). The strength of the (Co−C)-bond of **AdoCbl** has been calculated to be about 30 kcal/mol by using detailed kinetic analyses of its thermal decomposition in aqueous and glycerol solutions.[64,69,70] The quantitative determination of the homolytic (Co−C)-bond dissociation energy (BDE) of **AdoCbl** is hampered by cage effects and the presence of both "base-on" and "base-off" forms.[70] In a similar way, the slightly higher homolytic (Co−C)-BDE of **MeCbl** has been determined at 37 kcal/mol.[71] In the gas-phase the homolytic (Co-C)-BDEs of the "incomplete" organocorrinoids adenosylcobinamide (**AdoCbi**) and methylcobinamide (**MeCbi**) were determined in mass spectrometric experiments as amounting to 41.5 and 44.6 kcal/mol, respectively.[72]

The nucleotide coordinated "base-on" forms of some organocobalamins decomposed considerably faster than their (protonated) "base-off" forms, or than the related "incomplete" organocobinamides.[73] Therefore, the intramolecular coordination of the nucleotide was associated with labilizing the (Co−C)-bond of organometallic B_{12}-derivatives.[64,73] However, the contribution of the nucleotide coordination to the ease of homolysis of **AdoCbl** is relatively small: On the basis of available thermodynamic data concerning the coordination of the nucleotide in **AdoCbl** and of the homolysis product cob(II)alamin (**Cbl^{II}**), the coordination of the nucleotide was derived to weaken the (Co−C)-bond by only 0.7 kcal/mol.[56,74] With **MeCbl**, the intramolecular coordination of the nucleotide was determined to even increase the homolytic (Co−C)-BDE slightly by about 0.3 kcal/mol according to studies of the methyl-group transfer equilibrium between **MeCbl** /cob(II)inamide (**Cbi^{II}**) and methylcobinamide (**MeCbi**)/Cbl^{II}.[56]

The (Co−C)-bond of most organocorrinoids have long been known to be sensitive to visible light,[75,76] which induces cleavage of the (Co-C)-bond of, e.g., **AdoCbl** and **MeCbl**, very effectively.[77] Hence, by the use of visible irradiation, organocorrinoids are a convenient source for organic radicals.[78] Interestingly, (phenyl)alkynyl-Cbls are a striking exception and are inert to cleavage by visible light.[79]

Organocobalamins are also accessible by the reaction between Co(II)corrins and organic radicals. The penta-coordinated Co(II)-center of the persistent radicaloid **Cbl^{II}** fulfills all the structural criteria of a highly efficient "radical trap"[46]: The reactions of **Cbl^{II}** with alkyl radicals are indicated to occur with negligible restructuring of the cobalt corrin moiety and to furnish **AdoCbl** and other organo-Cbls directly by the "reverse homolytic" mode (radical recombination) with formation of the (Co−C)-bond.[46] This feature helps to rationalize the remarkably high reaction rate of **Cbl^{II}** with alkyl radicals (such as the 5′-deoxy-5′-adenosyl radical), as well as the diastereospecificity for the reaction at the β-face.[10,80] **Cbl^{II}** even traps the very short lived acetyl-radicals efficiently, providing an effective synthetic route to acetyl-cobalamin[81] by "slaving-in" the acetyl radical.[82]

20.3.1.2 The Nucleophile-Induced Heterolysis Mode

The second biologically broadly relevant mode of formation of the (Co−C)-bond is represented by the nucleophilic substitution reaction (see Fig. 20.5) of highly nucleophilic ("supernucleophilic") Co(I)-corrins[83,84] with alkylating agents. This heterolytic reaction type follows the mechanism of an SN_2-reaction without any evidence for the existence of a free methyl (or alkyl cation), and is particularly relevant in typical enzyme-catalyzed methyl-transfer reactions, as well as in the biosynthesis of adenosyl-corrinoids.[15,63,85,86]

Alkylation of cob(I)alamin (**Cbl^{I}**) normally proceeds via the "classical" bimolecular nucleophilic substitution (SN_2) mechanism at the corrin-bound Co(I) center, where **Cbl^{I}** acts as a strong nucleophile.[63,83,86,87] However, in certain cases alkylation with Co(I)-corrins occurs via a two-step ("inner-sphere") electron transfer path, where Co(I)-corrins act as strong reducing agents and the process involves Co(II)-corrin and radical intermediates.[74,87]

The nucleophilicity of Co(I)-corrins, like **Cbl^{I}**, is indicated to be virtually independent of the presence of the cobalt-coordinating (DMB) nucleotide: Both "complete" and "incomplete" Co(I)-corrins react with similar rates and with strong preference for their β-face, which, therefore, is the more nucleophilic of the two diastereo-faces of the corrin-bound Co(I)-ion.[39,87]

The nucleophile-induced demethylation of methyl-Co(III)-corrins is the biologically important basis for substrate methylation by cleavage of the (Co−C)-bond. It furnishes a strongly reducing Co(I)-corrin,[63] and, formally, represents a reductive *trans* elimination at cobalt.[39] Strong nucleophiles, such as thiolates de-methylate the "incomplete" **MeCbi^{+}**-ion approximately 1000 times faster than **MeCbl**,[88] reflecting the strong stabilization by the coordinated DMB-nucleotide in **MeCbl**, which amounts to about 4 kcal/mol.[39,56] This effect is of relevance also for enzymatic methyl-group transfer reactions involving protein bound methyl-Co(III)-corrins, where the coordinated histidine ligand displays a significant role.[63,89]

20.3.1.3 Radical-Induced (Co-C)-Bond Cleavage of Methylcorrinoids

A further mode of cleavage of the (Co−C)-bond of organometallic B_{12} derivatives operates via a radical-induced substitution at the cobalt-bound methyl group (see Fig. 20.5).[10,68,90] This type of a substitution reaction was observed, e.g., in the reaction of a malonyl-methyl-radical with **MeCbl**.[68] The thermodynamically very favorable and kinetically highly effective abstraction of the cobalt-bound methyl group of **MeCbl** by an alkyl radical has been suggested to represent a second biological role of methyl-corrinoids.[68] As a multitude of unusual biological (C−C)-bond forming reactions[91] and biosynthetic methylations at inactivated carbon centers by B_{12}- and S-adenosyl-methionine (SAM) dependent radical enzymes have recently been reported,[92–94] the "novel" radical methyl group abstraction mechanism with **MeCbl** experiences considerable interest lately (see section IV.A.2. below).

By, formally, a related reaction, the cobalt-bound methyl group of methylcorrinoids, such as **MeCbl**, is rapidly abstracted by Co(II)corrinoids, such as **Cbi^{II}**, giving **MeCbi** and **Cbl^{II}**.[56] This type of reaction does not involve free methyl radicals and, under appropriate conditions (aprotic solvents), it is not (even) sensitive to the presence of molecular oxygen.[95]

20.3.2 Redox-Chemistry of B_{12}-Derivatives

Under physiological conditions, vitamin B_{12}-derivatives exist as Co(III)-, Co(II)-, or Co(I)-corrins, each oxidation state possessing different coordination properties and correspondingly differing reactivity.[9,10,22,39] Electrochemistry has been used for determining the crucial redox-potentials in solution,[96] for the controlled electro-synthesis of organometallic B_{12}-derivatives,[97] for the generation of reduced forms of protein bound B_{12}-derivatives[98] and of electrode-bound B_{12}-derivatives for analytical applications.[99]

Axial coordination to the corrin-bound cobalt center depends on the formal oxidation state of the cobalt ion, and, as a rule, the number of axial ligands decreases with the cobalt oxidation state.[10,33,96] In the thermodynamically predominating forms of cobalt corrins, the diamagnetic Co(III) has two axial ligands bound (coordination number 6), the paramagnetic (low spin) Co(II) has one axial ligand bound (coordination number 5) and for the diamagnetic Co(I) axial ligands are not bound (coordination number 4), or only very weakly (see Fig. 20.6). Electron transfer reactions involving B_{12}-derivatives are, therefore, accompanied by a change in the number of axial ligands, which influences the thermodynamic and kinetic features of the redox-processes of cobalt corrins (reviewed in[96,97]).

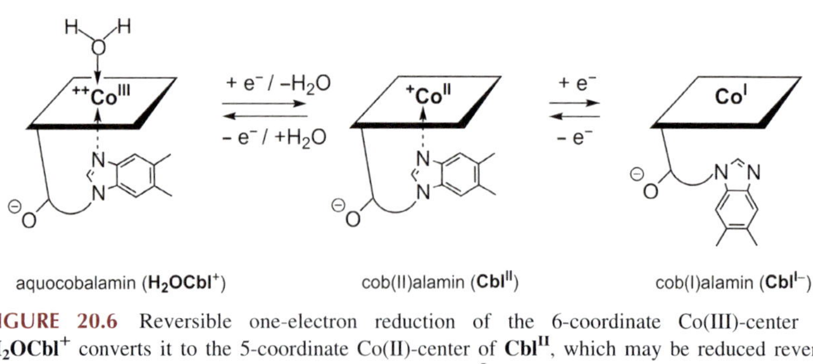

FIGURE 20.6 Reversible one-electron reduction of the 6-coordinate Co(III)-center of **H_2OCbl^+** converts it to the 5-coordinate Co(II)-center of **Cbl^{II}**, which may be reduced reversibly at more negative potential to the Co(I)-corrin **Cbl^{I}** with a 4-coordinate Co(I)-center. The reduction steps are accompanied by the loss of one axial ligand (oxidations steps involve the corresponding addition of one axial ligand).

Electrochemical studies of organometallic B_{12}-derivatives are generally complicated due to a rapid and irreversible loss of the organic ligand upon reduction.[96,100] Organo-corrinoids are, hence, labile to strong one-electron reducing agents, and the (Co–C)-bond is of **MeCbl** is weakened considerably by the reduction.[71,96] However, the standard potential of the typical Co(III)-/Co(II)-redox pair of organometallic B_{12}-derivatives is strikingly more negative than that of $\mathbf{Cbl^{II}/Cbl^{I}}$ and out of the reach of biological reductants.[96] Thus, the further reduction of organometallic Co(III)-corrins typically does not occur at the potentials needed for electro-generation of the highly nucleophilic Co(I)-corrins. Thus, the selective electrochemical production of Co(I)-corrins in the presence of suitable alkylating agents provides an efficient and selective preparative access to alkyl-corrinoids [97] and other complex organo-corrinoids.[47,55,101,102] However, organo-corrinoids with electron-withdrawing substituents in their cobalt-bound organic ligand are reduced more easily than **MeCbl**,[103,104] rendering it difficult to prepare them along the simple path via the strongly reducing Co(I)-corrins as nucleophilic intermediates.[105]

20.4 ENZYMATIC REACTIONS BASED ON ORGANOMETALLIC B_{12}-COFACTOR

Methyl- and adenosyl-corrinoids, such as **MeCbl** and **AdoCbl**, are often observable in functioning enzymes,[10] where such "complete" corrinoids may be bound in their characteristic "base-on",[106,107] "base-off/his-on,"[59,60,108] or "base-off" forms.[109] The catalytically equally important enzyme-bound protein-bound cofactor form $\mathbf{Cbl^{II}}$ has also been characterized by crystallography lately (see e.g.,[110,111]). However, the transient $\mathbf{Cbl^{I}}$-form in enzyme reactions has been observed only indirectly, for reasons of its thermodynamic instability and inaccessibility under typical physiological conditions.[15,86]

20.4.1 B_{12}-Dependent Methyl Transferases

B_{12}-dependent methyltransferases have widespread and important biosynthetic roles. Two basic biochemical mechanisms of methyl group transfer are now established.[86,92] B_{12}-dependent methionine synthases use nucleophile induced heterolytic methyl group transfer steps and occur in many organisms, including humans.[86] Related B_{12}-dependent methyltransferases operate broadly in one-carbon metabolism in microorganisms, most importantly in methanogenesis,[112] in acetogenesis[113] as pathway of anaerobic CO_2 fixation,[114,115] and in acetic acid catabolism to methane and CO_2 in some anaerobic microbes.[116,117] The reactivity of the nucleophilic Co(I)-forms and of the organometallic methyl-Co(III)-forms of the B_{12}- cofactors provide the basis for the catalysis of these enzymatic methyl-group transfer reactions.[9,10,63] Alternatively, bifunctional B_{12}- and SAM-dependent radical methyl transferases (now classified as "class B" radical SAM methyltransferases) were also recognized recently as a second important group of biosynthetic methylating enzymes,[92–94,118] as discussed below.

20.4.1.1 B_{12}-Dependent Methionine Synthase

B_{12}-dependent methionine synthase (MetH) is a widespread organometallic enzyme and is, probably, the most extensively studied B_{12}-dependent methyltransferase.[15,86,119] It is detailed here as a basic model for other methyl transferases that operate via a nucleophile induced methyl group transfer.[114] The enzyme MetH from *E. coli* is has been a particularly useful representative of B_{12}-dependent methionine synthase.[86,119] Methyl group transfer, catalyzed by MetH, follows basically a two-step ping-pong mechanism (see Fig. 20.7).

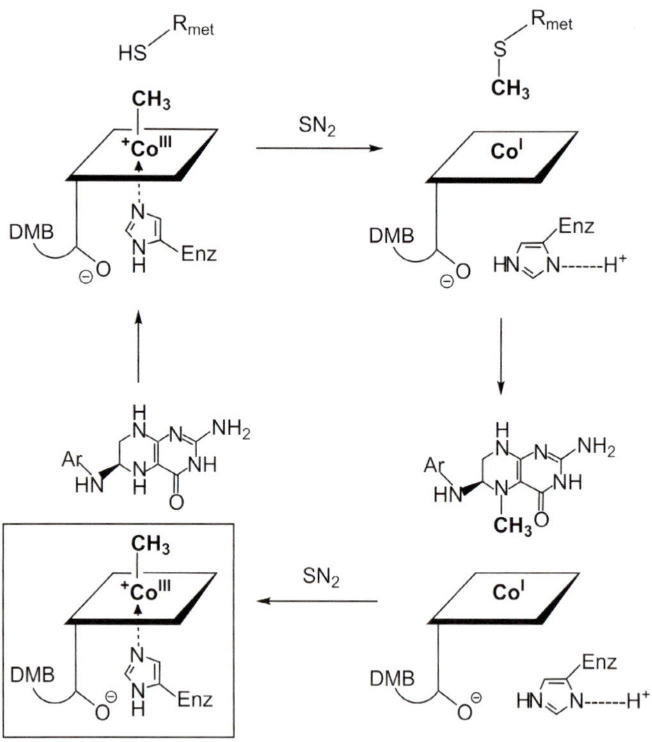

FIGURE 20.7 Methionine synthase (MetH) catalyzes the Cbl-dependent formation of methionine from homocysteine and demethylation of N^5-methyltetrahydrofolate to tetrahydrofolate involving the protein bound Cbls **MeCbl** in a "base-off/His-on"-state and **Cbl^I** (the imidazole ring symbolizes His759 of MetH). The heterolytic methyl group transfer occurs in a ping-pong mechanism via two nucleophilic substitution (SN_2) steps. The resting state with **MeCbl** bound "base-off/His-on" is highlighted at the bottom left.

In a first step, the protein-bound **MeCbl** is demethylated by activated homocysteine, furnishing protein-bound **Cbl^{I}** and methionine. In the second step, protein-bound **Cbl^{I}** is methylated by activated N-methyl-tetrahydrofolate, generating tetrahydrofolate and regenerating the protein-bound **MeCbl**.[114] The two steps proceed with an overall retention of configuration at the methyl group carbon, consistent with two SN_2-type nucleophilic displacement steps, each occurring with inversion of configuration.[63,120] These two methyl-group transfer steps are considered to be subject to the strict geometric control of SN_2 reactions, i.e., to require in-line arrangements of the incoming nucleophile, the transferred CH_3-group and of the leaving group. Although free methyl cations or radicals are excluded as intermediates, the methyl group transfer catalyzed by MetH involves heterolytic (and nucleophile-induced) cleavage/formation of the $(Co - CH_3)$-bond. Hence, in a formal sense, it represents a methyl "cation" transfer.

During turnover striking structural changes accompany the transitions of the enzyme MetH between its state with (tetra-coordinate) **Cbl^{I}** bound and the one with (hexa-coordinate) base-off/his-on **MeCbl** (see Fig. 20.8). The protein environment plays a crucial role in controlling substrate positions, as well as in providing access to the catalytic center.[114]

The X-ray crystal analysis of the B_{12}-binding domain of MetH provided the first insight into the three-dimensional structure of a B_{12}-dependent enzyme.[59,121] An eye-opening revelation of this work was the finding that the cobalt-coordinating DMB-nucleotide tail of the protein-bound organometallic cofactor **MeCbl** was displaced at cobalt by the histidine of a conserved His-Asp-Ser-triad and bound by the core of a "Rossmann fold" of the protein.[59,121] Consequently, in MetH the corrinoid cofactor is bound in a "base-off/His-on" mode. In various other B_{12}-dependent methyltransferases the methyl-Co(III)-corrinoid cofactor has been observed in a "base-off/His-on" binding mode, or even in a "base-off" form (i.e., without DMB-or His-coordination).[122]

The axial bond of the histidine of the His-Asp-Ser-triad to the protein-bound **MeCbl** in MetH helps to position the corrinoid cofactor for methyl group transfer.[121,123] A thermodynamic role of the histidine coordination in the methyl transfer reactions of MetH has also been discussed.[74,86,89,121] Indeed, a significant thermodynamic *trans*-effect of the DMB-coordination in **MeCbl** on heterolytic methyl group transfer reactions has been observed in aqueous solution.[39,56,74] The coordinating DMB-ligand stabilized **MeCbl**, opposing nucleophilic abstraction of the methyl group by roughly 4 kcal/mol.[56] The His-Asp-Ser-triad may, furthermore, function as the "relay" for H^{+}-uptake/release accompanying the enzymatic methylation / demethylation cycles.[123,124] It may, thus, fine-tune the bound corrinoid for enzyme catalysis: Weakening of the axial (Co-N)-bond activates the methyl group of

MeCbl for abstraction by a nucleophile, as well as adventitiously formed **Cbl**^{II} for re-reduction to protein-bound **Cbl**^{I}.

20.4.1.2 B_{12}- and S-Adenosylmethionine (SAM)-Dependent Radical Methyl Transferases

The incorporation of intact methyl groups, originating from methionine, in the course of the biosynthetic methylation at saturated carbon-positions of some antibiotics was a puzzling observation, first made in the late 1980s.[125] It was incompatible with a mechanism like the one of MetH, suggesting a new biological path for methylation. The direct, efficient and thermodynamically very favorable methylation of a primary (malonate-derived) carbon-radical by abstraction of the cobalt-bound methyl group of **MeCbl** provided a first model reaction.[68] Indeed, the eventual identification of a class of the abundant enzymes with protein signatures of "radical" SAM-enzymes[126] was consistent with the broad biosynthetic involvement of radicals.[127,128] Evidence for a large sub-class of B_{12}-dependent "radical" SAM-enzymes has subsequently been obtained, many (but not all) of which appear to be B_{12}-dependent methyltransferases.[92,94] Indeed, methylation of radicals by **MeCbl** is now an accepted mechanism for, e.g., the methyltransferase Fom3 in the course of the biosynthesis of fosfomycin (see Fig. 20.8).[94] The radical enzyme Fom3 methylates the fosfomycin precursor 2-hydroxyethylphosphonate.

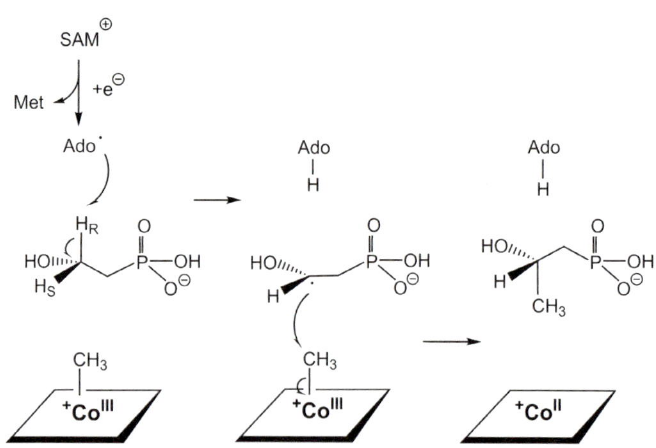

FIGURE 20.8 Fom3 is a **MeCbl-** and radical SAM-dependent methyl group transferase that catalyzes the methylation of 2-hydroxyethylphosphonate to (*S*)-2-hydroxypropylphosphonate. *Adapted from Kräutler B, Puffer B. Vitamin B12-derivatives: Organometallic catalysts, cofactors and ligands of bio-macromolecules. In: Kadish KM, Smith KM, Guilard R, editors. Handbook of Porphyrin Science. Vol 25. World Scientific; 2012. P. 133–265.*

Overall, the methylation by Fom3 occurs with loss of H_R to give (*S*)-2-hydroxypropyl-phosphonate and is thus indicated to involve stereochemical inversion at carbon-2 of the phosphonate. Fom3 was shown to consist of a B_{12}-binding domain and a "radical" SAM domain.[94,129] A coupled H-atom abstraction/methylation process was proposed for the formation of (*S*)-2-hydroxypropyl-phosphonate. This involves abstraction of an H-atom (H_R) from hydroxyethyl-2-phosphonate by an Ado-radical produced in the radical SAM domain and generation of the short lived hydroxyethyl-2-phosphonate radical. Rapid, stereo-selective methylation of the latter radical by **MeCbl** of the B_{12}-binding domain would lead to (*S*)-2-hydroxypropyl-phosphonate (see Fig. 20.8).[10,94] The observed stereochemical inversion (at C-2 of hydroxyethyl-2-phosphonate) is consistent with a pre-organized enzyme, in which the substrate (hydroxyethyl-2-phosphonate) is positioned between the radical generating SAM and the methylating B_{12}-domain that presents **MeCbl** from its β-face. For the biosynthesis of fosfomycin and of various other natural products (see for example Refs.[92,94,130]) radical methylation processes are now indicated, in which **MeCbl** and other methylcorrinoids would serve as direct methylating reagents. However, in other cases the class B radical SAM-methylases appear to methylate their substrates by still less established pathways.[92] A biochemical curiosity, at present, is the B_{12}- and SAM-dependent radical enzyme that catalyzes the (furane to oxetane) ring-contraction step of the bacterial antibiotic Oxetanocin A, in which a function for the B_{12}-cofactor is not yet implicated.[131]

20.4.2 Organometallic Chemistry of Enzymes Dependent on Coenzyme B_{12}

The classical **AdoCbl**-dependent enzymes rely on the reactivity of bound organic radicals, which are formed (directly or indirectly) by an H-atom abstraction by the 5′-deoxy-5′-adenosyl radical, the actual reactive agent that originates form the homolysis of the (Co—C)-bond of **AdoCbl**. The structurally highly sophisticated **AdoCbl** acts as a "pre-catalyst" (or catalyst precursor) and functions merely as reversible source for the 5′-deoxy-5′-adenosyl radical.[10,132] On the other hand, the question has also been a matter of discussion, to what degree the remaining Co(II)-corrin fragment $\mathbf{Cbl^{II}}$ is merely a "spectator", or takes part as a "conductor" in the **AdoCbl**-dependent isomerases.[133,134]

A class of exceptional ("non-classical") **AdoCbl**-dependent DNA-binding enzymes makes use of the light sensitive organometallic **AdoCbl** in an amazing photo-regulatory role of gene expression thanks to an efficient light-induced cleavage of the (Co-C)-bond of **AdoCbl** See below).[75,135,136]

20.4.2.1 Coenzyme B_{12}-Dependent Isomerases and Ribonucleotide Reductases

A B_{12}-dependent ribonucleotide reductase and ten B_{12}-dependent isomerases use **AdoCbl** (or an analogous AdoCba) as cofactors of radical enzymes.[10,17,21,62,106] These enzymes are five carbon skeleton mutases (methyl-malonyl-CoA mutase,[21] ethylmalonyl-CoA-mutase,[137] glutamate mutase,[138] methylene glutarate mutase,[16] and isobutyryl-CoA-mutase[139]) and diol and glycerol dehydratase,[106] ethanolamine ammonia lyase,[106] and two amino mutases (ornithine-4,5-aminomutase and D-lysine/L-β-lysine-5,6-aminomutase[17,140]). The **AdoCbl**-dependent enzymes are disproportionately distributed in living organisms. Only methylmalonyl-CoA mutase (MCM) is required for a functioning metabolism of humans and other mammals (see Fig. 20.9).[15]

The mentioned eleven coenzyme B_{12}-dependent enzymes make use of the protein-activated homolysis of the weak (Co—C)-bond of **AdoCbl**, furnishing **Cbl^{II}** and a 5′-deoxy-5′-adenosyl radical.[132] The latter tightly bound primary radical abstracts an H-atom from its substrate, inducing the further enzymatic transformations.[16] Indeed, **AdoCbl**-dependent enzymes perform chemical transformations that are difficult to achieve by typical "organic reactions." With the exception of the enzymatic ribonucleotide reduction,[141]

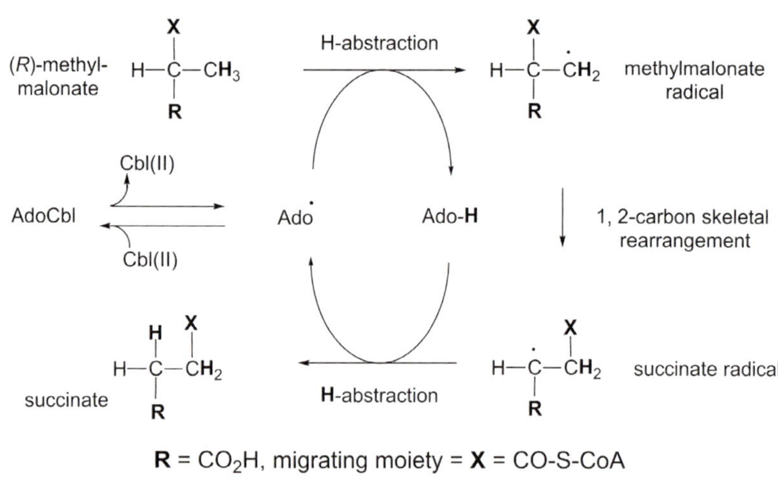

FIGURE 20.9 Methylmalonyl-CoA mutase converts (R)-methylmalonyl-CoA to succinyl-CoA and is induced by (Co-C)-homolysis of enzyme-bound **AdoCbl** and simultaneous formation **Cbl^{II}** and of the 5′-deoxy-5′-adenosyl radical (**Ado$^{\bullet}$**). The proposed mechanism involves abstraction of an H-atom from (R)-methylmalonyl-CoA (top, left) by **Ado$^{\bullet}$**, leading to the 2-methyl-malon-2′-yl-CoA radical (top, right); in this enzyme-bound radical the thioester moiety **X** undergoes a 1,2-migration (a carbon skeleton rearrangement) to the succin-3′-yl-CoA radical (bottom, right); H-atom abstraction by this radical from 5′-deoxyadenosine (**Ado-H**) provides succinyl-CoA (bottom, left) and regenerates **Ado$^{\bullet}$**, ready for recombination with **Cbl^{II}**. The sequence of steps of this reversible enzyme reaction is only shown in the "forward" direction.

the results of coenzyme B_{12}-catalyzed enzymatic reactions correspond to isomerisations with vicinal exchange of a hydrogen atom and of a group with heavy atom centers.

The homolysis of the (Co—C)-bond of the protein-bound **AdoCbl** needs to be activated and accelerated by a factor of about 10^{12} to agree with the observed reaction rates of the coenzyme B_{12}-dependent enzymes.[64,69] The deduced dramatic destabilization of the bound organometallic cofactor towards homolysis of the (Co—C)-bond and its mechanism are intriguing and still much discussed facets of the **AdoCbl**-dependent enzymes.[15,46,47,106,138] Covalent restructuring of the bound cofactor (except for the formation of the "base-off/His-on"-form in the carbon skeleton mutases) is not indicated. In addition, protein and solvent molecules can only weakly stabilize a radical center.[142] Destabilization of the (Co—C)-bond towards homolysis may come about largely from a protein- and substrate-assisted separation of the largely non-strained homolysis fragments, a 5′-deoxy-5′-adenosyl radical and $\mathbf{Cbl^{II}}$ (in either a "base-off/His-on" or "base-on" form), and strong binding by the protein of the separated fragments.[46,47,143,144] The existence in some of these enzymes of a binding interface (e.g., of an "adenosine-binding pocket") which does not allow for unstrained binding of the organometallic moiety, helps to support this picture.[108,145]

Interestingly, **AdoCbl**-dependent enzymes come in two structural classes, one of them with the B_{12}-cofactor bound "base-on", as found e.g., in diol-dehydratases and in B_{12}-dependent ribonucleotide reductase,[106,141,146] the other one, with the B_{12}-cofactor bound in a "base-off" (and "His-on") form, as discovered in methyl-malonyl-CoA mutases (MCM).[60] Fixed placement of the corrin moiety at the interfaces of the B_{12}-binding and substrate-binding/activating domains of MCM appears to be of high significance. The "regulatory triad" may not be involved in proton-transfer steps in the mutases, and it may conserve its structure largely during enzymatic turnover. Indeed, "electronic effects" of the axial *trans* ligand on the (Co—C)-bond homolysis in **AdoCbl** and **MeCbl** are less important.[39] The proper substrate to product rearrangement steps of **AdoCbl**-dependent enzymatic rearrangements are accomplished by tightly protein-bound radicals that are controlled in their reaction space.[10,132] The major functions of the protein-part of the enzyme concern not only the assistance in its proper reactions (by activation of protein-bound **AdoCbl**) but also the reversible generation of the radical intermediates and the protection of its protein environment from non-specific radical chemistry, a function classified as "negative catalysis".[142]

20.4.2.2 Coenzyme B_{12} as Light-Receptor in the Photo-Regulation of Gene Expression

In a striking twist from its known biological roles in radical enzymes, **AdoCbl** has joined the list of the light-sensing cofactors involved in photo-regulation of gene expression.[135,147,148] The **AdoCbl**-based photoreceptor CarH regulates biosynthesis of the photo-protecting carotenoids in the bacterium *Myxococcus xanthus* and is a representative of an abundant class of B_{12}-based bacterial photoregulators.[149] In very elegant crystallographic, mutational and mechanistic studies, the question has been clarified largely, how the organometallic B_{12}-cofactor **AdoCbl** could be repurposed to play a broadly relevant light-sensing gene-regulatory role.[136,150,151]

According to these studies, **AdoCbl** is not directly involved in DNA binding.[136] The mechanism of gene regulation by CarH relies on the modulation of the structure of the protein through interaction with the B_{12}-cofactor. Under low-light conditions CarH binds intact **AdoCbl** and forms a dimer-of-dimers-type tetramer. In this state CarH binds with high affinity to the promoter region of genes coding for carotenoid biosynthesis, inhibiting their transcription. Upon photolytic cleavage of the (Co−C)-bond of **AdoCbl**, the adenosyl group is lost as unreactive 4′,5′-anhydroadenosine,[150] and a conformational change of the protein leads to the disintegration of the protein, which can no longer bind strongly to DNA.[136] It remains to be established how the protein moiety "reprograms" the path of the light-triggered cleavage of the Co-C bond to the here observed unique ("heterolytic") mode, thus repurposing **AdoCbl** perfectly for effective photo-regulation.[75,152]

A situation related to the one now characterized in CarH,[136] was found in the **AdoCbl**-dependent photo-regulator AerR in the photosynthetic bacterium *Rhodobacter capsulatus.*[153] When carrying aquocobalamin, from aerobic photolysis of bound **AdoCbl**, the B_{12}-binding protein AerR binds CrtJ, the regulator of genes coding for tetrapyrrole biosynthesis.

20.4.3 B_{12}-Dependent Dehalogenases

The ability of some anaerobic microorganisms, using reduced corrinoids as cofactors, to dehalogenate haloalkanes and haloaromatics reductively, is a globally relevant biological feature.[154,155] A variety of environmentally relevant dehalogenation reactions have been disribed, such as of chloromethane,[156] of chloroethenes,[44] of trichloroethane[157] of hexachlorocyclohexane,[158] and of chlorinated phenols.[15,159] The anaerobic bacterium *Sulfurospirillum multivorans* dechlorinates tetrachloroethene by its B_{12}-dependent tetrachloroethene reductive dehalogenase (PCE).[44] The membrane bound dehalogenase PCE uses a reduced form of an unusual "complete" corrinoid cofactor, isolated as *nor*-pseudovitamin B_{12}[42] in order to reduce tetrachloroethene to trichloroethene (first) and to *cis*-dichloroethene.[44] *Nor*-pseudovitamin B_{12} from *D. multivorans* was about fifty times more active than **CNCbl** in an in vitro reduction of trichloroacetate[160] and addition of DMB into the growth medium inducing formations of *nor*-cobalamins (*nor*-Cbls) actually reduced bacterial growth.[45] In aqueous solution *nor*-cobamides (and *nor*-Cbls) have been shown to assemble to the B_{12} "base-on" forms with lesser preference than Cbls,[43] enabling redox reactions with *nor*-cobamides to occur at potentials less negative than those of the corresponding Cbl redox-couples.[161]

Subsequent X-ray analytical studies have revealed the structures of PCE of *Sulfurospirillum multivorans*[160] with the "base-off" Co(II)-form of *nor*-pseudovitamin B_{12} bound,[44] as well as of a dehalogenase associated with respiratory chlorophenol dechlorination, with the "base-off" form of **Cbl^{II}** bound.[162] In the case of PCE of *Sp. multivorans* the Co(I)-form of the enzyme is proposed to reduce the haloalkene via a long distance electron-transfer,[163] whereas in the latter situation the structural data suggest an "inner-sphere" reduction by **Cbl^{I}**[74] of the halophenol via a formal halogen-atom transfer.[162] Formation of organometallic intermediates, as considered earlier,[48,159,164] appears unlikely in these two enzymes. Interestingly, the B_{12}-binding region of both dehalogenases is structured similarly as the B_{12}-processing enzyme CblC[52,109] (see below).

20.5 COBALAMIN PROCESSING ENZYMES AND ANTIVITAMINS B_{12}

20.5.1 Human Cobalamin Processing Enzymes

Vitamin B_{12} (**CNCbl**), the most common form of B_{12}, has no direct physiological functions in humans and is transformed inside healthy human cells to the cofactors methylcobalamin (**MeCbl**) and coenzyme B_{12} (**AdoCbl**).[14] In patients with deranged Cbl metabolism who manifest signs of Cbl-deficiency despite of a proper supply with **CNCbl**, eight genes were identified.[165] Two of these genes are responsible for intracellular processing of Cbls. One (*cblC*) encodes for the cobalamin processing enzymes CblC,[166] the other one (*cblB*) for the **AdoCbl** synthesizing adenosyltransferase ACA.[167]

The *cblC* type Cbl disorders led to a malfunction of accumulation of methylmalonate and homocysteine. The genetic locus responsible for this was referred to as the MMACHC gene (for *m*ethyl*m*alonic *a*ciduria type *C* and *h*omo*c*ystinuria).[166] The translational product of this gene is the protein CblC that "tailors" different cobalamins to **Cbl^{II}**, which is processed further in the cell to the physiologically active cofactors **MeCbl** and **AdoCbl**. Depending on the Cbl substrate, CblC employs two different mechanisms to produce **Cbl^{II}** (Fig. 20.10).[168]

When **CNCbl** is bound to the enzyme, it catalyzes the reductive decyanidation by NADPH via flavin cofactors.[169] Alternative binding of alkylcobalamins (such as **MeCbl**) initiates the removal of the upper ligand *via* nucleophilic substitution by glutathione (GSH).[170] The crystal structure of human CblC in complex with **MeCbl** (but lacking GSH) allowed for detailed insights into the three-dimensional structure of this unique 26 kDa ptotein.[109] The substrate **MeCbl** is bound "base-off" with a five-coordinated cobalt-center

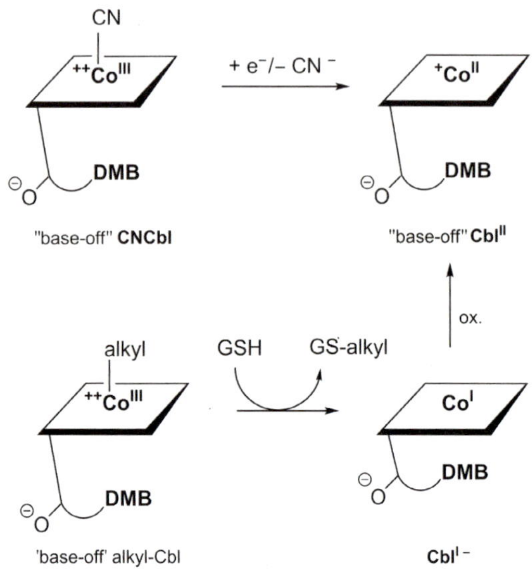

FIGURE 20.10 The dual performance enzyme CblC uses two mechanisms. The common product **Cbl**^{II} is formed in "base-off"-form either by reductive decyanation of ("base-off") **CNCbl** or by nucleophile substitution of ("base-off") alkylcobalamins (such as **MeCbl**) by glutathione (GSH), followed by oxidation of the directly resulting **Cbl**^{I}.

in a large cavity at the domain interface, with the nucleotide tail buried in a crevice of the N-terminal domain. In a more recent X-ray analysis of the complex of CblC with GSH and the "antivitamin B_{12}" difluorophenylethynylcobalamin (**F2PhEtyCbl**), which inhibits Cbl-processing (see below), the important further structuring of CblC by inclusion of GSH was revealed.[52]

The *cblB* type disorders are due to a non-functional enzyme ATP:Cbl^{I} adenosyltransferase (ACA), responsible for the biosynthesis of **AdoCbl** from the precursor corrinoid **Cbl**^{II}.[14,171] The enzyme-catalyzed adenosyl transfer is based on the intermediate formation of the nucleophilic **Cbl**^{I}, which attacks the 5′-carbon of correctly bound ATP. As the reduction of "base-on" **Cbl**^{II} to **Cbl**^{I} requires a reduction potential beyond the capacities of in vivo reducing agents, ACA activates **Cbl**^{II} for reduction by binding it in a four-coordinate "base-off" form. This was shown by crystallographic snapshots of the ACA structure with **Cbl**^{II} and ATP and the partially formed reaction product **AdoCbl**[171] bound in the active site. The remarkable enzyme catalyzed adenosyl transfer makes use of a customized protein environment that facilitates not only the reduction to **Cbl**^{I} but lowers the energy barrier for adenosylation (so far, the direct adenosylation of **Cbl**^{I} with ATP has not been successful).

20.5.2 Antivitamins B_{12}

Cbls that resist tailoring by CblC are not converted to **Cbl**^{II} (nor, subsequently, to **AdoCbl** and **MeCbl**) in the cell and they have been classified as "antivitamins B_{12}."[172] Indeed, "functional" Cbl-deficiency is induced in animals by the supply with such metabolically inert Cbls,[173] which would also be poisonous for humans. Two organometallic classes of Cbls have been suggested, so far, to be inert to tailoring by CblC and, hence, to represent potential "antivitamins B_{12}": (1) aryl-Cbls, which have a substitution and reduction inert organometallic bond, such as 4-ethylphenylcobalamin (**EtPhCbl**, the first representative of this class of organo-Cbls)[49] or phenylcobalamin (**PhCbl**),[50] and, for the same reason; (2) suitably substituted (hydrolysis-resistant) alkynyl-Cbls, such as 2-phenyl-ethynylcobalamin (**PhEtyCbl**),[51] and difluorophenyl-ethynylcobalamin (**F2PhEtyCbl**).[52] Due to acute interest in "antivitamins B_{12}"[172,174] several other phenylalkynyl-Cbls have been made subsequently.[54] The "antivitamins B_{12}" **EtPhCbl** and **PhEtyCbl** have been shown to bind to the three important human B_{12}-transporting proteins intrinsic factor, transcobalamin, and haptocorrin,[58] indicating these inert Cbls to induce functional B_{12}-deficiency when supplied orally.[172] Antivitamins B_{12}, when resistant to further metabolic use by bacteria and archaea that depend upon B_{12}, also have features of useful antibiotics.[172,174,175]

Suitably structured B_{12}-derivatives, having the B_{12}-specific cobalt-center replaced by, e.g., the homologous rhodium-ions, may also have the features of "antivitamins B_{12}".[172] Thus, the organometallic Rh-analogue of **AdoCbl**, adenosyl-rhodibalamin (**AdoRhbl**), has a structure very similar to **AdoCbl**. **AdoRhbl** inhibited a bacterial diol-dehydratase very effectively, as well as the growth of *Salmonella enterica*.[176]

20.6 CONCLUSION AND FUTURE PERSPECTIVES

The discovery of the organometallic nature of B_{12}-coenzymes in the Hodgkin labs[1] has opened the field of bio-organometallic chemistry. Nature makes use of the capacities of the organometallic B_{12}-catalysts in remarkable and multiple ways. In the cellular metabolism, the B_{12}-organometallic cofactors are to proteins, as well as to B_{12}-binding nucleotides, e.g., in bacterial B_{12}-riboswitches.[177] The structures of both of these classes of complexes will continue to give new insights.[148,178–180] Amazing new forms of gene expression, controlled by corrinoids, continue to appear on the scene.[148,181] Increasing insights into the dependency of important organisms from most kingdoms of life, give the complex and sparse corrinoids previously unrecognized and often very specific ecological roles.[35,36,182] Likewise, a range of hardly rationalized effects of B_{12}-deficiency in human and mammalian physiology[18,183] call for new and better diagnostic B_{12}-tools in medicine, possibly, as provided by "antivitamins B_{12}."[172] The organometallic corrinoids are unique compounds that extend the capacity for biological organometallic catalysis, for control of life processes and for new biomedical applications.[92,101,176,184–186] B_{12}-derivatives also have a remarkable potential for the development of novel applications based on "purely" chemical (organometallic and analytical) research.[187] Clearly, bio-structural, biochemical and chemical biological studies concerning the "most beautiful" B_{12}-cofactor[188] will go on in strengthening the demand in the interdisciplinary "B_{12}-fraternity" for a better understanding of the unique reactivity of organometallic B_{12}-derivatives and their role in biology,[10,148,168,189] which will surely continue to also fascinate its neighboring bioorganometallic field.

ACKNOWLEDGMENTS

I have enjoyed working with a group of dedicated and talented doctoral and postdoctoral coworkers, whose names are listed in the references. Over the years, our work in the B_{12}-field was supported by the European Commission and by generous and continuous support by the Austrian National Science Foundation (FWF).

LIST OF ABBREVIATIONS

ACA	ATP:**Cbl^{I}** adenosyltransferase
Ado	5′-deoxy-5′-adenosyl (group or radical)
AdoCbl	coenzyme B_{12} (adenosylcobalamin)
AdoCby	adenosylcobyrate
AdoRhbl	adenosylrhodibalamin
ATP	adenosine-triphosphate
BDE	bond dissociation energy
Cba	cob(III)amide
Cbi	cob(III)inamide
Cbi^{II}	cob(II)inamide
Cbl	cob(III)alamin (DMB-cob(III)amide)
Cbl^{II}	cob(II)alamin
Cbl^{I}	cob(I)alamin
Cby	cob(III)yric acid
CNCbl	vitamin B_{12} (cyanocob(III)alamin)
CN,H_2O-Cby	cyano, aquo-cobyrate
DMB	5,6-dimethylbenzimidazole
EtPhCbl	4-ethylphenylcobalamin
F2PhEtyCbl	2(2,4-difluorophenyl)-ethynylcobalamin
H_2OCbl^+	aquocobalamin (cation)
HOCbl	hydroxocobalamin
MCM	methylmalonyl-CoA mutase
MeCbl	methylcobalamin
MeCbi$^+$	methylcobinamide
MetH	B_{12}-dependent methionine synthase

MMACHC	methylmalonic aciduria type C and homocystinuria
NADH	nicotine-adenine-dinucleotide hydride
PCE	tetrachloroethene reductive dehalogenase
PhEtyCbl	2-phenylethynylcobalamin
SAM	S-adenosyl-methionine (also abbreviated as AdoMet)

REFERENCES

1. Hodgkin DC. X-ray analysis of complicated molecules. *Science* 1965;**150**(3699):979–88.
2. Rickes EL, Brink NG, Koniuszy FR, Wood TR, Folkers K. Crystalline vitamin B_{12}. *Science* 1948;**107**:396–7.
3. Smith EL, Parker LFJ. Purification of anti-pernicious anaemia factor. *Biochem. J.* 1948;**43** (1):R8–9.
4. Battersby AR. How nature builds the pigments of life - the conquest of vitamin B_{12}. *Science* 1994;**264**(5165):1551–7.
5. Eschenmoser A, Wintner CE. Natural product synthesis and vitamin-B_{12}. *Science* 1977;**196** (4297):1410–26.
6. Woodward RB. The total synthesis of vitamin B_{12}. *Pure Appl. Chem.* 1973;**33**(1):145–77.
7. Kräutler B, Arigoni D, Golding BT, editors. Vitamin B_{12} and B_{12}-Proteins. Weinheim: John Wiley VCH; 1998.
8. Banerjee R, editor. Chemistry and Biochemistry of B_{12}. New York, Chichester: John Wiley & Sons; 1999.
9. Brown KL. Chemistry and enzymology of vitamin B_{12}. *Chem. Rev.* 2005;**105** (6):2075–149.
10. Kräutler B, Puffer B. Vitamin B_{12}-derivatives: Organometallic catalysts, cofactors and ligands of bio-macromolecules. In: Kadish KM, Smith KM, Guilard R, editors. *Handbook of Porphyrin Science*, Vol 25. World Scientific; 2012. p. 133–265.
11. Warren MJ, Raux E, Schubert HL, Escalante-Semerena JC. The biosynthesis of adenosylcobalamin (vitamin B_{12}). *Nat. Prod. Rep.* 2002;**19**(4):390–412.
12. Obeid R. Vitamin B_{12} - Advances and Insights. CRC Press, Taylor & Francis Group; 2017.
13. Nielsen MJ, Rasmussen MR, Andersen CBF, Nexo E, Moestrup SK. Vitamin B_{12} transport from food to the body's cells-a sophisticated, multistep pathway. *Nature Reviews Gastroenterology and Hepatology*. 2012;**9**(6):345–54.
14. Banerjee R, Gherasim C, Padovani D. The tinker, tailor, soldier in intracellular B_{12} trafficking. *Curr. Opin. Chem. Biol.* 2009;**13**(4):484–91.
15. Banerjee R, Ragsdale SW. The many faces of vitamin B_{12}: Catalysis by cobalamin-dependent enzymes. *Ann. Rev. Biochem.* 2003;**72**:209–47.
16. Buckel W, Golding BT. Radical enzymes[a]. In: Chatgilialoglu CSA, editor. Encyclopedia of Radicals in Chemistry, Biology and Materials. John Wiley & Sons; 2012.
17. Frey PA, Hegeman AD, Reed GH. Free radical mechanisms in enzymology. *Chem. Rev.* 2006;**106**(8):3302–16.
18. Scalabrino G. The multi-faceted basis of vitamin B_{12} (cobalamin) neurotrophism in adult central nervous system: Lessons learned from its deficiency. *Progr. Neurobiol.* 2009;**88** (3):203–20.
19. Friedrich W. *Vitamins*. Berlin: Walter de Gruyter; 1988.
20. Green R, Miller JW. Vitamin B_{12}. In: Zempleni J, Suttie JW JFG, Stover PJ, editors. *Handbook of Vitamins*. 5th ed. Boca Raton, USA: CRC Press; 2014. p. 447–89.

21. Banerjee R. Radical carbon skeleton rearrangements: Catalysis by coenzyme B_{12}-dependent mutases. *Chem. Rev.* 2003;**103**(6):2083–94.
22. Pratt JM. Inorganic Chemistry of Vitamin B_{12}. New York: Academic Press; 1972.
23. Dolphin D, editor. B_{12}. New York, Chichester: John Wiley & Sons; 1982.
24. Mutti E, editor. Vitamin B_{12} - Chemical Aspects, Transport, Cause and Symptoms of Deficiency, Dietary Sources and Health Benefits. New York: Nova Publishers; 2015. BiochemistryResearch Trends.
25. Glusker JP. X-ray crystallography of B_{12} and cobaloximes. In: Dolphin D, editor. B_{12}, Vol I. New York: John Wiley & Sons; 1982. p. 23–106.
26. Widner FJ, Gstrein F, Kräutler B. Partial synthesis of coenzyme B12 from cobyric acid. *Helv. Chim. Acta.* 2017;**100**(9):e1700170.
27. Murtaza S, Butler PA, Kratky C, Gruber K, Kräutler B. A crystalline B_{12}-dimer from β-cyano-neocobyrate. *Chem. Eur. J* 2008;**14**:7521–4.
28. Eschenmoser A. Vitamin-B_{12} - experiments concerning the origin of its molecular-structure. *Angew. Chem. Int. Ed* 1988;**27**(1):5–39.
29. Murakami Y, Hisaeda Y, Kajihara A. Hydrophobic vitamin B12. I. Preparation and axial ligation behavior of hydrophobic vitamin B12r. *Bull. Chem. Soc. Japan.* 1983;**56**(12):3642–6.
30. Fischli A, Daly JJ. Cob(I)alamin as catalyst. 8. cob(I)alamin and heptamethyl Cob(I)yrinate during the reduction of α,β-unsaturated carbonyl derivatives. *Helv. Chim. Acta.* 1980;**63**(6):1628–43.
31. Kräutler B, Keller W, Hughes M, Caderas C, Kratky C. A crystalline cobalt(II)corrinate derived from vitamin $_{B1}$2: Preparation and X-ray crystalStructure. *J. Chem. Soc., Chem. Comm.* 1987;1678–80.
32. Randaccio L, Geremia S, Nardin G, Würges J. X-ray structural chemistry of cobalamins. *Coord. Chem. Rev.* 2006;**250**(11-12):1332–50.
33. Kratky C, Kräutler B. Molecular structure of B_{12} cofactors and other B_{12} derivatives In: Banerjee R, ed. *chemistry and biochemistry of* B_{12} 1999:9–41.
34. Kräutler B. B_{12}-nomenclature and a suggested atom-numbering. In: Kräutler B, Arigoni D, Golding BT, editors. Vitamin B_{12} and B_{12}-Proteins. Weinheim: Wiley VCH; 1998. p. 517–21.
35. Helliwell KE, Lawrence AD, Holzer A, et al. Cyanobacteria and eukaryotic algae use different chemical variants of vitamin B_{12}. *Curr. Biol.* 2016;**26**:999–1008.
36. Helliwell KE. The roles of B vitamins in phytoplankton nutrition: New perspectives and prospects. *New Phytologist.* 2017;**216**(1):62–8.
37. Croft MT, Warren MJ, Smith AG. Algae need their vitamins. *Eukaryotic Cell.* 2006;**5**(8):1175–83.
38. Hoffmann B, Oberhuber M, Stupperich E, et al. Native corrinoids from *clostridium cochlearium* are adeninylcobamides: Spectroscopic analysis and identification of pseudovitamin B_{12} and factor A. *J. Bacteriol.* 2000;**182**(17):4773–82.
39. Kräutler B. Organometallic chemistry of B_{12}-coenzymes. In: Sigel A, Sigel H, Sigel RKO, editors. *Metal-Ions in Life Sciences*, Vol 6. Cambridge, UK: RSC Publishing; 2009. p. 1–51.
40. Taga ME, Walker GC. Pseudo-B_{12} joins the cofactor family. *J. Bacteriol.* 2008;**190**(4):1157–9.
41. Kräutler B, Konrat R, Stupperich E, Färber G, Gruber K, Kratky C. Direct evidence for the conformational deformation of the corrin ring by the nucleotide base in vitamin-B_{12} - synthesis and solution spectroscopic and crystal-structure analysis of Co_{β}-cyano-imidazolyl-cobamide. *Inorg. Chem.* 31 1994;**33**(18):4128–39.

42. Kräutler B, Fieber W, Ostermann S, et al. The cofactor of tetrachloroethene reductive dehalogenase of *Dehalospirillum multivorans* is norpseudo-B_{12}, a new type of a natural corrinoid. *Helv. Chim. Acta.* 2003;**86**(11):3698–716.
43. Butler PA, Ebert M-O, Lyskowski A, Gruber K, Kratky C, Kräutler B. Vitamin B_{12} - A methylgroup without a job ?. *Angew. Chem. Int. Ed.* 2006;**45**(6):989–93.
44. Bommer M, Kunze C, Fesseler J, Schubert T, Diekert G, Dobbek H. Structural basis for organohalide respiration. *Science.* 2014;**346**(6208):455–8.
45. Keller S, Ruetz M, Kunze C, Kräutler B, Diekert G, Schubert T. Exogenous 5,6-dimethylbenzimidazole caused production of a non-functional tetrachloroethene reductive dehalogenase in *Sulfurospirillum multivorans. Environm. Microbiol.* 2014;**16**(11):3361–9.
46. Kräutler B, Keller W, Kratky C. Coenzyme B_{12}-chemistry: The crystal and molecular structure of cob(II)alamin. *J. Am. Chem. Soc.* 1989;**111**:8936–8.
47. Gschösser S, Hannak RB, Konrat R, et al. Homocoenzyme B_{12} and bishomocoenzyme B_{12}, covalent structural mimics for homolyzed, enzyme-bound coenzyme B_{12}. *Chem. Eur. J.* 2005;**11**:81–93.
48. McCauley KM, Pratt DA, Wilson SR, Shey J, Burkey TJ, van der Donk WA. Properties and reactivity of chlorovinylcobalamin and vinylcobalamin and their implications for vitamin B_{12}-catalyzed reductive dechlorination of chlorinated alkenes. *J. Am. Chem. Soc.* 2005;**127**(4):1126–36.
49. Ruetz M, Gherasim C, Fedosov SN, Gruber K, Banerjee R, Kräutler B. Radical synthesis opens access to organometallic aryl-cobaltcorrins - 4-ethylphenyl-cobalamin, a potential "antivitamin B_{12}". *Angew. Chem. Int. Ed.* 2013;**52**(9):2606–10.
50. Brenig C, Ruetz M, Kieninger C, Wurst K, Kräutler B. Alpha- and beta-diastereoisomers of phenylcobalamin from cobalt-arylation with diphenyliodonium chloride. *Chem. Europ. J.* 2017;**23**(41):9726–31.
51. Ruetz M, Salchner R, Wurst K, Fedosov S, Kräutler B. Phenylethynylcobalamin: A light-stable and thermolysis-resistant organometallic vitamin B_{12} derivative prepared by radical. *Angew. Chem. Int. Ed.* 2013;**52**(43):11406–9.
52. Ruetz M, Shanmuganathan A, Gherasim C, et al. Antivitamin B-12 inhibition of the human B-12-processing enzyme CblC: Crystal structure of an inactive ternary complex with glutathione as the cosubstrate. *Angew. Chem. Int. Ed.* Jun 2017;**56**(26):7387–92.
53. Chrominski M, Lewalska A, Gryko D. Reduction-free synthesis of stable acetylide cobalamins. *Chem. Comm.* 2013;**49**(97).11406–8.
54. Chrominski M, Lewalska A, Karczewski M, Gryko D. Vitamin B_{12} derivatives for orthogonal functionalization. *J. Org. Chem.* 2014;**79**(16):7532–42.
55. Gschösser S, Gruber K, Kratky C, Eichmüller C, Kräutler B. B_{12}-retro-riboswitches: Constitutional switching of B_{12} coenzymes induced by nucleotides. *Angew. Chem. Int. Ed.* 2005;**44**(15):2284–8.
56. Kräutler B. Thermodynamic trans-effects of the nucleotide base in the B_{12} coenzymes. *Helv. Chim. Acta.* 1987;**70**(5):1268–78.
57. Fedosov SN, Fedosova NU, Kräutler B, Nexo E, Petersen TE. Mechanisms of discrimination between cobalamins and their natural analogues during their binding to the specific B_{12}-transporting proteins. *Biochemistry.* 2007;**46**(21):6446–58.
58. Fedosov SN. Physiological and molecular aspects of cobalamin transport. In: Stanger O, editor. *Water Soluble Vitamins*, Vol 56. Springer; 2012. p. 347–68.
59. Drennan CL, Huang S, Drummond JT, Matthews RG, Ludwig ML. How a protein binds B_{12} - a 3.0-angstrom X-ray structure of B_{12}-binding domains of methionine synthase. *Science.* 1994;**266**(5191):1669–74.

60. Ludwig ML, Evans PR. X-Ray crystallography of B_{12} enzymes: Methylmalonyl-CoA mutase and methionine synthase. In: Banerjee R, editor. Chemistry and Biochemistry of B_{12}. New York, Chichester: John Wiley & Sons; 1999. p. 595–632.
61. Hannak RB, Konrat R, Schüler W, Kräutler B, Auditor MTM, Hilvert D. An antibody that reconstitutes the "base-on" form of B_{12} coenzymes. *Angew. Chem. Int. Ed.* 2002;**41**(19):3613–16.
62. Marsh ENG, Drennan CL. Adenosylcobalamin-dependent isomerases: New insights into structure and mechanism. *Curr. Opin. Chem. Biol.* 2001;**5**(5):499–505.
63. Matthews RG. Cobalamin-dependent methyltransferases. *Acc. Chem. Res* 2001;**34**(8):681–9.
64. Halpern J. Mechanisms of coenzyme B_{12}-dependent rearrangements. *Science.* 1985;**227**(4689):869–75.
65. Jensen MP, Halpern J. Dealkylation of coenzyme B_{12} and related organocobalamins: Ligand structural effects on rates and mechanisms of hydrolysis. *J. Am. Chem. Soc.* 1999;**121**(10):2181–92.
66. Craig PJ, Glockling F, editors. *The Biological Alkylation of Heavy Elements.* London: Royal Society of Chemistry; 1988.
67. Desimone RE, Penley MW, Charbonn L, et al. Kinetics and mechanism of cobalamin-dependent methyl and ethyl transfer to mercuric ion. *Biochim. Biophys. Acta.* 1973;**304**(3):851–63.
68. Mosimann H, Kräutler B. Methylcorrinoids methylate radicals - Their second biological mode of action? *Angew. Chem. Int. Ed.* 2000;**39**(2). 393-200.
69. Finke RG, Hay BP. Thermolysis of adenosylcobalamin - a product, kinetic, and Co-C5′ bond-dissociation energy study. *Inorg. Chem.* 1984;**23**(20):3041–3.
70. Finke RG. Coenzyme B_{12}-based chemical precedent for Co-C bond homolysis and other key elementary step. In: Kräutler B, Arigoni D, Golding BT, editors. Vitamin B_{12} and B_{12}-Proteins. Weinheim: Wiley-VCH; 1998. p. 383–402.
71. Martin BD, Finke RG. Methylcobalamins full-strength vs half-strength cobalt-carbon sigma-bonds and bond-dissociation enthalpies - a-greater-than-10(15) Co-CH_3 homolysis rate enhancement following one-antibonding-electron reduction of methylcobalamin. *J. Am. Chem. Soc.* 1992;**114**(2):585–92.
72. Kobylianskii I, Widner F, Kräutler B, Chen P. Co − C bond energies in adenosylcobinamide and methylcobinamide in the gas phase and in silico. *J. Am. Chem. Soc.* 2013;**135**(37):13648–51.
73. Grate JH, Schrauzer GN. Chemistry of cobalamins and related compounds. 48. sterically induced, spontaneous dealkylation of secondary alkylcobalamins due to axial base coordination and conformational-changes of the corrin ligand. *J. Am. Chem. Soc.* 1979;**101**(16):4601–11.
74. Kräutler B. B_{12} Coenzymes, the central theme. In: Kräutler B, Arigoni D, Golding BT, editors. Vitamin B_{12} and B_{12} Proteins. Weinheim: Wiley-VCH; 1998. p. 3–43.
75. Jones AR. The photochemistry and photobiology of vitamin B12. *Photochem. & Photobiol. Sci.* 2017;**16**(6):820–34.
76. Rury AS, Wiley TE, Sension RJ. Energy cascades, excited state dynamics, and photochemistry in Cob(III)alamins and ferric porphyrins. *Acc. Chem. Res.* 2015;**48**(3):860–7.
77. Sension RJ, Harris DA, Cole AG. Time-resolved spectroscopic studies of B-12 coenzymes: Comparison of the influence of solvent on the primary photolysis mechanism and geminate recombination of methyl-, ethyl-, n-propyl-, and 5'-deoxyadenosylcobalamin. *J. Phys. Chem. B.* 2005;**109**(46):21954–62.
78. Hisaeda Y, Tahara K, Shimakoshi H, Masuko T. Bioinspired catalytic reactions with vitamin B_{12} derivative and photosensitizers. *Pure Appl. Chem.* 2013;**85**(7):1415–26.

79. Miller NA, Wiley TE, Spears KG, et al. Toward the design of photoresponsive conditional antivitamins B-12: A transient absorption study of an arylcobalamin and an alkynylcobalamin. *Journal of the American Chemical Society*. 2016;**138**(43):14250–6.
80. Endicott JF, Netzel TL. Early events and transient chemistry in the photohomolysis of alkylcobalamins. *J. Am. Chem. Soc.* 1979;**101**:4000–2.
81. Kräutler B. Acetyl-cobalamin from photoinduced carbonylation of methyl-cobalamin. *Helv. Chim. Acta.* 1984;**67**(4):1053–9.
82. Fischer H. Unusual selectivities of radical reactions by internal suppression of fast modes. *J. Am. Chem. Soc.* 1986;**108**(14):3925–7.
83. Schrauzer GN, Deutsch E, Windgassen RJ. The nucleophilicity of vitamin B_{12s}. *J. Am. Chem. Soc.* 1968;**90**(9):2441–2.
84. Haglund J, Magnusson A-L, Ehrenberg L, Törnqvist M. Introduction of cob(I)alamin as an analytical tool: Application to reaction-kinetic studies of oxiranes. *Toxicol. Environ. Chem.* 2003;**85**(4-6):81–94.
85. Stich TA, Buan NR, Escalante-Semerena JC, Brunold TC. Spectroscopic and computational studies of the ATP : Corrinoid adenosyltransferase (CobA) from Salmonella enterica: Insights into the mechanism of adenosylcobalamin biosynthesis. *J. Am. Chem. Soc.* 2005;**127**(24):8710–19.
86. Matthews RG, Koutmos M, Datta S. Cobalamin-dependent and cobamide-dependent methyltransferases. *Curr. Op. Struct. Biol.* 2008;**18**(6):658–66.
87. Kräutler B, Caderas C. Complementary diastereoselective cobalt methylations of the vitamin-B_{12} derivative cobester. *Helv. Chim. Acta.* 1984;**67**(7):1891–6.
88. Hogenkamp HPC, Bratt GT, Sun S. Methyl transfer from methylcobalamin to thiols - a reinvestigation. *Biochemistry.* 1985;**24**(23):6428–32.
89. Dorweiler JS, Matthews RG, Finke RG. Providing a chemical basis toward understanding the histidine base-on motif of methylcobalamin-dependent methionine synthase: An improved purification of methylcobinamide, plus thermodynamic studies of methylcobinamide binding exogenous imidazole and pyridine bases. *Inorg. Chem.* 2002;**41**(24):6217–24.
90. Kräutler B, Dérer T, Liu PL, et al. Oligomethylene-bridged vitamin-B_{12} dimers. *Angew. Chem. Int. Ed.* 1995;**34**(1):84–6.
91. Galliker PK, Gräther O, Rümmler M, Fitz W, Arigoni D. New structural and biosynthetic aspects of the unusual core lipids from archaebacteria. In: Kräutler B, Arigoni D, Golding BT, editors. Vitamin B_{12} and B_{12}-Proteins. Weinheim: Wiley-VCH; 1998. p. 447–58.
92. Benjdia A, Balty C, Berteau O. Radical SAM enzymes in the biosynthesis of ribosomally synthesized and post-translationally modified peptides (RiPPs). *Front. Chem.* 2017;**5**:87
93. Fujimori DG. Radical SAM-mediated methylation reactions. *Curr. Op. Chem. Biol.* 2013;**17**(4):597–604.
94. Zhang Q, van der Donk W, Liu W. Radical-mediated enzymatic methylation: A tale of two SAMS. *Acc. Chem. Res.* 2012;**45**:555.
95. Kräutler B, Hughes M, Caderas C. Thermal methyl-group transfer between methylcobalt (III) corrinates and cobalt(II) corrinates - equilibration experiments with heptamethyl cobyrinates and cobalamins. *Helv. Chim. Acta.* 1986;**69**(7):1571–5.
96. Lexa D, Savéant JM. The electrochemistry of vitamin B_{12}. *Acc. Chem. Res.* 1983;**16**(7):235–43.
97. Kräutler B. Electrochemistry and organometallic electrochemical synthesis. In: Banerjee R, editor. Chemistry and Biochemistry of B_{12}. New York: John Wiley; 1999. p. 315–39.
98. Ragsdale SW. Enzymology of the acetyl-CoA pathway of CO_2 fixation. *Crit. Rev. Biochem. Mol. Biol.* 1991;**26**(3-4):261–300.

99. Hisaeda Y, Shimakoshi H. Bioinspired catalysts with B_{12} enzyme functions. In: Kadish KM, Smith KM, Guilard R, editors. *Handbook of Porphyrin Science*, Vol 10. 2010. p. 313–64.
100. Birke RL, Huang QD, Spataru T, Gosser DKj. Electroreduction of a series of alkylcobalamins: Mechanism of stepwise reductive cleavage of the Co − C bond. *J. Am. Chem. Soc.* 2006;**128**(6):1922–36.
101. Hunger M, Mutti E, Rieder A, Enders B, Nexo E, Kräutler B. Organometallic B_{12}-DNA-conjugate: Synthesis, structure analysis and studies of binding to human B_{12}-transporterProteins. *Chem. Eur. J.* 2014;**20**:13103–7.
102. Gschösser S, Kräutler B. B_{12}-retro-riboswitches: Guanosyl-induced constitutional switching of B_{12}-coenzymes. *Chem. Eur. J.* 2008;**14**(12):3605–19.
103. Tinembart O, Walder L, Scheffold R. Reductive cleavage of the Co,C-bond of [(methoxycarbonyl)methyl]cobalamin. *Ber. Bunsen-Ges. Physic. Chem. Chem. Phys.* 1988;**92** (11):1225–31.
104. Puchberger M, Konrat R, Kräutler B, Wagner U, Kratky C. Reduction-labile organo-cob (III)alamins via cob(II)alamin: Efficient synthesis and solution and crystal structures of [(methoxycarbonyl)methyl]cob(III)alamin. *Helv. Chim. Acta.* 2003;**86**(5):1453–66.
105. Zhou DL, Tinembart O, Scheffold R, Walder L. Influence of the axial alkyl ligand on the reduction potential of alkylcob(III)alamins and alkylcob(III)yrinates. *Helv. Chim. Acta.* 1990;**73**(8):2225–41.
106. Toraya T. Cobalamin-dependent dehydratases and a deaminase: Radical catalysis and reactivating chaperones. *Arch. Biochem. Biophys.* 2014;**544**:40–57.
107. Shibata N, Masuda J, Tobimatsu T, et al. A new mode of B_{12} binding and the direct participation of a potassium ion in enzyme catalysis: X-ray structure of diol dehydratase. *Structure.* 1999;**7**:997–1008.
108. Gruber K, Reitzer R, Kratky C. Radical shuttling in a protein: Ribose pseudorotation controls alkyl-radical transfer in the coenzyme B_{12} dependent enzyme glutamate mutase. *Angew. Chem. Int. Ed.* 2001;**40**(18):3377–80.
109. Koutmos M, Gherasim C, Smith JL, Banerjee R. Structural basis of multifunctionality in a vitamin B_{12}-processing enzyme. *J. Biol. Chem.* 2011;**286**(34):29780–7.
110. Bauer CB, Fonseca MV, Holden HM, et al. Three-dimensional structure of ATP: Corrinoid adenosyltransferase from *Salmonella typhimurium* in its free state, complexed with MgATP, or complexed with hydroxycobalamin and MgATP. *Biochem.* 2001;**40** (2):361–74.
111. Moore TC, Newmister SA, Rayment I, Escalante-Semerena JC. Structural insights into the mechanism of four-coordinate cob(II)alamin formation in the active site of the salmonella enterica ATP:Co(I)rrinoid adenosyltransferaseenzyme: Critical role of residues Phe91 and Trp93. *Biochem.* 2012;**51**(48):9647–57.
112. Sauer K, Thauer RK. The Role of corrinoids in methanogenesis. In: Banerjee R, editor. Chemisty and Biochemistry of B_{12}. New York, Chichester: John Wiley & Sons; 1999. p. 655–79.
113. Stupperich E. Corrinoid-dependent mechanism of acetogenesis from methanol. *Acetogenesis.* 1994;180–94.
114. Matthews RG, editor. *Cobalamin- and Corrinoid-Dependent Enzymes.* Cambridge: RSC Publishing; 2009. Sigel A, Sigel H, Sigel RKO, eds. Metal-Carbon Bonds in Enzymes and Cofactors, Metal Ions in Life Sciences; No. 6.
115. Appel AM, Bercaw JE, Bocarsly AB, et al. Frontiers, opportunities, and challenges in biochemical and chemical catalysis of CO2 fixation. *Chem. Rev.* 2013;**113**(8):6621–58.

116. Thauer RK, Mollerzinkhan D, Spormann AM. Biochemistry of acetate catabolism in anaerobic chemotropic bacteria. *Annu. Rev. Microbiol.* 1989;**43**:43–67.
117. Ragsdale SW. Metals and their scaffolds to promote difficult enzymatic reactions. *Chem. Rev.* 2006;**106**(8):3317–37.
118. Chan KKJ, Thompson S, O'Hagan D. The mechanisms of radical SAM/cobalamin methylations: An evolving working hypothesis. *ChemBioChem.* 2013;**14**(6):675–7.
119. Ludwig ML, Matthews RG. B_{12}-dependent methionine synthase: A structure that adapts to catalyze multiple methyl transfer reactions. *ACS Symposium Series: Structures and Mechanisms, from Ashes to enzymes.* 2002;**827**(827):186–201.
120. Zydowsky TM, Courtney LF, Frasca V, et al. Stereochemical analysis of the methyl transfer catalyzed by cobalamin-dependent methionine synthase from *Escherichia coli B. J. Am. Chem. Soc.* 1986;**108**(11):3152–3.
121. Ludwig ML, Drennan CL, Matthews RG. The reactivity of B_{12} cofactors: The proteins make a difference. *Structure.* 1996;**4**(5):505–12.
122. Svetlitchnaia T, Svetlitchnyi V, Meyer O, Dobbek H. Structural insights into methyltransfer reactions of a corrinoid iron-sulfur protein involved in acetyl-CoA synthesis. *Proc. Nat. Acad. Sci. USA.* 2006;**103**(39):14331–6.
123. Bandarian V, Ludwig ML, Matthews RG. Factors modulating conformational equilibria in large modular proteins: A case study with cobalamin-dependent methionine synthase. *Proc. Nat. Acad. Sci. USA.* 2003;**100**(14):8156–63.
124. Kräutler B, Kratky C. Vitamin B_{12}: The haze clears. *Angew. Chem. Int. Ed.* 1996;**35** (2):167–70.
125. Zhou P, Ohagan D, Mocek U, et al. Biosynthesis of the antibiotic thiostrepton - methylation of tryptophan in the formation of the quinaldic acid moiety by transfer of the methionine methyl-group with net retention of configuration. *J. Am. Chem. Soc.* 1989;**111** (18):7274–6.
126. Sofia HJ, Chen G, Hetzler BG, Reyes-Spindola JF, Miller NE. Radical SAM, a novel protein superfamily linking unresolved steps in familiar biosynthetic pathways with radical mechanisms: Functional characterization using new analysis and information visualization methods. *Nucl. Acids Res.* 2001;**29**(5):1097–106.
127. Frey PA, Magnusson OT. *S*-Adenosylmethionine: A wolf in sheep's clothing, or a rich man's adenosylcobalamin? *Chem. Rev.* 2003;**103**(6):2129–48.
128. Frey PA, Hegeman AD. *Enzymatic Reaction Mechanisms.* New York: Oxford University Press; 2007.
129. Woodyer RD, Li G, Zhao H, van der Donk WA. New insight into the mechanism of methyl transfer during the biosynthesis of fosfomycin. *Chem. Commun.* 2007;(4):359–61.
130. Zhang Q, Liu W. Complex biotransformations catalyzed by radical S-adenosylmethionine enzymes. *J. Biol. Chem.* 2011;**286**(35):30245–52.
131. Bridwell-Rabb J, Zhong A, Sun HG, Drennan CL, Liu H-w. A B12-dependent radical SAM enzyme involved in oxetanocin A biosynthesis. *Nature.* 2017;**544**(7650):322–6.
132. Marsh ENG, Patterson DP, Li L. Adenosyl radical: Reagent and catalyst in enzyme reactions. *ChemBioChem.* 2010;**11**(5):604–21.
133. Buckel W, Kratky C, Golding BT. Stabilization of methylene radicals by cob(II)alamin in coenzyme B_{12} dependent mutases. *Chem. Eur. J.* 2006;**12**:352–62.
134. Buckel W, Friedrich P, Golding BT. Hydrogen bonds guide the short-lived 5′-deoxyadenosyl radical to the place of action. *Angew. Chem. Int. Ed.* 2012;**51**(40):9974–6.

135. Ortiz-Guerrero JM, Polanco MC, Murillo FJ, Elias-Arnanz M, Padmanabhan S. Light-dependent gene regulation by a coenzyme B_{12}-based photoreceptor. *Proc. Natl, Acad. Sci. USA*. 2011;**108**(18):7565–70.
136. Jost M, Fernandez-Zapata J, Polanco MC, et al. Structural basis for gene regulation by a B12-dependent photoreceptor. *Nature*. 2015;**526**:536–41.
137. Erb TJ, Retey J, Fuchs G, Alber BE. Ethylmalonyl-CoA mutase from rhodobacter sphaeroides defines a new subclade of coenzyme B-12-dependent acyl-CoA mutases. *J. Biol. Chem.* 2008;**283**(47):32283–93.
138. Gruber K, Kratky C. Coenzyme B_{12} dependent glutamate mutase. *Curr. Op. Chem. Biol.* 2002;**6**(5):598–603.
139. Jost M, Born DA, Cracan V, Banerjee R, Drennan CL. Structural basis for substrate specificity in adenosylcobalamin-dependent isobutyryl-CoA mutase and related acyl-CoA mutases. *J. Biol. Chem.* 2015;**290**(45):26882–98.
140. Berkovitch F, Behshad E, Tang K-H, Enns EA, Frey PA, Drennan CL. A locking mechanism preventing radical damage in the absence of substrate, as revealed by the X-ray structure of lysine 5,6-aminomutase. *Proc. Nat. Acad. Sci. USA*. 2004;**101**(45):15870–5.
141. Stubbe J. Ribonucleotide reductases: The link between an RNA and a DNA world? *Curr. Opin. Struct. Biol.* 2000;**10**(6):731–6.
142. Rétey J. Enzymatic-reaction selectivity by negative catalysis or how do enzymes deal with highly reactive intermediates. *Angew. Chem. Int. Ed.* 1990;**29**(4):355–61.
143. Fukuoka M, Nakanishi Y, Hannak RB, Kräutler B, Toraya T. Homoadenosylcobalamins as probes for exploring the active sites of coenzyme B_{12}-dependent diol dehydratase and ethanolamine ammonia-lyase. *FEBS Journal*. 2005;**272**(18):4787–96.
144. Robertson WD, Wang M, Warncke K. Characterization of protein contributions to cobalt-carbon bond cleavage catalysis in adenosylcobalamin-dependent ethanolamine ammonia-lyase by using photolysis in the ternary complex. *J. Am. Chem. Soc.* 2011;**133**(18):6968–77.
145. Roman-Melendez GD, von Glehn P, Harvey JN, Mulholland AJ, Marsh ENG. Role of active site residues in promoting cobalt-carbon bond homolysis in adenosylcobalamin-dependent mutases revealed through experiment and computation. *Biochem.* 2014;**53**(1):169–77.
146. Sintchak MD, Arjara G, Kellogg BA, Stubbe J, Drennan CL. The crystal structure of class II ribonucleotide reductase reveals how an allosterically regulated monomer mimics a dimer. *Nature Struct. Biol.* 2002;**9**(4):293–300.
147. Pérez-Marín MC, Padmanabhan S, Polanco MC, Murillo FJ, Elías-Arnanz M. Vitamin B_{12} partners the CarH repressor to downregulate a photoinducible promoter in *Myxococcus xanthus*. *Mol. Microbiol.* 2008;**67**(4):804–19.
148. Bridwell-Rabb J, Drennan CL. Vitamin B-12 in the spotlight again. *Curr. Op. Chem. Biol.* 2017;**37**:63–70.
149. Elias-Arnanz M, Padmanabhan S, Murillo FJ. Light-dependent gene regulation in nonphototrophic bacteria. *Curr. Opin. Microbiol.* 2011;**14**(2):128–35.
150. Jost M, Simpson JH, Drennan CL. The transcription factor CarH safeguards use of adenosylcobalamin as a light sensor by altering the photolysis products. *Biochem.* 2015;**54**(21):3231–4.
151. Kutta RJ, Hardman SJO, Johannissen LO, et al. The photochemical mechanism of a B_{12}-dependent photoreceptor protein. *Nature Comm.* 2015;6.
152. Gruber K, Kräutler B. Coenzyme B12 repurposed for photo-regulation of gene expression. *Angew. Chem. Int. Ed.* 2016;**55**:5638–40.

153. Cheng Z, Li KR, Hammad LA, Karty JA, Bauer CE. Vitamin B_{12} regulates photosystem gene expression via the CrtJ antirepressor AerR in Rhodobacter capsulatus. *Mol. Microbiol.* 2014;**91**(4):649–64.
154. Jugder BE, Ertan H, Lee M, Manefield M, Marquis CP. Reductive dehalogenases come of age in biological destruction of organohalides. *Trends Biotechnol.* 2015;**33**(10):595–610.
155. Smidt H, de Vos WM. Anaerobic microbial dehalogenation. *Ann. Rev. Microbiol.* 2004;**58**:43–73.
156. Studer A, Stupperich E, Vuilleumier S, Leisinger T. Chloromethane: Tetrahydrofolate methyl transfer by two proteins from *Methylobacterium chloromethanicum* strain CM4. *Eur. J. Biochem.* 2001;**268**(10):2931–8.
157. Sun BL, Griffin BM, Ayala-del-Rio HL, Hashsham SA, Tiedje JM. Microbial dehalorespiration with 1,1,1-trichloroethane. *Science.* 2002;**298**(5595):1023–5.
158. Raina V, Rentsch D, Geiger T, et al. New metabolites in the degradation of α- and γ-hexachlorocyclohexane (HCH): Pentachlorocyclohexenes are hydroxylated to cyclohexenols and cyclohexenediols by the haloalkane dehalogenase LinB from *Sphingobium indicum* B90A. *J. Agric. Food Chem.* 2008;**56**(15):6594–603.
159. Krasotkina J, Walters T, Maruya KA, Ragsdale SW. Characterization of the B_{12}- and iron-sulfur-containing reductive dehalogenase from *Desulfitobacterium chlororespirans*. *J. Biol. Chem.* 2001;**276**(44):40991–7.
160. Neumann A, Siebert A, Trescher T, Reinhardt S, Wohlfarth G, Diekert G. Tetrachloroethene reductive dehalogenase of *Dehalospirillum multivorans*: Substrate specificity of the native enzyme and its corrinoid cofactor. *Arch. Microbiol.* 2002;**177**(5):420–6.
161. Diekert G, Gugova D, Limoges B, Robert M, Saveant J-M. Electroenzymatic reactions. Investigation of a reductive dehalogenase by means of electrogenerated redox cosubstrates. *J. Am. Chem. Soc.* 2005;**127**(39):13583–8.
162. Payne KAP, Quezada CP, Fisher K, et al. Reductive dehalogenase structure suggests a mechanism for B_{12}-dependent dehalogenation. *Nature.* 2015;**517**(7535):513–16.
163. Kunze C, Bommer M, Hagen WR, et al. Cobamide-mediated enzymatic reductive dehalogenation via long-range electron transfer. *Nat Commun.* 2017;8.
164. Holliger C, Wohlfarth G, Diekert G. Reductive dechlorination in the energy metabolism of anaerobic bacteria. *FEMS Microbiol. Rev* 1999;**22**(5):383–98.
165. Quadros EV. Advances in the understanding of cobalamin assimilation and metabolism. *British J. Haematol.* 2009;**148**(2):195–204.
166. Lerner-Ellis JP, Tirone JC, Pawelek PD, et al. Identification of the gene responsible for methylmalonic aciduria and homocystinuria, cblC type. *Nature Genetics.* 2006;**38**(1):93–100.
167. Dobson CM, Wai T, Leclerc D, et al. Identification of the gene responsible for the cblB complementation group of vitamin B12-dependent methylmalonic aciduria. *Human Molecular Genetics.* 2002;**11**(26):3361–9.
168. Gherasim C, Lofgren M, Banerjee R. Navigating the B_{12} road: Assimilation, delivery, and disorders of cobalamin. *J. Biol. Chem.* 2013;**288**(19):13186–93.
169. Kim J, Gherasim C, Banerjee R. Decyanation of vitamin B_{12} by a trafficking chaperone. *Proc. Natl. Acad. Sci. USA.* 2008;**105**(38):14551–4.
170. Hannibal L, Kim J, Brasch NE, et al. Processing of alkylcobalamins in mammalian cells: A role for the MMACHC (cblC) gene product. *Mol. Genet. Metab.* Aug 2009;**97**(4):260–6.

171. St Maurice MS, Mera P, Park K, Brunold TC, Escalante-Semerena JC, Rayment I. Structural characterization of a human-type corrinoid adenosyltransferase confirms that coenzyme B_{12} is synthesized through a four-coordinate intermediate. *Biochemistry.* 2008;**47**(21):5755–66.
172. Kräutler B. Antivitamins B_{12} - a structure- and reactivity-based concept. *Chem. Eur. J.* 2015;**21**:11280–7.
173. Mutti E, Ruetz M, Birn H, Kräutler B, Nexo E. 4-Ethylphenyl-cobalamin impairs tissue uptake of vitamin B_{12} and causes vitamin B_{12} deficiency in mice. *Plos One.* 2013;**8**(9): e75312.
174. Zelder F, Sonnay M, Prieto L. Antivitamins for medicinal applications. *ChemBioChem.* 2015;**16**:1264–78.
175. Guzzo MB, Nguyen HT, Pham TH, et al. Methylfolate trap promotes bacterial thymineless death by sulfa drugs. *Plos Pathog.* 2016;**12**(10):e1005949.
176. Widner FJ, Lawrence AD, Deery E, et al. Total synthesis, structure, and biological activity of adenosylrhodibalamin, the non-natural rhodium homologue of coenzyme B12. *Angew. Chem. Int. Ed.* 2016;**55**(37):11281–6.
177. Winkler WC, Breaker RR. Regulation of bacterial gene expression by riboswitches. *Ann. Rev. Microbiol.* 2005;**59**:487–517.
178. Peselis A, Serganov A. Structural insights into ligand binding and gene expression control by an adenosylcobalamin riboswitch. *Nat. Struct. Mol. Biol.* 2012;**19**(11):1182–4.
179. Johnson JE, Reyes FE, Polaski JT, Batey RT. B_{12} Cofactors directly stabilize an mRNA regulatory switch. *Nature.* 2012;**492**:133–7.
180. Sukumar N. Crystallographic studies on B_{12} binding proteins in eukaryotes and prokaryotes. *Biochimie.* 2013;**95**(5):976–88.
181. Serganov A, Nudler E. A Decade of riboswitches. *Cell.* Jan 17 2013;**152**(1-2):17–24.
182. Croft MT, Lawrence AD, Raux-Deery E, Warren MJ, Smith AG. Algae acquire vitamin B_{12} through a symbiotic relationship with bacteria. *Nature.* 2005;**438**(7064):90–3.
183. Chan W, Almasieh M, Catrinescu MM, Levin LA. Cobalamin-associated superoxide scavenging in neuronal cells is a potential mechanism for vitamin B-12-deprivation optic neuropathy. *Am. J. Pathol.* 2018;**188**(1):160–72.
184. Rownicki M, Wojciechowska M, Wierzba AJ, et al. Vitamin B-12 as a carrier of peptide nucleic acid (PNA) into bacterial cells. *Scientific Reports.* 2017;**7**(1):7644.
185. Prieto L, Rossier J, Derszniak K, et al. Modified biovectors for the tuneable activation of anti-platelet carbon monoxide release. *Chem. Comm.* 2017;**53**(51):4.
186. Mutti E, Hunger M, Fedosov S, Nexo E, Kräutler B. Organometallic DNA-B-12 conjugates as potential oligonucleotide vectors: Synthesis and structural and binding studies with human cobalamin-transport proteins. *ChemBioChem.* 2017;**18**(22):2280–91.
187. Zelder F, Alberto R. Vitamin B12 derivatives for spectroanalytical and medicinal applications. In: Kadish KM, Smith KM, Guilard R, editors. *Handbook of Porphyrin Science*, Vol 25. World Scientific; 2012. p. 84–132.
188. Stubbe J. Binding-site revealed of natures most beautiful cofactor. *Science.* 1994;**266**(5191):1663–4.
189. Gruber K, Puffer B, Kräutler B. Vitamin B_{12}-derivatives - enzyme cofactors and ligands of proteins and nucleic acids. *Chem. Soc. Rev.* 2011;**40**(8):4346–63.

Index

Note: Page numbers followed by "*f*" and "*t*" refer to figures and tables, respectively.

C

D

E

F

G

H

M

N

O

P

Q

T

U

V

Made in the USA
Monee, IL
12 September 2026

62977625R00283